Migrations and Dispersal of Marine Organisms

Developments in Hydrobiology 174

Series editor
K. Martens

Migrations and Dispersal of Marine Organisms

Proceedings of the 37th European Marine Biology Symposium held in Reykjavík, Iceland, 5–9 August 2002

Edited by

M.B. Jones[1], A. Ingólfsson[2], E. Ólafsson[3], G.V. Helgason[2], K. Gunnarsson[4] & J. Svavarsson[2]

[1] *School of Biological Sciences (Plymouth Environmental Research Centre), University of Plymouth, U.K.*
[2] *Institute of Biology, University of Iceland, Reykjavík, Iceland*
[3] *Department of Zoology, University of Stockholm, Sweden*
[4] *Marine Research Institute, Reykjavík, Iceland*

Reprinted from Hydrobiologia, volume 503 (2003)

Kluwer Academic Publishers
Dordrecht/Boston/London

Library of Congress Cataloging-in-Publication Data

European Marine Biology Symposium (37th : 2002 : Reykjavík, Iceland)
Migrations and dispersal of marine organisms: proceedings of the 37th European Marine Biology Symposium held in Reykjavík, Iceland, 5–9 August 2002 / edited by M.B. Jones ... [et al.].
p. cm. – (Developments in hydrobiology; 174)
"Reprinted from Hydrobiologia, volume 503 (2003)."
ISBN 1-4020-1736-7 (alk. paper)
1. Marine animals–Dispersal–Congresses. 2. Marine animals–Migration–Congresses.
I. Jones, M. B. (Michael B.), 1946- II. Hydrobiologia. III. Title. IV. Series. 2003062043

ISBN 1-4020-1736-7

Published by Kluwer Academic Publishers,
P.O. Box 17, 3300 AA Dordrecht, The Netherlands

Sold and distributed in North, Central and South America
by Kluwer Academic Publishers,
101 Philip Drive, Norwell, MA 02061, U.S.A.

In all other countries, sold and distributed
by Kluwer Academic Publishers Group,
P.O. Box 322, 3300 AH Dordrecht, The Netherlands

Printed on acid-free paper

Printed in the Netherlands

TABLE OF CONTENTS

OPEN SESSION

Preface

The 37th European Marine Biology Symposium, held in Reykjavík, Iceland, from 5–9th August 2002, was hosted jointly by the Marine Research Institute and the Institute of Biology of the University of Iceland. The Organizing Committee consisted of Agnar Ingólfsson, Jörundur Svavarsson and Guðmundur Víðir Helgason (Institute of Biology), Karl Gunnarsson (Marine Research Institute), Guðrún Marteinsdóttir, (affiliated with both institutions), and Emil Ólafsson (Department of Zoology, University of Stockholm). The host institutions acknowledge the support of the Ministry of Fisheries, the Ministry for the Environment, Landsbanki Islands and the Icelandic Research Council.

The theme of the symposium, Migrations and Dispersal of Marine Organisms, attracted about 170 participants from 25 countries, including several outside of Europe. The Organizing Committee invited 5 keynote speakers to give authoritative overviews of the themes of the symposium. In addition, there were about 75 oral presentations in 8 consecutive sessions and about 85 posters were displayed in two poster sessions. Most of the contributions centered around the main themes of the symposium, but there were also open sessions, following the tradition of previous EMBS meetings.

The Editorial Committee, lead by Malcolm Jones (School of Biological Sciences, University of Plymouth, UK), received 48 manuscripts based upon presentations given at the meeting. Members of the Editorial Committee took great care in approaching only scientists of international reputation to review the manuscripts, using 2–3 reviewers per submitted manuscript. We thank all our referees for their scrupulous attention to detail and for the high scientific acumen, freely given, they have contributed to the Proceedings Volume. The reviewing process resulted in 26 accepted papers. Time constraints, imposed by our publication date, was a factor in this high rejection rate, and we hope that many of the papers not included in this Proceedings Volume will appear at a later date in *Hydrobiologia* or elsewhere, as all authors benefited from the constructive reviewers' comments. In this work we would like to thank the editorial secretary, Björg Sigurðardóttir, for her invaluable help.

We have strived to maintain the high standards of previous EMBS Symposium Volumes, and hope that the current Proceedings will not only stimulate further research but also contribute positively to the continued valuable activity of the EMBS.

MALCOLM B. JONES
AGNAR INGÓLFSSON
EMIL ÓLAFSSON
GUÐMUNDUR VÍÐIR HELGASON
KARL GUNNARSSON
JÖRUNDUR SVAVARSSON

Guest Editors

Hydrobiologia **503:** xi–xii, 2002.
M.B. Jones, A. Ingólfsson, E. Ólafsson, G.V. Helgason, K. Gunnarsson &
J. Svavarsson (eds), Migrations and Dispersal of Marine Organisms.

List of participants

Australia
Schreider, Maria

Austria
Ott, Joerg

Belgium
Bonne, Wendy
De Troch, Marleen

Chile
Thiel, Martin

Finland
Hemmi, Anne
Kinnunen, Veijo
Kostamo-Liusvaara, Kirsi
Nappu, Niko
Ruuskanen, Ari

France
Denis, Francoise
Dupont, Lise
Lemaire, Christophe
Montaudouin, Xavier
Sauriau, Pierre-Guy
Thiebaut, Eric

Germany
Groben, René
Gutow, Lars
Kahle, Jens
Kruse, Inken
Liebezeit, Gerd
Normant, Monika
Tuschling, Kirsten

Greece
Aligizaki, Katerina
Megalofonou, Persefoni
Nicolaidou, Artemis

Ireland
Davenport, John
Watson, Douglas

Iceland
Ástþórsson, Ólafur
Böðvarsdóttir, Anna Rósa
Dongala Dombaxe, María
Einarsson, Haraldur A.
Einarsson, Sólmundur
Eiríksson, Þorleifur
Eydal, Agnes
Galan, Anton
Gíslason, Ástþór
Guijarro Garcia, Elena
Gunnarsson, Björn
Gunnarsson, Karl
Gunnlaugsson, Þorvaldur
Guðfinnsson, Hafsteinn
Guðmundsson, Guðmundur
Guðmundsson, Guðmundur
Halldórsson, Halldór P.
Helgason, Guðmundur Víðir
Ingólfsson, Agnar
Jaworski, Andrzej
Jónsdóttir, Ingibjörg
Júlíusdóttir, Stefanía
Marteinsdóttir, Guðrún
Ólafsson, Magnús Freyr
Ólafsson, Ólafur Patrick
Örnólfsdóttir, Erla Björk
Óskarsson, Guðmundur
Pálsson, Jónbjörn
Pálsson, Ólafur K.
Pétursdóttir, Gróa
Ragnarsson, Stefán
Sólmundsson, Jón
Steinarsdóttir, María Björk
Steingrímsson, Sigmar
Svavarsson, Jörundur
Valtýsson, Heiðar Þór
Víkingsson, Gísli A.
Þórisson, Konráð
Þorsteinsson, Vilhjálmur

Israel
Goren, Menachem

Italy
Alidani, Stefano
Cannizzaro, Leonardo
Castriota, Luca
Comaschi, Alessandra
Del Piero, Donatella
Fiorentino, Fabio
Merello, Stefania
Modigh, Monica
Nasci, Christina
Relini, Giulio
Relini, Lidia Orsi
Rinelli, Paola
Socal, Giorgio

Japan
Hanyu, Kazuhiro
Narita, Teruyoshi
Sekiguchi, Hideo

Mexico
Ardisson, Pedro
Signoret Poillon, Giséle
Signoret, Martha

Netherlands
Coyer, James
Hiddink, Jan Geert
Hoarau, Galice
Mekking, Suzanne
Peijnenburg, Katja
Pierrot-Bults, Annelise
Ysebaert, Tom

Norway
Beuchel, Frank
Christie, Hartvig
Falk-Petersen, Stig
Lekve, Kyrre
Norderhaug, Kjell Magnus
Salvanes, Anne Gro Vea

Poland
Sapota, Mariusz

Styczynska-Jurewicz, Ewa
Surosz, Waldemar
Szaniawska, Anna
Zmijewska, Maria Iwona

Portugal
Avila, Sergio
Barreiros, João
Brotas, Vanda
Cabecadas, Leonor
Daguin, Claire
Mendes, Jose Carlos
Monteiro, M. Teresa
Pearson, Gareth
Ribeiro da Cunha, Marina
Serrao, Ester
Silvert, William

Slovenia
Flander, Vesna

South Africa
McQuaid, Christopher
Porri, Francesca

Spain
Duran, Sandra
Macho, Gonzalo
Sánchez-Lizaso, José

Sweden
Berger, Rita
Ganning, Björn
Hjelmgren, Anna
Isæus, Martin
Kalvas, Arja
Kautsky, Lena
Neideman, Rasmus
Ólafsson, Emil
Råberg, Sonja
Strand, Malin
Svenson, Anders
Ullberg, Jörgen
Ungfors, Anette
Wenngren, Johan
Wikström, Sofia A.

U.S.A.
Carlton, Debby
Carlton, James
Cheng, Lanna
Piontkovski, Sergey
Scheltema, Amelie
Scheltema, Rudolf S.

Ukraine
Shulman, Gregory E.
Tokarev, Yuri

United Kingdom
Barnes, David K.A.
Bell, James
Brand, Andy
Burrows, Michael
Callaway, Ruth
Davis, Martin
Ferrero, Timothy
Gibson, Robin N.
Hartnoll, Richard
Hays, Graeme
Herborg, Leif-Matthias
Jenkins, Stuart
Johnson, Mark
Jones, Malcolm B.
Morritt, David
Paramor, Odette
Peck, Lloyd
Poloczanska, Elvira
Porter, Joanne
Ryland, John
Scott, Catherine Lousie
Sims, David
Southall, Emily
Tyler, Paul
Williams, Robert

Hydrobiologia **503:** 1–8, 2003.
M.B. Jones, A. Ingólfsson, E. Ólafsson, G.V. Helgason, K. Gunnarsson & J. Svavarsson (eds), Migrations and Dispersal of Marine Organisms.

Community assembly and historical biogeography in the North Atlantic Ocean: the potential role of human-mediated dispersal vectors

James T. Carlton
Maritime Studies Program, Williams College - Mystic Seaport, Mystic, CT 06355, U.S.A.
E-mail: jcarlton@williams.edu

Key words: invasions, invasive species, cryptogenic, marine biogeography

Abstract

Historical and modern migrations and dispersal of most marine organisms (intertidal, benthic, meiofaunal, planktonic, nektonic, or neustonic) are classically interpreted in terms of their natural dispersal potential. Exceptions are introduced species, largely recognized since the 19th century, known to have been transported by human activities. However, humans were transporting species along coastlines and across oceans for millennia and centuries prior to the advent of the first biological surveys. Thus, the presumptive natural distributions of many species may be questioned. Reviewed here are some basic concepts about invasions of non-native species. Human activities move species isolated in time and space from other oceans or continents, and thus human-mediated transport does not simply speed up natural dispersal processes. Both past and modern-day invasions are often overlooked, leading to an underestimation of the scale of invasion diversity and impact. Because vectors, donor regions, and recipient regions change over time, invasions will continue along long-standing but un-managed corridors. The impact of most invasions has never been studied and, therefore, it is not possible to conclude that most invasions have no impact, nor is it generally possible to say that invasions have become 'integrated' into a community or ecosystem in ecological time. Finally, invasions in the ocean are not limited to harbours and ports, but are found in a wide variety of marine habitats, ranging from the open ocean continental shelf to exposed rocky shores. The existence of human-mediated vectors has created extraordinary challenges to our understanding and interpretation of the ecology, biogeography, evolutionary biology, and conservation biology of marine communities.

Introduction

Historical and modern-day migrations and dispersal (and thus gene flow) of most marine organisms, whether intertidal, benthic, meiofaunal, planktonic, nektonic, neustonic, or other species, are classically interpreted in terms of their natural dissemination potential. This dissemination may be along continental or island corridors, across oceans, or between oceans. In turn, dispersal is influenced by the nature of coastal or oceanic currents, by the movements of continents and the history of ocean basins (that is, plate tectonics), by glacial histories at higher latitudes and the creation or dissolution of refugia, and by other small-scale or large-scale natural phenomena. Exceptions to these broad processes, that have led to the marine biota of a given region to be seen as shaped largely by natural means, are those species known to have been transported by human activities, activities that are often independent in space and time of natural barriers or corridors. Examples include fouling species attached to ship bottoms and a wide variety of taxa inside ships in ballast sands, with ballast rocks, and in more recent times, in ballast water.

Although we are aware of unquestioned invasions of exotic species due to such human activities, in general when we seek to elucidate the historical biogeography, evolutionary history, and genetics history of a regional biota (such as the coastal marine life of the North Atlantic Ocean) our working assumptions are that natural forces have by and large created, shaped and modified the histories of most species. Stated another way, a 'default' assumption in biogeographical analysis is that distributions should be first interpreted as a result of natural phenomena, rather than having been influenced or altered by human activity. Thus if we examine the amphiatlantic marine biota ranging from western and northern mainland Europe and the Faroe Islands to Iceland and Greenland, and then down the Labrador Sea to eastern Canada and south to the Gulf of Maine, we see (or perhaps we seek to see) a biota that is now, in large part, as it was since the retreat of the last Holocene glaciers. Only since between the 1830s and 1850s, with the advent of biological surveys, have we noted the appearance (whether by presumptive natural means or human-mediated vectors) of novel species on either side of the North Atlantic Ocean. This temporal marker may thus inadvertently lead to the assumption that the pre-1830s biota was not altered, or less altered, by the arrival of new taxa.

As Carlton (1999) noted however, humans were moving by watercraft along coastlines and across oceans for millennia and centuries (see also, for example, Labaree, 1957; Hourani, 1995; Schwartz, 2002) prior to the advent of the first biological surveys. In the North Atlantic Ocean, for example, Viking vessels were sailing back-and-forth across the Atlantic Ocean by 1000 AD (Petersen et al., 1992; Strasser, 1999). As almost all vessels have organisms associated with some part of their outer structure, we can presume that Viking vessels were transporting organisms on and in their hulls, thus planting some of the earliest potential seeds of post-glacial human-mediated transoceanic transport, independent of natural dispersal mechanisms. Here, I review briefly some basic considerations of the science of marine bioinvasions as it links to the assemblage of marine communities and the historical biogeography of the North Atlantic Ocean. I suggest that the presumptive natural distributions of many species should be brought into question, and that the existence of human-mediated vectors has created extraordinary challenges to our understanding and interpretation of the ecology, biogeography, evolutionary biology, and conservation biology of marine communities.

Introduced species

Introduced species are human-mediated biological invasions in historical times (Carlton, 1979; Ruiz et al., 1997; Ruiz et al., 2000; Grosholz, 2002). Natural bioinvasions are known as range expansions. Introductions are also known as alien, exotic, non-indigenous and invasive species, among many other terms. Species are only considered to be introduced when they reproduce in the wild and establish continuing generations. Species are not considered as being introduced to a region if they are simply released into a new region (such releases are *inoculations*) and there is no evidence of reproduction or establishment; nor are species considered introduced to a country if they have been imported but remain solely within mariculture or aquaculture establishments (even if occasional larvae, juveniles, or adults escape). Recent reviews of human-mediated invasion vectors are available (Carlton, 1999; Sandlund et al., 1999; Ruiz et al., 2000; Wonham et al., 2000; Carlton, 2001; Minchin & Gollasch, 2002; Läppakoski et al., 2002).

Invasions increase as the vectors that serve to engage, transport and release species increase. Thus, as human populations began continental, followed by transoceanic and then interoceanic movements, they transported a host of associated species carried intentionally (for many reasons) and accidentally in, on, and with humans and their associated belongings and foods. Indeed, classical maps of historical demography, especially those showing the origins, destinations, numbers, and rates of movements of peoples (see also King & Motulsky, 2002), may be good predictors of the origins, destinations, numbers, and rates of movements of many species of animals and plants around the world. The greatest voluntary migrations in human history were the movement of Europeans to North America in the 19th century; the greatest involuntary migrations were the transportation of Africans as slaves to the Americas between 1550 and 1860 (Oxford Atlas of the World, 2002). All of these movements were by ships moving across the North and South Atlantic Oceans. These ship movements are likely to have left strong biogeographical and genetic tracks criss-crossing the North and South Atlantic Oceans, as well as cascading elsewhere around the world.

Invasions also increase as vector speed increases. Over the past 100 years, the speeds of almost all transport vehicles (automobiles, trucks, trains, planes, and ships) have increased dramatically. The probability of survival of a given organism from point A to point B may thus be proportional to increased transport speed, depending on the species' physiological requirements, life history and similar considerations.

Given the steadily increasing number of transport events over millennia and centuries combined with the increasing rate of speed of such events, invasions should thus inexorably increase over time (in the absence of any attempts to control the vectors that move species). How many invasions have occurred, and when and where they are detected by biologists in the receiving countries, is of course another matter.

Misperceptions and myths in invasion science

Not surprisingly, as a result of this vast movement of organisms around the globe, a number of misperceptions, if not myths, have developed relative to the human-mediated movement of species by humans. Examples of these include:

1. Human activities simply speed up the natural processes of dispersal.
2. Most major invasions of the past are known.
3. Most species have been introduced by now.
4. Most new invasions are generally recognized.
5. Most invasions have little or no impact.
6. Invasions become 'naturalised' after some period of time.
7. Marine invasions are restricted largely to harbours and ports.

I will comment briefly on each of these below, as they might apply to the biogeography of the North Atlantic Ocean.

Human activities are simply speeding up the natural process of invasions

Human-mediated transport vectors bridge all natural dispersal barriers in time and space. Thus, such activities do not speed up the natural process of invasions: the vast majority of invasions involve species that would never have gained access to new regions, regardless of the amount of time. This misperception serves to underscore that the distinction between human-mediated invasions and natural invasions is not an arbitrary or artificial one. In the North Atlantic Ocean, the Gulf Stream and North Atlantic Drift serve to transport organisms from west to east. For cold-water species to gain natural access from Europe to North America would, in general, require transport in currents that flow to the south from Europe into warm subtropical and tropical waters, and then west to the Americas, for many months in hostile, oligotrophic open ocean waters. In contrast, vessels sailed in a matter of days and weeks directly from Europe to North America, remaining largely in colder waters.

Most invasions (or at least most major invasions) of the past are now known

The potential antiquity of invasions argues potently against this concept. We have little to no understanding of the scale and rate of introductions prior to the 19th century. By the 1500s, vessel traffic (largely for fishing and exploration) across the North Atlantic was in place and, by the 1600s, numerous voyages of colonization had commenced. Thus, by at least 350 years ago, the anthropogenic bioflow across the North Atlantic of marine species had commenced on a regular basis. These transport events may have involved at times more than 150 species per ship (Carlton, 1999).

Given this nautical bioflow, if only one species was successfully introduced from, for example, Europe to North America once every 5 years between 1600 and 1850, then some 50 species of protists, animals and plants that are now in the western Atlantic Ocean (and that we are confident represent a portion of the aboriginal and natural distributional picture of the marine world) are in fact introduced species. Consideration of introductions in the centuries prior to the 1600s would add to this number. That some of these earlier invasions may now be abundant and ecologically important taxa seems probable. Centuries of natural coastal dispersal following introduction could then result in broad distributions that would appear to mimic natural patterns. Thus, ancient invasions may no longer have the classic 'disjunct' distributions of many introduced species (wherein, for example, a species occurs from Scandinavia to Spain, and outside of this range only in Chesapeake Bay). Rather, after centuries of post-introduction dispersal, a species could spread between Canada and Cape Hatteras, if not further south, essentially fully 'filling in' its temperature range. Thus the Japanese alga *Polysiphonia harveyi* arrived on the North American Atlantic coast by the 1840s, and today occurs from Newfoundland to Georgia (Schneider and Searles, 1991; McIvor et al., 2001). Long-standing biases that 'type localities' may reflect where a species occurs naturally may serve to obscure and obfuscate potential candidates for species introduced on ship bottoms over the past 350–1000 years across the North Atlantic. For example, the abundant ship-fouling seasquirt (ascidian) *Molgula manhattensis* was described by DeKay from Long Island Sound (Manhattan, part of New York City) in 1843. Had the description of the European *Molgula tubifera* by Orsted in 1844 (now recognized as *M. manhattensis*; Berrill, 1950) appeared in 1842, we might more charitably re-consider *M. manhattensis* ' provenance. Such species as ship-fouling seasquirts, susceptible to transport by human vectors, are candidates for careful re-examination of their historical biogeography. Easily underestimated in this regard is the scale of biodiversity of entrainable organisms on and in the hulls of wooden vessels or, since the 1880s, in the ballast water of ships, as well as by many other vectors, such as the movement of species for aquaculture (mariculture) purposes. Carlton (1999) and Carlton et al. (1995) have presented partial lists of such candidate marine and estuarine taxa, many of whose modern-day distributions have rarely (if ever!) been considered in light of centuries of potential dispersal by ships and other means.

Table 1 presents a synthesized list of many of these taxa. Included are five kingdoms (Monera, Protoctista, Fungi, Plantae, and Animalia) and most non-protistan phyla. Various life-cycle stages (from larvae through juveniles to adults) may be entrained differentially by transport vectors. In addition, many taxa have resting stages that make them susceptible to long-distance transport, both my natural and human-mediated vectors (Carlton et al., 1995). Examples among freshwater, estuarine, and marine invertebrates include protozoans, hydrozoans, flatworms, nematodes, rotifers, bryozoans, cladocerans and copepods, among other taxa, that can form cysts, cocoons, and resting eggs. A particularly remarkable example of cyst formation is found in the sea anemone *Diadumene lineata* (=*Haliplanella luciae*). This small Asian orange-striped anemone is a member of the family Diadumenidae, perhaps the only anemones in the world capable of encystment. *Diadumene lineata* forms a thick mucous cyst (Kiener, 1972) in which the foodless anemone is seemingly impervious to changes in temperature and salinity for many months (J. T. Carlton, pers. obs.). This characteristic no doubt aids in long-term survival on ship bottoms. *Diadumene lineata* now occurs on both sides of the North Atlantic Ocean.

Species that are not clearly introduced or native are *cryptogenic species* (Carlton, 1996a). In classical evolutionary biogeography, cryptogenic species are considered native. Assigning species to a cryptogenic status does not imply that they are *probably* exotic, however. Cryptogenic species are essentially in a 'neutral' or holding status, awaiting proper investigation to determine their evolutionary and thus geographic origins. Data to resolve the provenance include paleontological, archeological, historical, genetic, biogeographic, ecological, and vector science evidence (the latter in terms of the potential of a given species to be transported by natural and/or anthropogenic vectors). There are many hundreds of cryptogenic species in North America and European waters. The relative number of cryptogenic marine species in Greenland, Iceland, the Faroe Islands, and other major regions between Europe and North America, remains unknown.

Most species have been introduced by now

A common assumption is that if a vector has been operating over a given route for a lengthy period of time, most species that could have been transported and successfully introduced would have done so already. In reality, however, as long as a vector is operating, invasions can and will continue, even if the vector has been in place for centuries. Carlton (1996b) addressed a series of hypotheses to elucidate why invasions continue to arrive over time on long-established routes. Broadly, these hypotheses invoke a combination of changes in vector rate, timing, direction, source, speed, and management with changes in the environment, including habitat alteration, chemical pollution and eutrophication, climate change, community changes resulting from overextraction fisheries or fisheries methods, and introduced species themselves (a process that may involve facilitation of further invasions, known as *invasional meltdown*, Simberloff & Von Holle, 1999). The holistic result of all of these processes is a region that changes through time in its susceptibility or resistance to invasions. Thus, invasions continue across the North Atlantic on the bottoms of ships or in ballast water, despite the fact that ships have been traversing these waters for centuries.

In some cases, of course, these invasions may simply be the result of the acquisition of a new invasion on one side of the ocean, with the species in question then carried to the other side. An example is the invasion of the Asian shore crab *Hemigrapsus sanguineus* in North America in the 1980s; it was then transported by ships (either in ballast water or ship fouling) to Europe by the late 1990s (Breton et al., 2002). In turn, the Asian shore crab *Hemigrapsus penicillatus* invaded Europe in the late 1980s

Table 1. A partial synthetic list of marine, estuarine, and maritime organisms whose distributions have been or are susceptible to alteration by human-mediated transport

Monera (bacteria)	Mysidacea (mysids)
Granuloreticulosa (foraminiferans)	Cumacea (cumaceans)
Rhizopoda (amoebae and allies)	Copepoda (copepods)
Actinopoda (radiolarians)	Ostracoda (ostracods)
Ciliophora (ciliate 'protozoans', including meiofaunal species)	Decapoda (shrimps, crabs, hermit crabs, etc.)
Dinoflagellida (dinoflagellates)	Euphausiacea (krill)
Bacillariophyta (diatoms)	Insecta (insects)
Protoctista (other protists, including a vast number of other photosynthetic or heterotrophic organisms)	Tardigrada (tardigrades)
Placozoa (placozoans)	Acarina (marine mites)
Porifera (sponges)	Pycnogonida (sea spiders)
Hydrozoa (hydroids)	Gastropoda (snails, seaslugs and pteropods)
Scyphozoa (jellyfish)	Bivalvia (clams, mussels, oysters, etc.)
Anthozoa (sea anemones)	Polyplacophora (chitons)
Ctenophora (comb jellies)	Cephalopoda (squids and octopuses)
Platyhelminthes (flatworms, trematodes, cestodes)	Phoronida (phoronids)
Gnathostomulida (gnathostomulids)	Brachiopods (lamp shells)
Nemertea (nemerteans)	Ectoprocta (bryozoans)
Rotifera (rotifers)	Kamptozoa (entoprocts)
Priapulida (priapulans)	Chaetognatha (arrow worms)
Gastrotricha (gastrotrichs)	Asteroidea (seastars)
Kinorhyncha (kinorhynchs)	Ophiuroidea (brittle stars)
Loricifera (loriciferans)	Holothuroidea (sea cucumbers)
Nematoda (nematodes)	Echinoidea (sea urchins and sand dollars)
Oligochaeta (oligochaetes)	Hemichordata (hemichordates, enteropneusts)
Polychaeta (polychaetes)	Ascidiacea (seasquirts)
Sipuncula (sipunculans)	Thaliacea (salps and doliolids)
Echiura (echiurans)	Larvacea (larvaceans)
Cirripedia (barnacles)	Osteichthyes (fish)
Cladocera (water fleas)	Rhodophyta (red algae)
Copepoda (copepods)	Phaeophyta (brown algae)
Amphipoda (amphipods)	Chlorophyta (green algae)
Isopoda (isopods)	Potamogetonaceae, Hydrocaritaceae (seagrasses)
Tanaidacea (tanaids)	Anthophyta (other plants such as marsh grasses)
	Fungi (especially lignicolous species)

(Gollasch, 1999) and transatlantic shipping will now bring it to the United States (although where and when this will occur is not predictable). Thus, North America still awaits the arrival of the New Zealand barnacle *Elminius modestus* (which invaded Europe in the 1940s) and the arrival of the Japanese brown seaweed *Sargassum muticum* (which invaded Europe in the 1970s). After some 60 and 30 years, respectively, neither species has appeared in the western Atlantic Ocean. The Japanese green alga *Codium fragile tomentosoides* was first collected in 1900 in Europe but was not found in North America until 1957 (Carlton & Scanlon, 1985).

On the other hand, native European species continue to appear in American waters. In 1987, the European bryozoan *Membranipora membranacea* was first collected in the Gulf of Maine (Berman et al., 1992). In the 1980s, the common fouling European ascidian *Ascidiella aspersa* was first detected in New England (Carlton, 2000b), another example of a species that would appear to have been available for transatlantic colonization for centuries. As noted earlier, invasion rates may increase if the frequency and speed of vectors increase, leading to increased survivorship during the dispersal event, and thus the arrival of larger numbers of individuals potentially capable of reproducing. Alternatively, the recipient environment may shift in ways that would increase its susceptibility to invasions of species that were previously unable to become established (Carlton, 1996b, 2000a). Stachowicz et al. (2002) argued that certain recent invasions (including *Ascidiella aspersa*) in New England have been mediated by warming temperatures.

Most new invasions are generally recognized

In fact, invasions often go unnoticed and even larger species (of 15 cm length or height or larger) may go unrecognized, depending upon the availability of local taxonomic expertise, the lack of exploration of certain habitats, and so on. The relatively large Japanese seastar *Asterias amurensis*, which invaded Tasmania in the 1980s, was misidentified for some years as the native seastar *Uniophora granifera* (Buttermore et al., 1994). The Caribbean barnacle *Chthamalus proteus* may have gone unnoticed for nearly 20 years in the Hawaiian Islands (Southward et al., 1998). Much more often overlooked, however, are invasions of smaller species; indeed, the smaller the species, the more likely the invasion will either go entirely unnoticed, or be considered a previously undetected native species. The recognition of invasions as inversely correlated with body size is referred to as the 'smalls rule' of invasion ecology (Carlton, in Wyatt & Carlton, 2002).

Most invasions have little or no impact

This misperception is derived from the so-called 'tens rule' (Williamson, 1996), which has been cited widely as meaning that only 10% of all invasions have an ecological impact. However, as originally defined (see Williamson, 1996) the 'tens rule' means that a 'rough' range of 5–20% (*not* strictly 10%) of invasions become *pests*, defined as species 'with a negative economic effect'. It is important to note that the definition of pest is without quantitative boundaries. Data on the relative pestiferous economic status of invasions are largely derived from selected sets of terrestrial plants, and omit most taxa and many regions of the world. Most importantly, the 'tens rule' does not apply to the percentage of invasions having *ecological or other environmental* impacts; nor, of course, could any such conclusions yet be drawn, as perhaps 95–99% of all invasions, depending upon the taxonomic group, have not been experimentally studied to demonstrate causative processes (versus correlative patterns). Rather, for many introductions, most judgements on 'impact' are based on correlative, qualitative, and anecdotal assumptions and presumptions. A logical default is to assume that, since invasions are utilizing resources (space, food, or otherwise), the more likely outcome is that there has been some level of alteration (impact) on the utilization of those resources by the previously resident species (whether earlier invasions or native species). Experimental studies are required to determine if these alterations are statistically significant, or if they may have long-term, cascading biological or ecological consequences.

Of course, invasions may occur in a spatial sense where no other species is believed to have existed previously, or, for example, in a trophic sense, where no other species in a particular feeding guild was believed to exist (such as a filter feeder or grazer). It is important to emphasize that occupation of presumptive previously unutilized space, or the arrival of a novel filter feeder or grazer in a habitat should *not* be described as the invasion of an 'empty (or open or vacant) niche'. A niche describes *all* aspects of a species' role and function in the economy of nature: unutilized 'bare space', or consuming some resource fraction not apparently consumed by an existing species, does not an 'empty niche' make (Herbold & Moyle, 1986).

Invasions become 'naturalized' after some period of time

There is little empirical evidence or theoretical basis for setting a time limit after which an introduced species has become integrated or 'naturalized' into a community, nor are there rigorous definitions of what such integration would entail (or how it could be measured!), other than the establishment of self-reproducing wild populations. While 'naturalization' may simply imply the latter, establishment of populations of a non-native animal and plant 'in the wild' does not imply community integration, nor does an arbitrary passage of time (such as one or two human generations).

Marine invasions largely occur in harbours and ports

An investigatory bias in understanding the extent of both historical and modern-day invasions may be a continuing focus on harbours and ports to assess invasion diversity. Carlton (2002) provided evidence for invasions in a wide variety of marine habitats, including open ocean neritic shelf habitats (both pelagic and benthic), kelp beds, sandy beaches and coral reefs, as well as invasions in mangrove communities, salt marshes and supratidal communities (including the supralittoral fringe and coastal dunes). In some instances, invasions on open coast shores may be significant, such as the colonization of the rocky intertidal in parts of South Africa by the Mediterranean mussel *Mytilus galloprovincialis* (Griffiths et al., 1992).

Examples of non-harbour, non-port invasions in the North Atlantic Ocean include the predominance of two major European intertidal consumers, the snail *Littorina littorea* and the crab *Carcinus maenas* on the rocky shore of New England, the invasion of the European flatworm *Convoluta convoluta* in the open waters of the Gulf of Maine, and the establishment of the Asian seaweed *Codium fragile tomentosoides* and the Asian seasquirt *Styela clava* in open subtidal, hard-bottom environments of the western Atlantic from eastern Canada to the mid-Atlantic coast of the United States (Carlton, 2002).

Conclusions

Critical breakthroughs in our understanding of the historical biogeography (the assembling) of shallow-water marine communities across the North Atlantic will include our ability to dissect out the role of human-mediated vectors in transporting species in the half-millennium since the 1500s (and earlier invasions as well, if possible). Lacking this understanding, we remain looking through a largely clouded window on how coastal, shallow-water communities were structured prior to the advent of human exploration, commerce, and colonization.

References

Berman, J., L. Harris, W. Lambert, M. Buttrick & M. Dufresne, 1992. Recent invasions of the Gulf of Maine: three contrasting ecological histories. Conserv. Biol. 6: 435–441.

Berrill, N. J., 1950. The Tunicata with an account of the British species. Ray Society, London. 354 pp.

Breton G., M. Faasse, P. Noël & T. Vincent, 2002. A new alien crab in Europe:*Hemigrapsus sanguineus* (Decapoda: Brachyura: Grapsidae). J. Crust. Biol. 22: 184–189.

Buttermore, R. E., E. Turner & M. G. Morrice, 1994. The introduced northern Pacific seastar *Asterias amurensis* in Tasmania. Mem. Queensland Museum 36: 21–25.

Carlton, J. T., 1979. Introduced invertebrates of San Francisco Bay. In Conomos, T. J. (ed.), San Francisco Bay: The Urbanized Estuary. American Association for the Advancement of Science, Pacific Division, San Francisco: 427–444.

Carlton, J. T., 1996a. Biological invasions and cryptogenic species. Ecology 77: 1653–1655.

Carlton, J. T., 1996b. Pattern, process, and prediction in marine invasion ecology. Biol. Conserv. 78: 97–106.

Carlton, J. T., 1999. The scale and ecological consequences of biological invasions in the world's oceans. In Sandlund, O. T., P. Johan Schei & Å. Viken (eds), Invasive Species and Biodiversity Management. Kluwer Academic Publishers, Dordrecht: 195–212.

Carlton, J. T., 2000a. Global change and biological invasions in the oceans. In Mooney, H. A. & R. J. Hobbs (eds), Invasive Species in a Changing World. Island Press, Covelo CA: 31–53.

Carlton, J. T., 2000b. Quo Vadimus Exotica Oceanica?: Marine Bioinvasion Ecology in the Twenty-First Century. In Pederson, J. (ed.), Marine Bioinvasions: Proceedings of the First National Conference. Massachusetts Institute of Technology, MIT Sea Grant College Program, MITSG 00-2, Cambridge, Massachusetts: 6–23.

Carlton, J. T., 2001. Introduced species in U.S. coastal waters: environmental impacts and management priorities. Pew Oceans Commission, Arlington, Virginia. 28 pp.

Carlton, J. T., 2002. Bioinvasion Ecology: Assessing Invasion Impact and Scale. In Leppäkoski, E., S. Gollasch & S. Olenin (eds), Invasive Aquatic Species of Europe. Distribution, Impacts, and Management, Kluwer Academic Publishers, Dordrecht, The Netherlands: 7–19.

Carlton, J. T. & J. A. Scanlon, 1985. Progression and dispersal of an introduced alga: *Codium fragile* ssp. *tomentosoides* (Chlorophyta) on the Atlantic coast of North America. Bot. mar. 28: 155–165.

Carlton, J. T., D. M. Reid & H. van Leeuwen, 1995. Shipping Study. The role of shipping in the introduction of non-indigenous aquatic organisms to the coastal waters of the United States (other than the Great Lakes) and an analysis of control options. The National Sea Grant College Program/Connecticut Sea Grant Project R/ES-6. Department of Transportation, United States Coast Guard, Washington, D.C. and Groton, Connecticut. Report Number cg-D-11-95. Government Accession Number AD-A294809. xxviii + 213 pages and Appendices A-I (122 pages).

Gollasch, S., 1999.The Asian decapod *Hemigrapsus penicillatus* (de Haan, 1833) (Decapoda, Grapsidae) introduced in European waters, status quo and future perspective. Helgol. Meeresunters. 52: 359–366.

Griffiths, C. L., P. A. R. Hockey, C. Van Erkom Shurink & P. J. Le Roux, 1992. Marine invasive aliens on South African shores: implications for community structure and trophic functioning. S. Afr. J. mar. Sci.12: 713–722.

Grosholz, E., 2002. Ecological and evolutionary consequences of coastal invasions. Trends Ecol. Evol. 17: 22–27.

Herbold, B. & P. B. Moyle, 1986. Introduced species and vacant niches. Am. Nat. 128: 751–760.

Hourani, G. F., 1995. Arab Seafaring. 2nd edn. Princeton University Press. 189 pp.

Kiener, A., 1972. Contribution a l'ecologie, la physiologie et l'ethologie de l'actinie *Diadumene luciae* (Verrill). Bull. Soc. Zool. France 96: 581–603.

King, M.-C. & A. G. Motulsky, 2002. Mapping human history. Science 298: 2342–2343.

Labaree, B. W., 1957. How the Greeks sailed into the Black Sea. Am. J. Archaeol 61: 29–33.

Leppäkoski, E., S. Gollasch & S. Olenin (eds), 2002. Invasive Aquatic Species of Europe. Distribution, Impacts, and Management, Kluwer Academic Publishers, Dordrecht, The Netherlands. 583 pp.

McIvor, L., C. A. Maggs, J. Provan & M. J.Stanhope. 2001. *rbc*L sequences reveal multiple cryptic introductions of the Japanese red alga *Polysiphonia harveyi*. Mol. Ecol. 10: 911–919.

Minchin, D. & S. Gollasch, 2002. Vectors – how exotics how get around. In Leppäkoski, E., S. Gollasch & S. Olenin (eds), Invasive Aquatic Species of Europe. Distribution, Impacts, and Management, Kluwer Academic Publishers, Dordrecht, The Netherlands: 183–192.

Oxford Atlas of the World, 2002. Oxford Atlas of the World. 10th edn. Oxford University Press: 304 pp.

Petersen, K. S.., K. L. Rasmussen, J. Heinemeler & N. Rud. 1992. Clams before Columbus? Nature 359: 679.

Ruiz, G. M., J. T. Carlton, E. D. Grosholz & A. H. Hines, 1997. Global invasions of marine and estuarine habitats by non-indigenous species: mechanisms, extent, and consequences. Am. Zool. 37: 621–632.

Ruiz, G. M., P. Fofonoff, A. H. Hines & E. D. Grosholz, 1999. Non-indigenous species as stressors in estuarine and marine communities: assessing invasion impacts and interactions. Limnol. Oceanogr. 44 (3, part 2): 950–972.

Ruiz, G. M., P. W. Fofonoff, J. T. Carlton, M. J. Wonham & A. H. Hines, 2000. Invasion of coastal marine communities in North America: apparent patterns, processes, and biases. Ann. Rev. Ecol. Syst. 31: 481–531.

Sandlund, O. T., P. J. Schei & Å. Viken (eds), 1999. Invasive Species and Biodiversity Management. Kluwer Academic Publishers, Dordrecht. 431 pp.

Schneider, C. W. & R. B. Searles, 1991. Seaweeds of the South-eastern United States. Cape Hatteras to Cape Canaveral. Duke University Press, Durham and London. 554 pp.

Schwartz, M., 2002. Early evidence of reed boats from southeast Anatolia. Antiquity 76: 617–618.

Simberloff, D. & B. Von Holle, 1999. Positive interactions of nonindigenous species: invasional meltdown? Biol. Invasions 1: 21–32.

Southward, A. J., R. S. Burton, S. L. Coles, P. R. Dando, R. DeFelice, J. Hoover, P. E. Parnell, T. Yamaguchi & W. A. Newman, 1998. Invasion of Hawaiian shores by an Atlantic barnacle. Mar. Ecol. Prog. Ser. 165: 119–126.

Stachowicz, J. J., J. R. Terwin, R. B. Whitlatch & R. W. Osman, 2002. Linking climate change and biological invasions: ocean warming facilitates nonindigenous species Invasions. Proc. natl. Acad. Sci. 99: 15497–15500.

Strasser, M., 1999. *Mya arenaria* – an ancient invader of the North Sea coast. Helgol. Meeresunters. 52: 309–324.

Williamson, M., 1996. Biological invasions. Chapman & Hall, London. 244 pp.

Wonham, M. J., J. T. Carlton, G. M. Ruiz & L. D. Smith, 2000. Fish and ships: relating dispersal frequency to success in biological invasions. Mar. Biol. 136: 1111 – 1121.

Wyatt, T. & J. T. Carlton, 2002. Phytoplankton introductions in European coastal waters: why are so few invasions reported? pp. 41–46. In CIESM (Commission Internationale pour l'Exploration Scientifique de la mer Mediterranee) Workshop Monographs no. 20, 136 pp. Monaco.

Hydrobiologia **503:** 9–19, 2003.
M.B. Jones, A. Ingólfsson, E. Ólafsson, G.V. Helgason, K. Gunnarsson & J. Svavarsson (eds), *Migrations and Dispersal of Marine Organisms*.

Dispersal at hydrothermal vents: a summary of recent progress

Paul A. Tyler[1] & Craig M. Young[2]
[1]*School of Ocean and Earth Science, University of Southampton SOC, Southampton SO14 3ZH, U.K.*
[2]*Oregon Institute of Marine Biology, University of Oregon, P.O. Box 5389, Charleston, OR 97420, U.S.A.*
E-mail: pat8@soc.soton.ac.uk

Key words: dispersal, larvae, seeps, vents, Vestimentifera

Abstract

The discovery of hydrothermal vents along the Galapagos Rift in 1977 opened up one of the most dynamic and productive research themes in marine biology. In the intervening 25 years, hydrothermal vent faunas have been described from the eastern, northeastern and western Pacific, the mid-Atlantic Ridge and the Indian Ocean in the region of the Rodriguez Triple Junction. In addition, there is evidence of hydrothermal signals from the Gakkel Ridge in the Arctic, the central and southwest Indian Ridges and the Scotia Arc in Antarctica. Although often perceived as a continuous linear structure, there are many discontinuities that have given rise to separate biogeographic provinces. In addition, the intervening 25 years have seen a massive increase in our understanding of the biological processes at hydrothermal vents. However, how vents are maintained, and how new vents are colonised has been relatively poorly understood until recently. This review addresses the known larval development of vent-endemic invertebrates. The distribution of larvae in relation to the hydrothermal plume, and the ocean ridge in general, are discussed and the experimental evidence of larval longevity and transport are discussed using such variables as gene flow and larval development rates. The concept of larval dispersal along the mid-ocean ridge is discussed in relation to dispersal barriers and relates the known biogeography of hydrothermal vent systems to both local and evolutionary processes.

Introduction

The discovery of hydrothermal vents along the Galapagos Ridge in 1977 forced marine biologists to reassess the energy available for primary production in marine ecosystems. Up to that time, with a few minor exceptions, all energy for primary production was believed to derive ultimately from the sun in the form of photosynthetic primary production. With the discovery of hydrothermal vents, and subsequently cold seeps, a new energy source was discovered that originated beneath the seabed. In the case of hydrothermal vents, this energy source was hydrogen sulphide used by chemolithoautotrophic bacteria for primary production, the products of which could be transferred to supply the nutritional requirements of the host organism. The discovery of cold seeps further demonstrated that chemical energy could be used for significant primary production, in this case, being in the form of methane and hydrogen sulphide. At vents, hydrogen sulphide was available as a product of seawater passing through the fractured basalt of the mid ocean ridge, being heated, reacting chemically with the surrounding rock and all the sulphate being reduced inorganically to hydrogen sulphide (Van Dover, 2000). At seeps, hydrogen sulphide is made available by the bacterial reduction of sulphate using methane and the release of carbon dioxide and hydrogen sulphide (Sibuet & Olu, 1998). Hydrogen sulphide is also made available by the biological degradation of organic matter. Thus at seeps (and some vents) it is not uncommon for the hosts of thiotrophs and methanotrophs to live side by side, and, in some cases, bacteria capable of methanotrophy and thiotrophy can be found in the same host. In addition, seeps may be fuelled by particulate organic matter from pelagic production or by allochthonous organic matter carried into the sea by large rivers.

Chemolithoautotrophy was not unknown to marine ecologists, as chemosynthesis is known to be carried out by bacteria at the interface of anoxic and oxic conditions such as in the Black Sea (Sorokin, 1964) and in sediment, visible as a black layer (Gray, 1981). However, vents and seeps were the first expression of major ecosystems being fuelled by chemosynthesis.

In the very first publications describing hydrothermal vents, the importance of vent maintenance through reproduction, dispersal and recruitment was recognised. Corliss et al. (1979) noted "there are many questions to be answered about these animal communities. One concerns how they locate and colonize new vents. It is clear that an individual vent area has a finite lifetime". In the intervening 25 years, there has been a wealth of literature on the physiology, ecology, evolution and biogeography of hydrothermal vents (for reviews see Tunnicliffe, 1991; Childress & Fisher, 1992; Van Dover, 2000; Van Dover et al., 2002) and also on cold seeps (Sibuet & Olu, 1998), but study of the reproductive biology, dispersal and recruitment has been patchy (Tyler & Young, 1999). One of the most difficult aspects of studying the life-history biology of vent organisms is the necessity to examine temporal processes, requiring visits to the remote sampling sites of vents at different times of the year. This has had major logistical and fiscal constraints. However, with programmes such as the EU-funded AMORES programme and the US-funded LARVE program considerable progress has been made in the last few years.

In this mini-review, we address a number of aspects that have led to our understanding of the dispersal at hydrothermal vents. We have concentrated on hydrothermal vents as they form part of the linear mid-ocean ridge systems and have a number of features that are comparable.

Distribution of known vent sites

The known distribution of vents is by no means complete. Initial discoveries of vents were along the Galapagos Rift (Fig. 1). These were quickly followed by vent discoveries along the East Pacific Rise at 13° and 21° N, and 17–19° S, in the Guaymas Basin of the Gulf of California, and more recently at 9° N (Table 1) (Lonsdale, 1977; Corliss et al., 1979; Spiess et al., 1980; Hekinian et al., 1983; Lonsdale & Becker, 1985; Haymon et al., 1991). At the same time, vents were being discovered along the Juan de Fuca Ridge, the Explorer Ridge and the Gorda Ridge in the Northeast Pacific (Normark et al., 1982; Tunnicliffe et al., 1986; Rona et al., 1990). During the mid-1980s, there was extensive and successful exploration of the back-arc basins of the western Pacific, the first being the Manus basin (Both et al., 1986) (Table 1).

Major vent sites were not discovered in the Atlantic until the dredging of the TAG hydrothermal sites in 1985 (Rona et al., 1986). Subsequently, seven sites along the northern part of the Mid-Atlantic Ridge (MAR), south of the Azores have been discovered and described (Desbruyères et al., 2000), as well as sites to the north of Iceland, although no fauna has been described from these last sites. Very recently, hydrothermal plumes have been observed south of Ascension Island on the southern MAR (C.R. German, pers. comm.). In the Atlantic, hydrothermal vents are generally widely spaced compared to the EPR. In the southern hemisphere, hydrothermal vents with fauna occur at the Rodriguez Triple Junction in the Indian Ocean (Hashimoto et al., 2001; Van Dover et al., 2001) and vent plumes have been discovered along the Southwest and Southeast Indian Ridge (Table 1) (German et al., 1998; Scheirer et al., 1998). In the Atlantic sector of the southern ocean, vent plumes occur along the back-arc basin of the Scotia Arc (German et al., 2000). Lastly, in the Arctic, there is evidence of hydrothermal signals along the Knipovitch and Gakkel Ridges (Table 1), although a true vent fauna has not been sampled (see Edmonds et al., 2003).

Biogeographic implications of vent discovery and first indications of dispersal

The discovery and faunal descriptions of vent sites gave rise to the description of biogeographic 'provinces' and also gave the first indications of dispersal (Van Dover et al., 2002). The biogeographic province identified first was that of the East Pacific Rise (Fig. 1), dominated by the tube worm *Riftia pachyptila* Jones, the bivalves *Calyptogena magnifica* Boss & Turner, and *Bathymodiolus thermophilus* Kenk & Wilson, together with a variety of smaller species especially gastropod and polychaetes (Hessler & Smithey, 1983; Fustec et al., 1987).

With the discovery of vents in the Northeast Pacific, the dominant tubeworm was *Ridgeia piscesae*, together with the polychaete *Paralvinella sulfincola* Desbruyeres & Laubier (Tunnicliffe et al., 1985, 1986). Species of *Calyptogena* and *Bathymodiolus*

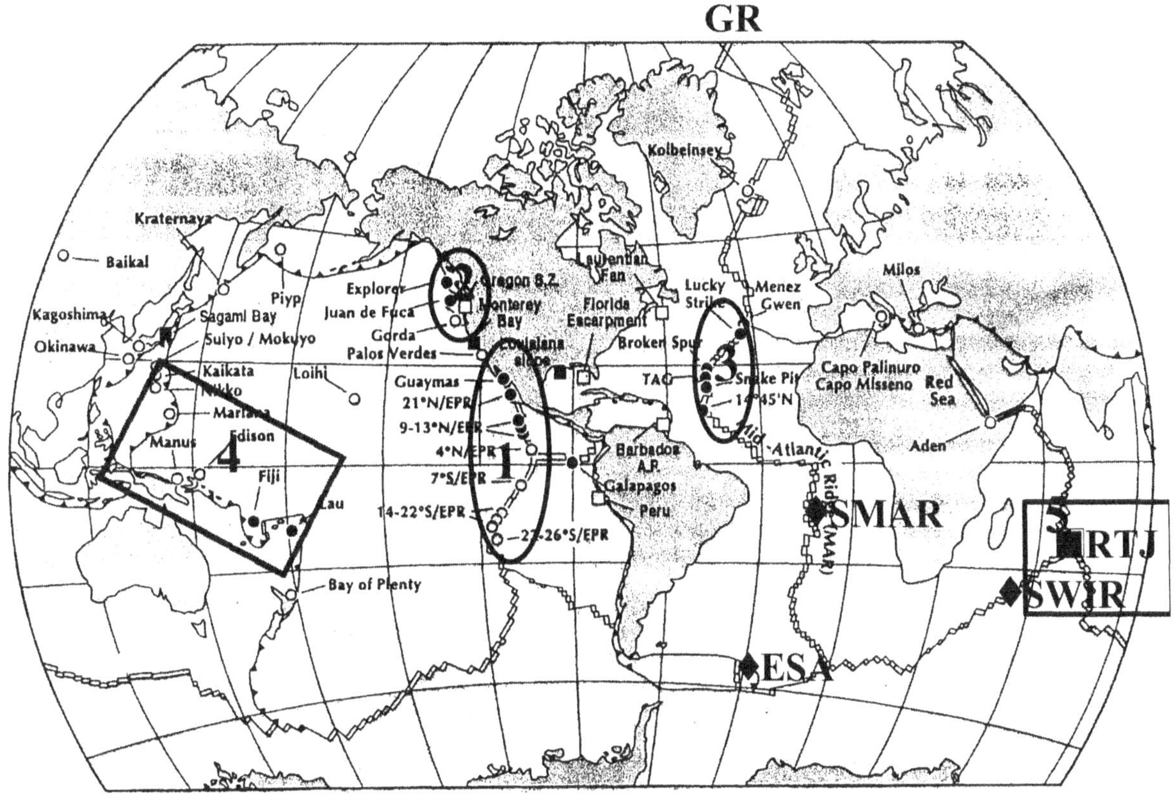

Figure 1. Distribution of the main hydrothermal vent sites in the world ocean. Recent discoveries of vents or plume signals are identified, RTJ: Rodriguez Triple Junction; Kairei and Edmond hydrothermal vents; Sites with plume signals include Southwest Indian Ridge (SWIR), Gakkel Ridge (GK), Southern Mid-Atlantic Ridge (SMAR) and East Scotia Arc (ESA). The known biogeographic provinces are 1. East Pacific; 2. NE Pacific, 3. Atlantic (Azores); 4. Western Pacific; 5. Indian Ocean (Figure adapted from Desbruyères & Segonzac, 1997).

were thought to be absent but there is evidence they occur occasionally, but not to the extent of the EPR, the most notable site being Clam Acres on the Juan de Fuca Ridge. These differences give rise to the Northeast Pacific geographical province. Although the MOR of the Northeast Pacific had been attached to that of the East Pacific in the geological past, separation by seafloor spreading had isolated the respective communities such that dispersal could not maintain a metapoluation and speciation occurred between the two biogeographic provinces (Tunnicliffe, 1988; Tunnicliffe & Fowler, 1996).

More profound differences among vent sites were observed with the discovery and description of vents in the Atlantic (see Desbruyères et al., 2000 for review). These sites are dominated by the shrimp *Rimicaris exoculata* Williams & Rona and the bivalve *Bathymodiolus azoricus* von Cosel, Comtet & Krylova at the shallower sites, and *B. puteoserpentis* von Cosel, Metivier & Hashimoto at the deeper sites (Desbruyères et al., 2000, 2001; Gebruk, 2002). Discovery of species of *Bathymodiolus* suggested that this genus is one of the most widespread of vent and seep genera; it is found in the EPR and the Atlantic, as well as the Indian Ocean and is also common at cold seeps in the Gulf of Mexico (MacDonald et al., 1990). Vestimentiferan tube worms are not found at any of the Atlantic vents sites although vestimentiferan species are found at cold seeps in the Gulf of Mexico and the eastern Atlantic (Kennicutt et al., 1985; Dando et al., 1992).

Although within the same ocean as the EPR, the back-arc basins of the western Pacific form their own potential biogeographic province (Fig. 1) dominated by species of *Bathymodiolus* but also have dominants such as gastropods (Desbruyères et al., 1994).

The discovery of a hydrothermal vent fauna in the Indian Ocean gave an opportunity to determine if the Atlantic or Pacific faunas had the greater contribution. Initial observations suggested that the dominant species (as determined by biomass) were shrimp, closely related to the Atlantic fauna, and anemones, displaying a close faunal affinity to the western Pacific fauna.

Table 1. Discovery and biogeography of known hydrothermal vents deeper than 500 m

Biogeographic province	Site	First recognised*	Depth (m)
Eastern Pacific	Galapagos	1977	2500
	EPR 21° N	1979	2600
	EPR 13° N	1981	2500
	Guaymas	1980	2000
	9° 50′ N	1989	2500
	17°/19° S	1980	2600–2800
	>25° S to SE Indian Ridge: unexplored		
North-East Pacific	Juan de Fuca	1981	2200
	Explorer	1984	1850
	Gorda	1988	2800
Mid-Pacific	Loihi	1987	1000–1200
Western Pacific	Manus	1986	2190
Back-arc basins	Marianus	1987	3660
	Fiji	1987	2000–2700
	Lau	1985	1707–1887
	Okinawa	1988	
Mid-Atlantic Ridge	Menez Gwen	1994	850
	Lucky Strike	1992	1700
	Rainbow	1996	2260–2350
	Broken Spur	1993	2900
	TAG	1985	4000
	Snake Pit	1985	3480
	Logachev	1994	3000
	South MAR	2001	3000–3500
East Scotia Ridge		1999	2600
Indian Ocean	Central Ridge	1986	3200
	RTJ	1995	2450
	SWIR	1997	3100–4100
	SEIR	1996	2500–2800
Arctic	Gakkel	2001	3500–4500
	Knipovitch	2000	3200

*Either the vent was observed or there was evidence from hydrothermal signals.

Careful examination of the species diversity suggested a close faunal affinity to the Pacific (Hashimoto et al., 2001; Van Dover et al., 2001).

These observations of biogeographic provinces demonstrate that the fauna of hydrothermal vents is not panmictic and that separation does occur. This separation may be as a result of seafloor spreading and that larval dispersal is insufficient to counteract the separation by geological processes.

How far will a larva disperse?

The simplest way to calculate the dispersal of a larva would be to multiply the advective process by the

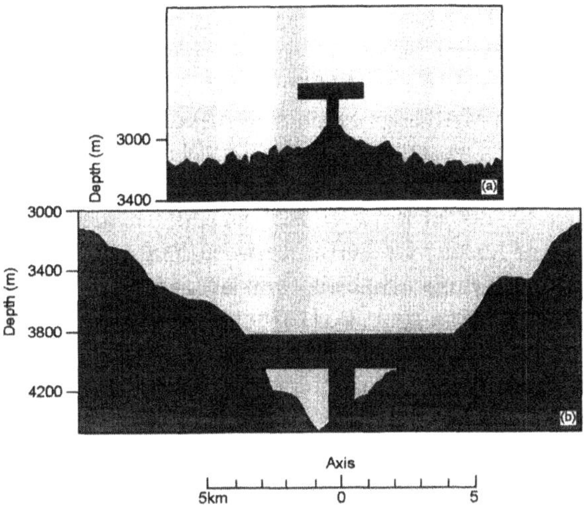

Figure 2. Diagrammatic representation of the constraint of hydrothermal plumes in the axial summit graben of (a) the East Pacific Rise; (b) the Mid-Atlantic Ridge.

length of larval life. In the 'normal' (non-vent and non-cold seep) deep sea this may be valid as the larva will develop to competency while being transported and, owing to the vastness of the sedimentary abyssal plains, have little problem settling on a suitable site, although the deep-sea may appear as a more heterogenous habit if one takes into account food falls (animal and plant), bioturbated areas and phytodetrital patches. At vents, however, the problem is much more complex. Mid-ocean ridge systems are linear, vents are widely spaced and often ephemeral, and advective process may not keep a larva at or near the MOR, let alone near a vent. Losses can occur by hydrographic processes either over the vent, at transform faults along the MOR, or by large geological features such as Iceland that sit directly over the MAR. To understand dispersal at vents, we have to consider the advective processes in relation to the local and distant seabed topography, and the larval life of a species.

Geological and hydrographic factors affecting dispersal

One of the most noticeable hydrographic features at vents is the hot water rising from the vent, the 'smoker'. Plume dynamics are complicated and beyond the scope of this review (see Van Dover, 2000). However, the rising water is likely to transport eggs or zygotes away from the vent into the water column. The rising water is called the 'buoyant plume' and, when it reaches a level of neutral density, it advects horizontally as the 'effluent layer'. At most vent sites the effluent layer is about 200–300 m above the vent, although, in the case of megaplumes found in the Northeast Pacific the effluent layer may be 1000 m above the vent site (Baker et al., 1987). The lateral transport of the effluent layer is determined by local topography. Along the East Pacific Rise, the summit axial graben is very shallow (at most a few 10s m deep) (MacDonald, 1982). In this situation, the plume rises above the local topography and can be advected in any direction (Fig. 2), either along the ridge axis or away laterally into the ocean interior (Kim et al., 1994; Kim & Mullineaux, 1998). Conversely, in the Atlantic, the axial summit graben may be as deep as 1000 m and the effluent layer is constrained by the sides of the summit graben (Fig. 2) (McDonald, 1982). In this case, flow is along the axis of the ridge, often constrained towards one graben wall by Coriolis force. Water can only escape this constraint when it reaches one of the transform faults that cut the MAR in many places (Fig. 1). The site at Rainbow is an anomaly in the Atlantic as it is in a transform fault, whereas the other sites tend to be towards the centre of segments.

Transform faults may impede dispersal of larvae along the mid-ocean ridge by the lateral loss of larvae from the ridge. Conversely, they may transport larvae from one segment to the next. Other topographic dispersal constraints include large land masses such as Iceland that, presumably, prevent the dispersal of vent larvae north and south of the island. A second example from the Atlantic is the Azores Rise that forms a relatively shallow sill that may form a bathymetric barrier to larval dispersal. A constraint to larval dispersal may arise along the Chile Rise where the southeastern end of the Rise forms a 'dead end' against the southern tip of South America. Lastly, some segments of ridge such as the East Scotia Ridge in the south Atlantic are so isolated that colonisation may have been difficult, if it ever occurred!

There are still relatively few data for the hydrographic characteristics of flow along the mid-ocean ridge that may transport larvae (see Tyler & Young, 1999) (Table 2). The most recent studies are those of Marsh et al. (2001) who deployed a current meter for 4 months, 175 m above the seabed at 9° N on the EPR. Current velocities were variable and periodicities were semi-diurnal, diurnal and fortnightly, with residual flows trending along the ridge axis. These data suggested that the furthest distance a particle could travel from the point source was ~100 km SSE, whilst to the

Table 2. Measured and inferred flow close to mid-ocean ridges (cm s^{-1})

Site	Flow
Juan de Fuca	5–20
East Pacific Rise	'few'
10° N	1–2
13° N	4.2–5.2
Axial seamount	4
Mid-Atlantic Ridge	20 (Variable)

NNW it was 47 km. These authors concluded from the flow data (and larval life-see below) that most larvae would be retained within a 'few tens of kilometers' of their release. In the Atlantic, Khripounoff et al. (2001) deployed current metres at the Rainbow hydrothermal vent field and showed variable current speeds between 9 and 19 cm s^{-1} at 15 mab and between 6 and 17 cm s^{-1} at 310 mab.

Evidence of larval dispersal from field sampling

Observations of larval development were initially indirect. The types of larval shell in molluscs (prodissoconch and dissoconch in bivalves, protoconch in gastropods) were indicative of the period of time that the larvae spent in the plankton (Lutz et al., 1980; Mullineaux et al., 1996). Of particular interest is the larval shell of species of *Bathymodiolus*. Interpretation of the large size of the larval shell suggests that larvae spend considerable time in the plankton and may even by planktotrophic. There is strong evidence that recruitment is seasonal (Comtet & Desbruyères, 1998). The field sampling of larvae close to hydrothermal vents has become more sophisticated in recent years. Mullineaux et al. (1995) used a MOCNESS plankton sampler to sample the water column over the Juan de Fuca Ridge whilst at the same time measuring the light attenuation (Δc), a measure of the hydrothermal plume. Along-axis and transverse tows demonstrated that larvae of vent gastropods were significantly associated with the plume (21 individuals per 1000 m^3 in the plume compared to 1.4 individuals per 1000 m^3 in the open ocean). Larvae of *Calyptogena* occurred exclusively in the plume (0.5 individuals per 1000 m^3).

In the Atlantic, Herring & Dixon (1998) sampled for larvae of vent shrimp (particularly *Rimicaris exoculata*). By sampling with a Rectangular Midwater Trawl (RMT 1+8) over a depth range of 2000–3050 m, these authors found that vent shrimp were widely dispersed round the Broken Spur field, and extended into the next MAR segment and the Atlantis fracture zone beyond. Density of larvae declined both horizontally and vertically (from an average of 14.83 individuals per 100 000 m^3 between 2750 and 3050 m to an average of 1.53 per 100 000 m^3 between 2000 and 2500 m) from the Broken Spur site with larvae being found up to 100 km from the site (Herring & Dixon, 1998). This would allow them to colonise adjacent fields. These field observations support genetic data (Creasey et al., 1996; Shank et al., 1998) that suggest high gene flow in *Rimicaris exoculata*.

Evidence of larval development from laboratory and field experimentation

One of the most difficult aspects of understanding larval development and dispersal of vent organisms is the laboratory culture of larvae and their use in laboratory and field experimentation. The main problems are the same for those of 'normal' deep sea taxa in that the larvae have to be cultured at ambient temperature and pressure, requiring special pressure vessels.

The first successful culture of a chemosynthetically-related metazoan was the larval development of the vestimentiferan *Lamellibrachia* sp. from the cold seeps of the Gulf of Mexico (Young et al., 1996). More recently, Marsh et al. (2001) have presented data on the larval development of *Riftia pachyptila*. Embryonic development was conducted in both the laboratory and in the field, and the rate of development between the two showed no significant difference. Development was through a lecithotrophic trochophore larva that at 34 d had two ciliated bands but no mouth, apical tuft or other features (Marsh et al., 2001; Young, pers. obs.). These authors calculated the growth rate and larval life by determining the respiration rate of the larvae and how long it took the larva to consume its energy reserves. The estimated larval life span was ~38 d and the survival of larvae decreased almost linearly from >95% at day 5 to approaching zero at day 40 (Marsh et al., 2001). When these data are combined with the flow data (see above), the net dispersal of a larva of *Riftia pachyptila* was ~100 km along the mid-ocean ridge. In a parallel experiment, Pradillon et al. (2001) determined the larval survival of *Alvinella pompejana* Desbruyeres & Laubier at different temperatures from 2 to 27 °C at ambient pressure. De-

Table 3. Pressure/temperature effects on the first larval stage of *Mirocaris fortunata* (from Tyler & Dixon, 2000)

	Replicate		
	1	2	3
At 10 °C			
1 atm	Very active	Very active	Very active
150 atm	Very active	Active (L)	Very active
250 atm	Active (S)	Active (S)	Active (S)
300 atm	Dead	Low active	Low active
At 20 °C			
1 atm	Dead	Dead	Dead
150 atm	Low active	Very active	Very active
250 atm	Active (L)	Very active	Active (L)
300 atm	Active (S)	Dead (*)	Dead

Responses: very active: larva moves by whole body jerks with no stimulation; active: legs active without stimulation (L) or whole body moves with stimulation (S); low active: very slow response to stimulation; dead: no response; *lipid granules displaced.

velopment only occurred at 10 and 14 °C, suggesting that the warmer waters of the vent were necessary for larval development. In subsequent experiments these authors subjected larvae to a period of cold (2 °C) and them 10 °C water, followed by 2 °C. Pradillon et al. (2001) interpreted these data as the cold non-vent temperatures delaying development but once vent temperatures were encountered embryonic development was rapid.

In the Atlantic, Tyler & Dixon (2000) used pressure/temperature tolerances of larvae of *Mirocaris fortunata* to determine potential dispersal. Hatched zoeae 1 were released naturally in the laboratory and subjected to pressures of 1, 150, 250 and 300 atm at 10 and 20 °C for 24 h. Resulting data (Table 3) was interpreted as suggesting that larvae of *M. fortunata* could ascend the water column as far as the permanent thermocline (~200 m) but 1 atm at 20 °C was lethal. The position of bresiliid shrimp larvae in the water column is of interest as there is evidence that the larvae feed on phytoplankton from surface production (Allen Copley et al., 1998) either in the water column or on material that has sunk to the deep-sea bed. However, there is no evidence of seasonality of reproduction in vent shrimps (Ramirez et al., 2000).

The experimental work described in this section has demonstrated that embryonic experimentation using vent organisms is tractable and provides a valuable way forward in the future.

Evidence of dispersal from molecular analysis

There has been considerable interest in the use of molecular techniques for providing information on the dispersal of vent organisms. This aspect has been reviewed by Vrijenhoek (1997) and Tyler & Young (1999) and is presented in Table 4. From these data, we can ascertain that most vent species have an N_m greater than 1 [N_m is a measure of the number of migrants per generation (Vrijenhoek, 1997)]. The N_m is $\gg 1$ for a number of species including the alvinellid polychaetes but particularly high for *Rimicaris exoculata*. Analysis of the N_m with distance between populations (Vrijenhoek, 1997) suggests that the tubeworm *Riftia pachyptila* appears to conform to the 'stepping stone model' where there is a steady change in genetic composition between adjacent populations within a metapopulation. Conversely, in *Bathymodiolus thermophilus* and *Rimicaris exoculata*, the data conform to the 'island model', which suggests that all the populations within a metapopulation both contribute to, and recruit from what is effectively a large single pool of larvae.

Where does the study of vent species dispersal and biogeography go from here?

In the 25 years since the discovery of hydrothermal vents, there has been considerable effort in finding new vent sites, describing the very high proportion of new species and understanding the way vent ecosystems function (reviewed by Van Dover, 2000). However, because of fiscal and logistical constraints, including the problems of weather, most of the known vent sites are found at tropical and subtropical latitudes. From the data available, it has been possible to identify a number of biogeographic provinces (see above and Table 1 and Fig. 1). If we are to understand the extent of dispersal it is important to extend our knowledge by exploration of potential vents sites both to northern and southern higher latitudes.

In the Arctic, there is evidence of hydrothermal activity along the Gakkel Ridge north of Siberia and under almost permanent sea ice (Edmonds et al., 2003). The ridge may support a vent fauna unique to the Arctic, as the Gakkel Ridge is apparently separated from the MAR north of Iceland by the exceptionally long transform fault. However, the latest charts from this region suggest this is a bend in the MOR rather than a transform fault (P. Hunter, SOC, pers. com.).

Table 4. Gene flow in hydrothemal vent species (from Tyler & Young, 1999)

Species	Location	Method	Gene flow/N_m	Ref.
Paralvinella grasslei Desbruyères & Laubier	11, 13, 21° N EPR,Galapagos, Guyamas	Allozyme	High/3.4	6, 12
Alvinella pompejana Desbruyères & Laubier	13, 21° N EPR	Allozyme	High/5.7	6, 12
Alvinella pompejana Desbruyères & Laubier	13° N EPR	Rest. Analysis	High/5.2	14
Alvinella caudata Desbruyères & Laubier	13, 21° N EPR	Allozyme	High/6.7	6, 12
Riftia pachyptila Jones	Galapagos; 21° N EPR	Allozyme	High	2
Riftia pachyptila Jones	9, 11, 13, 21° N EPR, Galapagos, Guyamas	Allozyme	High/5.4	4, 12
Ridgeia piscesae Jones	JdeF, Explorer, Gorda	Allozyme	High/3.3	9
	Gorda	DNA	High/3.3	13
Tevnia jerichonana Jones	9, 11, 13° N EPR	Allozyme	High/2.4	12, 13
Oasisia alvinae Jones	9, 11, 13, 21° N EPR	Allozyme	High/1.2	13
Bathymodiolus sp.	Lau, Fiji, 13° N	Allozyme	High	5
B. thermophilus Kenk & Wilson	13° N, Galapagos	Allozyme	Low	1
B. thermophilus Kenk & Wilson	9, 11, 13° N, Galapagos	Alloz/DNA	High/5.5	7
Calyptogena magnifica Boss & Turner	9, 21° N, 18° S, Gal	Alloz/DNA	High/11.7	10
Lepetodrilus elevatus McLean	9, 11, 13, 21° N Gal	Allozyme	High/1.8	11
L. e. galriftensis McLean	9, 13, 21° N EPR	Allozyme	High/1.4	11
L. pustulosus McLean	9, 11, 13° N EPR, Gal	Allozyme	High/2.5	11
Eulepetopsis vitrea McLean	9, 11, 13, 20° N, Gal	Allozyme	High/1.0	11
Ventiella sulfuris Barnard & Ingram	11° N EPR, Galapagos	Allozyme	High on EPR/0.3	3
Rimicaris exoculata Williams & Rona	TAG, Broken Spur	Allozyme	High/250	8

Based on 1. Grassle (1985); 2. Bucklin (1988); 3. France et al. (1992); 4. Black et al. (1994); 5. Moraga et al. (1994); 6. Jollivet et al. (1995); 7. Craddock et al. (1995); 8. Creasey et al. (1996); 9. Southward et al. (1996); 10. Karl et al. (1996); 11. Craddock et al. (1997); 12. Vrijenhoek (1997); 13. Black et al. (1998); 14. Jollivet et al. (1998). For definition of N_m see text. See Figure 1 for sites.

The MAR north of Iceland is separated from the main MAR south of Iceland by the land mass of Iceland. It is almost certain that the Arctic, and maybe the whole region north of Iceland, will form a separate vent (and non-vent) biogeographic province.

In the southern hemisphere, nothing is known of hydrothermal vent activity along the mid-ocean ridge from the Easter island microplate, south of Australia to the Southeast Indian Ridge. There is evidence of a hydrothermally-influenced community on the Pacific-Antarctic Ridge between 37° 30′ S and 110° 30′ W, comprising filter feeders and a species of *Bathymodiolus* (Stecher et al., 2002). More is known of hydrothermal activity in the Atlantic and Indian sectors of the Southern Ocean. The Kairei and Edmond sites have been discovered and their fauna analysed (Hashimoto et al., 2001; Van Dover et al., 2001). There is evidence of vent sites from the Central Indian Ridge, the Southwest Indian Ridge, the Southeast Indian Ridge, the MAR south of Ascension Island and from the Scotia Arc (Herzig & Pluger, 1988; German et al., 1998; German et al., 2000; German, pers. obs 2002). The fauna of these vent sites is, as yet, unsampled let alone described. The Scotia Arc is the most isolated piece of mid-ocean ridge in the world ocean and is ~10 million years old. It is known to have venting in segments 2 and 9 (German et al., 2000). Examining its geographic position, it is not easy to predict what fauna may be found there. Tectonically, it is connected to the southern part of the MAR although there are numerous long transform faults along this segment. Conversely, it is connected hydrographically to the Pacific with a major water flow from the Pacific, through the Drake Passage and across the Scotia Sea.

Conclusions

For the last 25 years, discoveries at hydrothermal vents have caused great excitement for the scientifically-interested public and have posed intellectual challenges for the scientists analysing vent ecology. One of the most challenging aspects of vent biology has been an understanding of the processes that maintain current vents and colonise new vents. In parallel with this has been the development of our understanding of the biogeography of hydrothermal vents. Recent developments have led to a breakthrough in the ability to culture larvae of vent invertebrates. These data, in conjunction with physiological methods and an analysis of the flow regimes in and around vents, have allowed the prediction of the dispersal distances of vent-originated larvae. In future, it is necessary to explore for new vent sites to determine if larval dispersal has allowed these sites to be colonised or whether geological or hydrographic factors limit dispersal, allowing speciation at different places on the MOR leading to separate biogeographic provinces.

Acknowledgements

The authors wish to thank the organisers of EMBS37 for their kind invitation to participate in the conference in Iceland. Supported in part by NSF grants OCE-0118733 and OCE-9619606.

References

Allen Copley, C. E., P. A. Tyler & M. S. Varney, 1998. Lipid profiles of hydrothermal vent shrimps. Cah. Biol. Mar. 39: 229–231.

Baker, E. T., G. J. Massoth & R. E. Feely, 1987. Cataclysmic hydrothermal venting on the Juan de Fuca Ridge. Nature 329: 149–151.

Black, M. B., R. A. Lutz & R. C. Vrijenhoek, 1994. Gene flow among vestimentiferan tube worm (*Riftia pachyptila*) populations from hydrothermal vents of the eastern Pacific. Mar. Biol. 120: 33–39.

Black, M. B., A. Trivedi, P. A. Y. Maas, R. A. Lutz & R. C. Vrijenhoek, 1998. Population genetics and biogeography of vestimentiferan tube worms. Deep-Sea Res. II 45: 365–382.

Both, R. E. A., 1986. Hydrothermal chimneys and associated fauna in the Manus back-arc basin, Papua New Guinea. EOS, Transactions, American Geophysical Union 67: 489–490.

Bucklin, A., 1988. Allozymic variability of *Riftia pachyptila* populations from the Galapagos Rift an 21 degree N hydrothermal vents. Deep-Sea Res. I 35: 1759–1768.

Childress, J. J. & C. R. Fisher, 1992. The biology of hydrothermal vent animals: Physiology, biochemistry, and autotrophic symbioses. Ocean. mar biol. Ann. Rev 30: 337–441.

Comtet, T. & D. Desbruyères, 1998. Population structure and recruitment in mytilid bivalves from the Lucky Strike and Menez Gwen hydrothermal vent fields (37° 17′ N and 37° 50′ N on the Mid-Atlantic Ridge). Mar. Ecol. Prog. Ser. 163: 165–177.

Corliss, J. B., J. Dymond, L. I. Gordon, J. M. Edmond, R. P. von Herzen, R. D. Ballard, K. Green, D. Williams, A. Bainbridge, K. Crane & T. H. Andel, 1979. Submarine thermal springs on the Galapagos Rift. Science 203: 1073–1083.

Craddock, C., W. R. Hoeh, R. A. Lutz & R. C. Vrijenhoek, 1995. Extensive gene flow among mytilid (*Bathymodiolus thermophilus*) populations from hydrothermal vents of the eastern Pacific. Mar. Biol. 124: 137–146.

Craddock, C., R. A. Lutz & R. C. Vrijenhoek, 1997. Patterns of dispersal and larval development of archaeogastropod limpets at hydrothermal vents in the eastern Pacific. J. exp. mar. Biol. Ecol. 210: 37–51.

Creasey, S., A. D. Rogers & P. A. Tyler, 1996. Genetic comparison of two populations of the deep-sea vent shrimp *Rimicaris exoculata* (Decapoda: Bresiliidae) from the Mid-Atlantic Ridge. Mar. Biol. 125: 473–482.

Dando, P. R., A. J. Southward, E. C. Southward, D. R. Dixon, A. Crawford & M. Crawford, 1992. Shipwrecked tube worms. Nature 356: 667.

Desbruyères, D., A. M. Alayse-Danet & S. Ohta, 1994. Deep-sea hydrothermal communities in southwestern Pacific back-arc basins (the North Fiji and Lau Basins): Composition, microdistribution and food web. Mar. Geol. 116: 227–242.

Desbruyères, D. & M. Segonzac, 1997. Handbook of Deep-sea Hydrothermal Vent Fauna. IFREMER. 279 pp.

Desbruyères, D., A. Almeida, M. Biscoito, T. Comtet, A. Khripounoff, N. Le Bris, P.-M. Sarradin & M. Segonzac, 2000. A review of the distribution of hydrothermal vent communities along the northern Mid-Atlantic Ridge: dispersal vs. environmental controls. Hydrobiologia 440: 201–216.

Desbruyères, D., M. Biscoito, J. C. Caprais, A. Colaco, T. Comtet, P. Crassous, Y. Fouquet, A. Khripounoff, N. Le Bris, K. Olu, R. Riso, P.-M. Sarradin, M. Segonzac & A. Vangriesheim, 2001. Variations in deep-sea hydrothermal vent communities on the Mid-Atlantic Ridge near the Azores plateau. Deep-Sea Res. I 48: 1325–1346.

Edmonds, H., P. J. Michael, E. T. Baker, D. P. Connelly, J. E. Snow, C. H. Langmuir, H. J. B. Dick, R. Mühe, C. R. German & D. W. Graham, 2003. Discovery of abundant hydrothermal venting on the ultraslow-spreading Gakkel ridge in the Arctic Ocean. Nature 421: 252–256.

France, S. C., R. R. Hessler & R. C. Vrijenhoek, 1992. Genetic differentiation between spatially-disjunct populations of the deep-sea, hydrothermal vent-endemic amphipod *Ventiella sulfuris*. Mar. Biol. 114: 551–559.

Fustec, A., D. Desbruyères & S. K. Juniper, 1987. Deep-sea hydrothermal vent communities at 13° N on the East Pacific Rise: microdistribution and temporal variation. Biol. Oceanogr. 4: 121–164.

Gebruk, A., 2002. Biology of hydrothermal systems. KMK Press, Moscow. 543 pp (in Russian).

German, C. R., E. T. Baker, C. Mevel & K. Tamaki, 1998. Hydrothermal activity along the southwest Indian ridge. Nature 395: 490–493.

German, C. R., R. A. Livermore, E. T. Baker, N. I. Bruguier, D. P. Connelly, A. P. Cunningham, P. Morris, I. P. Rouse, P. J. Statham & P. A. Tyler, 2000. Hydrothermal plumes above the East Scotia Ridge: an isolated high-latitude back-arc spreading centre. Earth Planet. Sci. Lett. 184: 241–250.

Grassle, J. P., 1985. Genetic differentiation in populations of hydrothermal vent mussels (*Bathymodiolus thermophilus*) from the Galapagos Rift and at 13° N on the East Pacific Rise. Bull. Biol. Soc. Wash. 6: 429–442.

Gray, J. S., 1981. The Ecology of Marine Sediments. Cambridge University Press. 185 pp.

Hashimoto, J. and others, 2001. Hydrothermal vents and associated biological communities in the Indian Ocean. InterRidge News 10: 21–22.

Haymon, R. M., D. J. Fornari, M. H. Edwards, S. Carbotte, D. Wright & K. C. MacDonald, 1991. Hydrothermal vent distribution along the East Pacific Rise crest 9° 09′–54′ N and its relationship to magmatic and tectonic processes on fast-spreading mid-ocean ridges. Earth Planet. Sci. Lett. 104: 513–534.

Hekinian, R., M. Fevrier, F. Avedick, P. Cambon, J. L. Charlou, H. Needham, J. Raillard, J. Boulegue, L. Merlivat, A. Moinet,

S. Manganini & J. Lange, 1983. East Pacific Rise near 13° N: geology of new hydrothermal fields. Science 219: 1321–1324.

Herring, P. J. & D. R. Dixon, 1998. Extensive deep-sea dispersal of postlarval shrimp from a hydrothermal vent. Deep-Sea Res. I 45: 2105–2118.

Herzig, P. M. & W. L. Pluger, 1988. Exploration for hydrothermal activity near the Rodriguez Triple Junction, Indian Ocean. Can. Mineral. 26: 721–736.

Hessler, R. R. & W. M. Smithey, 1983. The distribution and community structure of megafauna at the Galapagos Rift hydrothermal vents. In Rona, P. A., K. Bostrom, L. Laubier & K. L. Smith Jr. (eds), Hydrothermal Processes at Seafloor Spreading Centres. Plenum, New York.

Jollivet, D., D. Desbruyèeres, F. Bonhomme & D. Moraga, 1995. Genetic differentiation of deep-sea hydrothermal vent alvinellid populations (Annelida: Polychaeta) along the East Pacific Rise. Heredity 74: 376–391.

Jollivet, D., L. J. R. Dixon, D. Desbruyères & D. R. Dixon, 1998. Ribosomal (rDNA) variation in a deep sea hydrothermal vent polychaete, *Alvinella pompejana*, from 13° N on the East Pacific Rise. J. mar. biol. Ass. U.K. 78: 113–130.

Karl, S. A., S. Schutz, D. Desbruyères, R. Lutz & R. C. Vrijenhoek, 1996. Molecular analysis of gene flow in the hydrothermal vent clam (*Calyptogena magnifica*). Mol. Mar. Biol. Biotech. 5: 193–202.

Kennicutt II, M. C., J. M. Brooks, R. R. Bidigare, R. R. Fay, T. L. Wade & T. J. McDonald, 1985. Vent-type taxa in a hydrocarbon seep region on the Louisiana slope. Nature 317: 351–353.

Khripounoff, A., A. Vangriesheim, P. Crassous, M. Segonzac, A. Colaco, D. Desbruyères & R. Barthelemy, 2001. Particle flux in the Rainbow hydrothermal vent field (Mid-Atlantic Ridge): dynamics, mineral and biological composition. J. mar. Res. 59: 633–656.

Kim, S. L., L. S. Mullineaux & K. R. Helfrich, 1994. Larval dispersal via entrainment, into hydrothermal plumes. J. Geophys. Res. 99: 12 655–12 665.

Kim, S. L. & L. S. Mullineaux, 1998. Distribution and near-bottom transport of larvae and other plankton at hydrothermal vents. Deep-Sea Res. II 45: 423–440.

Lonsdale, P., 1977. Clustering of suspension-feeding macrobenthos near abyssal hydrothermal vents at oceanic spreading centres. Deep-Sea Res. I 24: 857–863.

Lonsdale, P. & K. P. Becker, 1985. Hydrothermal plumes, hot springs and conductive heat flow in the Southern Trough of Guaymas Basin. Earth Planet. Sci. Lett. 73: 211–225.

Lutz, R. A., D. Jablonski, D. C. Rhoads & R. D. Turner, 1980. Larval dispersal of a deep-sea hydrothermal vent bivalve from the Galápagos Rift. Mar. Biol. 57: 127–133.

MacDonald, I. R., J. F. Reilly II, N. L. Guinasso Jr., J. M. Brooks, R. S. Carney, W. A. Bryant & T. J. Bright, 1990. Chemosynthetic mussels at a brine-filled pockmark in the northern Gulf of Mexico. Science 248: 1069–1099.

MacDonald, K. C., 1982. Mid-ocean ridges: fine scale tectonic, volcanic, and hydrothermal processs within the plate boundary zone. Ann. Rev. Earth Planet. Sci. 88: 119–131.

Marsh, A. G., L. S. Mullineaux, C. M. Young & D. T. Manahan, 2001. Larval dispersal potential of the tubeworm *Riftia pachyptila* at deep-sea hydrothermal vents. Nature 411: 77–88.

Moraga, D., D. Jollivet & F. Denis, 1994. Genetic differentiation across the western Pacific populations of the hydrothermal vent bivalve *Bathymodiolus* spp. and the eastern Pacific (13° N) population of *Bathymodiolus thermophilus*. Deep-Sea Res. I 41: 1551–1567.

Mullineaux, L., P. H. Wiebe & E. T. Baker, 1995. Larvae of benthic invertebrates in hydrothermal vent plumes over the Juan de Fuca Ridge. Mar. Biol. 122: 585–596.

Mullineaux, L. S., S. L. Kim, A. Pooley & R. A. Lutz, 1996. Identification of archaeogastropod larvae from a hydrothermal vent community. Mar. Biol. 124: 551–560.

Normark, W. R., J. E. Lupton, J. W. Murray, R. A. Koski, D. A. Clague, J. L. Morton, J. R. Delaney & H. P. Johnson, 1982. Polymetallic sulfide deposits and water-column tracers of active hydrothermal vents on the southern Juan de Fuca Ridge. Mar. Technol. Soc. J. 16: 46–52.

Pradillon, F., B. Shillito, C. M. Young & F. Gaill, 2001. Developmental arrest in vent worm embryos. Nature 413: 698–699.

Ramirez-Llodra, E. Z., P. A. Tyler & J. T. Copley, 2000. Reproductive biology of three caridean shrimp, *Rimicaris exoculata*, *Chorocaris chacei* and *Mirocaris fortunata* (Caridea: Decapoda), from hydrothermal vents. J. mar. biol. Ass. U.K. 80: 473–484.

Rona, P. A., G. Klinkhammer, T. A. Nelson, J. H. Trefry & H. Elderfield, 1986. Black smokers, massive sulphides and vent biota at the Mid-Atlantic Ridge. Nature 321: 33.

Rona, P. A., R. P. Deninger, M. R. Fisk, K. J. Howard, K. D. Klitgord, J. S. McClain, G. R. McMurray, G. L. Taghon & J. C. Wiltshire, 1990. Major off-axis hydrothermal activity on the northern Gorda Ridge. Geology 18: 493–496.

Scheirer, D. S., E. T. Baker & K. T. M. Johnson, 1998. Detection of hydrothermal plumes along the Southeast Indian Ridge near the Amsterdam-St. Paul Plateau. Geophys. Res. Lett. 25: 97–100.

Shank, T. M., R. A. Lutz & R. C. Vrijenhoek, 1998. Molecular systematics of shrimp (Decapoda; Bresiliidae) from deep-sea hydrothermal vents: I. Enigmatic 'small orange' shrimp from the Mid-Atlantic Ridge are *Rimicaris exoculata*. Mol. mar. biol. Biotech. 7: 88–96.

Sibuet, M. & K. Olu, 1998. Biogeography, biodiversity and fluid dependence of deep-sea cold-seep communities at active and passive margins. Deep-Sea Res. II 45: 517–567.

Sorokin, Y. I., 1964. On the primary production and bacterial activities in the Black Sea. J. Cons. Int. Explor. Mer 29: 41–60.

Southward, E. C., V. Tunnicliffe, M. B. Black, D. R. Dixon & L. J. R. Dixon, 1996. Ocean-ridge segmentation and vent tubeworms (Vestimentifera) in the NE Pacific. In MacLeod, C. J., P. A. Tyler & C. L. Walker (eds), Tectonic, Magmatic, Hydrothermal and Biological Segmentation of Mid-Ocean Ridges. Geological Society of London, London.

Spiess, R., K. C. MacDonald, T. Atawater, R. D. Ballard, A. Carranza, D. Cordoba, C. Cox, V. Diaz Garcia, J. Francheteau, J. Guerrero, J. Hawkins, R. M. Haymon, R. Hessler, T. Juteau, M. Kastner, R. Larson, B. Luyendyk, J. MacDougall, S. Miller, W. R. Normark, J. Orcutt & C. Rangin, 1980. East Pacific Rise: hot springs and geophysical experiments. Science 207: 1421–1433.

Stecher, J., M. Türkay & C. Borowski, 2002. Faunal assemblages on the Pacific-Antarctic Ridge near the Foundation Seamount Chain (37° 30′ S, 110° 30′ W). Cah. Biol. Mar. 43: 271–274.

Tunnicliffe, V., 1988. Biogeography and evolution of hydrothermal-vent fauna in the eastern Pacific Ocean. Proc. r. Soc. Lond. B, 233: 347–366.

Tunnicliffe, V., 1991. The biology of hydrothermal vents: Ecology and Evolution. Ocean. mar. biol. Ann. Rev. 29: 319–407.

Tunnicliffe, V., S. K. Juniper & M. E. de Burgh, 1985. The hydrothermal vent community of Axial seamount, Juan de Fuca Ridge. Bull. Biol. Soc. Wash. 6: 453–464.

Tunnicliffe, V., M. Botros, M. E. De Burgh, A. Dinet, H. P. Johnson, S. K. Juniper & R. E. McDuff, 1986. Hydrothermal vents of Explorer Ridge, Northeast Pacific. Deep-Sea Res. I 33: 401–412.

Tunnicliffe, V. & C. M. R. Fowler, 1996. Influence of sea-floor spreading on the global hydrothermal vent fauna. Nature 379: 531–533.

Tyler, P. A. & C. M. Young, 1999. Reproduction and dispersal at vents and cold seeps. J. mar. biol. Ass. U.K. 79: 193–208.

Tyler, P. A. & D. R. Dixon, 2000. Temperature/pressure tolerance of the first larval stage of *Mirocaris fortunata* from Lucky Strike hydrothermal vent field. J. mar. biol. Ass. U.K. 80: 739–740.

Van Dover, C. L., 2000. The ecology of deep-sea hydrothermal vents. Princeton University Press, Princeton, New Jersey. 424 pp.

Van Dover, C. L., C. R. German, K. G. Speer, L. M. Parson & R. C. Vrijenhoek, 2002. Evolution and biogeography of deep-sea vent and seep invertebrates. Science 295: 1253–1257.

Van Dover, C. L., S. E. Humphris, D. Fornari, C. M. Cavanaugh, R. Collier, S. K. Goffredi, J. Hashimoto, M. D. Lilley, A. L. Reysenbach, T. M. Shank, K. L. Von Damm, A. Banta, R. M. Gallant & R. C. Vrijenhoek, 2001. Biogeography and ecological setting of Indian Ocean hydrothermal vents. Science 294: 818–823.

Vrijenhoek, R. C., 1997. Gene flow and genetic diversity in naturally fragmented metapopulations of deep-sea hydrothermal vent animals. J. Hered. 88: 285–293.

Young, C. M., E. Vazquez, A. Metaxas & P. A. Tyler, 1996. Embryology of vestimentiferan tube worms from deep-sea methane/sulphide seeps. Nature 381: 514–516.

Hydrobiologia **503:** 21–28, 2003.
M.B. Jones, A. Ingólfsson, E. Ólafsson, G.V. Helgason, K. Gunnarsson & J. Svavarsson (eds), Migrations and Dispersal of Marine Organisms.

Spread of the Chinese mitten crab (*Eriocheir sinensis* H. Milne Edwards) in Continental Europe: analysis of a historical data set

L.-M. Herborg[1], S. P. Rushton[2], A. S. Clare[1] & M. G. Bentley[1]
[1]*School of Marine Science & Technology, Ridley Building, University of Newcastle-upon-Tyne, Newcastle upon Tyne, NE1 7RU, U.K.*
[2]*Centre for Life Science Modelling, School of Biology, Porter Building, University of Newcastle-upon-Tyne, Newcastle upon Tyne, NE1 7RU, U.K.*
E-mail: leif-matthias.herborg@ncl.ac.uk

Key words: invasive species, *Eriocheir sinensis*, crustacean, migration, catadromous, Europe

Abstract

The Chinese mitten crab, *Eriocheir sinensis* (H. Milne Edwards), is an invasive species that lives as an adult predominantly in freshwater but migrates seawards to breed. It has spread via ballast water and/or intentional introduction to Continental Europe, Southern France, U.S.A. (San Francisco Bay) and the United Kingdom. Analysis of detailed historic data from the outbreak in Europe was digitised and analysed using Geographical Information Software. This revealed that there were two separate invasions in Northern Europe and Southern France, with an average range expansion during the peak period of 562 km/year from 1928–1938 (Northern Europe) and 380 km/year from 1954–1960 in Southern France. Size class distribution data from the lower estuary of the River Elbe (Germany) (1932–1936) illustrate migration patterns to and from the estuary over the year. Marking experiments determined that the mean rate of downstream migration for adults was 11.5 km/day (SD 3.54; n=7), up to a maximum of 18.1 km/day. The carapace width of upstream-migrating animals increased by 3 mm/100 km. The peak period for upstream migration was March to July, followed by the downstream season from July to September. This data set, extracted from historic references, represents one of the most complete pictures of the life cycle and spreading behaviour of this alien invader.

Introduction

The Chinese mitten crab, *Eriocheir sinensis* (H. Milne Edwards), originates in the Far East, its native range extending from Hong Kong (≈22° N) to the border with North Korea (≈40° N) (Hymanson et al., 1999). It is catadromous returning to the sea to reproduce once or rarely twice and then die. During the upstream migration of juveniles from estuaries, mitten crabs can reach rivers, lakes, and ponds as far as 1200 km from the coast (Peters, 1933). There they grow and mature. The age at which *E. sinensis* reaches sexual maturity varies from 1–3 years in China (Jin et al., 2001) to 3–5 years in Europe (Schubert, 1938). The Chinese mitten crab returns, when sexually mature at the end of its life cycle, to estuaries for reproduction in late summer. After mating, females move into the more marine parts of the estuary and release their larvae in early spring. Similar to *E. japonica* (de Haan), *E. sinensis* releases up to three batches of eggs (Peters, 1933; Kobayashi & Matsuura, 1999; Kobayashi, 2001). After one reproductive season, the life cycle of most *E. sinensis* draws to an end and they die (Peters, 1938c). Larval development consists of 6 pelagic stages (1 prezoea, 5 zoea) lasting ca. 90 days at 12 °C (Anger, 1991; Montu et al., 1996). After settlement; juvenile crabs migrate upstream. The pelagic larval stages are the most likely vector of spread for *E. sinensis* in ship's ballast water (Peters, 1933; Cohen & Carlton, 1997).

This invasive species has several impacts on humans and biodiversity. Firstly, it interferes with recreational and commercial fishing; in the San Francisco Bay area, the presence of mitten crabs has lead to commercial shrimp trawlers abandoning certain fish-

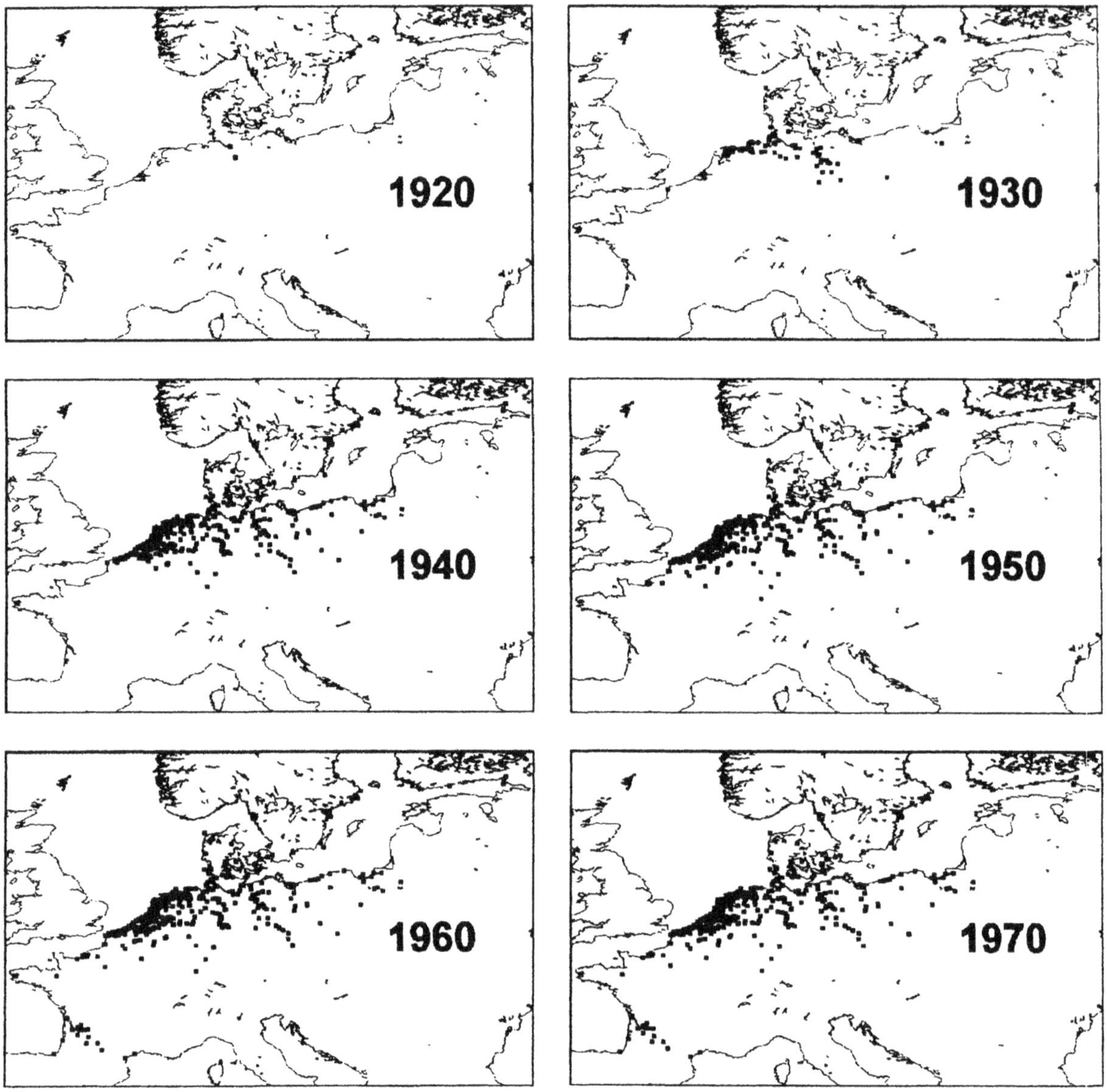

Figure 1. Cumulative sightings of *Eriocheir sinensis* from the literature across Continental Europe between 1912 and 1970.

ing areas. In the same area, there have been numerous complaints from recreational fishermen who blame the crabs for causing an increased loss of bait (Veldhuizen & Stanish, 1999). Similar problems have been recorded in Europe where Ingle (1986) reported that, in 1981, Dutch fishermen suffered serious net damage from mass occurrence of the mitten crab. Secondly, the burrow-digging habit of *E. sinensis* can cause serious river bank erosion. This is observed mainly in tidally-influenced areas, or other stretches of rivers with fluctuations in water level. Burrows are made in river banks with a step gradient, and the necessary structural strength to allow burrowing. They can be located anywhere between the high-water mark, under reed banks to below the low-water mark with the highest occurrence around mid- water level (up to $30/m^2$) (Peters, 1933, 1938b; Dutton & Conroy, 1998). Thirdly, *E. sinensis* may compete for resources with local freshwater fish and invertebrates. Overlap of dietary and habitat preferences with the economical important introduced signal crayfish (*Pacifastacus leniusculus,* Dana) occurs in San Francisco Bay (Veldhuizen & Stanish, 1999). In Britain, the already threatened native crayfish (*Austropotamobius pallipes,* Lereboullet) could face similar pressure (Clark et al., 1998). Large numbers of mitten crabs occurring in some areas, their large size and aggressive behaviour, in combination with overlap in feeding and habitat selection, gives rise to concerns about the possible impacts of current and future invasions (Clark & Rainbow, 1996).

The sight of thousands of crabs migrating overland around weirs in Europe in the 1930s was a major factor in raising public awareness. The scale of the occurrence of this crustacean in rivers and lakes, especially during migrations has caused sufficient public interest to engender many publications documenting its spread. Some of these provide much data on population distribution, migration patterns and spread particularly in Germany (Peters, 1933, 1938a; Panning, 1933, 1938a, b). The aim of this study was to analyse the invasion of *E. sinensis* in continental Europe, using literature contemporary with the major outbreak in.

Materials and methods

The spread characteristics of the Chinese mitten crab during two invasive events in Northern Europe (ca. 1910–1950) and in Southern France (ca. 1950–1960) were quantified. This was achieved by digitising all sightings (n=646) from the literature (Fig. 1) and using Geographic Information System (GIS) GRASS 4.0 software to measure the total length of river occupied from the estuary to the furthest upstream record for each year. Therefore, if in year 1 crabs were found 40 km upstream from the estuary in a particular river and in year 2 were 60 km upstream from the estuary in the same river, the average annual distance of upstream migration is 30 km/year. This unit of measurement was chosen since each population found further upstream than any previously observed ones must originate from the estuary due to its migration behaviour.

Size distribution data of the mitten crab populations over several years were collected and a wide area of Germany assessed during the 1930s' invasion (for population distributions see Peters, 1933; Panning, 1938a; for migration, see Peters, 1933; Panning, 1938b). This information from the major invasions in northern Europe has been analysed to describe the life-history characteristics of this species. Data from the Lower Elbe from 1932–6 were pooled for each month of the year to obtain average monthly size-class distributions of *E. sinensis*. The data (n=8736) were derived from various sampling methods: small specimens (as small as 2 mm) were collected in large numbers from the banks of the river Elbe throughout the year; larger crabs were collected on an occasional basis from the catches of fishermen using bottom trawls, gill nets and eel buckets. Size classes within the population in the Elbe were determined using modal class progression analysis (Bhattacharya's method) in the FiSAT stock assessment package. Using this method, different cohorts within the monthly size-frequency distribution data could be identified, and a mean and standard deviation derived. It should be noted that all size measurements used in the historic literature are in mm carapace length (cl), rather than the carapace width used in more recent literature. Therefore, carapace length is used during the analysis; this can be transformed into carapace width using the formula derrived by Peters (1933):

$$\text{carapace width (mm)} = 1.13\times \text{carapace length (mm)}.$$

Migratory behaviour was assessed using catch (in kg) from crabs collected in various types of traps placed at weirs to catch mitten crabs during migration. These were installed in the 1930s to reduce the mitten crab population which was a major pest by this time. Based on the structure of these traps, the only crabs caught were those migrating across or around the weir (Panning, 1938b). Panning (1938b) used carapace marking experiments to determine the downstream migratory speed of *E.sinensis* over various distances. The recatch occurred either in the crab traps around a weir or by fishermen.

Results

Spread across Continental Europe and Southern France

The first record of *E. sinensis* was in 1912 in a tributary of the River Weser in North Germany (Peters, 1933); it was also found in the River Elbe from 1914 onwards (both rivers run into the North Sea only 60 km apart). From there, it reached the Baltic Sea via the Kiel Canal in 1927 (Peters, 1938a). Despite the fact that the number of mitten crabs in the Baltic remained much lower than in the river systems along the North Sea, the species spread as far as Vyborg (Russia) and Finland by 1933. In the Netherlands, the first crab was recorded in 1931 and the population had spread into most rivers by 1936 (Kamps, 1937). It reached Belgium in 1933

(Peters, 1938a and citations within; Leloup, 1937). France was invaded by 1930 (Hoestlandt, 1945) and Denmark by 1927 (Rasmussen, 1987).

An important finding was also the extent of spread upstream in some rivers. *E. sinensis* had migrated along the tributaries of the Elbe as far as Prague in the Czech Republic (700 km inland) by 1932 (Peters, 1938a). In the same year, it had reached 512 km along the Rhine and, by 1934, it had spread 464 km upstream in the Oder to Breslau. Analysis revealed that a separate invasion occurred along the river Gironde (1954–1960) in Southern France. Previously, *E. sinensis* had spread only as far as Le Havre (1943) along the French Channel coast (Hoestlandt, 1959; Vincent, 1996). In 1954, specimens were caught in Nantes in the River Loire estuary and the same year near Bordeaux in the estuary of the Gironde. Migrating upstream in the Gironde, the mitten crab reached the artificial lagoons along the Mediterranean coast via canals by 1959, a distance of 504 km (Petit & Mizoule, 1974). These measurements gave an average total distance of upstream migration of 48 km/year between 1912–1927, 562 km/year for 1928–1939 and 69 km/year in 1940–1955 for Northern Europe, and 104 km/year for Southern France (1954–1960) (Fig. 2). The distance recorded upstream for the northern European and southern French river systems in each year of the invasion is shown in Figure 2. In Northern Europe the Chinese mitten crab spread slowly over a 15-year period (1912–1927) but then spread rapidly from 1928 to 1939, slowing afterwards. Part of the apparent slowing in the spread of mitten crabs in Northern Europe after 1939 might be related to the reduced monitoring effort during the war. In Southern France, records show no establishment phase but a rather sharp increase in spread between 1954 and 1960.

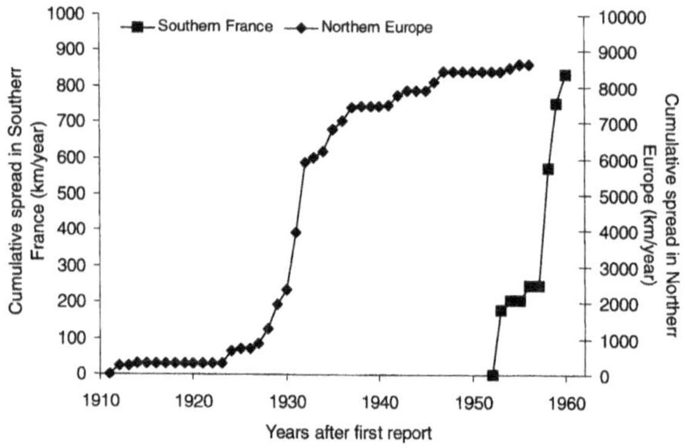

Figure 2. The cumulative total length of river (in km) occupied by *Eriocheir sinensis* from the estuary to the furthest upstream record for each year, measured using Geographic Information System (GIS) GRASS 4.0 software. The spread in Northern Europe (■) is plotted separately from the spread in Southern France (♦). Note the different scales of distance of river invaded used for the two invasions.

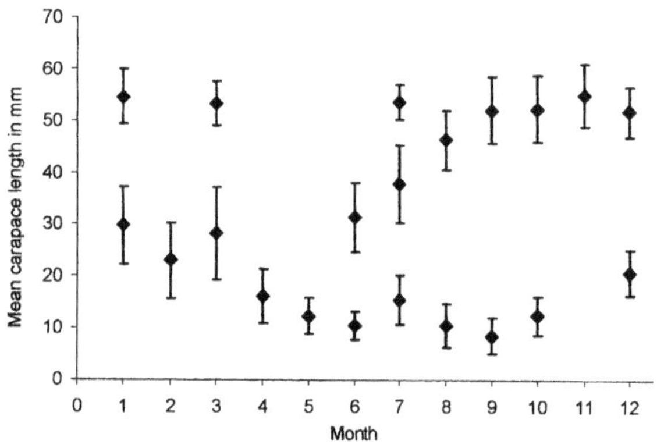

Figure 3. The size-class distribution (± 1 SD) of *Eriocheir sinensis* in the lower Elbe per month as determined using modal progression analysis (Bhattacharya's method) in the FiSAT stock assessment package.

Observations on population distribution and migratory pattern

The size-class distribution (Fig. 3) of mitten crabs (both male and female) in the lower Elbe shows the presence of at least two different cohorts during spring: a larger/adult class (54.6 mm ± 5.3 mm cl for January) disappearing in May, and a smaller/juvenile class (29.8 mm ± 7.6 mm cl for January) decreasing in size considerably by May (12.4 mm ± 3.4 mm cl). The smaller cohort remained over the whole year, fluctuating in mean size between 8.6 mm ± 3.5 mm cl (September) and 20.7 mm ± 4.4 mm cl (December). In June, the larger cohort was found again in the estuary with a smaller carapace length (31.5 mm ± 6.7 mm), increasing gradually until December to 51.7 mm ± 4.9 mm cl.

The mean carapace length of upstream-migrating *E. sinensis,* measured using animals caught in traps at weirs along the Weser and Elbe, showed an average carapace length increase of 3 mm per 100 km upstream from the estuary (Fig. 4) (Panning, 1938b). This can be used to estimate the annual growth rate as 6.6 mm over the annual migration distance of 225 km (according to Panning, 1938b). The annual growth rate estimated by (Panning, 1938a) suggests an increase of about 12 mm per year up to year 4. Despite the discrepancy of these growth rates, they are the only available estimates for *E. sinensis* in Europe. The peak downstream migration period at three locations (estuarine, 310 km upstream,

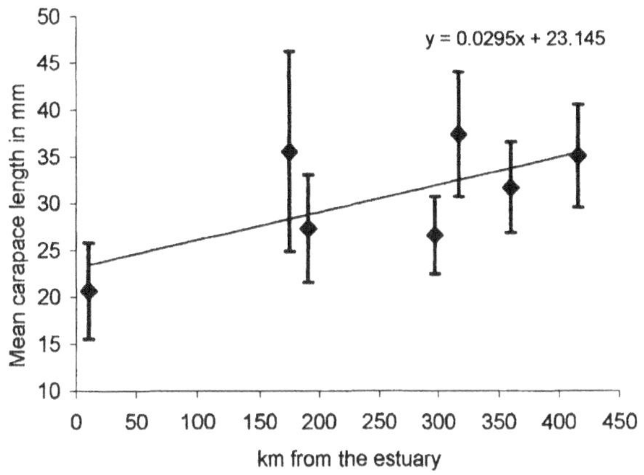

Figure 4. The mean carapace length (in mm) of *Eriocheir sinensis* migrating upstream along the Rivers Elbe and Weser (Germany).

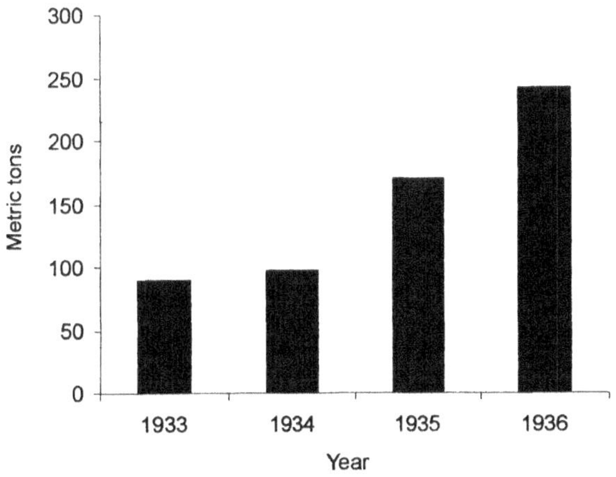

Figure 5. The annual catch of *Eriocheir sinensis* (in metric tons) from the River Weser and Elbe (Germany) during the invasion of Northern Europe (1933–1936).

410 km upstream) was August–October. Absence of a difference in migration period between these sites can be explained using downstream marking experiments. These showed that mature animals can cover large distances (400 km or more) within this 3-month migration period, since analysis revealed an average downstream migration speed of 11.5 km/day (SD± 3.54) (Panning, 1938b).

Using the catch data from the rivers Elbe and Weser, population numbers can be estimated for the mid 1930s; the peak catch occurred in 1936 with an annual catch of mitten crabs totalling 242 tonnes in Germany alone (Panning, 1938b) (Fig. 5). Based on an average weight of 55 g per individual from collections in the river Thames by Herborg (unpubl. data), total catch can be estimated to be the equivalent of 4.4 million crabs. Considering this is only an estimation for two rivers out of many in Europe this species has invaded, the total population must have been very large.

Discussion

The most obvious feature of these results is the similar pattern in expansion in length of river colonised in both Southern France and Northern Europe. In both river systems there is evidence of a sigmoidal relationship between length of river invaded and time. Whilst the pattern may have been similar, comparison of the rates of spread between Northern Europe and Southern France indicates a much faster rate of spread during the exponential phase of the spread in Northern Europe. These differences probably reflect differences in the structure of the two river systems. In Northern Europe, river systems are extensive, especially along the North Sea coast where the Rhine, Weser and Elbe give access to an extended network of tributaries as well as to canals linking to other rivers. The outbreak in Southern France was limited by the nature of the Gironde system, which is less extensive. Towards the south, the Pyrenees form a natural barrier and, to the north of the Gironde, relatively few large rivers run into the Bay of Biscay.

The presence of a 15-year period of slow spread in Germany at the early stage of invasion can be interpreted as an 'establishment' phase which is observed commonly during invasions (Drake & Williamson, 1986). Absence of a establishment phase in Southern France may be due to a lack of data, or perhaps that the Chinese mitten crab population, at this time already present for over 40 years in Europe, was well acclimatised to European environmental conditions. Another factor associated with a sharp population increase of *E. sinensis* is periods of low river flow as observed for the U.K. in 1989–1992 (Attrill & Thomas, 1996).

The pattern of spread of the Chinese mitten crab in Europe compares well with the spread of the European shore crab (*Carcinus maenas*, L.) along the Californian coast. Both species have a similar pelagic larval period in their life cycles (ca. 60 days in 20 °C) (Anger, 1991; Cohen et al., 1995). Observations of the invasion rates of *C. maenas* also showed fluctuations (between 16 and 63 km/year) in the rate of spread along the coast in different invasive events (Grosholz & Ruiz, 1996 and citations within). Grosholz (1996), analysing the rate of spread for 10 invasions of the marine environment, reported an average rate of spread for marine invaders of 50.7 km/year with individual values between 12 and 115 km/year. Estimates of the rate of spread for three invertebrate species (*Balanus improvisus,* Darwin,*Potamopyrgus antipodarium,* J.E Gray, *Marenzelleria viridis,* Verrill) in the Baltic Sea range between 30 and 480 km/year (Leppaekoski & Olenin, 2000). The sigmoidal rate of spread has also been described in other invasions: the spread of red deer (*Cervus elaphus,* L.) on the South Island of New Zealand described a 40-year establishment phase before the area of invasion increased rapidly, flattening off again after ca. 20 years (Clarke, 1971).

Whilst the spread of *E. sinensis* in Northern Europe seems to originate from the Weser area, the question arises as to the origin of the Chinese mitten crab at this site. An introduction via ballast water from China is the most likely vector, since there has been frequent ship traffic from Hamburg (River Elbe) and Bremen (River Weser) to China since the end of the 19th century. The typical steamer on this route had 15 ballast water tanks, each holding 90 tonnes, with an intake grid of 15 × 460 mm, large enough not only for larvae but also juvenile mitten crabs. Peters (1933) mentions the case of two adult mitten crabs (40–50 mm) which were found in the sea chest of one of these vessels during its salvage. Introduction of the Chinese mitten crab is considered as one of the very first known cases of a species being transported via ballast water (Carlton, 1985).

It is likely that invasion of Southern France arose in a similar way. The arrival of mitten crabs in Southern France demonstrated a significant jump in spread along the coast. Before 1954, the year the Chinese mitten crab appeared in the Gironde and the Loire, it had only been reported from the Seine, a distance of 855 and 205 km along the coast, respectively. Considering this distance, as well as the prevailing surface transport going northwards (Ellien et al., 2000; Alvarez et al., 2001), larval transport along the coast is highly unlikely. The most likely vector is human-aided transport, most likely via ship ballast water or in association with the intensive oyster aquaculture along this coast. Furthermore, the long time gap between the invasion of the northern Atlantic coast of France, and that of the Bay of Biscay, also suggests that a chance ship transport across an expansion barrier was the mechanism of introduction rather than natural spread.

The size-frequency distribution of crabs in the Elbe estuary over the year shows the recruitment, departure and arrival of different cohorts of animals according to their life cycle. At the beginning of the year, a large and small cohort was observed. The larger cohort comprised animals in excess of 43 mm, the minimum size for sexual maturity of both sexes, and are most likely animals from the previous autumn's mating season which are either ovigerous females or part of the small fraction which survives the mating season (Peters, 1933). The size reduction of the juvenile cohort from March to April coincided with the peak upstream migration period as determined by migration rates (data not shown). The carapace length of juveniles leaving the estuary may indicate that the smaller cohort observed in the Elbe (January–March) consisted of year-class 1 and 2, whereas from May onwards, only smaller crabs of the year 1 cohort remained in the estuary and the year 2 cohort started their migration further upstream.

Fluctuation of the mean size of the juvenile cohort during the summer and autumn months could be caused by two separate factors. Firstly, the settlement of megalopa larvae from the water column in July–September (Schubert, 1938) at a length of 5mm (Schnakenbeck, 1933) and, secondly, continuous growth of the previous year class. The larger size class appearing in the Elbe as early as July may be explained as being either animals which remained in the lower part of the estuary, or juveniles migrating downstream to the estuary to become sexually mature. Further increase in this cohort is caused by adult crabs arriving downstream during the peak migratory period (August–October). Analysis of catch numbers along the Elbe and Weser showed that August to October is the peak downstream migration period in the River Elbe (data not shown).

Recent invasions of Chinese mitten crabs in Great Britain (Attrill & Thomas, 1996; Clark et al., 1998; Herborg et al., 2002) and San Francisco Bay (Rudnick et al., 2003) show similar patterns of spread to those analysed here. There is also evidence that the invasion process is ongoing, as recent findings of *E. sinensis* in the Black Sea (Gomui et al., 2002) and the Sea of Azov (Murina & Antonovsky, 2001) underline. The study presented highlights the potential of major invasions of *E.sinensis*. There has been little description in the historic work of the ecological impact of invasions such as those discussed here. Nevertheless, considering the size of the population at its peak, ecological impacts may well have been considerable. What is required now is some quantification of these impacts, in order that threats to biodiversity can be assessed and requirements for management of the invasion problem identified.

Acknowledgements

LMH would like to thank the Faculty of Science, Agriculture and Engineering, University of Newcastle (Swales Studentship) and the Esmeé Fairbairn Foundation (RG 01-1816) for their financial support.

References

Alvarez, P., L. Motos, A. Uriarte, J. & Egana, 2001. Spatial and temporal distribution of European hake, *Merluccius merluccius* (L.), eggs and larvae in relation to hydrographical conditions in the Bay of Biscay. Fish. Res. 50: 111–128.

Anger, K., 1991. Effects of Temperature and salinity on the larval development of the Chinese mitten crab *Eriocheir sinensis* (Decapoda, Grapsidae). Mar. Ecol. Prog. Ser. 72: 103–110.

Attrill, J. M. & R. M. Thomas, 1996. Long-term distribution patterns of mobile estuarine invertebrates (Ctenophora,Cnideria, Crustacea: Decapoda) in relation to hydrological parameters. Mar. Ecol.- Progr. Ser. 143: 25–36.

Carlton, J. T., 1985. Transoceanic and interoceanic dispersal of coastal marine organisms: the biology of ballast water. Oceanogr. mar. Biol. 23: 313–371.

Clark, P. & P. Rainbow, 1996. The Chinese mitten crab in the Thames catchment. 000208370, Environment Agency Report.

Clark, P. F., P. S. Rainbow, R. S. Robbins, B. Smith, W. E. Yeomans, M. Thomas & G. Dobson, 1998. The alien Chinese mitten crab *Eriocheir sinensis* (Crustacea: Decapoda: Brachyura), in the Thames catchment. J. mar. biol. Ass. U.K. 78: 1215–1221.

Clarke, C. M. H., 1971. Liberations and dispersal of red deer in northern South Island districts. New Zeal. J. For. Sci. 1: 194–207.

Cohen, A. N. & J. T. Carlton, 1997. Transoceanic transport mechanisms: introduction of the Chinese mitten crab, *Eriocheir sinensis*, to California. Pacif. Sci. 51: 1–11.

Cohen, A. N., J. T. Carlton & M. C. Fountain, 1995. Introduction, dispersal and potential impacts of the Green Crab *Carcinus maenas* in San Francisco Bay, California. Mar. Biol. 122: 225–237.

Drake, J. A. & M. Williamson, 1986. Invasions of natural communities. Nature 319: 718–719.

Dutton, C. D. & C. Conroy, 1998. Effects of burrowing Chinese mitten Crabs (*Eriocheir sinensis*) on the Thames Tideway, Environment Agency Report.

Ellien, C., E. Thiebaut, A. S. Barnay, J. C. Dauvin, F. Gentil & J. C. Salomon, 2000. The influence of variability in larval dispersal on the dynamics of a marine metapopulation in the eastern Channel. Oceanol. Acta. 23: 423–442.

Gomui, M. T., B. Alexandrov, N. Shadrin & T.Y. Zaitsev, 2002. The Black Sea – a recipient, donor and transit area for alien species. In Leppaekoski, E. S. Gollasch & S. Olenin (eds), Invasive Aquatic Species of Europe. Distribution, Impacts and Management. Kluwer Academic Publishers: 341–350.

Grosholz, E. D., 1996. Contrasting rates of spread for introduced species in terrestrial and marine systems. Ecology 77: 1680–1686.

Grosholz, E. D. & G. M. Ruiz, 1996. Predicting the impact of introduced marine species: lessons from the multiple invasions of the European green crab *Carcinus maenas*. Biol. Conserv. 78: 59–66.

Herborg, L.-M., M. G. Bentley & A. S. Clare, 2002. First confirmed record of the Chinese mitten crab (*Eriocheir sinensis*) from the River Tyne, United Kingdom. J. mar. biol. Ass. U.K. 82: 921–922.

Hoestlandt, H., 1945. Le crabe chinois (*Eriocheir sinensis* Mil. Ed.) en Europe et principalement en France. Ann Épiphy. XI: 223–233 (in French).

Hoestlandt, H., 1959. Répartition actuelle du crabe chinois (*Eriocheir sinensis* H. Milne Edwards) en France. Bul. Franc. Pisicul. 194: 5–13 (in French).

Hymanson, Z., J. Wang & T. Sasaki, 1999. Lessons from the home of the Chinese mitten crab. IEP Newsletter 12: 25–32.

Ingle, R. W., 1986. The Chinese mitten crab *Eriocheir sinensis* a continuous immigrant. Lond. Nat. 65: 101–105.

Jin, G., Z. Li, & P. Xie, 2001. The growth patterns of juvenile and precocious Chinese mitten crabs, *Erocheir sinensis* (decapoda, grapsida), stocked in freshwater lakes of China. Crustaceana 74: 261–273.

Kamps, L. F., 1937. De Chineesche wolhand krab in Nederland. Akad. Proef. Groningen: 1–112 (in Dutch).

Kobayashi, S., 2001. Fecundity of the Japanese mitten crab *Eriocheir japonica* (de Haan). Benth. Res. 56: 1–7.

Kobayashi, S. & S. Matsuura, 1999. Reproductive ecology of the Japanese mitten crab *Eriocheir japonica* (de Haan): a review. Jap. J. Benth. 54: 24–35.

Leloup, E., 1937. Contributions à l'étude de la faune belge. VII. - La propagation du crabe chinois en Belgique pendant l'année 1936. Bull. Mus. Hist. Natur. Belg. 13: 1–7 (in French).

Leppaekoski, E. & S. Olenin, 2000. Non-native species and rates of spread: lessons from the brackish Baltic Sea. Biol. Invas. 2: 151–163.

Montu, M., K. Anger & C. deBakker, 1996. Larval development of the Chinese mitten crab *Eriocheir sinensis* H Milne-Edwards (Decapoda: Grapsidae) reared in the laboratory. Helgolaender Meeresunters. 50: 223–252.

Murina, V. V. & A. G. Antonovsky, 2001. Chinese crab, *Eriocheir sinensis* is an invader into the basin of the Sea of Azov. Ekol. morya 55: 37–39 (in Russian).

Panning, A., 1933. Gestaltkundlicher Teil. In Peters, N., A. Panning & W. Schnackenbeck (eds), Die chinesische Wollhandkrabbe (*Eriocheir sinensis*) in Deutschland. Zool. Anz. (Leipzig): 1–58 (in German).

Panning, A., 1938a. Die Verteilung der Wollhandkrabbe über das Flußgebiet der Elbe nach Jahrgängen. Mitt. Hamb. Zool. Mus. Inst. 47: 65–82 (in German).

Panning, A., 1938b. Über die Wanderung der Wollhandkrabbe. Markierungsversuche. Mitt. Hamb. Zool. Mus. Inst. 47: 32–49 (in German).

Peters, N., 1933. Lebenskundlicher Teil. In Peters, N., A. Panning & W. Schnakenbeck, (eds), Die chinesische Wollhandkrabbe (*Eriocheir sinensis* H.Milne-Edwards) in Deutschland. Zool. Anz. (Leipzig): 59–155 (in German).

Peters, N., 1938a. Ausbreitung und Verbreitung der chinesischen Wollhandkrabbe (*Eriocheir sinensis* H.M.-Edw) in Europa im Jahre 1933 bis 1935. Mitt. Hamb. Zool. Mus. Inst. 47: 1–31 (in German).

Peters, N., 1938b. Neue Untersuchungen über die Erdbauten der Wollhandkrabbe. Mitt. Hamb. Zool. Mus. Inst. 47: 129–139 (in German).

Peters, N., 1938c. Zur Fortpflanzungsbiologie der Wollhandkrabbe (*Eriocheir sinensis* H.M.-Edw). Mitt. Hamb. Zool. Mus. Inst. 47: 112–128 (in German).

Petit, G. & R. Mizoule, 1974. En douze ans le 'crabe Chinois' n'a pu réussir son implantation dans les lagunes du Languedoc. Vie Milieu Ser. C Biol. Terr. 23: 181–186 (in French).

Rasmussen, E., 1987. Status over uldhåndskrabbens (*Eriocheir sinensis*) udbredelse og forekomst i Danmark. Flora og Fauna 93: 51–58 (in Danish).

Rudnick, D. A., K. Hieb, K. F. Grimmer & V. H. Resh, 2003. Patterns and processes of biological invasion: the Chinese mitten crab in San Francisco Bay. Basic appl. Ecol. 4 (in press).

Schubert, K., 1938. Häutung, Wachstum und Alter der Wollhandkrabbe. Mitt. Hamb. Zool. Mus. Inst. 47: 83–104 (in German).

Veldhuizen, T. C. & S. Stanish, 1999. Overview of the life history, distribution, abundance, and impact of the Chinese mitten crab, *Eriocheir sinensis*, California Department of Water Resources, Environmental Services Office.

Vincent, T., 1996. Le crabe Chinois *Eriocheir sinensis* H. Milne-Edwards, 1854 (Crustacea, Brachyura) en Seine-Maritime, France. Ann. Inst. océanogr. 72: 155–171 (in French).

Hydrobiologia **503:** 29–35, 2003.
M.B. Jones, A. Ingólfsson, E. Ólafsson, G.V. Helgason, K. Gunnarsson & J. Svavarsson (eds), Migrations and Dispersal of Marine Organisms.

Characterising invasion processes with genetic data: an Atlantic clade of *Clavelina lepadiformis* (Ascidiacea) introduced into Mediterranean harbours

Xavier Turon[1], Isabel Tarjuelo[1], Sandra Duran[1] & Marta Pascual[2]
[1]*Department of Animal Biology*
[2]*Department of Genetics, Faculty of Biology, University of Barcelona, 645, Diagonal Ave, 08028 Barcelona, Spain*
E-mail: xaviert@bio.ub.es

Key words: cryptic colonisation, invasions, harbours, COI gene, mtDNA, ascidians

Abstract

We studied the genetic structure of populations of the Atlanto-Mediterranean ascidian *Clavelina lepadiformis* (Müller, 1776). A 369 bp segment of the COI mitochondrial gene was sequenced in Mediterranean and Atlantic populations from inside harbours, marinas and fjords (interior populations), and from the open rocky littoral (exterior populations). Previous work identified genetic differences between *C. lepadiformis* inhabiting Mediterranean harbours and the Mediterranean rocky littoral, however, the origin of these two clades remained speculative. Here we compared the Mediterranean populations with four Atlantic populations (two interior and two exterior). Gene differentiation and maximum likelihood analyses showed that the Atlantic forms were not divided into interior and exterior clades, and were closely related to the interior clade in the Mediterranean. The results support the hypothesis that both clades evolved allopatrically in the two seas, and that a recent colonisation of Mediterranean marinas from the Atlantic was caused by ship-hull transport. Colonisation of habitats by new genetic variants, morphologically indistinguishable from local populations, may be common among benthic invertebrates, and only genetic tools can uncover these cryptic invasions.

Introduction

Colonisation of new geographical areas by marine species is often caused by transport on ships' hulls and in ballast waters, with harbours and marinas often being the entrance gates for new invasions (Carlton & Geller, 1993). This is especially true of invertebrate species with short-lived larvae, whose opportunities for long-distance dispersal may rely, in many cases, on rafting on human-made floating objects. Ascidians belong to this latter category due to their larval characteristics (Svane & Young, 1989), and the role of harbours and marinas in the introduction of non-indigenous ascidian species is increasingly recognised (Monniot et al., 1985, 1991; Monniot & Monniot, 1994; Lambert & Lambert, 1998). In the Mediterranean Sea, non-indigenous ascidian species have been reported recently in harbours and marinas (Brunetti, 1979; Monniot, 1981; Zibrowius, 1991).

The contribution of maritime transport to benthic assemblages may go undetected if the coloniser is a genetic variant or a cryptic species that is morphologically indistinguishable from native populations/species. Such cryptic invasions may be common, with colonisers confined, at first, to marginal marine habitats such as harbours and lagoons, and eventually spreading to open habitats. While ecophenotypic variability may provide the first clue that a new variant has arrived in a region, only molecular tools can identify reliably these genetic variants and uncover cryptic colonisation of new areas. Indeed, there are instances of genetic discontinuities between supposedly conspecific populations inhabiting open littoral and marginal marine habitats (Dalby, 1997; Boisselier-Dubayle &

Gofas, 1999) that suggest the arrival of new genetic variants in the latter type of habitat.

Clavelina lepadiformis (Müller, 1776) is an Atlanto-Mediterranean species whose distribution ranges from Scandinavia (Millar, 1966) to the Aegean Sea in the eastern Mediterranean (Koukouras et al., 1995). Wirtz & Martins (1993) and Wirtz (1998) reported this species from the Azores and Madeira, and suggested that it may have been introduced by ships, spreading subsequently from island to island. In the Mediterranean, *C. lepadiformis* has been reported from harbours and shallow rocky littoral habitats (Harant, 1927; Turon, 1987; Ramos, 1988). In Atlantic waters, it is known from rocky bottoms down to 50 m, but is also present in sheltered environments including wharfs, harbours, fjords and shallow estuarine bays (Berrill, 1950; Millar, 1966; Vázquez, 1993). Here we use the term 'exterior populations' for those inhabiting open littoral rocky shores, and 'interior populations' for those inside harbours, marinas or fjords. A recent study of *C. lepadiformis* based on mitochondrial DNA sequence data (Tarjuelo et al., 2001) showed that Mediterranean populations inhabiting harbours (interior form) were genetically distinct from those thriving in the open habitats (exterior form). It was concluded that the two forms could be cryptic species. A biological study of both forms (De Caralt et al., 2002) revealed noticeable differences in parameters such as growth, biological cycles, chemical defence production and juvenile survival. Therefore, although the two clades were morphologically identical, genetic and ecophenotypic divergences were observed.

The origin of the two clades (interior and exterior), however, remained speculative. Two hypotheses were advanced by Tarjuelo et al. (2001): firstly, parapatric speciation occurred in the Atlantic between populations in open shores and those in sheltered, marginal marine environments. The form evolving in these marginal habitats would have been preadapted to colonise man-made harbours and constructions. The two clades could have colonised the corresponding Mediterranean habitats, possibly after the Messinian salinity crisis of the Mediterranean (late Miocene, between 5 and 6 MY, Maldonado, 1985). The second hypothesis was that the two clades diverged allopatrically from an original population split in the Mediterranean and Atlantic basins since the Miocene, and that the interior form represents the Atlantic clade that has colonised recently Mediterranean harbours through ship traffic (Tarjuelo et al., 2001).

Here, we examine the origin and relationships between the two Mediterranean clades by adding to the existing genetic database representatives of interior and exterior populations of the Atlantic. Specifically, we address the following questions: is there a genetic divergence between interior and exterior populations in the Atlantic, as is the case in the Mediterranean? What are the relationships between the Atlantic populations and the two Mediterranean clades?

Materials and methods

We studied 73 individuals of *Clavelina lepadiformis* belonging to two exterior and two interior populations from Atlantic shores, and three exterior and three interior populations from the Spanish Mediterranean (Fig. 1). The Mediterranean populations studied were: Cadaqués and Tossa de Mar (exterior populations), Estartit and Port Ginesta (interior populations), and Blanes (which had both exterior and interior populations, Fig. 1). The Atlantic populations sampled were: Smatiolmarma, at the Gullmarsfjorden (Sweden), Vigo (NW Spain), Sesimbra (Portugal), and Sao Miguel Island (Azores). The Portuguese mainland and Azorean populations sampled were from exterior habitats, while the Vigo sampling point was inside the Baiona Port; the Swedish locality was inside a fjord. The Mediterranean locations were the same as in Tarjuelo et al. (2001).

We also sequenced two specimens of *C. dellavallei* (Zirpolo, 1925) from Cabo de Gata (SE of Spain) and one specimen of *C. oblonga* Herdman, 1880 from Faial Island (Azores) as outgroups.

Mitochondrial DNA was extracted from ethanol-preserved samples using the protocol for *Drosophila* sp. described in Latorre et al. (1986). We used specific primers designed for a region of cytochrome oxidase I in this species (Tarjuelo et al., 2001). The amplification yielded a fragment of ca. 500 bp that, after alignment and trimming, reduced to 369 positions for the analyses. No gaps were needed in the alignment.

Amplification was performed in a 20 μl total reaction volume with 0.4 μl of each primer (25 μM), 3 μl dNTP's (1 mM), 2 μl 10X buffer containing 15 mM $MgCl_2$ (Promega), 1U Taq polymerase (Promega) and 1 μl DNA. A single soak at 96 °C for 3 min was followed by 35 cycles (denaturation at 95 °C for 55 s, annealing at 47 °C for 55 s, and extension at 72 °C for 1.5 min), and a final extension at 72 °C for 5 min, on a Perkin-Elmer 9600 machine. The same primers were

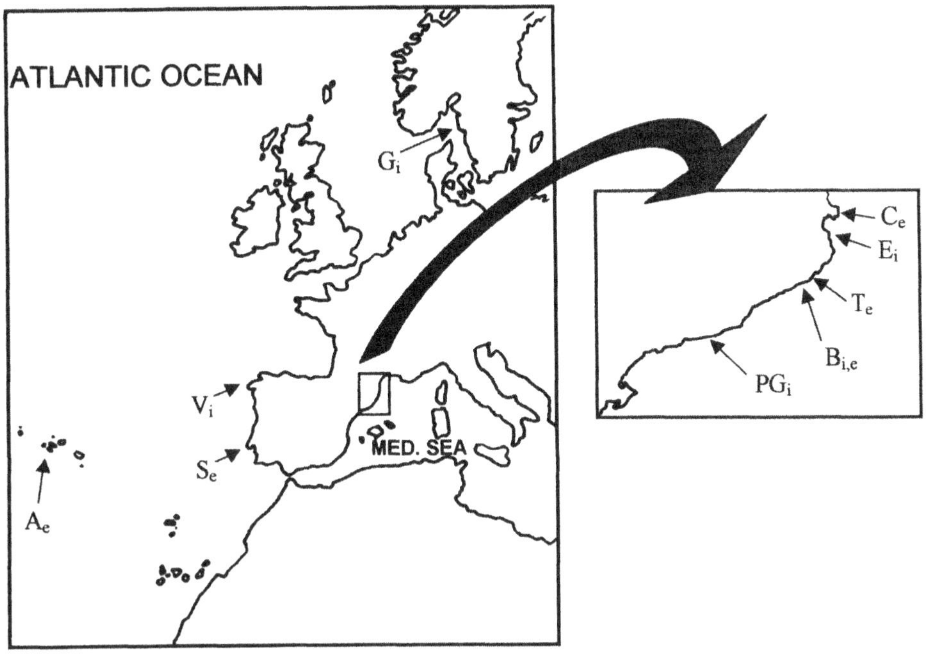

Figure 1. Map showing the position of the populations sampled. Codes of localities: (A) Azores Islands, (B) Blanes, (C) Cadaqués, (E) Estartit, (G) Gullmarsfjorden, (PG) Port Ginesta, (S) Sesimbra, (T) Tossa, (V) Vigo. Subscripts e and i mean exterior and interior types of habitat, respectively.

used for the sequencing reaction, and the sequences were obtained using the ABI Big-Dye Ready-Reaction kit of Perkin Elmer on an ABI Prism 9700 automated sequencer belonging to the Scientific and Technical Services of the University of Barcelona.

Nucleotide diversity within and between populations (π, Nei, 1987, equation 10.5; Lynch & Crease, 1990), divergence between populations (*Dxy*, Nei, 1987, equation 10.20), and haplotype diversity (Nei, 1987, equation 8.4) were measured. Genetic divergence was estimated by measuring $Gamma_{ST}$ (γ_{ST}) (Nei, 1982). The significance of estimates of genetic differentiation was tested with a permutation (randomisation) test. Gene flow was calculated using the standard relationship $\gamma_{ST} = 1/(1+2N_{ei}m)$ to obtain $N_{ei}m$, the effective number of migrants, where N_{ei} is the inbreeding effective population size and m is the migration rate per generation. These estimates are contingent on the assumptions of the island model of population structure (Wright, 1951). The population genetic parameters were estimated and analysed using DnaSP v.3.97 (Rozas & Rozas, 1999) and Arlequin v. 2000 (Schneider et al., 2000).

Phylogeographic information in the data was studied by maximum likelihood analysis of the sequences. To select the best-fit model of nucleotide substitution for the COI data set, the hierarchical likelihood ratio tests approach implemented in Modeltest 3.0 (Posada & Crandall, 1998) was applied to a NJ cladogram obtained from the data. A haplotype tree was then estimated under the maximum likelihood criterion (ML; Felsenstein, 1981), with the parameter estimates obtained under the best-fit model. A heuristic search with 1000 replicates of random stepwise addition and tree bisection-reconnection (TBR) branch-swapping was implemented. As we used mid-point rooting, there was no need to force the ingroup to be monophyletic. Confidence in the resulting nodes was assessed by 1000 bootstrap replicates (Felsenstein, 1985). Phylogenetic analyses were implemented in PAUP* (Swofford, 1998).

Type sequences are available from GenBank (Mediterranean haplotypes, accession # AF368352 and AF368353; Atlantic haplotypes, accession # AY211529, AY211530 and AY211531).

Results

Thirteen different haplotypes were found in the samples (Table 1). Overall, there were 32 variable sites (8.67%), of which only 5 yielded non-synonymous changes. 14 fixed nucleotide positions

Table 1. Clavelina lepadiformis. Haplotype frequencies, nucleotide diversity within populations (π) and number of polymorphic sites (PS) in the populations studied. The codes used in Figure 1 to designate the sampling points are indicated in parentheses, e and i denote exterior and interior populations, respectively

C.lepadiformis	Haplotypes															
Population	I	II	III	IV	V	VI	VII	VIII	IX	X	XI	XII	XIII	Total	π	PS
Cadaques (C), e	2	1	-	-	-	-	-	-	-	-	-	-	-	3	0.0022	1
Tossa (T), e	7	-	1	7	-	-	-	-	-	-	-	-	-	15	0.0024	3
Blanes (B), e	4	-	-	-	2	1	3	-	-	-	-	-	-	10	0.0027	3
Estartit (E), i	-	-	-	-	-	-	-	3	7	1	-	-	-	11	0.0017	2
Blanes (B), i	-	-	-	-	-	-	-	4	11	-	-	-	-	15	0.0012	1
Port Ginesta (PG), i	-	-	-	-	-	-	-	6	2	-	1	-	-	9	0.0022	3
Vigo (V), i	-	-	-	-	-	-	-	-	5	-	-	-	-	5	0	0
Sesimbra (S), e	-	-	-	-	-	-	-	-	-	-	-	1	-	1	0	0
Gullmarsfjorden (G), i	-	-	-	-	-	-	-	-	2	-	-	-	-	2	0	0
Açores (A), e	-	-	-	-	-	-	-	-	1	-	-	-	1	2	0.0162	6
Totals	13	1	1	7	2	1	3	13	28	1	1	1	1	73		

(3.79%, of which 2 represented non-synonymous changes) discriminated the Mediterranean exterior populations from the group formed by the Mediterranean interior and Atlantic populations. There were no coincident haplotypes between exterior and interior Mediterranean populations, or between exterior Mediterranean and Atlantic populations, while haplotype IX was shared and was the most abundant both in the interior Mediterranean and the Atlantic samples (Table 1). The nucleotide diversity (π) pooling Mediterranean exterior populations was 0.0037, in Mediterranean interior populations it was 0.0018 and in Atlantic populations it was 0.0063. The values of haplotype diversity were 0.728, 0.550, and 0.378, respectively. When Mediterranean exterior populations were compared with Mediterranean interior+Atlantic samples, the average number of nucleotide differences between these two groups was 20.048 (5.42%), and the average number of nucleotide substitutions per site (Dxy) was 0.054.

Table 2 shows pairwise values of γ_{ST} between the studied populations and the corresponding estimates of gene flow ($N_{ei}m$, subject to the island model of population structure). Again, two groups formed by Mediterranean exterior and Mediterranean interior+Atlantic samples were evident, as reflected by low genetic divergence and high gene flow within groups, and high genetic divergence and low gene flow between groups. Within Mediterranean exterior populations, average γ_{ST} was 0.143±0.035 (mean±SE), and within Mediterranean interior+Atlantic populations it was 0.192±0.035. In contrast, average γ_{ST} value between the two groups was 0.891±0.015. The values of genetic differentiation between the two groups of populations were significant (permutation test at p=0.05) in all pairwise comparisons except for the comparisons of Cadaqués with Gullmarsfjorden and Azores (possibly as a result of low sample size in the latter two populations). Conversely, none of the pairwise measures within groups was significant except the comparison between Port Ginesta and Vigo. The estimates of gene flow were lower in between-group comparisons ($N_{ei}m$ = 0.063±0.010) than in within-group pairs ($N_{ei}m$ =3.23±0.77 and 5.31±1.97 for exterior Mediterranean and interior Mediterranean+Atlantic groups, respectively).

A maximum likelihood analysis was performed on the 13 haplotypes found. Two haplotypes found in the congeneric *C. dellavallei* from the Mediterranean and one sequence of *C. oblonga* from the Azores Islands were used as outgroups. The hierarchical comparison between likelihood scores of each model of evolution showed that Kimura's three-substitution-type model (Kimura, 1981) with unequal base frequencies was the best-fitting model.

The results of the ML haplotype tree show (Fig. 2) that Mediterranean interior and Atlantic haplotypes cluster together in a monophyletic clade with 88% bootstrap support, while the Mediterranean exterior haplotypes formed another group, with the same support.

Table 2. Clavelina lepadiformis. Gamma$_{ST}$ (γ_{ST})values (above the diagonal) and gene flow estimates (*Nm*, below the diagonal) for the different populations studied (nps: no polymorphic sites). The Sesimbra datum was not included as it consisted of a single sequence. Comparisons between the two recognized clades (exterior Mediterranean and interior Mediterranean+Atlantic) are marked by double-line boxes. (e) and (i) denote exterior and interior populations, respectively

	Cad. (e)	Tossa (e)	Blanes (e)	Blanes (i)	P. Gin. (i)	Estartit (i)	Vigo (i)	Gullm. (i)	Açores (e)
Cadaques (e)		0.13	0.09	0.92	0.91	0.91	0.98	0.97	0.85
Tossa (e)	3.29		0.21	0.92	0.90	0.90	0.89	0.79	0.73
Blanes (e)	4.54	1.87		0.93	0.91	0.92	0.93	0.87	0.80
Blanes (i)	0.04	0.04	0.03		0.13	0.02	0.08	0.04	0.32
P. Ginesta (i)	0.05	0.05	0.05	3.23		0.08	0.29	0.17	0.32
Estartit (i)	0.05	0.05	0.04	28.49	5.85		0.12	0.06	0.30
Vigo (i)	0.01	0.06	0.04	5.50	1.25	3.58		Nps	0.42
Gullmarsfj. (i)	0.02	0.13	0.08	11.69	2.46	7.26	nps		0.33
Azores (e)	0.09	0.18	0.13	1.07	1.05	1.16	0.70	1.00	

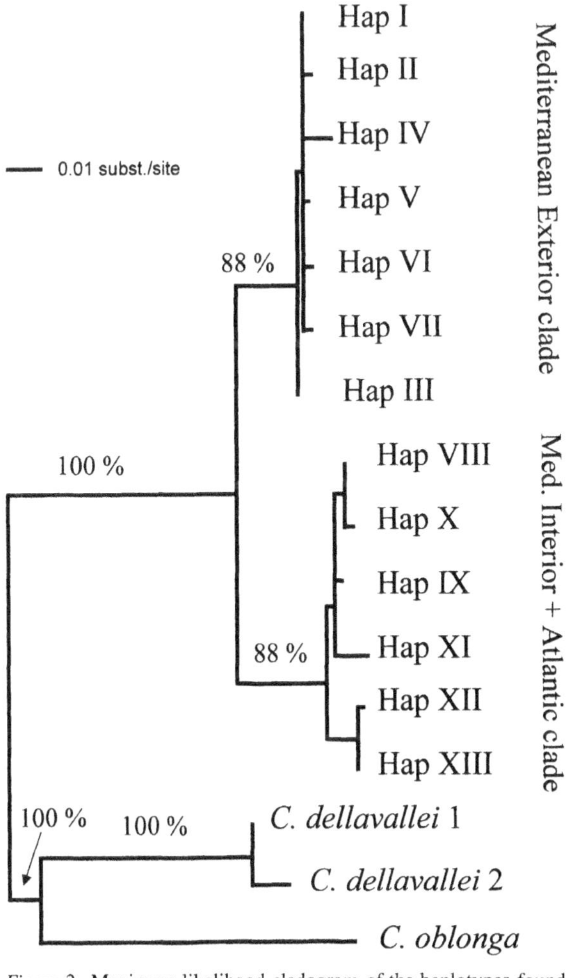

Figure 2. Maximum likelihood cladogram of the haplotypes found in *Clavelina lepadiformis* and the outgroups. Codes of haplotypes as in Table 1. Bootstrap support for the clades is indicated. See text for details.

Discussion

Present results provide strong evidence that the interior Mediterranean form of *Clavelina lepadiformis* originated from the Atlantic clade. The high genetic similarity between the Atlantic interior populations (Vigo and Sweden) and the exterior ones (Azores, Portugal) indicates that there is only one clade in the Atlantic. This Atlantic clade thrived subsequently in an environment (inside of Mediterranean harbours) that was more similar to the turbid and richer Atlantic waters than the clearer and more oligotrophic open waters of the Mediterranean. Its presence in harbours suggests a colonisation related to ship traffic between ports. Low genetic divergence, and high gene flow estimates between Atlantic and interior Mediterranean populations, suggests strongly that ship traffic favours genetic exchange overwhelming the dispersal ability of the short-lived (hours to days, Tarjuelo, 2001) larvae of this ascidian.

The favoured hypothesis, that the two clades evolved allopatrically in the Mediterranean and Atlantic basins, is in agreement with other instances of genetic discontinuities in conspecific populations or congeneric species-pairs of benthic invertebrates between the north-east Atlantic and Mediterranean (Saavedra et al. 1993; Quesada et al., 1995; Côrte-Real et al., 1996; Panacciulli et al., 1997; McFadden, 1999). The presence of two clades of the brittle star *Ophiothrix* in the Mediterranean could, similarly, be the result of allopatric divergence in the Atlantic and Mediterranean, followed by a second invasion of the Mediterranean by the Atlantic clade, as suggested by Baric & Sturmbauer (1999).

The level of nucleotide diversity, as well as the number of haplotypes found, is similar in the two main clades recognized, indicating an ancient separation that has allowed the two forms to undergo genetic differentiation. As the Mediterranean basin suffered a strong drying out period during the Messinian (between 5 and 6 million years ago), it seems likely that the species first originated in the Atlantic and colonised the Mediterranean after the Messinian crisis. This time scale would provide ample time for genetic differentiation of the two clades, followed by a recent invasion of Mediterranean marinas by the Atlantic clade. There are insufficient data to establish a precise time frame for this invasion, but Harant (1927) reported the presence of *C. lepadiformis* in harbours of the western Mediterranean. Although we are not making formal taxonomic distinctions, both genetic and biological data (De Caralt et al., 2002) indicate that the two clades might be given species status. As the original species description was based on Scandinavian specimens, the original name would belong to the Atlantic form and a new name could be given to the exterior Mediterranean form.

The extent of cryptic introductions of invertebrate species may be important and deserves further attention. Increased recreational sailing in the Mediterranean, and the proliferation of marinas in the past decades, provide additional points for step-by-step colonisation of alien forms, even with low natural dispersal abilities. In terms of conservation, not only is the arrival of non-indigenous species important, but so too are the invasions of new genetic variants. We believe that more studies of populations inhabiting marginal marine environments, supposedly conspecific with local species, may reveal more instances of genetic exchange through harbour traffic. Genetic tools are a useful method to investigate these cryptic invasions, as they allow the identification of sources, patterns of dispersal and degree of gene flow with local forms.

Acknowledgements

We thank Peter Wirtz, Malin Werner and Gonzalo Macho for providing samples, David Posada for advice with the analyses, and Julio Rozas for the latest version of DnaSP. This study was funded by projects MAR98-1004, BOS2000-0295 and REN2001-2312 of the Spanish Ministry of Science and Technology, and by the Interreg IIIA program of the European Union.

References

Baric, S. & C. Sturmbauer, 1999. Ecological parallelism and cryptic species in the genus *Ophiothrix* derived from mitochondrial DNA sequences. Mol. Phyl. Evolut. 11: 157–162.

Berrill, N. J., 1950. The tunicata. With an account of the British species. Ray Society, London. 354 pp.

Boisselier-Dubayle, M. C. & S. Gofas, 1999. Genetic relationships between marine and marginal marine populations of *Cerithium* species from the Mediterranean Sea. Mar. Biol. 135: 671–682.

Brunetti, R., 1979. *Polyandrocarpa zorritensis* (Van Name, 1931) a colonial ascidian new to the Mediterranean record. Vie Milieu 29: 647–652.

Carlton, J. T. & J. B.Geller, 1993. Ecological roulette: the global transport of nonindigenous marine organisms. Science 261: 78–82.

Côrte-Real, H. B., J. P. Hawkins & J. P. Thorpe, 1996. Population differentiation and taxonomic status of the exploited limpet *Patella candei* in the Macaronesian islands (Azores, Madeira, Canaries). Mar. Biol. 125: 141–152.

Dalby, J. E., 1997. Dimorphism in the ascidian *Pyura stolonifera* near Melbourne, Australia, and its evaluation through field transplant experiments. PSZNI: Mar. Ecol. 18: 253–271.

De Caralt, S., S. López-Legentil, I. Tarjuelo, M. J. Uriz & X. Turon, 2002 Contrasting biological traits of *Clavelina lepadiformis* (Ascidiacea) populations from inside and outside harbours in the western Mediterranean. Mar. Ecol. Prog. Ser. 244: 125–137.

Felsenstein, J., 1981. Evolutionary trees from DNA sequences: a maximum likelihood approach. J. Mol. Evol. 17: 368–376.

Felsenstein, J., 1985. Confidence limits on phylogenies: an approach using the bootstrap. Evolution 39: 783–791.

Harant, H., 1927. La faune Ascidiologique de Banyuls et Cette. Essai de revision des Ascidies de la Méditerranée occidentale. Annl. Inst. océanogr., Paris 4: 209–251.

Kimura, M., 1981. Estimation of evolutionary distances between homologous nucleotide sequences. Proc. natn. Acad. Sci. U.S.A. 78: 454–458.

Koukouras, A., E.Voultsiadou-Koukoura, T. Kevrekidis & D.Vafidis,1995. Ascidian fauna of the Aegean sea with a check list of eastern Mediterranean and Black sea species. Annl. Inst. océanogr., Paris 71: 19–34.

Lambert, C. C. & G. Lambert, 1998. Non-indigenous ascidians in southern California harbors and marinas. Mar. Biol. 130: 675–688.

Latorre, A., A. Moya & F. Ayala, 1986. Mitochondrial DNA polymorphism of *Drosophila subobscura*. Proc. natn. Acad. Sci. U.S.A. 83: 8649–8653.

Lynch, M. & T. J. Crease, 1990. The analysis of population survey data on DNA sequence variation. Mol. Biol. Evol. 7: 377–394.

Maldonado, A., 1985. Evolution of the Mediterranean Basins and a detailed reconstruction of the Cenozoic paleoceanography. In Margalef, R. (ed.), Western Mediterranean. Pergamon Press, Oxford: 17–60.

McFadden, C. S., 1999. Genetic and taxonomic relationships among Northeastern Atlantic and Mediterranean populations of the soft coral *Alcyonum coralloides*. Mar. Biol. 133: 171–184.

Millar, R. H., 1966. Marine invertebrates of Scandinavia. 1. Tunicata Ascidiacea. Scandinavian University Books, Oslo. 123 pp.

Monniot, C., 1981. Apparition de l'ascidie *Microcosmus exasperatus* dans les ports méditerranéens. Tethys 10: 59–62.

Monniot, C. & F. Monniot, 1994. Additions to the inventory of eastern tropical atlantic ascidians: arrival of cosmopolitan species. Bull. mar. Sci. 54: 71–93.

Monniot, C., F. Monniot & P. Laboute, 1985. Ascidies du port de Papeete (Polynèsie française): relations avec la milieu naturel et apports intercontinentaux par la navigation. Bull. Mus. natn. Hist., Paris 7A: 481–495.

Monniot, C., F. Monniot F & P. Laboute, 1991. Coral Reef Ascidians of New Caledonia. O.R.S.T.O.M., Paris. 247 pp.

Nei, M., 1982. Evolution of human races at the gene level. In Bonne-Tamir, B., T. Cohen & R. M. Goodman (eds), Human Genetics, Part A. The Unfolding Genome. Alan R. Liss, New York: 167–181.

Nei, M.,1987. Molecular Evolutionary Genetics. Columbia University Press, New York. 512 pp

Pannacciulli, F. G., D. D. Bishop & S. J. Hawkins, 1997. Genetic structure of populations of two species of *Chthamalus* (Crustacea: Cirripedia) in the north-east Atlantic and Mediterranean. Mar. Biol. 128: 73–82.

Posada, D. & K. A. Crandall, 1998. Modeltest: testing the model of DNA substitution. Bioinformatics 14: 817–818.

Quesada, H., C. Zapata & G. Alvarez, 1995. A multilocus allozyme discontinuity in the mussel *Mytilus galloprovincialis*: the interaction of ecological and life-history factors. Mar. Ecol. Prog. Ser. 116: 99–115.

Ramos, A. A., 1988. Ascidias litorales del Mediterráneo Ibérico: faunística, ecología y biogeografía. Ph.D. Thesis. University of Alicante. 405 pp.

Rozas, J. & R. Rozas, 1999. DnaSP version 3.0: an integrated program for molecular population genetics and molecular evolution analyses. Bioinformatics 15: 174–175.

Saavedra, C., C. Zapata, A. Guerra & G. Alvarez G. 1993. Allozyme variation in European populations of the *Ostrea edulis*. Mar. Biol. 115: 85–95.

Schneider, S., D. Roessli & L. Excoffier. 2000. Arlequin: a software for population genetics data analysis. Ver 2000. Genetics and Biometry Lab., Dept. of Anthropology, University of Geneva. 111 pp.

Svane, I. & C. M. Young. 1989. The ecology and behaviour of ascidian larvae. Oceanogr. mar. biol. Ann. Rev. 27: 45–90.

Swofford, D. L., 1998. PAUP*: Phylogenetic Analysis using Parsimony and other Methods. Sinauer Associates, Sunderland.

Tarjuelo, I., 2001. Reproductive strategies in colonial ascidians: relationships with other life-history traits and genetic structure. Ph.D. Thesis. University of Barcelona. 159 pp.

Tarjuelo, I., D. Posada, K. A. Crandall, M. Pascual & X. Turon, 2001. Cryptic species of *Clavelina* (Ascidiacea) in two different habitats: harbours and rocky littoral zones in the northwestern Mediterranean. Mar. Biol. 139: 455–462.

Turon, X., 1987. Estudio de las ascidias de las costas de Cataluña e Islas Baleares. Ph.D. Thesis. University of Barcelona. 353 pp.

Vazquez, E., 1993. Estudio faunístico, autoecológico y biogeográfico de los tunicados bentónicos de la ria de ferrol (Galicia). Ph.D. Thesis. University of Santiago de Compostela. 443 pp.

Wirtz, P. & H. R. Martins, 1993. Notes on some rare and little known marine invertebrates from the Azores, with a discussion of the zoogeography of the region. Arquipélago 11A: 55–63.

Wirtz, P., 1998. Twelve invertebrate and eight fish species new to the marine fauna of Madeira, and a discussion of the zoogeography of the area. Helg. Meeres. 52: 197–207.

Wright, S., 1951. The genetical structure of populations. Ann. Eugenics 15: 323–352.

Zibrowius, H., 1991. Ongoing modification of the Mediterranean marine fauna and flora by the establishment of exotic species. Mésogée 51: 83–107.

Hydrobiologia **503:** 37–44, 2003.
M.B. Jones, A. Ingólfsson, E. Ólafsson, G.V. Helgason, K. Gunnarsson & J. Svavarsson (eds), Migrations and Dispersal of Marine Organisms.

Shallow-water hydrothermal vents in the Mediterranean sea: stepping stones for Lessepsian migration?

Anna Maria De Biasi[1] & Stefano Aliani[2]
[1]*Centro Interuniversitario di Biologia Marina, Via N. Sauro, 4 I-57127 Leghorn, Italy*
[2]*Institute for Physical Oceanography CNR, Forte S. Teresa, I-19036 Pozzuolo La Spezia, Italy*
E-mail: a.debiasi@cibm.it

Key words: hydrothermal vents, temperature, thermophilous algal species, Milos, Greece

Abstract

The present study was carried out in Palaeochori Bay, south eastern coast of Milos, where outflow of hot gas bubbles and hydrothermal water seepage was observed. The study tests hypotheses about relationships between seawater temperature and species with warm-water affinities in areas close to shallow-water vents. We predicted that if temperature plays a major role in influencing abundance of species with warm-water affinities, then there should be more thermophilous species in areas with higher seawater temperature. Time series of seawater temperature were recorded in Palaeochori Bay and Pollonia Bay where hot emission was not observed. Temperature, water pressure and conductivity were measured every 30 min from 19 June 1996 to 16 June 1997. Fourteen qualitative samples of benthic flora were collected by SCUBA diving in six rocky sites. Photosamples of benthic assemblages were collected using an UW camera equipped with a wide-angle lens (15 mm) and an electronic flash unit. A constant area of 0.7 m^2 was sampled. Quantitative estimates of abundance of algal species were performed as percent cover. This paper shows that patterns of fluctuation in seawater temperature in Palaeochori were different from the control where no venting was found. In Palaeochori, algal assemblages were dominated by species with warm-water affinities that were not found elsewhere. These results are consistent with the hypothesis that temperature plays a major role in influencing abundance of species with warm-water affinities. According to our results, hydrothermal vents are potential oases that experience biological-physical coupling not yet clearly understood. They are important in affecting migration of alloctonous thermophilous species suggesting a possible role of vent areas as stepping stones for species migration.

Introduction

In recent years, intrusion of warm-water species through the Suez Canal was recorded in the Mediterranean Sea and these species were called 'Lessepsian' from the name of the builder of the canal. This migration increased after enlargement of the canal which is at present 2.5 times its size of 1975, 4 times its size in 1956, and 14 times its size in 1869 when it opened. Apart from higher fluxes due to increase in the canal cross section, the new channel by-passes the Bitter Lakes. The latter previously represented a strong barrier for migration because of their extremely high salinity. This scenario of faunal/floristic exchange is helped by a general warming of the deep Mediterranean (Roether et al., 1996; Boscolo & Bryden, 2001) and shallow waters (Cerrano et al., 2000). Exotic and thermophilic species can take advantage of this new situation. In fact, after becoming very abundant in the eastern part of the Mediterranean, species with warm-water affinity spread slowly over the Mediterranean and were recorded in the northernmost areas (Aliani & Meloni, 1999) where they found suitable habitats in 'warm oases' along the coastline (Astraldi et al., 1995).

In marine hydrothermal vents, hot fluids emitted from the seafloor create some of the most extreme environments on Earth, sometimes unsustainable for life.

In shallow water venting is characterized typically by massive free gas release and hot water efflux, and extends from the intertidal down to more than 200 m water depth (Dando et al., 1995). Temperatures at shallow vents, due to the lower boiling point, are not as high as in the widely-studied deep sea vents, which can exceed 300 °C (Tunnicliffe et al., 1990), but sediment temperatures over 120 °C had been recorded (Dando et al., 1995) as well as sediment temperature series over 80 °C for months (Aliani et al., 1998). Temperature is one of the most important environmental factors affecting the hydrothermal vent ecosystems and seems to be a major limiting factor for bacterial production (Chevaldonné et al., 1991). Knowledge of the venting temperatures may explain some of the changes in composition and distribution of the biological communities (Tunnicliffe et al., 1990; Chevaldonné et al., 1991; Dando et al., 1995). Although dilution rates of the heat are poorly known (Johnson et al., 1988; Chevaldonné et al., 1991), the consequences of hot venting on the surrounding habitat are remarkable (Tarasov et al., 1999). Hydrothermalism in the Aegean has been widely studied (Dando et al., 1995; Thiermann et al., 1997, Gamenick et al., 1998, Morri et al., 1999) and several of studies concentrated on the very active area to the south east of Milos Island (Kiklades Archipelago), probably the most active site in Milos. In this area, from 3 to about 100 m depth, there is a spectacular outflow of gas bubbles (CO_2, CH_4, H_2, H_2S) and hydrothermal water seepage from the seabed.

The present study, carried out in Palaeochori Bay situated in the south east coast of Milos, tested hypotheses about relationships between seawater temperature and species with warm-water affinities in areas close to shallow-water vents. We predicted that if temperature plays a major role in influencing the abundance of species with warm-water affinities, then there should be more thermophilous species in areas with higher seawater temperature.

Materials and methods

Study area

The study was carried out off Milos Island located in the Aegean Sea one of the most seismically active region on earth (Bath, 1983; Makkropoulos & Burton, 1984) (Fig. 1). Milos, Kos, Yali and Santorini are distributed as drops on a line along the northern part of the Sea of Crete from East to West. Studies on the biota were made in Palaeochori Bay. Hot gaseo-hydrothermal vents coming out of the sandy bottom sustain a peculiar ecosystem based upon chemosynthetic bacteria that are organized on the sea bed as concentric rings of bacterial mats of different species. The sediment temperature in these areas can be extremely high (~85 °C) and even swimming very close to these vents could produce skin burn. These peculiar bacterial-based ecosystems are like glades within the seagrass meadows that largely cover the sandy bottoms in the bay. *Posidonia oceanica* was dominant all over the Bay and *Cymodocea nodosa*, which was more tolerant to the venting, could be found closer to the border of the bacterial mats rings. However, both were absent where sediment temperature was 2.5 °C higher than ambient (Aliani et al., 1998).

Temperature time series

Time series of seawater temperature were recorded at about 8 m depth in Palaeochori Bay at about 40 m distance from a vent field (36° 40.32′ N 24° 31.49′ E). Parallel control measurements were collected in a nearby bay in the summer (36° 40.11′ N 24° 30.37′ E) and in Pollonia Bay for the remaining part of the year. Control areas were considered free of vent activity after visual survey. Temperature sensors (Sensor Fenwall GB 32JM19 accuracy ±0.05 °C) mounted in a Aanderaa WLR7 depth gauge collected synchronised data series every 30 min. The instrument also measured water pressure and conductivity. A 1-year time series of seawater temperature was collected starting from 19 June 1996 at 07:30 GMT. Gaps in the series were on 26 and on 28 September 1996 (Palaeochori and control instruments, respectively) due to intermediate data download and instruments check. The series ended on 16 June 1997 at 01:00 GMT. Gaps in the data set were filled with symbols (NaN) that were not considered in the statistics (e.g. calculation of average) but kept the time series synchronised.

Benthic assemblages

Fourteen qualitative samples of benthic flora were collected by SCUBA divers scraping rocks with hands, knives or by a hammer in six rocky sites: CR, VS, SR, ST, E, S (Fig. 1). The shoal CR, positioned off Palaeochori Bay, was made by volcanic rock where continuous vent activity was observed. It was explored between 26 and 30 m depth. The shoal VS located about 1.8 km offshore from Palaeochori Bay, is a

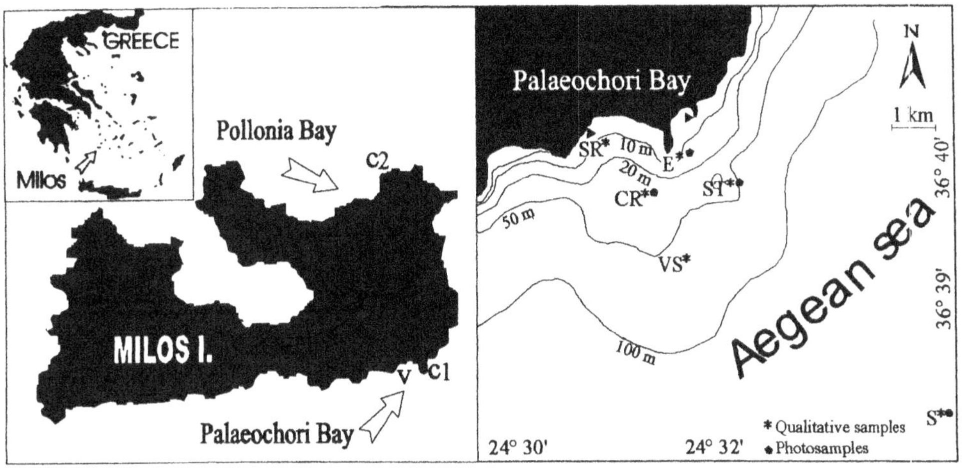

Figure 1. Study area. C1, C2, V are currentmeters. CR, VS, SR, ST, E, S indicate sites where algae were sampled or photographed.

Table 1. Average values, minimum and standard deviation of seawater temperature in Palaehori Bay (P) and control (C)

	Mean			Min			Stdev		
	P	C	P-C	P	C	P-C	P	C	P-C
Jun-96	21.6	21.7	–0.1	20.2	20	0.2	1.14	1.26	–0.1
Jul-96	23.3	23.4	–0.1	21.7	21.6	0.1	0.96	1.02	–0.1
Aug-96	23.5	23.7	–0.2	22.2	22.4	–0.2	0.53	0.57	0.0
Sep-96	23.8	23.3	0.5	22.3	21.8	0.5	0.41	0.6	–0.2
Oct-96	21.7	20.8	0.9	19.8	17.6	2.2	0.78	1.33	–0.6
Nov-96	19.0	18.4	0.6	17.6	16.4	1.2	0.58	0.73	–0.2
Dec-96	17.2	16.2	0.9	16.1	13	3.1	0.36	0.81	–0.5
Jan-97	16.3	15.2	1.0	14.9	12.4	2.5	0.33	1.07	–0.7
Feb-97	15.5	14.5	1.0	14.6	11.2	3.4	0.25	1.13	–0.9
Mar-97	15.2	14.5	0.8	14.4	12.1	2.3	0.2	1.15	–1.0
Apr-97	15.4	15.3	0.0	14.8	11.6	3.2	0.31	1.26	–1.0
May-97	17.9	19.0	–1.1	15.5	14.7	0.8	1.67	2.17	–0.5
Jun-97	21.0	21.5	–0.5	19.7	19.7	0.0	0.44	0.96	–0.5

rocky platform emerging from a sandy floor at 44 m depth. Vent emissions were observed from the central part of the shoal occupied by a sandy depression and from the sediments surrounding the shoal. Samples were collected between 45 and 81 m depth. Site SR is inside the Bay with a depth ranged from a few to 12 m. Samples were collected on very shallow substrata down to 12 m. ST is a rocky shoal made by two basaltic pillars located about 900 m east off Spathi Point where no thermal activity was observed. It was investigated only at 9 m depth. Site E was located at the base of Spathi Point closing the east side of the bay. At this site, only one sample was collected at 11 m depth. The shoal S is situated about 5.4 km offshore to the south-eastern side of the bay where there were two emerged rocky pinnacles named Vrakhos Ktenia whose upper portion was made of granite; the remaining part was composed of black basaltic rock. Samples were collected between 15 and 40 m depth. For a more detailed description of these sites see Cocito et al. (2000).

Photosamples of benthic assemblages were collected at four (CR, E, ST, S) different sites (Fig. 1) in June 1996. At each site, images were taken at four stations with different slope, i.e. (sub)horizontal (H, 0° to 30°), inclined (I, 30° to 70°), (sub)vertical (V, 70° to 90°) and overhanging (>90°). Three replicates were collected at each station. Frame still photographs were collected using an UW camera equipped with a wide-angle lens (15 mm) and a flash. The frame (30×30cm) was maintained parallel to the lens at a distance of 60 cm by a metal stick connected to the camera stirrup, thus sampling a constant area of 0.7 m^2. Quantitative estimates of abundance of algal species were performed as percent cover. Visual estimation of large species were performed, by snorkelling in September 1996 and June 1997 also in other parts of the islands (two bays close to Palaeochori and two sites to the north of the Island).

Results and discussion

Temperature time series

No extreme values of seawater temperature were recorded in the series from Palaeochori, even though some of the bottom sediment temperatures reached

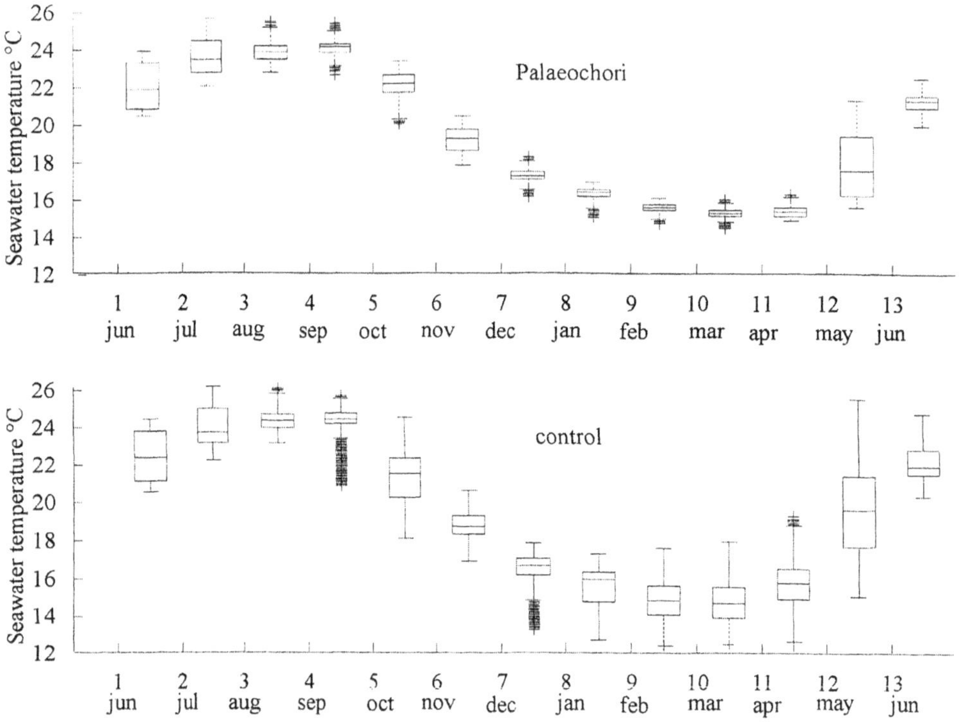

Figure 2. Plot of seawater temperature in Palaeochori and in the control. The box has lines at the lower quartile, median, and upper quartile values. The whiskers are lines extending from each end of the box to show the extent of the rest of the data. Outliers are data with values beyond the ends of the whiskers.

120 °C at about 40 m distance from the sensor. Some statistical properties of temperature series are summarised in Figure 2. In summer 1996, seawater temperatures in Palaeochori were very similar to the control site and the mean temperature was slightly colder (0.1 °C) than the control. At the end of summer, the monthly mean temperature inside Palaeochori became warmer than the control, being 0.5 °C higher in September and about 1 °C higher in mid winter. From spring to summer, the mean temperature inside Palaeochori became lower than the control (Table 1). A trend similar to the previous year may be inferred but, unfortunately, the series ended before summer. Differences in monthly temperature minima were compared and were greatest (3.4 °C) in February 1997. These differences were only partly dependant on the lower mean temperature of the active bay in winter. In fact, high standard deviations in the control suggested that strong, maybe episodic, events produced abrupt changes in the water temperature at the control site that did not take place in Palaeochori.

A possible hypothesis to explain these differences is that the series were from different water masses (different oceanographic properties or local vein of fresh water). To answer these questions we compared the series using conductivity data but, unfortunately, sediment particles affected the conductivity cell geometry and the final quality of the data set was quite poor. Superimposing short sections of salinity data were encouraging but they were not able to completely answer our question, as episodic events might not be included in the sections. If the temperature series referred to the same water mass, they must oscillate at a common frequency. We performed an analysis of the series in the frequency domain. Cross correlation between series was good ($r = 0.93$) and coherence analyses showed good relationships with a small time lag between series. Temperature fluctuations in the two series were very similar in frequency, so we reject the hypothesis of different water masses being involved. Available conductivity data were not different between Paleaochori and controls and temperature was the main property of seawater that changed. Differences in temperature cannot be explained by direct transfer of heat from vent-heated sediment because dilution of vent water in seawater was very strong (Johnson et al., 1988). Insulation in shallow water is probably the major source of heat in summer but,

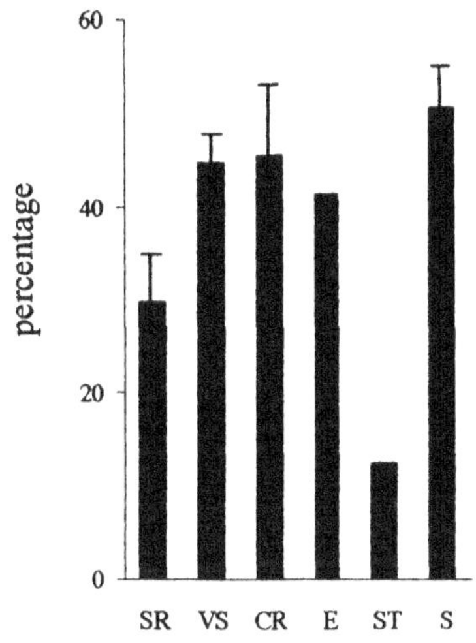

Figure 3. Percentage thermophilous species from samples. SR, VS, CR, E, ST, S sampling sites (see Fig. 1 for their location). Bars=standard error.

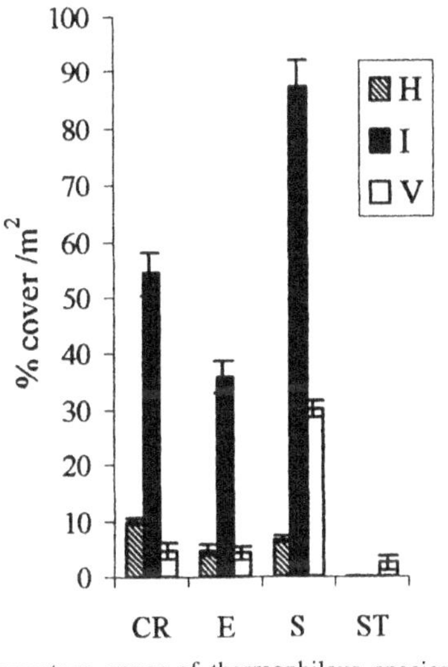

Figure 4. Percentage cover of thermophilous species from photosamples. See Figure 1 for site (CR, E, ST, S) location. Bars=standard error. H=horizontal, I=inclined, V=vertical.

in winter, a different source can be hypothesised and venting is the macroscopic difference between sites. At present, we cannot explain the reasons for the difference in temperature but it is an evidence that seawater temperature in Palaeochori Bay was higher and more stable over time than elsewhere in Milos.

Algal assemblages

A total of 83 algal species was identified from the fourteen qualitative samples. The list included 36 Rhodophyceae, 33 Fucophyceae, 13 Chlorophyceae and 1 Tribophycee [see Sartoni & De Biasi (1999) for the complete check list]. The list includes species not recorded previously in Milos Island (including *Amphiroa cryptarthrodia*, *Amphiroa verruculosa* Kützing, *Beckerella mediterranea*, *Botryocladia boergesenii*, *Corallophila cinnabarina*) and other species remarkable from both taxonomic and biogeographical views (among others, *Tribonema marinum*, *Cutleria* cf. *chilosa*, *Amphiroa rubra*). Noteworthy, was the presence of a large number of species (40%) with warm-water affinities (Table 2). They were widely recorded in all investigated sites where hot emissions were detected and where evident venting was not observed (Fig. 3). Benthic assemblages from overhanging sites included mainly calcareous encrusting algae not identified at species level and not considered hereafter. Data from photographs partly confirmed these results, although this last method provided a less complete inventory of algal species. Small size and cryptic species cannot be identified in the photosamples as well as epiphytes of larger taxa. However, photosamples provided additional information about distribution and abundance of conspicuous species. Algae with warm-water affinities were found in pictures from all sites (Fig. 4) and cover values were very different between different sites; *Beckerella mediterranea* and *Stypopodium schimperi* had high cover values in Palaeochori Bay.

The red alga *Beckerella mediterranea* is an apparently endemic species known from many islands in the Southern Archipelago, Peloponnesus and Crete (Athanasiadis, 1987). In Milos, it was found in vent and non-vent sites between 10 and 35 m depth mainly on inclined substrata. In the shoal S, *B. mediterranea* was conspicuous particularly on inclined substrata at 15 m depth where it was very abundant (51±5% cover) and able to shape the physiognomy of the epibenthic community. In Milos, the brown alga *Stypopodium schimperi*, had a wide depth distribution, ranging between 15 and 41 m depth. It was found on substrata with different slopes and exposure, but cover values were very different in vent and non-vent sites. In non-vent sites, and outside the Bay, it had a patchy distribution with low cover values (from 2 to 5% cover). At vent sites, this species was widely distributed with cover values over 25%. *S. schimperi*, being a large very conspicuous and abundant species, can be detected easily by snorkelling, but was not found during our surveys around the island outside the Palaeochori area, nor during previous work in Milos by other groups. This species is an example of the Lessepsian

Table 2. List of the species not recorded previously in Milos. Species with warm-water affinities are marked with an asterisk (*)

RHODOPHYCEAE
Amphiroa cryptarthrodia Zanardini
Amphiroa rubra (Philippi) Woelkerling
Beckerella mediterranea H. Huvé (*)
Botryocladia boergesenii J. Feldmann
Corallophila cinnabarina (Grateloup *ex* Bory de Saint-Vincent) R.E. Norris
Cruoria cruoriaeformis (P.L. *et* H.M. Crouan) Denizot
Drachiella minuta (Kylin) Maggs*et* Hommersand
Gelidium pusillum (Stackhouse) Le Jolis
Gulsonia nodulosa (Ercegovic) J. Feldmann *et* G. Feldmann
Halopithys incurva (Hudson) Batters (*)
Halydiction mirabile Zanardini
Hydrolithon farinosum (J.V. Lamouroux) Penrose *et* Chamberlain v. *chalicodityum* (W.R. Taylor) Serio
Jania adhaerens J. V. Lamouroux (*)
Laurencia microcladia Kützing (*)
Mesophyllum lichenoides (J. Ellis) Lemoine
Osmundaria volubilis (Linnaeus) R.E. Norris
FUCOPHYCEAE
Asperococcus bullosus J.V. Lamouroux
Cladosiphon cylindricus (Sauvageau) Kylin
Cladostephus spongiosus (Hudson) C. Agardh f.*verticillatus* (Lighfoot) Prud'homme van Reine
Cutleria cf. *chilosa* (Falkenberg) P.C. Silva
Cystoseria brachycarpa J. Agardh emend. Giaccone
Cystoseira schiffneri Hamel f. *latiramosa* (Ercegovic) Giaccone
Spatoglossum solieri (Chauvin *ex* Montagne) Kützing
Stilophora tenella (Esper) P.C. Silva
Stypopodium schimperi (Buchinger *ex* Kützing) M. Verlaque *et* Boudouresque (*)
Taonia atomaria (Woodward) J. Agardh
Zanardinia prototypus (Nardo) Nardo
Zonaria tournefortii (J.V. Lamouroux) Montagne (*)
CHLOROPHYCEAE
Codium bursa (Linnaeus) C. Agardh
Microdictyon tenuius Decaisne *ex* J. Gray (*)
TRIBOPHYCEAE
Tribonema marinum J. Feldmann (*)

migration process (Verlaque & Boudouresque, 1991; Verlaque, 1994). *Stypopodium* was first reported along the Syrian coastline [Mayhoub & Billard, 1991 as *S. zonale* (Lamouroux) Papenfuss]. Nowadays, it is very abundant there, replacing former algal assemblages in many cases. Verlaque & Boudouresque (1991) recorded this species in Cyprus and along the Turkish coasts, within a narrower bathymetric range (5–10 m depth) than in Palaeochori Bay. Our data are not sufficient to fully describe this pattern of invasion. We can only hypothesize a different susceptibility or resistance to invasion of these recipient regions. Furthermore, bias in the quantity or quality of existing data can never be completely excluded (Ruitz et al., 2000). Our study clearly showed that *S. shimperi*, despite recently introduced, was the most important and conspicuous species characterizing the epibenthic assemblages. Neither *Beckerella mediterranea* nor *Stypopodium shimperi* were observed outside Palaeochori Bay. The visual census did not allow us to check for the presence of small or cryptic species, so we cannot maintain that they were absent in other parts of the island, we can confirm that they were not as abundant as in Paleaochori. Some clues of the effect of vents on the macrobenthos have previously been suggested by different authors who investigated sessile epifauna in Palaeochori Bay. Morri et al. (1999) maintained that biodiversity was proportionally higher at sites closest to hydrothermal vents. In particular, species richness of sponges was consistently higher at vent sites than at the corresponding non-vent sites. Pansini et al. (2000) pointed out that the occurrence of some species such as *Cliona copiosa*, *C. rhodensis*, *Axinella cannabina* together with horny sponges (*Dysidea*, *Ircinia*, *Sarcotragus*, *Spongia*) suggests a warm-water affinity of the sponge fauna at Palaeochori Bay. By contrast, the occurrence of *Geodia cydonium* only at vent sites appears consistent with the fact that this species grows better in areas influenced by hydrothermal waters (Morri et al., 1994). With few exceptions (*Drachiella minuta*, *Halydiction mirabile*, *Hydrolithon farinosum* v. *chalicodityum*), our list does not include species recorded only at vent sites, suggesting that vent activity did not select an exclusive flora. Species found have been recorded at 'normal sites' in the Mediterranean Sea and cannot be called vent-obligate species according to the definition reported by Barry et al. (1996). Our study, as well as previous recording benthic communities in the area, was carried out at a large spatial scale. None planned a sampling strategy to include 'distance from the vent' as a main variable for sample collection. So, the direct vent effect at a small spatial scale (few meters apart) cannot be demonstrated for macroepibenthos as it has been for soft bottoms (Thiermann et al., 1994; Dando et al., 1995; Thiermann et al., 1997).

Conclusions

Patterns of fluctuation in seawater temperature, where vent emissions are evident, were different from the control where no venting was found. Our data do not demonstrate clearly that hot venting is involved directly in the differences recorded. We can affirm that seawater temperature in Palaeochori Bay was higher and more stable over time than in a similar area without venting. In Palaeochori, algal assemblages were dominated by species with warm-water affinities. In addition, the Lessepsian *S. shimperi*, despite being recently introduced, was abundant and can be considered a resident. Its absence outside the bay suggests that Palaeochori offers suitable environmental conditions not only for its settlement but also for survival. In addition, we report some evidence for the occurrence of seawater temperature anomalies at hot vent sites and the parallel abundance of thermophilous species. Our view is that the presence of seawater temperature anomalies and the record of species with warm-water affinities cannot be by chance. It is well known that many variables and ecological constraints are involved in the processes influencing invasion patterns, making inferences about the invasion success difficult to predict. The best forecasting tool to predict the invasiveness of non-native species (*sensu* Boudouresque & Verlaque, 2002) is its invasiveness elsewhere in the world. Our study has shown that peculiar patterns of invasiveness occurred in Palaeochori and that these deserve further investigation. Hydrothermal vents are potential oases that experience biological-physical coupling not yet clearly understood. They are important in affecting migration of alloctonous thermophilous species suggesting a possible role of vent areas as stepping stones for species migration.

Acknowledgements

This study was part-funded by the EC under contract MAS 95-0021. We thank G. Sartoni (University of Florence, Italy) for taxonomic identification of algae, C.N.Bianchi (ENEA, La Spezia, Italy) for scientific suggestions and A. Peirano, S. Cocito (ENEA, La Spezia, Italy) for collecting samples.

References

Aliani, S. & R. Meloni, 1999. Dispersal strategies of benthic species and water current variability in the Corsica Channel (Western Mediterranean). Sci. mar. 63: 137–145.

Aliani, S., L. Amici, P. R. Dando & R. Meloni, 1998. Time series of water pressure and bottom temperature in a marine shallow water hydrothermal vent of Milos Island (Aegean sea): preliminary results. Rapp. Comm. Int. Mèr Médit. 35: 46–47.

Astraldi, M., C. N. Bianchi, G. P. Gasparini & C. Morri, 1995. Climatic fluctuations, current variability and marine species distribution: a case study in the Ligurian Sea (North-Western Mediterranean). Oceanol. Acta 18: 139–149.

Athanasiadis, A., 1987. A survey of the seaweeds of the Aegean Sea with taxonomic studies on species of the tribe Antithamnieae (Rhodophyta). University of Gothenburg, Sweden. 174 pp.

Barry, J. P., H. G. Greene, D. L. Orange, C. H. Baxter, B. H. Robison, R. E. Kochevar, J. W. Nybakken, D. L. Reed & C. M. McHugh, 1996. Biologic and geologic characteristics of cold seeps in Monterey Bay, California. Deep Sea Res. 43: 1739–1762.

Bath, M., 1983. The seismology of Greece. Techtonophysics 98: 168–208.
Boscolo, R. & H. Bryden, 2001. Causes of long-term changes in Aegean deep water. Oceanol. Acta 24: 519–527.
Boudouresque, C. F. & M. Verlaque, 2002. Biological pollution in the Mediterranean Sea: invasive versus introduced macrophytes. Mar. Poll. Bull. 44: 32–38.
Cerrano, C., G. Bavestrello, C. N. Bianchi, R. Cattaneo-Vietti, S. Bava, C. Morganti, P. Picco, G. P. Sara, S. Schiapparelli, A. Siccardi & F. Sponga, 2000. A catastrophic mass-mortality episode of gorgonians and other organisms in the Ligurian Sea (North-western Mediterranean), summer 1999. Ecol. Lett. 3: 284–293.
Chevaldonné, P., D. Desbruyères & M. Le Haitre, 1991. Time series of temperature from three deep-sea hydrothermal vent sites. Deep Sea Res. 38: 1417–1430.
Cocito, S., C. N. Bianchi, C. Morri & A. Peirano, 2000. First survey of sessile communities on subtidal rocks area with hydrothermal vents: Milos island, Agean Sea. Hydrobiologia 426: 113–121.
Dando, P. R., J. A. Hughes, Y. Leah, S. J. Niven, R. J. Taylor & C. Smith, 1995. Gas venting rates from submarine hydrothermal areas around the island of Milos, Hellenic Volcanic Arc. Cont. Sh. Res. 15: 913–929.
Gamenick, L., M. Abbiati & O. Giere, 1998. Field distribution and sulphide tollerance of *Capitella capitata* (Annelida: Polychaeta) around shallow water hydrothermal vents off Milos (Aegean Sea). A new sibling species? Mar. Biol. 130: 447–453.
Johnson, K. S., J. J. Childress & C. L. Beehler, 1988. Short-term temperature variability in the Rose Garden hydrothermal vent field: an unstable deep-sea environment. Deep Sea Res. 35: 1711–1721.
Makropoulos, C. & P. W. Burton, 1984. Greek tectonics seismicity. Techtonophysics 106: 275–304.
Mayhoub, H. & C. Billard, 1991. Contribution à la connaissance d'un *Stypopodium* (Dictyotales, Phaeophyceae) installé récemment sur les côtes syriennes. Cryptog. Algol. 12: 125–136.
Morri, C., F. Cinelli & C. N. Bianchi, 1994. Sessile epifauna gigantism an a submarine cave with sulphur springs. Cave Diving 6: 4–9.
Morri, C., C. N. Bianchi, S. Cocito, A. Peirano, A. M. De Biasi, S. Aliani, M. Pansini, M. Boyer, F. Ferdeghini, M. Pestarino & P. Dando, 1999. Biodiversity of marine sessile epifauna at an Aegean island subject to hydrothermal activity: Milos, Eastern Mediterranean Sea. Mar. Biol. 135: 729–739.
Pansini, M., C. Morri & C. N. Bianchi, 2000. The sponge community of a subtidal area with hydrothermal vents: Milos Island, Aegean Sea. Estuar. coast shelf Sc. 51: 627–635.
Roether, W., B. Manca, B. Klein, D. Bregant, D. Georgopulos, V. Kovacecic & A. Lucchetta, 1996. Recent changes in Eastern Mediterranean deep water. Science 227: 333.
Ruiz, G. M., P. W. Fofonoff, J. T. Carlton, M. J. Wonham & A. H. Hines, 2000. Invasion of coastal marine communities in north America: apparent patterns, processes, and biases. Ann. Rev. Ecol. Syst. 31: 481–531.
Sartoni, G. & A. M. De Biasi, 1999. A survey of the marine algae of Milos Island, Greece. Cryptog. Algol 20: 271–283.
Tarasov, V. G., A. V. Gebruk, V. M. Shulkin, G. M. Kamenev, V. I. Faldeev, V. N. Kosmynin, V. V. Malakhov, D. A. Starynin & A. I. Obzhirov, 1999. Effect of shallow water hydrothermal venting on the biota of Matupi Harbour (Rabaul Caldera, Ner Britain Island, Papua New Guinea). Cont. Shelf. Res. 19: 79–116.
Thiermann, F., I. Akoumianaki, A. Huges & O. Giere, 1997. Benthic fauna of a shallow water gashydrothermal vent area in the Aegean Sea (Milos. Greece). Mar. Biol. 128: 149–159.
Thiermann, F., R. Windoffer & O. Giere, 1994. Selected meiofauna around shallow water hydrothermal vents off Milos (Greece): ecological and ultrastuctural aspects. Vie Milieu 44: 215–226.
Tunnicliffe, V., J. F. Garrett & P. H. Johnson, 1990. Physical and biological factors affecting the behaviour and mortality of hydrothermal vent tubeworms (vestimentiferans). Deep Sea Res. 37: 103–125.
Verlaque, M., 1994. Inventaire des plantes introduites en Méditerranée: origines et répercussions sur l'environnement et les activités humaines. Oceanol. Acta 17: 1–23.
Verlaque, M. & C. F. Boudouresque, 1991. *Stypopodium schimperi* (Buchinger *ex* Kützing) Verlaque *et* Boudouresque comb. nov. (Dictyotales, Fucophyceae), algue de Mer Rouge récemment apparue en Méditerranée. Cryptog. Algol. 12: 195–211.

Hydrobiologia **503:** 45–48, 2003.
M.B. Jones, A. Ingólfsson, E. Ólafsson, G.V. Helgason, K. Gunnarsson & J. Svavarsson (eds), Migrations and Dispersal of Marine Organisms.

Local population persistence as a pre-condition for large-scale dispersal of *Idotea metallica* (Crustacea, Isopoda) on drifting habitat patches

Lars Gutow
Alfred Wegener Institute for Polar and Marine Research, Biologische Anstalt Helgoland, Box 180, 27483 Helgoland, Germany
E-mail: lgutow@awi-bremerhaven.de

Key words: Idotea metallica, extinction, dispersal, habitat destruction, food quality

Abstract

Idotea metallica establishes self-sustaining populations exclusively on objects drifting at the sea surface. Large-scale transport of drift material with surface currents results in an efficient dispersal of the species. Two types of drifting objects are utilised (biotic and abiotic), providing quite different conditions of life. Ephemeral biotic substrata (mainly uprooted macroalgae) may be used for transport and food, however, resulting habitat destruction from feeding must be a major threat for local population persistence of *I. metallica*. Abiotic substrata or wood represent efficient vectors for long-distance dispersal due to their resistance to biodegradation, but do not provide food for this herbivorous species. In laboratory experiments, the spatially-limited conditions of drifting substrata were simulated in microcosms. *Idotea metallica* established persistent populations on both types of substrata. On abiotic substrata, however, where the animals were fed only on *Artemia* larvae, high variations in density and a reduced intrinsic rate of population growth increased the risk of population extinction. *Idotea metallica* avoids habitat destruction by limited feeding on macroalgae. In contrast, the coastally distributed congener *Idotea baltica* destroyed algal habitats by feeding about 10 times faster than *I. metallica*.

Introduction

Idotea metallica Bosc maintains persistent populations in the Mediterranean Sea (Dow & Menzies, 1958; Abelló & Frankland, 1997) and off the east coast of North America (Locke & Corey, 1989). Individuals occur in the entire North Atlantic (Naylor, 1957; Tully & McGrath, 1987; Davoult et al., 1999; Franke et al., 1999), and occasionally in the South Atlantic (Moreira, 1972) and the Pacific (Van der Baan & Holthuis, 1969). *Idotea metallica* disperses efficiently as self-sustaining local populations on drift material (Gutow & Franke, 2003). The persistence of these often abiotic drifting objects enables even species with low natural dispersal capacity to be passively transported over long distances with surface currents (Barnes, 2002).

Depending on the substrata, inhabiting animals are exposed to fundamentally different conditions. Due to their resistance to biodegradation, persistent objects (plastic bottles, remains of fishing nets and wood) represent efficient vectors for long-distance dispersal. Ephemeral substrata (mainly drifting, uprooted macroalgae), however, are exposed to a high risk of destruction particularly by the feeding activities of inhabiting animals. On drifting macroalgae, isopods feed on both their substrata and zooplankton from the surrounding water, whereas on persistent abiotic substrata, plant food is limited (if not entirely absent).

For *I. metallica*, efficiency of dispersal is linked to the persistence of local populations. The present paper, therefore, examines the risks of local population extinction for *I. metallica* through investigations on demographic stochasticity under laboratory conditions and environmental aspects such as food quality and risk of habitat destruction.

Materials and methods

All experiments were conducted at a constant temperature of 16 °C and an artificial 16:8 light–dark cycle.

Effect of food quality on population dynamics

Microcosm experiments were carried out in 5 l vessels with aerated seawater. One experimental group was run with 100 g of the brown alga *Fucus vesiculosus* L. as the biotic substratum. In the other group, 100 g of a plastic fishing net served as a persistent non-biotic substratum. The vessels were each stocked initially with 20 animals (10 males and 10 females) of about 45 days old. Each treatment was run in 5 replicates over 24 weeks. Freshly-hatched *Artemia* nauplii served as food and were offered *ad libitum*. Water was exchanged once to twice a week. Every 14 days, animals in the vessels were counted. At the same time, algae were replaced by fresh fronds but the nets remained unchanged.

The intrinsic rates of population growth (r) were calculated for the initial period of exponential growth (week 0-6) using the following equation:

$$r = \frac{\ln N_2 - \ln N_1}{t_2 - t_1},$$

where N = the number of animals and t = time (in weeks). In order to record not simply reproductive rates of initially-introduced animals but to include mortality as antagonist of reproduction over an appropriate period into the calculation, population densities were estimated as 3-point running means.

Rate of substratum destruction

To determine the rate of substratum destruction, two sets of experiments were carried out with *I. metallica* and results compared with the coastally distributed, benthic congener *Idotea baltica* (Pallas). Ten adult animals (5 males and 5 females) were each placed into 20 × 20 × 6 cm plastic boxes. The boxes were supplied with 1 l of aerated seawater and a branch of *Fucus vesiculosus* or *Ascophyllum nodosum* (L.) (the two dominant drift algae species in the North Sea), respectively. The water was exchanged daily and hatched juveniles and bitten off pieces of algae were removed. Animals that died during the experimental period were replaced. The fresh weight of the algal branch was measured every second day to 0.1 g. Five

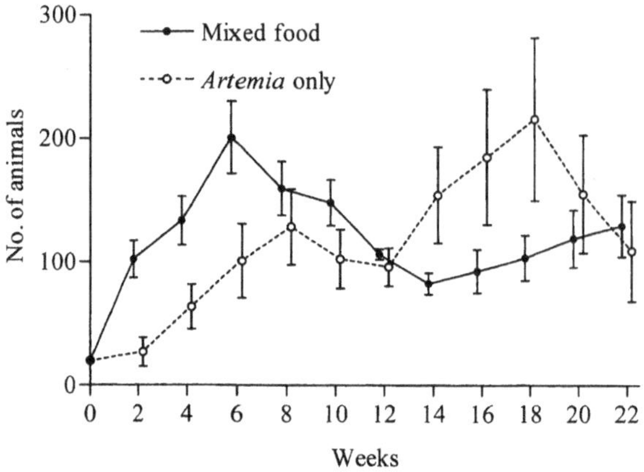

Figure 1. Number of animals of *Idotea metallica* in microcosms under different food conditions. Error bars denote standard error ($n = 5$).

experimental units (replicates) each were run for 10 days with and without *Artemia* larvae as an additional source of animal food.

Results

Effect of food quality on population dynamics

Idotea metallica developed persistent populations under both feeding conditions (Fig. 1). None of the experimental populations went extinct within 24 weeks. An initial overshoot in population density was followed by a period of marked oscillations around the value of the capacity starting in week 12 for each treatment. During the initial growth phase (week 0–6), populations grew significantly faster (Student's t-test, $P = 0.029$) under mixed food conditions ($r = 0.38 \pm 0.06^*$ wk^{-1}) than without plant food ($r = 0.24 \pm 0.10^*$ wk^{-1}). Food quality had no effect (Student's t-test, $P = 0.255$) on the capacities of the microcosms (calculated as mean population densities from week 12 on) for *I. metallica* (mixed food: 105.6 ± 30.2 ind.; without algae: 152.6 ± 80.4 ind.). Under mixed food conditions, oscillations (expressed as standard deviation of log-transformed population densities) were weaker (0.30 ± 0.18) than in the absence of plant food (0,68 ± 0.31) (Student's t-test, $P < 0.05$).

Rate of substratum destruction

Both species caused a significant (F-test, P always < 0.01) loss of algal fresh weight (Fig. 2). A 3-factorial ANOVA (comparing the slopes of linear regression

Table 1. Results of the 3-factorial ANOVA on the rates of habitat destruction ($g * ind.^{-1} * d^{-1}$)

	Source of variation	df	SS	MS	*F*	*P* value
(a)	Main effects					
	Factor 1 (isopod species)	1	0.0185	0.0185	123.907	<0.001
	Factor 2 (algae species)	1	0.0003	0.0003	1.954	0.172
	Factor 3 (presence of *Artemia*)	1	0.0008	0.0008	5.308	0.028
(b)	First-order interaction					
	Factors 1 × 2	1	0.0000	0.0000	0.268	0.608
	Factors 1 × 3	1	0.0010	0.0011	7.109	0.012
	Factors 2 × 3	1	0.0001	0.0001	0.564	0.458
(c)	Second-order interaction					
	Factors 1 × 2 × 3	1	0.0002	0.0002	1.357	0.253
	Within subgroups (error)	32	0.0048	0.0002		
	Total	39	0.0257	0.0007		

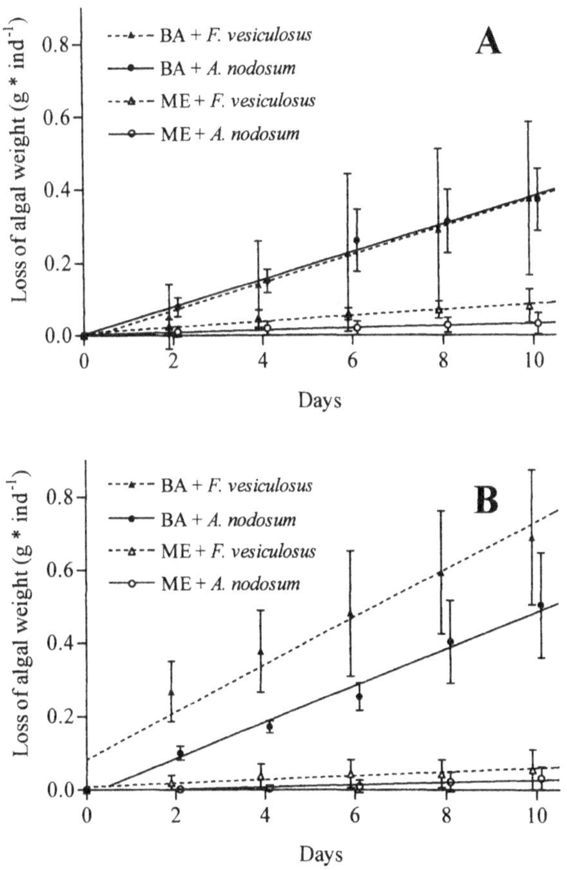

Figure 2. Rate of habitat destruction plotted as cumulative loss of algal freshweight with (A) and without (B) *Artemia* as additional source of food (ME = *Idotea metallica*; BA = *Idotea baltica*). Error bars denote standard error ($n = 5$).

in Fig. 2) revealed that *I. baltica* reduced algal fresh weight faster than *I. metallica* (Table 1). For each species, the rate of habitat destruction was independent of algal species. In *I. metallica,* habitat destruction proved to be independent of the presence/absence of additional zooplanktonic food while in *I. baltica* it increased in its absence. As a result, *I. baltica* reduced algal fresh weight about 6 times faster than *I. metallica* when *Artemia* nauplii were available (0.038 ± 0.016 and 0.006 ± 0.004 $g*ind^{-1}*d^{-1}$, respectively), but about 14 times faster when this food was absent (0.057 ± 0.018 and 0.004 ± 0.004 $g*ind^{-1}*d^{-1}$, respectively).

Discussion

Idotea metallica disperses as self sustaining local populations on drifting objects. Thus, the efficiency of dispersal is linked closely to population persistence on these objects. In general, animal populations face two sources of local extinction: an external risk caused by the variability of the environment, and an internal risk based on demographic stochasticity (Caughley, 1994).

Among environmental threats, habitat destruction is by far the most serious one. For *I. metallica,* destruction of a local habitat patch would lead to the complete extinction of a local population. Habitat destruction is most serious in biotic substrata such as drifting macroalgae. Present experiments revealed that *I. metallica* had limited feeding on macroalgae, indicating that its habitat is not likely to be destroyed quickly by feeding. Its congener *I. baltica*, however, destroyed macroalgal patches about 10 times faster, thus reducing the persistence time of macroalgal patches dramatically (especially when the high densities of *I. baltica* on drifting algae are considered). Franke et al. (1999) found up to about 85 specimens of *I. baltica* (considering only animals longer than 10 mm) per kg drift algae, while *I. metallica* occurred with a maximum density of 1 animal per kg. For *I. baltica*, rapid loss of algal substratum is not a problem as this coastally distributed species (Naylor, 1955) may assume a benthic life style after patch destruction (Locke & Corey, 1989). Even if additional planktonic animal food is absent, *I. metallica* hardly feeds on macroalgae, emphasizing the importance of patch persistence for this species.

Even though abiotic objects represent more persistent substrata, in microcosms with mixed feeding conditions, additional plant food increased the intrinsic rate of population increase (r) of *I. metallica* significantly for the initial phase of exponential population growth (week 0–6). Furthermore, population size varied stronger around the capacity (week 12–24) in the absence of plant food. Both a lower population growth rate and a higher variability in population size increase the stochastic risk of population extinction (Hanski, 1999).

In conclusion, although abiotic substrata represent a persistent habitat for *I. metallica* they do not provide optimum food supply.

References

Abelló, P. & R. J. Frankland, 1997. Population characteristics of the neustonic isopod *Idotea metallica* (Crustacea, Isopoda, Idoteidae) in the western Mediterranean (June 1993). Sci. mar. 61: 409–414.

Barnes, D. K. A., 2002. Invasions by marine life on plastic debris. Nature 416: 808–809.

Caughley, G., 1994. Directions in conservation biology. J. anim. Ecol. 63: 215–244.

Davoult, D., J.-M. Dewarumez, C. Luczak & A. Migné, 1999. Nouvelles signalisations d'espèces benthiques sur les côtes francaises de la Manche orientale et de la Mer du Nord. Cah. Biol. mar. 40: 121–127.

Dow, T. G. & R. J. Menzies, 1958. The pelagic isopod *Idotea metallica* in the Mediterranean. Pubbl. Staz. Zool. Napoli 30: 330–336.

Franke, H.-D., L. Gutow & M. Janke, 1999. The recent arrival of the oceanic isopod *Idotea metallica* Bosc off Helgoland (German Bight, North Sea): an indication of a warming trend in the North Sea? Helgoländer Meeresunters. 52: 347–357.

Gutow, L. & H.-D. Franke, 2003. Metapopulation structure of the marine isopod *Idotea metallica*, a species associated with drifting habitat patches. Helgol. mar. Res. 56: 259–264.

Hanski, I., 1999. Metapopulation Ecology. Oxford University Press, New York. 266 pp.

Locke, A. & S. Corey, 1989. Amphipods, isopods, and surface currents: a case for passive dispersal in the Bay of Fundy, Canada. J. Plankton Res. 11: 419–430.

Moreira, P. S., 1972. Species of marine Isopoda (Crustacea, Peracarida) from Southern Brazil. Bolm. Inst. Oceanogr., S Paulo 21: 163–179.

Naylor, E., 1955. The ecological distribution of british species of *Idotea* (Isopoda). J. anim. Ecol. 24: 255–269.

Naylor, E., 1957. The occurrence of *Idotea metallica* in British waters. J. mar. biol. Ass. U.K. 36: 599–602.

Tully, O. & D. McGrath, 1987. The status of *Idotea metallica* Bosc (Isopoda) in Irish waters. Ir. Nat. J. 22: 190–192.

Van der Baan, S. M. & L. B. Holthuis, 1969. On the occurrence if Isopoda in the surface plankton in the North Sea near the lightship "Texel". Neth. J. Sea Res. 4: 354–363.

Hydrobiologia **503:** 49–57, 2003.
M.B. Jones, A. Ingólfsson, E. Ólafsson, G.V. Helgason, K. Gunnarsson & J. Svavarsson (eds), Migrations and Dispersal of Marine Organisms.

Rafting of benthic macrofauna: important factors determining the temporal succession of the assemblage on detached macroalgae

Martin Thiel[1,2]
[1]*Facultad Ciencias del Mar, Universidad Católica del Norte, Larrondo 1281, Coquimbo, Chile*
E-mail: thiel@ucn.cl
[2]*Centro de Estudios Avanzados en Zonas Áridas (CEAZA), Coquimbo, Chile*

Key words: rafting, reproduction, dispersal, benthos, macroalgae, debris

Abstract

Rafting on biotic and abiotic substrata has been reported for many benthic marine invertebrates. Here, I describe important characteristics of common floating substrata and review published studies examining the succession of the assemblage on detached macroalgae to identify the most important factors determining this succession. Floating substrata differ in survival time (i.e. before they disintegrate) and in food value, with abiotic materials having high survival times but low food value and some biotic substrata (seagrasses, fresh wood) having high food values but short survival times. Large macroalgae with pneumatocysts may combine high survival times and high food values. Substratum survival and food value have consequences for the species composition of the rafting macrofauna. In general, suspension feeders dominate rafting assemblages on abiotic substrata, while grazing and boring species abound on macroalgae and wood. The succession of the rafting assemblage can be subdivided into three phases during which different processes predominate. During the initial phase some of the original colonizers (if present) disappear from the substratum, either due to active emigration or predation. This short, initial phase is followed immediately by the colonization phase, during which new organisms colonize the rafting substratum. Colonization may be rapid and intense in the beginning but then slows down. Towards the end of the colonization phase, some rafting organisms may reproduce and their offspring recruit within the parental raft. Results from two long-term studies confirmed that the proportion of species with direct development increased with duration of rafting. These successional changes will be most pronounced for large biotic substrata, rendering these particularly suitable for long-distance dispersal of organisms with direct development.

Introduction

Rafting on natural and anthropogenic floating material has been inferred as an important dispersal mechanism for many littoral marine invertebrates (Ingólfsson, 1995; Ó Foighil et al., 1999, 2001; Castilla & Guiñez, 2000). Organisms have been found rafting on a wide variety of different substrata of natural (wood, seagrasses, macroalgae, volcanic pumice, corals) and anthropogenic (plastics, tar balls, manufactured wood) origin. Floating substrata may differ not only in origin but also with respect to quality (food value, floatation potential, etc.). Some substrata may be much more suitable for rafting organisms than others. In particular, large macroalgae, which might continue to grow after detachment, combine a relatively high food value with high floatation potential.

Rafting assemblages may also be highly diverse (Hobday, 2000a), and strong competition for both space and food resources may exist among rafting organisms. Some species possess traits that make them much more suitable for rafting than others. For example, peracarid crustaceans, and other brooding marine invertebrates, are typically abundant on floating macroalgae (Hobday, 2000a; Castilla & Guiñez, 2000). These species may persist better than others during the voyage on floating substrata because their offspring may recruit directly to the parental raft, leading to successional changes of the rafting assemblage during the voyage. While many studies report the composition of the rafting assemblage at some moment after the start of the voyage (e.g. Helmuth et al., 1994; Ingólfsson, 1995; Hobday, 2000a), relatively little is known about the temporal succession of rafting assemblages. This succession depends not only on the biology of the rafting organisms, but also on the properties and history of the floating substrata. Since the outcome of the succession of the rafting assemblage during the voyage on floating substrata determines which species may arrive in new habitats, this process is of crucial importance when discussing rafting as a dispersal mechanism. Here, I identify some characteristics of the main substrata commonly harbouring rafting organisms and review several studies that report the temporal succession of the rafting assemblage. Rather than serve as an exhaustive review, this contribution represents a step in identifying some of the important processes and mechanisms involved in the succession of rafting assemblages.

Materials and methods

Recent publications were reviewed to identify and characterize the most ubiquitous floating substrata in the world's oceans. Particular emphasis was placed on characteristics of importance for potential rafting organisms. Species succession on abiotic floating substrata has been examined in a variety of different studies, but relatively little is known about this process on floating macroalgae. To obtain an estimate of successional changes of the rafting assemblage, three experimental studies, which investigated the succession of associated organisms on macroalgae after detachment from the primary substratum, were evaluated and examined for the existence of common patterns. All three studies were conducted with algal species that possess pneumatocysts and are commonly found floating at the sea surface. The first (Ingólfsson, 1998) examined the temporal succession of organisms associated with fronds of *Ascophyllum nodosum* (L.), while the second (Edgar, 1987) and third (Vásquez, 1993) examined the fauna associated with holdfasts of the giant kelp *Macrocystis pyrifera* (L.) (Table 1). I have evaluated these studies with respect to the total abundance of organisms associated with the detached macroalgae and the proportion of individuals with direct development. Furthermore, I mention briefly some of the important observations on the successional changes made by these authors.

Results and discussion

Floating substrata

A wide diversity of floating materials originating from human activities is found in the world's oceans. These items are reported frequently from seashores, even on remote islands (e.g. Barnes, 2002). Surveys have revealed that plastic items usually dominate floating marine debris, but glass bottles or metal containers are also found commonly in shore-based surveys (e.g. Garrity & Levings, 1993; Williams & Tudor, 2001). Besides these packaging materials, various other substrata originating from human activities have been reported, including tar balls and manufactured wood (Table 2). Sea-based surveys report similar materials, but plastic bags and fishing gear appear less abundant in shore-based surveys. Floating marine litter is most common at mid and low latitudes, close to major human populations (Matsumura & Nasu, 1997). Typically, these items are most abundant in coastal waters, but are also found in the open ocean. Among the most important biotic substrata are macroalgae, which are commonly reported from coastal areas, but have also been observed in the open ocean (Kingsford, 1992, 1995; Helmuth et al., 1994; Ingólfsson, 1995, 2000; Hobday, 2000b, c; Hirata et al., 2001).

Floating substrata differ widely in their nutritive value for rafting organisms. Macroalgae and seagrasses may be of high nutritional value for grazers, and animal remains may also be of high nutritional value depending on the proportion of organic tissue. Most floating litter (except for manufactured wood or other organic litter) present no nutritive value for rafting organisms, which is also true for other abiotic substrata such as volcanic pumice or buoyant coral (Table 2).

Survival estimates for floating items differ widely. Some abiotic substrata, such as pumice or some plastic litter, may float for years (Jokiel, 1989; Ebbesmeyer & Ingraham, 1992). Floating kelp has been inferred to remain afloat for more than 1 y (e.g. Dayton et al., 1984), while other authors suggest a lifetime of only a few days (Harrold & Lisin, 1989). Water temperatures and nutrient availability may play an important role in determining longevity of floating macroalgae (e.g. Edgar, 1987). Hobday (2000b) examined the aging rates of *M. pyrifera* in tethered plants and concluded that most rafts would disintegrate after <100 d, at least at water temperatures >20 °C. Surprisingly little is known about floating times of wood,

Table 1. Data of the three studies examining the fauna associated with macroalgae detached from the primary substratum and tethered at the sea surface or in the water column

Macroalgal species	Algal part	Study site	Water depth (m)	Duration (d)	Reference
Ascophyllum nodosum	fronds	Iceland	0	20	Ingólfsson (1998)
Macrocystis pyrifera	holdfast	Tasmania	15	191	Edgar (1987)
Macrocystis pyrifera	holdfast	California	5–10	110	Vásquez (1993)

Table 2. Common floating substrata and some important properties, relevant to rafting; information on food value and floating potential inferred from primary literature (see 'Reference' section)

Substratum	Sources	Biotic/abiotic	Food value	Floating potential
Macroalgae*	natural	biotic	high	high
Seagrass plants	natural	biotic	high	low
Sea beans	natural	biotic	intermediate	very high
Natural wood*	natural	biotic	intermediate	high
Animal remains	natural	biotic/abiotic	high/low	low
Volcanic pumice	natural	abiotic	low	very high
Buoyant coral	natural	abiotic	low	very high
Plastic*	anthropogenic	abiotic	none	high
Glass	anthropogenic	abiotic	none	intermediate
Metal	anthropogenic	abiotic	none	intermediate
Manufactured wood	anthropogenic	biotic	intermediate	high
Organic litter	anthropogenic	biotic	intermediate/low	low
Tar balls	natural/anthrop.	abiotic	low	intermediate

* Most commonly reported floating substrata.

but anecdotal information and dense accumulations of adult-sized marine invertebrates on wooden items, indicate that these can float for a long time. Since some floating items may remain afloat for long time periods (months), there exists a potential for the rafting assemblage to undergo successional change.

Floating substrata can be divided into two major categories: substrata that carry abundant and diverse epibiota before starting their voyage, and substrata that start out without any epibiota. The former comprises substrata that previously had been attached in coastal habitats where they may have become colonized by a wide variety of benthic organisms. Among these substrata, macroalgae, which may become detached during storms (e.g. Witman, 1987; Seymour et al., 1989), typically predominate. Substrata originating from anchored installations, such as harbours or aquaculture materials (e.g. buoys), may also be inhabited by fouling organisms at the moment of detachment. In contrast to these substrata, most marine litter, volcanic pumice and wood are not colonized by marine fauna before starting their voyage. In summary, biotic substrata tend to start their voyage with passengers from many different faunal and floral taxa, while most abiotic substrata start in a 'clean' state.

Temporal succession of the rafting assemblage

Although the succession of rafting assemblages is best studied by following floating items in time and space, tracking free-floating items across the ocean surface imposes considerable logistical constraints (see e.g. Harrold & Lisin, 1989). Consequently, most studies that follow the succession of rafting assemblages over long time periods (> a few days) use tethered substrata, except for two studies of free-floating macroalgae during the first few hours after detachment (Kingsford & Choat, 1985; Kingsford, 1992). Usually, tethering is done by tying floating items to weights, typically anchored in coastal habitats close to potential source populations. Furthermore, these experiments are conducted in sheltered bays to reduce the risk of losing tethered items.

There exists a vast literature on the species succession of the fouling assemblage on abiotic (typically artificial) substrata (for recent reviews see e.g. Wahl, 1997; Svane & Petersen, 2001). In general, these substrata become colonized by suspension-feeding organisms such as sponges, hydrozoans, bryozoans and ascidians together with a diverse associated fauna (e.g. Greene & Schoener, 1982; Relini et al., 2000; Holloway & Connell, 2002). This pattern, with a dominance of clonal suspension feeders, is in accordance with the results obtained from abiotic floating substrata collected in the flotsam (Winston et al., 1997; Barnes & Sanderson, 2000). While the main pattern of the successional changes on abiotic substrata (e.g. marine litter) is relatively well known, little is known about the species succession on floating biotic substrata. Some studies describe the successional changes in wood (e.g. Si et al., 2000; Tuente et al., 2002), but the succession of the rafting assemblages on macroalgae has been little studied despite their abundance and importance as floating substrata (e.g. Helmuth et al., 1994; Ingólfsson, 1995, 2000; Hobday, 2000c; Hirata et al., 2001; Smith, 2002).

In the three studies that examined the succession of the macrofauna organisms on macroalgae after detachment, some general trends were observed. The macroalgae remained intact for the duration of the respective experiments, but Vásquez (1993) and Edgar (1987) remarked that kelp holdfasts started to disintegrate after 110 d and 191 d, respectively. Edgar (1987) attributed this to the destructive activity of boring isopods, which increased in abundance during the experiment.

The temporal succession of the macrofaunal assemblage on detached macroalgae followed some general patterns. Numbers of species and total individuals remained very high or even increased after detachment (Fig. 1). In the *Ascophyllum*-study, abundance of associated organisms increased rapidly (Fig. 1A), primarily due to colonization by blue mussels and harpacticoid copepods via the water column (Ingólfsson, 1998). In the two holdfast studies, a small initial decrease in individual abundance was observed but, after approximately 10 d, numbers of associated organisms remained stable (Fig. 1B, C). In general, total species and individual numbers remained relatively high throughout the duration of the experiments, but some species and individual exchange occurred after deployment in the water column. The associated fauna of the kelp holdfasts comprised species from all major phyla (Arthropoda, Mollusca, Echinodermata and Annelida). Shortly after becoming detached, some of the original inhabitants disappeared from the substratum. Vásquez (1993) observed that, initially, ophiurids dominated the associated fauna, but their densities decreased strongly after deployment of the holdfasts in the water column. Molluscs and polychaetes also decreased in abundance shortly after deployment. This may be an active process where animals leave the floating substratum to fall or swim back to benthic habitats, or it may be a passive process where some organisms are selectively preyed upon by fish (Kingsford & Choat, 1985; Davenport & Rees, 1993; Shaffer et al., 1995; Deudero & Morales-Nin, 2000; Ingólfsson & Kristjánsson, 2002). Following this initial loss, either a strong increase of individual numbers on substrata can be observed or a stable phase occurs where eventual losses are balanced by new colonization. When substrata remain in the water column for relatively long time periods (>40 d), local reproduction and recruitment may occur, leading to a dominance of organisms with direct development on the floating substrata. Peracarid crustaceans were among the most abundant macrofauna organisms in the holdfasts and occurred also on the *Ascophyllum* fronds. In the study by Vásquez (1993), peracarids increased dramatically towards the end of the experiment, comprising >90% of all associated organisms at the final sampling date (110 d after deployment). He remarked that "it is interesting to note that both of these groups (isopods and amphipods) are brooders and reproduce inside the holdfasts". Similarly, Edgar (1987) mentioned that "the gradual population increases of holdfast species without pelagic larval stages during the first three months furthermore indicate that much of the fauna can complete all stages of their life cycles without leaving holdfasts". Many organisms that established strong local populations after considerable time periods were the original inhabitants of these biotic substrata. In all three studies, species numbers remained relatively stable during the experiment, suggesting a balance between immigration and extinction events.

As the kelp holdfasts (Edgar, 1987; Vásquez, 1993) were deployed at some depth below the sea surface, this may diminish substantially the possibility of encounter with other substrata and organisms floating at or near the water surface. Colonization by new inhabitants as reported in these two studies may, therefore, be substantially lower than that experienced by kelp plants floating at the sea surface. Ingólfsson (1998) remarked that tethered algal fronds frequently caught other free-floating algae, suggesting

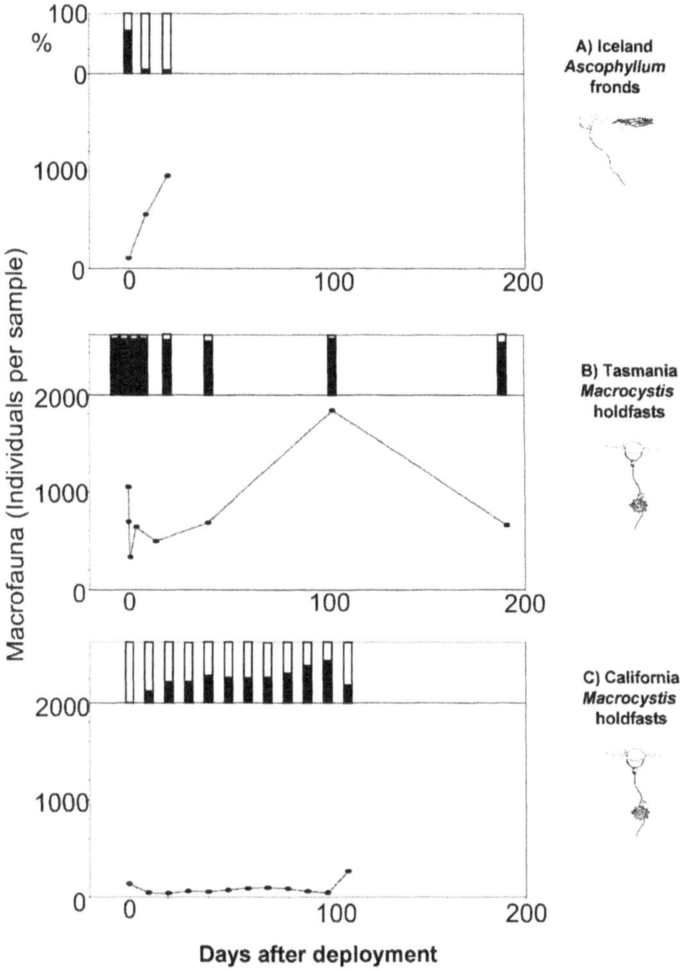

Figure 1. Abundance of macroinvertebrates in macroalgae separated from benthic habitats and exposed in the water column; columns in top figures show percentage of organisms with direct development (black) and with pelagic or unknown developmental mode (white); data taken from (A) Ingólfsson (1998); (B) Edgar (1987); (C) Vásquez (1993).

that some exchange between associated macrofauna from tethered algae and from free-floating algae may have occurred. These observations indicate that colonization of floating substrata via contact with other substrata may be important – a process inferred by other authors (Gutow & Franke, 2003). Nevertheless, the holdfast studies showed clearly the importance of local reproduction and recruitment in floating substrata. Other studies have also demonstrated that some organisms can complete their life cycle on floating substrata (e.g. Gutow & Franke, 2001).

General model for the temporal succession of rafting assemblages

Based on the results from the three studies reviewed, the successional changes of the associated fauna on any floating substratum may be subdivided into three distinct phases of different duration (Fig. 2).

Initial phase of floating

Substrata attached in coastal habitats before detachment (macroalgae, mangrove trees, buoys and anthropogenic installations) may harbour a diverse assemblage of associated fauna. After detachment, these substrata lose some of the original inhabitants because they either are not capable of holding onto floating items or because they are selectively eaten by predat-

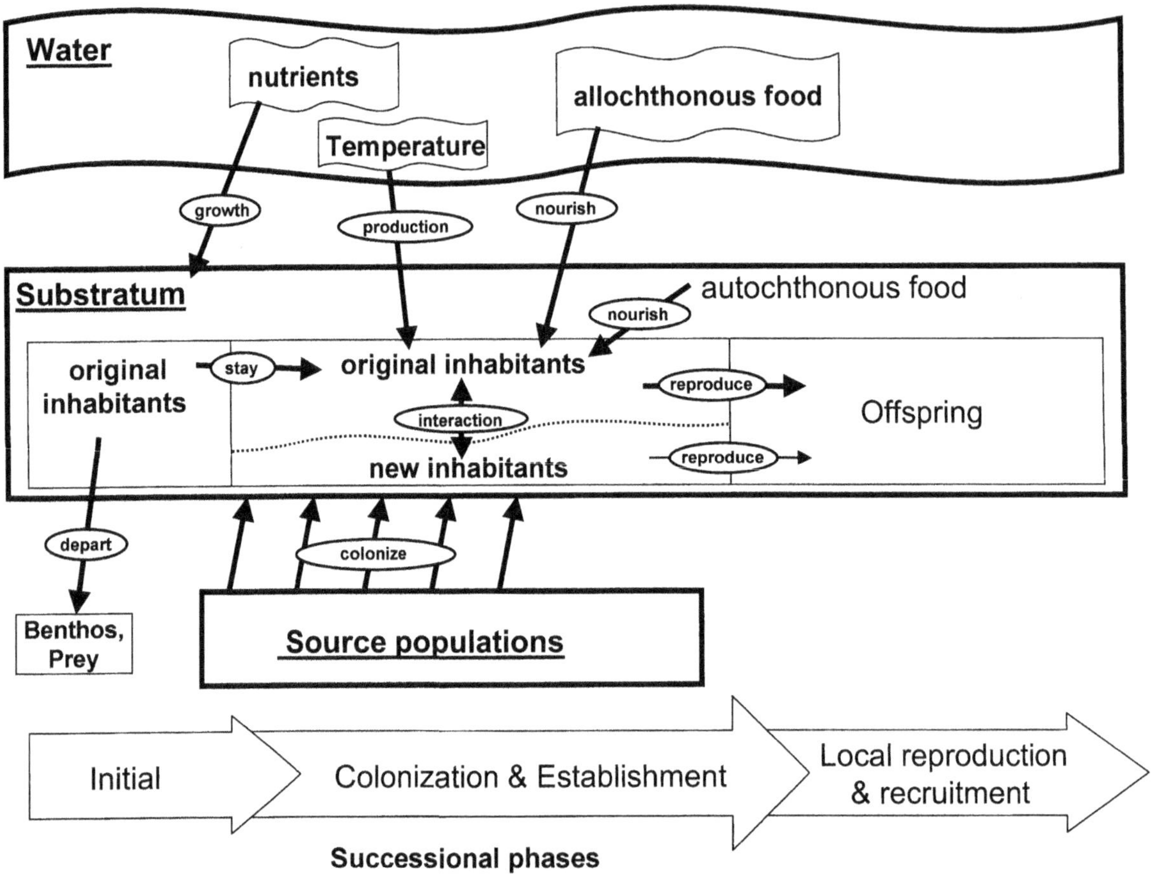

Figure 2. Schematic representation of the main factors and processes involved in the temporal succession of the rafting assemblage. Three distinct phases distinguished during the succession are indicated in the lower part of the figure.

ors (Fig. 2). Kingsford & Choat (1985) observed that many organisms, such as large, mobile invertebrates, actively abandoned kelp plants either immediately or soon after these had detached from the bottom and ascended to the water surface. Other organisms that live permanently attached to these substrata, or that inhabit protected structures such as burrows, will remain on these substrata. Many species of intermediate sizes (polychaetes, small bivalves, peracarid crustaceans) remain on the substrata after this initial phase.

Colonization and establishment phase of rafting fauna

Shortly after the start of the voyage, floating substrata become colonized by many different organisms, either arriving as larvae or as subadult stages. Following settlement on a floating item, a wide variety of biotic interactions are possible among rafting organisms, many of which might compete directly or indirectly for space and food. Species with a size advantage (e.g. original inhabitants) may suppress, or actively eliminate, small stages of other species. In contrast to these negative interactions, some species (e.g. settling macroalgae) may provide a habitat or food, thereby facilitating establishment of other organisms. The successional changes of the rafting assemblage will depend upon two main factors related to the characteristics of the surrounding water body: (1) the vicinity of floating material to source populations of potential colonizers, and (2) the dependence of colonizers on allochthonous food resources (Fig. 2).

Many studies of the colonization of benthic assemblages demonstrated the importance of larval supply (Roughgarden et al., 1987; Morgan, 2001). Similar mechanisms can be expected for floating substrata, but the influence of larval supply on the succession of the rafting assemblage is unknown. Other species colonize floating substrata in the subadult and adult stage.

For example, Brooks & Bell (2001) reported that floating algae that passed through a seagrass meadow were rapidly colonized by mobile peracarids. In coastal habitats, free-floating algae become quickly colonized by a variety of marine invertebrates (Kingsford & Choat, 1985). With increasing distances from suitable source populations (i.e. away from coastal habitats), floating substrata may be less prone to becoming colonized by littoral macrofauna.

Availability of food resources for new colonizers will also play an important role during the successional process. If the substratum itself does not provide food for colonizers (as is the case with abiotic substrata), colonizers will depend on food resources available in the surrounding water or growing on the raft. Therefore, suspension-feeding organisms (barnacles, bryozoans, hydrozoans) typically dominate abiotic substrata (Winston, 1982; Ye & Andrady, 1991; Minchin, 1996; Barnes & Sanderson, 2000). On biotic substrata, particularly those that continue to grow after detachment, colonizers may also feed on their substratum, i.e. on autochthonous food resources (e.g. Gutow, 2003). These substrata are more prone to destruction by their inhabitants than abiotic substrata. For example, rafting organisms that feed on their substratum may substantially contribute to the destruction of their home (Gutow & Franke, 2003; Gutow, this issue).

During this colonization and establishment phase, the importance of biotic interactions amongst rafting organisms is likely to increase. Colonizers may compete for space and food resources, or they may prey on each other. Nudibranch gastropods may feed on bryozoans or hydrozoans, omnivorous crustaceans may prey on larval or juvenile stages of other invertebrates and fish attracted to floating items may feed on associated macroinvertebrates (Hunter & Mitchell, 1967; Safran & Omori, 1990; Kingsford, 1992, 1995; Ingólfsson & Kristjánsson, 2002). The outcome of these biotic interactions during the colonization and establishment phase will determine which organisms persist on floating substrata. Successful raft organisms should be competitively superior to others, have only minor food requirements from their substratum, and be defended from predators (chemically or by crypsis).

Local reproduction and recruitment

The floating time of some substrata exceeds the life time of many of their invertebrate passengers. Some organisms may overcome this hurdle by reproducing within their floating habitat. In particular, species that produce propagules that can immediately colonize the natal float will gain important advantages over organisms that depend on pelagic larval stages, since the former are capable of establishing stable populations on floating substrata. Species with asexual reproduction (fission) or with direct development should, thus, be favoured during this later phase of rafting. Local reproduction and recruitment may gain particular importance on substrata that harbour original inhabitants with direct development and float away from potential coastal source populations. This late successional phase (sensu Tsikhon-Lukanina et al., 2001) may not be reached when floating substrata lack species with direct development, because they started their journey in a clean state and could only be colonized by species with pelagic larvae.

Outlook

Studies reviewed provide important insights into the successional changes occurring on macroalgae after detachment from the primary substratum. Due to logistic constraints, results are based on substrata anchored in coastal habitats, often in close vicinity to potential source populations. To achieve a sound understanding of the dynamics of rafting assemblages, it appears essential to track floating substrata during their voyage. Experiments investigating whether competition for food or space among rafting organisms is important could be conducted in the laboratory or in the field. Furthermore, it is important to establish whether rafting organisms arriving in new habitats are capable of successful colonization. Finally, biogeographic and molecular studies should be conducted to reveal whether rafting is an important process shaping the distributional pattern and genetic structure of the populations of coastal organisms (see e.g. Sponer & Roy, 2002).

Acknowledgements

I thank J. Varela for the drawings, and J. Vásquez and two anonymous reviewers for helpful comments on the manuscript. The English of the final version was kindly revised by J. Long. Special thanks to L. Gutow who convinced me with his thoughtful comments to follow the recommendations from the reviewers. Funding was provided by FONDECYT 1010356; the presentation of this study at the 37th EMBS was made possible by travel funds from Universidad Católica del Norte.

References

Barnes, D. K. A., 2002. Invasions by marine life on plastic debris. Nature 416: 808–809.

Barnes, D. K. A. & W. G. Sanderson, 2000. Latitudinal patters in the colonization of marine debris. In Herrera Cubilla, A. & J. B. C. Jackson (eds), Proceedings of the 11th International Bryozoology Association Conference. Smithsonian Tropical Research Institute, Balboa, R.P.: 145–153.

Brooks, R. A. & S. S. Bell, 2001. Mobile corridors in marine landscapes: enhancement of faunal exchange at seagrass/sand ecotones. J. exp. mar. Biol. Ecol. 264: 67–84.

Castilla, J. C. & R. Guiñez, 2000. Disjoint geographical distribution of intertidal and nearshore benthic invertebrates in the Southern Hemisphere. Rev. Chil. Hist. Nat. 73: 583–603.

Davenport, J. & E. I. S. Rees, 1993. Observations on neuston and floating weed patches in the Irish Sea. Estuar. coast. shelf Sci. 36: 395–411.

Dayton, P. K., V. Currie, T. Gerrodette, B. Keller, R. Rosenthal & D. van Tresca, 1984. Patch dynamics and stability of some California kelp communities. Ecol. Monogr. 54: 253–289.

Deudero, S. & B. Morales-Nin, 2000. Occurrence of *Polyprion americanus* under floating objects in western Mediterranean oceanic waters, inference from stomach content analysis. J. mar. biol. Ass. U.K. 80: 751–752.

Ebbesmeyer, C. C. & J. W. Ingraham, Jr., 1992. Shoe Spill in the North Pacific. EOS, Trans. am. Geophys. Union 73: 361–368.

Edgar, G. J., 1987. Dispersal of faunal and floral propagules associated with drifting *Macrocystis pyrifera* plants. Mar. Biol. 95: 599–610.

Garrity, S. D. & S. C. Levings, 1993. Marine debris along the Caribbean coast of Panama. Mar. Poll. Bull. 26: 317–324.

Greene, C. H. & A. Schoener, 1982. Succession on marine hard substrata: a fixed lottery. Oecologia 55: 289–297.

Gutow, L., 2003. Local population persistence as a pre-condition for large-scale dispersal of *Idotea metallica* (Crustacea: Isopoda) on drifting habitat patches. Hydrobiologia, 503 (Dev. Hydrobiol. 174): 45–48.

Gutow, L. & H. D. Franke, 2001. On the current and possible future status of the neustonic isopod *Idotea metallica* Bosc in the North Sea: a laboratory study. J. Sea Res. 45: 37–44.

Gutow, L. & H. D. Franke, 2003. Metapopulation structure of the marine isopod *Idotea metallica* – a species associated with drifting habitat patches. Helgol. mar. Res. 56: 259–264.

Harrold, C. & S. Lisin, 1989. Radio-tracking rafts of giant kelp: local production and regional transport. J. exp. mar. Biol. Ecol. 130: 237–251.

Helmuth, B, R. R. Veit & R. Holberton, 1994. Long-distance dispersal of a SubAntarctic brooding bivalve (*Gaimardia trapesina*) by kelp rafting. Mar. Biol. 120: 421–426.

Hirata, T., J. Tanaka, T. Iwami, T. Ohmi, A. Dazai, M. Aoki, H. Ueda, Y. Tsuchiya, T. Sato & Y. Yokohama, 2001. Ecological studies on the community of drifting seaweeds in the south-eastern coastal waters of Izu Peninsula, central Japan. I. Seasonal changes of plants in species composition, appearance, number of species and size. Phycol. Res. 49: 215–229.

Hobday, A. J., 2000a. Persistence and transport of fauna on drifting kelp (*Macrocystis pyrifera* (L.) C. Agardh) rafts in the Southern California Bight. J. exp. mar. Biol. Ecol. 253: 75–96.

Hobday, A. J., 2000b. Age of drifting *Macrocystis pyrifera* (L.) C. Agardh rafts in the Southern California Bight. J. exp. mar. Biol. Ecol. 253: 97–114.

Hobday, A. J., 2000c. Abundance and dispersal of drifting kelp *Macrocystis pyrifera* rafts in the Southern California Bight. Mar. Ecol. Prog. Ser. 195: 101–116.

Holloway, M. G. & S. D. Connell, 2002. Why do floating structures create novel habitats for subtidal epibiota? Mar. Ecol. Prog. Ser. 235: 43–52.

Hunter, J. R. & C. T. Mitchell, 1967. Association of fishes with flotsam in the offshore waters of Central America. Fish. Bull. 66: 13–29.

Ingólfsson, A., 1995. Floating clumps of seaweed around Iceland: natural microcosms and a means of dispersal for shore fauna. Mar. Biol. 122: 13–21.

Ingólfsson, A., 1998. Dynamics of macrofaunal communities of floating seaweed clumps off western Iceland: a study of patches on the surface of the sea. J. exp. mar. Biol. Ecol. 231: 119–137.

Ingólfsson, A., 2000. Colonization of floating seaweed by pelagic and subtidal benthic animals in south-western Iceland. Hydrobiologia 440: 181–189.

Ingólfsson, A. & B. K. Kristjánsson, 2002. Diet of juvenile lumpsucker *Cyclopterus lumpus* (Cyclopteridae) in floating seaweed: effects of ontogeny and prey availability. Copeia 2002: 472–476.

Jokiel, P. L., 1989. Rafting of reef corals and other organisms at Kwajalein Atoll. Mar. Biol. 101: 483–493. K

ingsford, M. J., 1992. Drift algae and small fish in coastal waters of northeastern New Zealand. Mar. Ecol. Prog. Ser. 80: 41–55.

Kingsford, M. J., 1995. Drift algae: a contribution to near-shore habitat complexity in the pelagic environment and an attractant for fish. Mar. Ecol. Prog. Ser. 116: 297–301.

Kingsford, M. J. & J. H. Choat, 1985. The fauna associated with drift algae captured with a plankton-mesh purse seine net. Limnol. Oceanogr. 30: 618–630.

Matsumura, S. & K. Nasu, 1997. Distribution of floating debris in the North Pacific Ocean: Sighting Surveys 1986–1991. In Coe, J. M. & D. B. Rogers (eds), Marine Debris: Sources, Impacts, and Solution. Springer, New York: 15–24.

Minchin, D., 1996. Tar pellets and plastics as attachment surfaces for lepadid cirripedes in the North Atlantic Ocean. Mar. Poll. Bull. 32: 855–859.

Morgan, S. G., 2001. The larval ecology of marine communities. In Bertness, M. D., S. D. Gaines & M. E. Hay (eds), Marine Community Ecology. Sinauer Associates, Sunderland, Massachusetts: 159–181.

Ó Foighil, D., B. A. Marshall, T. J. Hilbish & M. A. Pino, 1999. Trans-pacific range extension by rafting is inferred for the flat oyster *Ostrea chilensis*. Biol. Bull. 196: 122–126.

Ó Foighil, D., R. Jennings, J.-K. Park & D. A. Merriwether, 2001. Phylogenetic relationships of mid-oceanic ridge and continental lineages of *Lasaea* spp. (Mollusca: Bivalvia) in the northeastern Atlantic. Mar. Ecol. Prog. Ser. 213: 165–175.

Relini, G., M. Relini, L. Relini Orsi & M. Montanari, 2000. An offshore buoy as a small artificial island and a fish aggregating device (FAD) in the Mediterranean. Hydrobiologia 440: 65–80

Roughgarden, J., S. Gaines & H. Possingham, 1987. Recruitment dynamics in complex life-cycles. Science 241: 1460–1466.

Safran, P. & M. Omori, 1990. Some ecological observations on fishes associated with drifting seaweed off Tohoku coast, Japan. Mar. Biol. 105: 395–402.

Seymour, R. J., M. J. Tegner, P. K. Dayton & P. E. Parnell, 1989. Storm wave induced mortality of giant kelp, *Macrocystis pyrifera*, in southern California. Estuar. coast. shelf Sci. 28: 277–292.

Shaffer, J. A., D. C. Doty, R. M. Buckley & J. E. West, 1995. Crustacean community composition and trophic use of the drift vegetation habitat by juvenile splitnose rockfish *Sebastes diploproa*. Mar. Ecol. Prog. Ser. 123: 13–21.

Si, A., C. G. Alexander & O. Bellwood, 2000. Habitat partitioning by two wood-boring invertebrates in a mangrove system in tropical Australia. J. mar. biol. Ass. U.K. 80: 1131–1132.

Smith, S. D. A., 2002. Kelp rafts in the Southern Ocean. Global Ecol. Biogeogr. 11: 67–69.

Sponer, R. & M. S. Roy, 2002. Phylogeographic analysis of the brooding brittle star *Amphipholis squamata* (Echinodermata) along the coast of New Zealand reveals high cryptic genetic variation and cryptic dispersal potential. Evolution 56: 1954–1967.

Svane, I. & J. K. Petersen, 2001. On the problems of epibioses, fouling and artificial reefs, a review. Mar. Ecol. 22: 169–188.

Tsikhon-Lukanina, E. A., O. G. Reznichenko & G. G. Nikolaeva, 2001. Ecology of invertebrates on the oceanic floating substrata in the Northwest Pacific Ocean. Oceanology 41: 525–530.

Tuente, U., D. Piepenburg & M. Spindler, 2002. Occurrence and settlement of the common shipworm *Teredo navalis* (Bivalvia: Teredinidae) in Bremerhaven harbours, northern Germany. Helgol. mar. Res. 56: 87–94.

Vásquez, J. A., 1993. Effects on the animal community of dislodgement of holdfasts of *Macrocystis pyrifera*. Pac. Sci. 47: 180–184.

Wahl, M., 1997. Living attached: Aufwuchs, fouling, epibiosis. In: Nagabhushanam, R. & M.-F. Thompson (eds), Fouling Organisms of the Indian Ocean: Biology and Control Technology. Oxford & IBH Publ Co, New Delhi: 31–84.

Williams, A. T. & D. T. Tudor, 2001. Litter burial and exhumation: spatial and temporal distribution on a cobble pocket beach. Mar. Poll. Bull. 42: 1031–1039.

Winston, J. E., 1982. Drift Plastic – an expanding niche for a marine invertebrate?. Mar. Poll. Bull. 13: 348–351.

Winston, J. E., M. R. Gregory & L. M. Stevens, 1997. Encrusters, epibionts, and other biota associated with pelagic plastics: a review of biogeographical, environmental, and conservation issues. In Coe, J. M.& D. B. Rogers (eds), Marine Debris: Sources, Impacts, and Solution. Springer, New York: 81–97.

Witman, J. D., 1987. Subtidal coexistence: storms, grazing, mutualism, and the zonation of kelp and mussels. Ecol. Monogr. 57: 167–187.

Ye, S. & A. L. Andrady, 1991. Fouling of floating plastic debris under Biscayne Bay exposure conditions. Mar. Poll. Bull. 22: 608–613.

Hydrobiologia **503:** 59–67, 2003.
M.B. Jones, A. Ingólfsson, E. Ólafsson, G.V. Helgason, K. Gunnarsson & J. Svavarsson (eds), Migrations and Dispersal of Marine Organisms.

Hitch-hiking on floating marine debris: macrobenthic species in the Western Mediterranean Sea

Stefano Aliani[1] & Anne Molcard[1,2]
[1]*CNR, Istituto Scienze Marine - Sezione La Spezia, Forte Santa Teresa, 19036 Pozzuolo di Lerici (SP), Italy*
[2] *RSMAS-MPO University of Miami, 4600 Rickenbacker Causeway Miami, FL, 33149-1098, U.S.A.*
E-mail: stefano.aliani@sp.ismar.cnr.it

Key words: dispersal, rafting, flotsam, benthos, Lagrangian model

Abstract

Marine litter has been defined as solid materials of human origin discarded at sea, or reaching the sea through waterways. The effect of marine debris on wildlife, tourism and human health is well documented and there is considerable scientific literature about plastic litter in the sea and over the seabed, mostly highlighting the possible impact on marine mammals and tourism. Dispersal of marine and terrestrial organisms on floating objects has biogeographical and ecological interest. For some species, extension of their geographical range is more likely to be related to transport of mature individuals on floating rafts than to the active or passive dispersal of reproductive propagules. Variability and variety of rafting materials has increased dramatically in recent years and marine litter has been used widely as a raft by 'hitch-hiking' species. This paper reports on the benthic invertebrates living on marine debris transported by wind and surface currents over the western Mediterranean Sea. Plastics accounted for the major item of debris because of poor degradability, however glass, cans, fishing nets and polyurethane containers, were also found. Macro-benthos living on raft material comprised mainly molluscs, polychaetes and bryozoans. Large fish were found commonly below large plastic bags. Estimations of the distances that may be covered by hitch-hiking species and the contribution of rafting to the theoretical dispersal of species is provided.

Introduction

Typical marine invertebrates have life histories that include at least one dispersal stage. Tropical and warm-temperate species (e.g. >150 species of prosobranch gastropods) have a very long planktonic development and their teleplanic larvae provide evidence of very long range dispersal potential (Scheltema, 1988). However, there are exceptions to the traditional hypothesis 'longer larval period – wider distribution' (Thorson, 1950; Mileikovski, 1971) and alternative modes of dispersal have been considered (Zibrowius, 1983; Scheltema, 1986). For example, for littoral gastropods along the east coast of North America, there is no clear relationship between the length of development reported from laboratory experiments and geographical range. Apparently, dispersal along continental coastlines may be accomplished by stepwise progression as long as there are no ecological barriers such as ocean basins (Scheltema, 1989).

Brooding species, or species with a short-lived larval phase, may also be widely distributed and, for some of them (mainly living on hard substrata), extension of geographical range may be achieved through transport of mature individuals by floating rafts, rather than through the active or passive dispersal of reproductive propagules (Highsmith, 1985). The so-called 'Rockall paradox' is one example (Johannesson, 1988). Evidence from hydroids showed that wide oceanic distances are covered more by rafting hydroids than by long-distance dispersal of planulae (Cornelius, 1992; Boero & Bouillon, 1993). Colonies of the ascidian Botrylloides sp., attached to *Zostera* leaves in California, were found 200 times further away, and had greater recruitment success, compared with swimming larvae (Worchester, 1994).

Theories about the role of rafting in species distribution have become refined for several taxa living on hard substrata. Following Guppy's (1917) monograph on trans-Atlantic flotsam, rafting has been invoked as a main way of increasing the range of hydroids, (Ralph, 1961) and remote populations of cnidarians (Cornelius, 1992). Jackson (1986) proposed that rafting is the only reasonable explanation for the existence of the vast majority of (benthic) clonal species on Indo-Pacific oceanic islands and argued that, as medusal life was too short to disperse, species used flotsam (i.e. by rafting). Analysis of surface currents and reported drift patterns of pumice were used to estimate rafting of tropical corals (Jokiel, 1989); floating corals, due to air trapped, wander the sea for months (DeVantier, 1992).

The variety and availability of rafting material has increased dramatically in recent years with the spread of human population (Barnes, 2002). Ship hulls, glass, plastic bottles and even items of footwear can be used as rafts by hitch-hiking species. Prior to the spread of humans, available rafts were volcanic pumices, macroalgae, seagrasses, trees and seeds. Today marine litter is also available, i.e. solid material of human origin that is discarded at sea, or reach the sea through waterways or domestic and industrial outfalls (National Academy of Science, 1975 cited in Rees & Pond, 1995). Monitoring marine debris has been undertaken for several years (Rees & Pond, 1995), including assessment of the distribution and abundance of floating plastic in the Gulf of Mexico by aerial surveys (Lecke-Mitchell & Mullin, 1992), and by ship observers in the Eastern Mediterranean Sea (McCoy, 1988). Most papers focused mainly on the impact of debris on the marine environment in terms of pollution, aesthetic damage to tourism, or risks for turtles, cetaceans and fish.

In this paper, we identify the floating objects in the Ligurian Sea from Corsica to the Ligurian Coast (Western Mediterranean) and we asses their potential as ways of extending dispersal range of the benthic invertebrates living on them. Macrobenthic species were identified and the time needed for these benthic species to cover the distance from Corsica to Ligurian Coast by hitch hiking on flotsam was inferred from a numerical simulation of a transport model.

Materials and methods

Study area

The Ligurian Sea is situated at the north eastern border of the Western Mediterranean and is connected to the southern basin (Tyrrhenian Sea) across the Corsica Channel. The eastern and northern borders are the Tuscan and Ligurian coasts, and the western border is open toward the Provençal region and the Gulf of Lions. The major large-scale feature of the deep and surface layers of the Ligurian Sea is a cyclonic circulation active all year round, more intense in winter than in summer. Climatic forcing can greatly change the intensity of fluxes but the general pattern can be considered permanent (Astraldi & Gasparini, 1992). Southern waters, filling the Ligurian Sea, occur in two main currents running along each side of northern Corsica. The West Corsica Current runs along the western side of Corsica, and the warm and salty Tyrrhenian Current passes trough the Corsica Channel between Capraia and Corsica (Artale et al., 1994). The two waters merge to the north of Corsica and flow along the Ligurian coast toward the Gulf of Lions (Astraldi et al., 1995). Species living offshore in the Tyrrhenian basin can be carried northward by the Tyrrhenian current across the Corsica Channel, providing a forced passage for the fauna. Changes in the water fluxes due to climatic variability at this point can control faunal exchanges from the warm Tyrrhenian basin to the colder Ligurian Sea (Astraldi et al., 1995; Aliani & Meloni, 1999).

The oceanographic cruise

Visual sightings of floating objects were made in the Ligurian Sea from 10 to 15 July 1997 from the top deck of the R/V Urania. The type and GPS position of all objects on the sea surface during favourable light and sea conditions while underway and at each station were recorded. Sightings were made close to the Ligurian Coast (areas A,B,C, in Fig. 1), and close to Corsica and the coast of Tuscany (areas D,E,F, in Fig. 1). Samples of floating objects were collected after 15 min of searching by the ship's tender. Immediately after retrieval, samples were fixed in a buffered formaldehyde-seawater solution (final concentration: 5%) and the macrobenthic species identified in the laboratory. Cluster analysis was performed on the presence/absence species × samples matrix.

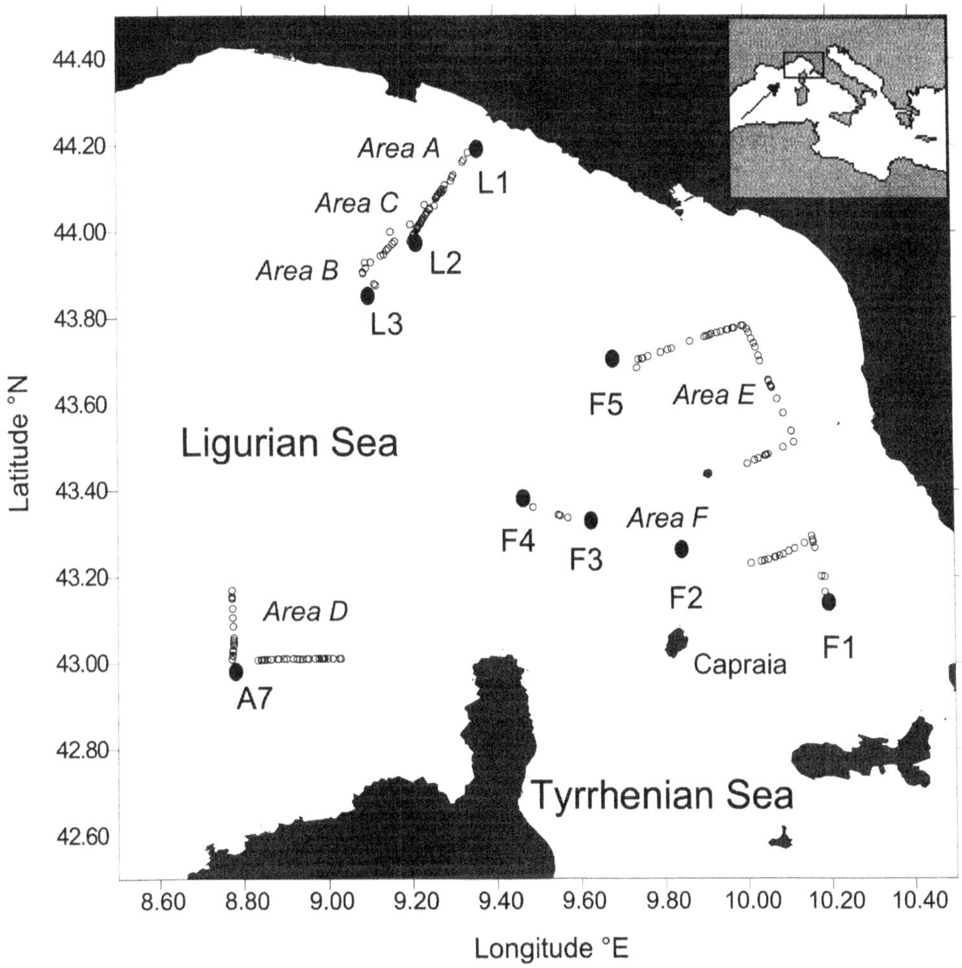

Figure 1. Map of the explored part of the Ligurian Sea. Small open circles indicate sightings and filled circles indicate sampling stations. Multiple sightings and replicates of samples are indicated with a single circles. Coordinates are expressed in degrees and decimals.

Transport models

A Lagrangian model (Falco et al., 2000) simulating the motion of independent particles in a turbulent flow was implemented to study dispersion of passive tracers by simulating a high number of particles. The model assumed that the velocity field can be decomposed in two distinct components: a large-scale mean flow U and a turbulent mesoscale field u'. The tracer particles are advected through two separate processes, the advection by U and the turbulent transport by u', which is characterised by some simple transport parameters such as the variance σ^2 and the turbulent decorrelation timescale T_L.

$$dx = (U + u')dt,$$

$$du' = -(1/T_L)u'dt + (\sigma^2/T_L)^{1/2}dw,$$

where dw is a random increment from normal distribution with zero mean and second order moment. This equation states that turbulent velocity following particles is a linear Gaussian-Markov process, characterised by an exponential autocovariance with e-folding timescale T_L (Risken, 1989).

For the mean flow, we used the output of a general circulation model applied to the Mediterranean Sea (Demirov & Pinardi, 2002) forced by a perpetual (monthly mean) surface forcing for a 7-year period, and averaged in a unique climatological surface mean current field (Fig. 2a). The spatial resolution of the model is 1/8 deg corresponding to 12 km grid resolution, that does not allow the resolution of eddy dynamics due to the high energetic mesoscale variability present in the Ligurian Sea. The mean flow can be underestimated and these small mesoscale effects

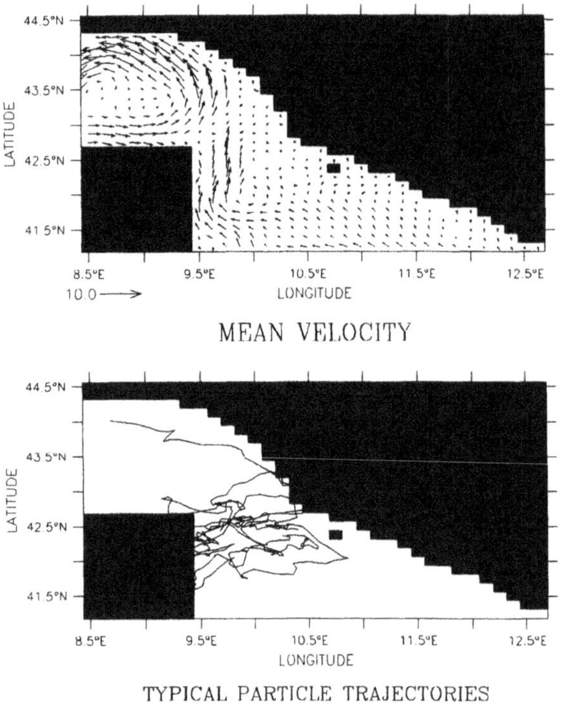

Figure 2. (a, above) Surface mean velocity obtained from a OGCM at 1/8deg resolution, units cm/s; (b, below) Typical trajectories obtained from a Lagrangian model.

were taken into account in the turbulent flow by the variance. The computational domain, where we apply the Lagrangian model, shown in Figure 1b, extends between 8.5° E and 12.5° E in longitude and between 41.25° N and 44.5° N in latitude. Values for σ^2 and T_L are taken from observations (Rupolo, 1993), and are set to 50 cm^2/s^2 and 1.5 d, respectively. Reflection boundary conditions were used at the closed boundaries, which represent the simplest way to guarantee the well-mixed conditions (Thompson, 1987). At the southern and northern boundaries, particles were free to exit following the current.

Results

Sightings

We recorded 169 sightings of different objects from Corsica to the Ligurian coast (Fig. 3). The average distance between sightings (on the transect direction) was 0.7 km and debris density was estimated to be of the order of 14–25 items per km^2 (Aliani et al., 2003). At these sites, we were able to identify 260 different objects, mostly plastic bags (at different stages of degradation) and plastic debris (hard substrata) with some Styrofoam, bottles, wood and fishing gear. All were equally distributed both in the northern (Liguria, area A+B+C) and southern part of the basin (Corsica, area D+E+F). For stations close to the Ligurian coast, dolphins, sea turtles and large fish were sighted below the larger objects.

Samples

The type of objects found in the samples was similar amongst the sightings, and included mainly man-made objects such as plastic litter, Styrofoam and a small percentage of natural floats such as pieces of wood, leaves and algae. All samples were colonised and a total of 22 macrobenthic species was found in 14 samples (Table 1). Some species were common to most flotsam. The most frequent species (>5 findings) were the lepadomorph barnacle *Lepas pectinata,* found at 12 out of 14 stations and the isopod *Idotea metallica* (9 stations). In some cases, juvenile or early life stages were the only fauna on the flotsam. In other cases, a complex population with individuals of different size classes was found. *Idotea metallica* was found preferentially on larger samples. The polychaetes *Spirobranchus polytrema* and *Nereis falsa* (7 findings each) occurred as adults and juveniles. Three records of *Phtisica marina* were found in stations with natural debris (e.g. seagrass leaves, algae) in Corsica and Liguria. Small specimens of the nudibranch mollusc *Doto* sp. were found at three stations and their distribution was limited to the southern stations close to Corsica (west and north), none was found in the Ligurian Sea. The hydroid *Obelia dichotoma* had a similar pattern. Hydroids (and bryozoans) accounted for a high number of species (5 and 4 species, respectively). Different unidentified egg masses were also found in some samples and some species had reproductive structures.

The plot of the cumulative number of species in all samples showed that the species-samples curve reached an asymptote after 10 samples, indicating exhaustive sampling and adequate evaluation of total species richness (Fig. 4). Curve for the Ligurian Sea, reached an asymptote after 4 samples and the curve for Corsica reached the same after 6 samples, suggesting difference in species richness between the two sites. These curves also suggested that little contribution to total richness from new samples from either Liguria or Corsica can be expected.

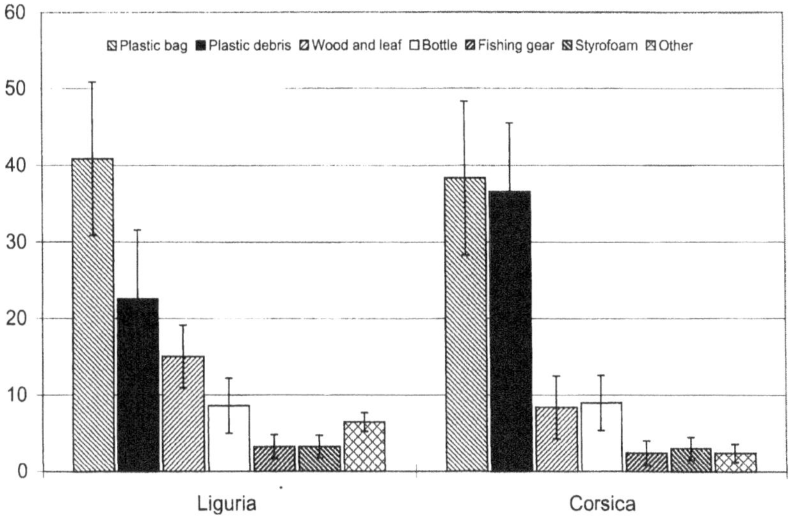

Figure 3. Percentage of major types of floating debris sighted in Liguria and Corsica.

Table 1. List of sessile macrofauna found on rafting objects. Stations are labelled with a letter and number code (e.g. L1, F3...); multiple samples from the same station with additional letter or numbers in brackets

Area	A		B		C		F						E	D
	L1(A)	L1(C)	L2(C1)	L2(C2)	L3(B)	L3(C)	F1	F2	F3(A)	F3(C)	F3(B)	F4	F5	A7
Algae *ind.*	-	-	-	-	-	-	-	-	+	-	-	-	-	-
Arbacia lixula (L.)	-	-	-	-	-	-	-	-	-	-	+	-	-	-
Bowerbankia gracilis Leidy	-	-	-	-	-	-	-	+	-	-	-	-	-	-
Callopora lineata (L.)	-	-	-	-	-	+	-	-	-	-	-	-	-	-
Clytia hemisphaerica (L.)	-	-	+	-	-	-	-	-	-	-	-	-	-	-
Cymodocea nodosa (Ucria) Asch.	+	-	-	-	-	-	-	-	-	-	-	-	-	-
Cystoseira sp.	-	-	-	-	-	-	-	-	-	-	-	+	-	-
Doto sp.	-	-	-	-	-	-	-	-	+	-	+	-	-	+
Electra posidoniae Gautier	-	-	+	-	-	-	-	-	-	-	-	-	-	-
Eudendrium sp.	+	-	-	-	-	-	-	-	-	-	-	-	-	-
Fiona pinnata (Eschscholtz)	-	-	-	-	-	-	-	-	+	-	-	-	+	-
Fosliella farinosa (Lam.)	-	-	-	-	-	-	-	-	-	+	+	-	-	-
Gonothyrea loveni Allm.	-	-	-	-	-	-	-	-	-	+	-	-	-	-
Idotea metallica Bosc	+	+	+	+	-	+	-	-	+	+	+	-	-	+
Laomedea angulata Hincks	-	-	+	-	-	-	-	-	-	-	-	-	-	-
Lepas pectinata (Spengler)	-	+	+	+	+	+	+	+	+	-	+	+	+	+
Membranipora membranacea (L.)	-	-	-	-	-	-	-	-	+	-	-	-	-	-
Nereis falsa Qautrefages	-	+	+	+	-	-	-	-	+	+	+	+	-	-
Obelia dichotoma (L.,)	-	-	-	-	-	-	-	-	-	-	+	+	-	+
Phtisica marina Slabber	+	-	-	-	-	-	-	-	+	-	+	-	-	-
Posidonia oceanica Delile	-	-	+	-	-	-	-	-	-	-	-	-	-	-
Spirobranchus polytrema (Phil.)	-	+	+	+	-	-	-	-	+	+	+	-	-	+

Cluster Analysis of the species × station matrix separated 4 groups (similarity 57%) (Fig. 5). They were accounted by two single samples (F4 and L1a) with species living on rafting plants (*Cymodocea* and *Cystoseira*) and two larger groups. One group (F1,F2,F5,L3b), included samples with few species

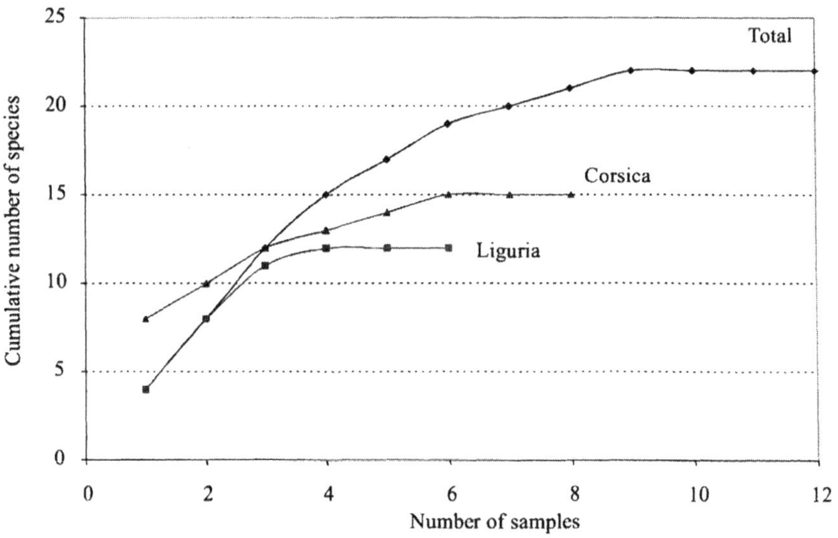

Figure 4. Cumulative number of species plotted against the number of samples in Liguria (square), Corsica (triangle) and against total number of samples (diamond).

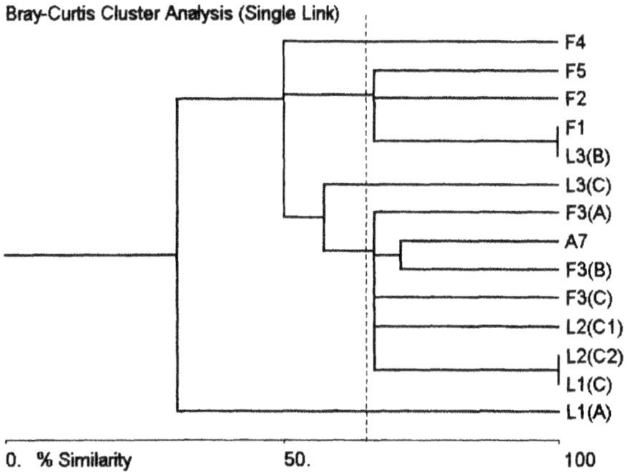

Figure 5. Cluster analysis on the species x stations matrix. The dotted line indicates the 57% similarity.

and the other (F3,A7,L1,L2,L3) had higher richness. Samples of these groups were spread over the basin without any apparent gradient and geographical pattern.

Numerical simulation

In the first experiment, 1000 particles were released in the western part of the Corsica channel (42.5° N 9.625° E), corresponding to the starting point of our observation of floating objects in the 1997 cruise. Figure 2b shows some typical trajectories obtained from the transport model and the pattern compare favourably with real drifter data (see web site

Table 2. Travel days and percentage of particles reaching region F3, and reaching the Ligurian coast when launched from F3

Travel days	Reaching F3	Reaching coast
10	1%	5%
20	5%	19%
30	11%	33%
40		40%
50		45%

doga.ogs.trieste.it and follow the drifter data links, Poulain). Figure 6 shows the diffusion clouds at 3, 30 and 60 d. The higher concentration can be found in the channel even though the clear northern advection mainly due to Tyrrhenian Current along the eastern side of Corsica allows some particles to reach the northern part of the Ligurian Sea. Figure 6 also depicts the time evolution of the percentage of particles able to reach the Ligurian Coast (all the particles that are crossing the 44° N parallel are considered landed in the Ligurian Coast). We first assumed that particles in the first experiment were hypothetical dispersal larvae and we set their life time to 30 d. Our model shows that only 3% of the total number of particles released were able to reach the Ligurian coast within this time.

Later on, we propose that some individuals use marine debris for transportation, and live and reproduce on these rafts. Length of larval life is not a limiting factor for adults settled on rafts, so we con-

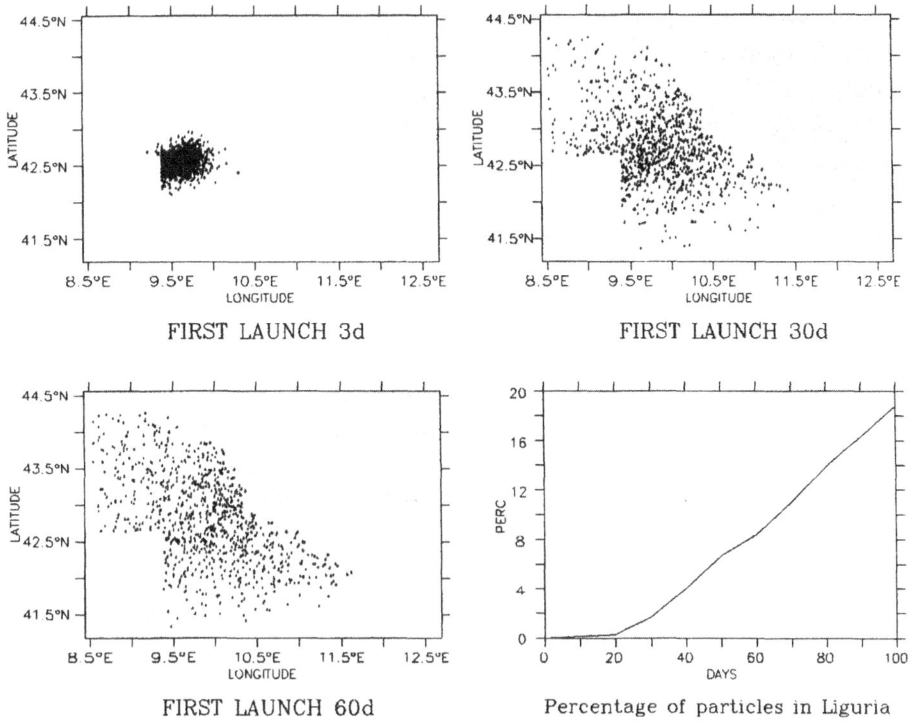

Figure 6. Dispersion of simulated particles and time evolution (lower right panel) of percentage that reach the Ligurian coast.

sidered their life time as infinite. In this case, we used the curve of Figure 6 for longer periods (>30 d), taking into account the probability for the benthic species passing through the Corsica Channel of finding a suitable object to use as a raft. In the second experiment, we focused on this probability. Area F is the closer region to the launching point where some floating objects were observed. Table 2 assumed the percentage of marine debris that had reached that region starting from the first point of release. After 30 d, 11% reached this region. We simulated 1000 new particles (rafting objects) starting at position F3 (43.33° N 9.64° E), and determined the percentage arriving in the Ligurian coast. The probability of reaching the coast increased, because the starting point was closer and the life time of the animal was almost infinite. After 30 d (Table 2), 33% of the animals that had taken the raft were found near the coast; after 50 d 45%.

Discussion

Our results indicated that the same type of floating object is spread over the Ligurian sea from Corsica to the eastern Ligurian coast. The amount of debris is probably higher as large-sized objects are more likely to be observed (McCoy, 1988; Dufault & Whitehead, 1994). Mainly man-made (plastic, fishing tools, Styrofoam) objects were found. Prior to the spread of humans, volcanic pumices, trees and seeds were the most common rafts (Guppy, 1917; DeVantier, 1992); however new types of rafts are now available, and these are very abundant, largely indestructible and almost continuously distributed (Barnes, 2002).

Some of this flotsam is completely submerged and remains just below the surface where it is transported by currents. Others (Styrofoam, bottles) have a ‘free board’ and are transported more by the wind than by currents. This has important implications for separating two different types of rafts whose directions and velocities of dispersal are not necessarily the same but share the same sea surface. Most of our flotsam was plastic bags or plastic debris, and was completely submerged. Samples included all types of objects and the benthic species on them were distributed without any marked difference in species composition between sites. Despite active discussion about the introduction of southern species in the Ligurian sea (Morri & Bianchi, 2001), no alien species was found on our rafts. The list included species reported previously in the

Mediterranean Sea as offshore fouling species (Bellan Santini,1970; Relini 1976; Relini et al., 1977; Aliani & Meloni, 1999; Relini et al., 2000). Most rafted species may be included in the 'list of superwanderers' (Cornelius, 1992; Aliani et al., 1998) that use rafts as their major means of dispersion (Cornelius, 1992). As Barnes (2002) pointed out, the distance from the mainland does not seem to influence the proportion of debris colonised and these species are widely distributed offshore in the Mediterranean Sea.

In contrast to these common widespread species, we encountered also single individuals. For these species, hitch hiking on rafts may offer a final opportunity to survive for a larvae close to the end of its planktonic period (Toonen & Pawlik, 1994) or just an alternative artificial substratum (Tursi et al., 1985). Their contribution to global biodiversity is not well understood. Another group of species, separated by Cluster Analysis, included those found rarely on plastic objects but commonly on seagrass leaves and on wood. They are typically reported as epibionts of seagrass and we found them as reproductive adults much further than the distance covered by short-living larvae. Rafting on leaves is a very effective means of dispersal for these species.

Species can jump to new rafts in the open sea as some unidentified eggs found on the rafts were too young to suggest a long dispersal history. Accumulation of floating objects in strips by Langmuir circulation (Nimmo Smith & Thorpe, 1999) may help in transferring to a new raft or exchanges of gametes between close rafts. A more complex community will probably evolve from the unidentified eggs and seasonal fluctuations in community composition may also occur.

Our numerical experiments showed that there is a limited chance for benthic invertebrates to travel long distances alone without mortality due to ecological constraints (Cowen et al., 2000). Using floating objects as a means of transportation (where they can also reproduce), increases their probability of survival. As plastic litter is everywhere, the probability of finding a raft is higher than hypothezised.

In conclusion, a continuum of species and rafts exist in the Ligurian sea. Rafts have different ages according to the dispersal processes that brought them offshore or according to recruitment processes occurring in far or nearby rafts. For these individuals, rafting objects extend their potential dispersal range beyond their theoretical extension.

Acknowledgements

We thank Annalisa Griffa for useful discussion and suggestions, and Encho Demirov for supplying the OGCM output. We are also grateful to S. Geraci, coordinator of the cruise, and to M. Faimali, F. Garaventa, M. Andrenacci, E. Canepa, E. Schiano for valuable help in the field work. Special thanks to the Captain and the crew of the R/V Urania. This paper forms part of the framework of the Italian Research Project SINAPSI. We thank two anonymous referees for greatly improving the manuscript with their suggestions.

References

Aliani, S., C. De Asmundis, R. Meloni, M. Borghini & G. P. Gasparini, 1998. Transport of benthic species in the Sicily channel: preliminary observations. In Piccazzo, M. (ed.), Atti XII Congresso Associazione Italiana Oceanologia e Limnologia: 173–182.

Aliani, S. & R. Meloni, 1999. Dispersal strategies of benthic species and water current variability in the Corsica Channel (Western Mediterranean) Sci. mar. 63: 137–145.

Aliani, S., A. Griffa & A.Molcard, 2003. Floating debris in the Ligurian Sea, North Western Mediterranean. Mar. Poll. Bull. 46: 1142–1149.

Artale, V., M. Astraldi, G. Buffoni & G. P. Gasparini, 1994. Seasonal variability of the gyre-scale circulation in the Northern Tyrrhenian sea. J. Geophys. Res. 99: 14127–14137.

Astraldi, M. & G. P. Gasparini, 1992. The seasonal characteristics of the circulation in the North Mediterranean Basin and their relationships with atmospheric climatic conditions J. Geophys. Res. 97: 9531–9540.

Astraldi, M., C. N. Bianchi, G. P. Gasparini & C. Morri, 1995. Climatic fluctuations, current variability and marine species distribution: a case study in the Ligurian Sea (North Western Mediterranean). Oceanol. Acta 18: 139–149.

Barnes, D. K. A., 2002. Invasions by marine life on plastic debris. Nature 416: 808–809.

Bellan Santini, D., 1970. Salissures biologiques de substrats vierges artificiels immergés en eau pure, durant 26 mois, dans la règion de Marseille Mediterranée Nord occidentale) I. Etude qualitative. Tethys 2: 335–356.

Boero, F. & J. Bouillon, 1993. Zoogeography and life cycle patterns of Mediterranean hydromedusae (Cnidaria). Biol. J. linn. Soc. 48: 239–266.

Cornelius, P. F., 1992. Medusa loss in leptolid Hydrozoa (Cnidaria), hydroid rafting and abbreviated life cycles among their remote-island faunae: an interim review. Sci. mar. 56: 245–261.

Cowen, R. K., K. M. M. Lwiza, S. Sponaugle, C. B. Paris & D. B. Olson, 2000. Connectivity of marine populations: open or closed? Science 287: 857–859.

Demirov, E. & Pinardi, N., 2002. Simulation of the Mediterranean Sea circulation from 1979 to 1993: Part I. The interannual variability. J. mar. Syst. (33–34): 23–50.

De Vantier, L. M., 1992. Rafting of tropical marine organisms on buoyant coralla. Mar Ecol. Prog. Ser. 86: 301–302.

Dufault, S. & H. Whitehead, 1994. Floating marine pollution in "the Gully' on the continental slope, Nova Scotia, Canada. Mar. Poll. Bull. 28: 489–493.

Falco, P., A. Griffa, P. M. Poulain & E. Zambianchi, 2000. Transport properties in the Adriatic Sea as deduced from drifter data. J. Phys. Oceanogr. 30: 2055–2071.

Guppy, H. B., 1917. Plants, seeds and currents in the West Indies and Azores. Williams and Norgate. London 1: 531.

Highsmith, R. C., 1985. Floating and algal rafting as potential dispersal mechanisms in brooding invertebrates. Mar. Ecol. Prog. Ser. 25: 169–179.

Jackson, J. B. C., 1986. Modes of dispersal of clonal benthic invertebrates: consequences for species' distribution and genetic structure of local populations. Bull. mar. Sci. 39: 588–606.

Johannesson, K., 1988. The paradox of Rockhall, why is a brooding gastropod (*Littorina saxatilis*) more widespread than one having a planktonic larval dispersal stage (*L. littorea*)? Mar. Biol. 99: 507–513.

Jokiel, P. L., 1989. Rafting of reef corals and other organisms at Kwajalein Atoll. Mar. Biol. 101: 483–493.

Lecke-Mitchell, K. & K. Mullin, 1992. Distribution and abundance of large floating marine plastic in the north-central gulf of Mexico. Mar. Poll. Bull. 24: 598–601.

Mc Coy, F., 1988. Floating megalitter in the Eastern Mediterranean. Mar. Poll. Bull. 19: 25–28.

Mileikovski, S. A., 1971. Types of larval development in marine bottom invertebrates, their distribution and ecological significance, a re-evaluation. Mar. Biol. 10: 193–213.

Morri, C. & C. N. Bianchi, 2001. Recent changes in biodiversity in the Ligurian Sea (NW Mediterranean): is there a climatic forcing? In Faranda, F. M., L. Guglielmo & G. Spezie (eds), Mediterranean Ecosystems: Structures and Processes. Springer Verlag: 375–385.

Nimmo Smith, W. A. M. & S. A. Thorpe, 1999. Dispersion of buoyant material by Langmuir circulation and a tidal current. Mar. Poll. Bull. 9: 824–829.

Ralph, P. M., 1961. New Zealand Thecate hydroids. Part V. The distribution of the New Zealand Thecate hydroids. Trans. Roy. Soc. N. Z. (Zool. ser.). 1: 103–111.

Rees, G. & K. Pond, 1995. Marine litter monitoring programmes – a review of methods with special reference to national surveys. Mar. Poll. Bull. 30: 103–108.

Relini, G., 1976. Fouling on different material immersed at a depth of 200 m in the Ligurian Sea. Proc. 4th Int. Congr. On Marine Corrosion. Juan les Pins: 431–443.

Relini, G., C. N. Bianchi, G. Diviacco & R. Rosso, 1977. Fouling di alcune piattaforme offshore dei mari italiani VI: Anfipodi e Policheti. Boll. Mus. Ist. Biol. Univ. Genova 45: 105–121.

Relini, G., M. Relini & M. Montanari, 2000. An offshore buoy as a small artificial island and a fish-aggregating device (FAD) in the Mediterranean. Hydrobiologia 40: 65–80.

Risken, H., 1989. The Fokker-Planck Equation: Methods of Solutions and Applications. Springer-Verlag 1: 472.

Rupolo, V., 1993. Studio delle caratteristiche principali della circolazione del mar Tirreno attraverso l'analisi di dati lagrangiani e la formulazione di un modello. Tesi di Laurea Università La Sapienza, Roma.

Scheltema, R. S., 1986. Alternative modes of dispersal. Bull. mar. Sci. 39: 310–312.

Scheltema, R. S., 1988. Initial evidence for the transport of teleplanic larvae of benthic invertebrates across the East Pacific barrier. Biol. Bull. 174: 145–152.

Scheltema, R. S., 1989. In Reproduction, genetics and distribution of marine organisms. In Ryland, J. S. & P. A. Tyler (eds), Olsen and Olsen Holdstedvj, Denmark: 186–187.

Thompson, D. J., 1987: Criteria for the selection of stochastic models of particle trajectories in turbulent flows. J. Fluid Mech. 180: 529–556.

Thorson, G., 1950. Reproduction and larval ecology of marine bottom invertebrates. Biol. Rev. 25: 1–45.

Toonen, R. J. & J. R. Pawlik, 1994. Foundations of gregariousness. Nature 370: 511–512.

Tursi, A., A. Matarrese, L. Scalera Liaci, G. Costantino, R. Cavallo & E. Cecere, 1985. Colonizzazione di substrati duri artificiali immersi in una biocenosi coralligena ed in un posidonieto. Oebalia 11: 401–416.

Worchester, S. E., 1994. Adults vs larval swimming: dispersal and recruitment of a botryllid ascidian on eelgrass. Mar. Biol. 121: 309–317.

Zibrowius, H., 1983. Extension de l'aire de repartition favorisee par l'homme chez les invertebres marins. Oceanis 9: 337–353.

Hydrobiologia **503**: 69–76, 2003.
M.B. Jones, A. Ingólfsson, E. Ólafsson, G.V. Helgason, K. Gunnarsson & J. Svavarsson (eds), Migrations and Dispersal of Marine Organisms.

Diurnal, horizontal and vertical dispersal of kelp-associated fauna

Nina Mari Jørgensen & Hartvig Christie
Norwegian Institute for Nature Research, P.O. Box 736, Sentrum, 0105 Oslo, Norway
E-mail: hartvig.christie@nina.no

Key words: kelp, fauna, colonization, artificial substratum, diurnal activity, dispersal

Abstract

The kelp *Laminaria hyperborea* is host for a rich fauna of mobile invertebrates. Dispersal patterns of these animals may be crucial for their availability to visual predators like fish, which are known to search for food in these productive habitats. Diurnal, horizontal and vertical dispersal within and out of the kelp forest were studied by analysing colonization of artificial substrata placed among kelps. The species composition of the fauna was different on three parts of the kelp: lamina, stipes (with epiphytes) and holdfast. The artificial substrata were colonized mainly by species associated with kelp. More species and individuals colonized the artificial substrata at night than during the day. The most abundant faunal groups on the kelps and the artificial substrata were amphipods and gastropods. Both groups dispersed at a significantly higher rate at night than during the day. Rapid horizontal dispersal out of the kelp forest was found. The artificial substrata were also colonized quickly by kelp fauna at all vertical levels inside and above the kelp forest. However, species associated with the kelp holdfast tended to disperse close to the bottom, while stipe-associated fauna moved through all parts of the kelp forest and even above the canopy layer. A high dispersal rate appears common amongst the mobile species living on kelp and seems to be advantageous, despite the risks involved in emigrating from the habitat and being exposed to predators. Higher activity at night may reduce availability to predators.

Introduction

Kelp, *Laminaria hyperborea* (Gunn.) Foslie, is distributed widely in the north-eastern Atlantic (Kain, 1971), and forms large forests along the Norwegian coast (Sjøtun et al., 1993). The high primary productivity (Mann, 1972) and very heterogeneous habitat may explain the high species diversity and abundance of invertebrates in kelp forests (Christie et al., 2003). This fauna is dependent on the macro-algal primary production (Hagermann, 1966; Duggins et al., 1989; Duggins & Eckman, 1994; Cruz-Rivera & Hay, 2000) and, in turn, supports fish (Nelson, 1979; Kennelly, 1983, 1991; Holmlund et al., 1990) and seabirds (Bustenes et al., 1997). In coastal areas where kelp forests are abundant, the kelp-associated fauna may represent a large food resource.

Mobile faunal densities exceeding 100 000 individuals per m^2 are common (Christie et al., 2003) and amphipods and gastropods are the most abundant and diverse groups. Most species are highly mobile and move frequently between kelp plants (Norderhaug et al., 2002; Waage-Nielsen et al., 2003). High mobility may be favourable for the maintenance of fauna in unstable habitats (Kuno, 1981). Kelp forests must be considered as unstable habitats as they are regularly disturbed and fragmented by storms and kelp mortality. Kelp forests are also harvested (Christie et al., 1998). Continuous dispersal can make it possible for a limited habitat to support higher fauna densities (Winder, 1990). Dispersal may reduce competition for space, while rapid dispersal gives a competitive advantage when new habitats are occupied (regrowth of kelp and epiphytic algae). However, mobility may have a negative effect by increasing the exposure of the kelp fauna to predators (such as fish). Exposure to fish predation may be reduced if dispersal takes place mainly at night.

Kelp forests form well-defined, three-dimensional patches on two-dimensional hard bottom surfaces (Jensen, 1999), consisting of a dense lamina canopy and stipes overgrown with epiphytic red algae. The

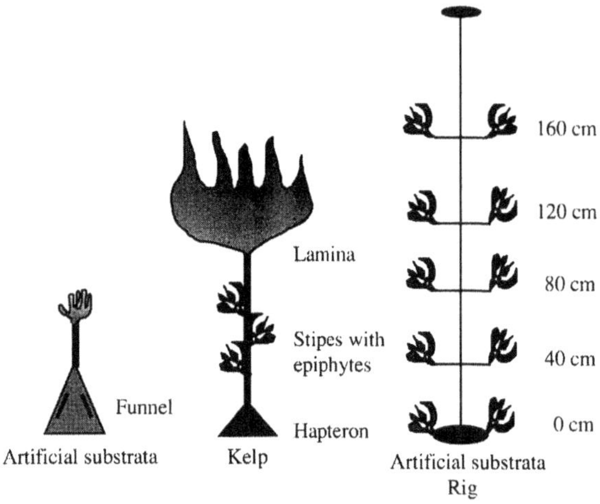

Figure 1. Artificial substrata used in the experiment.

most numerous mobile animals are gastropods and tube-building amphipods (Christie et al., 2003). If they live in close association with the kelp, dispersal from the dense kelp forest is unlikely to offer advantages. Kelp species move frequently between plants in dense kelp forests (Norderhaug et al., 2002; Waage-Nielsen et al., 2003), and the animals colonize new kelp growth (Christie et al., 1998). Migration between the plants in a kelp forest may also lead to loss or dispersal out of the forest. Dispersal out of sea grass beds is known to take place (Virnstein & Curran, 1986) and the epifauna in various habitats make nocturnal vertical migrations (Alldredge & King, 1977, 1985; Stretch, 1985; Sanchez-Jerez et al., 1999).

The aim of this study was to characterise the medium-scale dispersal patterns of the mobile kelp fauna. The study focused on the extent to which the fauna disperses horizontally and vertically out of the kelp forest, and on the extent of nocturnal and diurnal dispersal. Dispersal patterns were studied by examining the rates at which artificial substrata (kelp forest habitat mimics) were colonized by kelp fauna. Comparisons were made between mimics exposed during the day and at night, and between mimics inside, above and some horizontal distance away from the kelp forest.

Materials and methods

This study was carried out in June and September 1996 in a kelp forest at the outer islands of the Romsdal region on the west coast of Norway (62° 47′ N, 6° 32′ E). Two similar sites 0.6 km apart were used, where the kelp forest was undisturbed. The canopy height was 120 cm and the average kelp density was 10 adult kelp per m^2. The depth was about 5 m. The kelp forest community was dominated by mobile invertebrates associated mainly with the holdfasts (hapteron) and epiphytes (red algae) on the kelp stipes. These data are described more in detail by Christie et al. (2003). The fauna on natural kelp plants and on artificial substrata was sampled in three closely related experiments to assess dispersal patterns. Two different substrata were constructed (called 'funnel' and 'rig', Fig. 1) to mimic the holdfasts and epiphytes, respectively.

To study diurnal and horizontal dispersal, a funnel with eight slits, filled with hemp rope and with a textile brush on top (as described in Norderhaug et al., 2002) was used. Funnels were anchored to the bottom. For the study of vertical dispersal, a rig was constructed using a thin line weighted down with a stone and held up straight in the water by a buoy. Artificial substrata, each made of a combination of hemp rope, cotton and latex, were attached to the rope at 40-cm intervals, adding up to 5 levels. All substrata imitated a habitat of approximately 350–400 ml (displacement volume).

To compare the fauna that colonized the mimics with the natural kelp fauna, kelp plants were sampled in sections (lamina, upper, middle and lower stipes, and holdfast of each plant). Three plants were collected at each site and on each experiment date (lamina was only sampled in September). Each kelp plant section and artificial substratum was enclosed carefully in a separate bag *in situ* by SCUBA divers, and the bags were sealed quickly to avoid disappearance of mobile animals. Each sample was washed and sieved (250 μm), and animals were counted and identified to species or higher taxa. As the experiment focused on mobile fauna, hydroids, bryozoans, serpulid polychaets and other sessile groups were only recorded as present/absent.

Diurnal migration was studied at one site, and the experiment was repeated in June and September. Three replicate mimics (funnels) were placed inside the kelp forest and exposed for 10 h, either during the day or at night. Horizontal dispersion was investigated at the other site in June and September. Three replicate funnels were placed inside the kelp forest and at each of three distances (1 m, 4 m and 7 m) from the edge of the kelp forest on a sandy bottom. Funnels were collected after four days.

The vertical dispersal experiment was conducted at both sites in September, and 4 replicate rigs were used at each site. Rigs were lowered down from a

Table 1. The most common species and faunal groups found on kelp and artificial substrata in this study. Abundance is given for species and number of individuals/number of species is given for faunal groups. Faunal composition is shown for each part of the kelp, for all mimics in each of the experiments on diurnal and horizontal dispersal, while the animals colonizing the rig are split in three vertical levels. 'Others' includes the faunal groups Echinodermata, Ascidiacea and Osteichthyes

	Kelp Lamina $n = 6$	Kelp Stipes $n = 9$	Kelp Holdfast $n = 9$	Day/night $n = 12$	Horizontal $n = 24$	Rig Bottom $n = 8$	Rig Medium $n = 24$	Rig Upper $n = 8$
Polychaeta total	105/5	1034/7	2516/18	8/6	14/2	36/6	43/8	6/3
Idotea granulosa (Rathke)	1	8	0	0	101	0	31	21
Munna kroeyri (Goodsir)	1	352	360	13	0	0	9	0
Isopoda total	11/3	459/6	521/2	23/3	199/6	32/4	89/7	42/5
Stenothoidae indet	4	828	101	8	7	38	139	68
Apherusa bispinosa (Bate)	9	44	3	427	315	233	1052	550
Atylus swammerdami (Milne-Edwards)	0	0	0	0	159	0	10	2
Lembos websterii (Bate)	0	39	3035	10	30	19	22	1
Corophium bonnelli (G O Sars)	0	3	934	0	0	106	106	0
Jassa falcata (Montagu)	15	5673	2209	24	16	26	97	13
Gammarus locusta (L)	0	0	0	141	94	0	16	6
Dexamine thea (Boeck)	7	859	542	16	91	49	379	410
Gammaridea total	56/8	8970/14	7525/19	682/14	1117/18	1014/24	3842/32	1532/22
Caprellidea total	112/4	495/5	245/5	16/2	267/2	37/6	50/6	6/3
Other Crustacea total	33/3	142/8	805/10	191/6	145/9	42/8	87/12	10/5
Lacuna vincta (Montagu)	2206	7084	113	186	375	120	2609	1189
Rissoidea spp.	121	15972	5815	815	992	3791	7345	739
Gastropoda total	2433/7	24106/15	8269/17	1092/12	1404/12	4778/11	12031/14	2174/7
Mytilus edulis (L)	9	2642	279	50	35	105	165	9
Turtonia minuta (Fabricius)	0	125	69	10	19	51	55	0
Bivalvia total	9/3	4310/9	2929/10	67/4	60/5	184/10	256/11	11/3
Others total	2/1	209/5	356/8	19/4	6/3	21/3	22/3	0

Table 2. Results of statistical analysis of total number of individuals and species found in the two experiments: diurnal activity (1) and horizontal dispersal (2). In the horizontal dispersal experiments the interaction between season and distance regarding number of individuals was significant

	Season (1)			Time (1)			Season (2)			Distance (2)			Interaction (2)		
	df	*F*	*p*	df	*F*	*p*	df	*F*	*p*	df	*F*	*p*	df	*F*	*p*
No. individ.	1	87.3	<0.001	1	37.3	<0.001	1	19.4	<0.001	3	6.6	0.003	3	6.9	0.003
No. species	1	188.8	<0.001	1	35.8	<0.001	1	123.7	<0.001	3	0.99	0.42	3	2.7	0.07

boat haphazardly within the kelp forest, approximately 2 m apart. A diver inspected the rigs to ensure they were placed properly. The five vertical levels of the mimics were one at the bottom, two alongside the stipes/epiphytes, one at the kelp canopy level, and one 40 cm above the kelp. After 4 days, a diver collected the individual vertical mimics on each of the rigs. The scope of the experiments was limited to some extent because divers were needed to do the sampling, and there were large numbers of animals to identify in each sample (almost 10 000 individuals on each kelp plant).

ANOVA tests were used to identify differences in dispersal rates between species and faunal groups. Season (June and September), hour of the day (day and night), horizontal distance and vertical level were used as factors. In the experiment on vertical dispersal, rig was included as an error factor. Thus, in the test there has been considered a dependency between the levels on each rig. The data were transformed when necessary. All tests were performed in the SPLUS (6.0) computer package. MDS (Multi Dimensional Scaling, Shepard, 1962; Kruskal, 1964 a, b) ordinations based on similarity matrices were calculated using the PRIMER (Plymouth Routines In Multivariate Ecological Research, 5.2.1) computer package to identify patterns in the samples. The Bray–Curtis similarity index was used (Bray & Curtis, 1957) and the data were square-root transformed.

Table 3. Results of statistical analysis of the of total number of individuals and species, and number of individuals of the most common faunal groups, amphipods and gastropods, found in the vertical dispersal experiment

	Site			Level		
	df	*F*	*p*	df	*F*	*p*
No. individ.	1	4.1	0.08	4	15.6	<0.001
No. species	1	4.2	0.04	4	29.0	<0.001
No. of amphipods	1	0.09	0.75	4	6.24	<0.001
No. of gastropods	1	7.11	0.01	4	7.99	<0.001

Results

A total of 20 314 individuals of mobile macrofauna colonized the artificial substrata, and 69 127 animals were recorded on the reference kelp plants sampled. The most numerous kelp-associated faunal groups are shown by number of individuals/number of species in Table 1, together with the abundance of the most characteristic species in the study. Table 1 lists the fauna found in kelp parts and in the artificial substrata of each of the three experiments, and shows that kelp-associated animals colonized the mimics in all experiments. For a more thorough taxonomical breakdown of the fauna associated with kelps, see Christie et al. (2003). Table 1 also gives the proportion of animals on the vertical levels of the kelp and the rig, showing different vertical distributions among the species.

Colonization occurred during the day and at night, with more animals colonizing the kelp mimics at night than during the day (Fig. 2). The difference was significant when measured both as number of individuals and as number of species (Table 2). The number of individuals was higher in September and relatively higher nocturnal activity was also recorded in September.

The amphipods *Apherusa bispinosa* (Bate), *Corophium* spp, and species of the family Ischyroceridae (mainly *Jassa falcata* Montagu) showed particularly high activity at night. The two most abundant gastropods in the kelp forest, *Rissoa parva* (da Costa) and *Lacuna vincta* (Montagu), were also the most common gastropods on the artificial substrata. Amphipods showed a greater tendency for nocturnal activity (significant at 0.001 level) than the gastropods (significant at 0.03 level). There were twice as many gastropods in the night-time samples as in the daytime ones, whereas there were 11 times as many amphipods in the night-time samples.

The kelp fauna colonized artificial substrata placed up to 7 m from the kelp forest in both June and September within four days (Table 1, Fig. 3), but the abundance decreased with distance from the kelp forest. There were no significant differences in the number of species with distance from the kelp forest (Table 2), but ANOVA tests revealed that abundance on the mimics on the sandy bottom was significantly different from that on similar mimics within the kelp forest (Table 2).

Many animals that colonized the mimics placed outside the kelp forest belonged to the most abundant gastropods and amphipods of the kelp fauna, as found in the diurnal experiment presented above. However, some exceptions occurred, e.g. the amphipod *Atylus swammerdami* (Milne-Edwards) which was common in the mimics on sandy bottoms but not found as part of the kelp-associated fauna. Also *A. bispinosa* was more common on the mimics than on the kelp plants (Table 1).

A large number of animals colonized the artificial substrata at all vertical levels inside the kelp forest and 40 cm above the kelp canopy (Table 1, Fig. 4). Numbers of species and individuals were generally high at all levels of the rig, but ANOVA tests showed significant differences between levels in both number of species and number of individuals (Table 3). The number of species was especially high at the lowest level, and the number of individuals was also highest here. The numbers of gastropods and amphipods are presented in Figure 4. The distribution of these groups also showed significant differences between vertical levels (Table 3). Gastropods were most abundant at the bottom level and amphipods at the upper levels (Fig. 4).

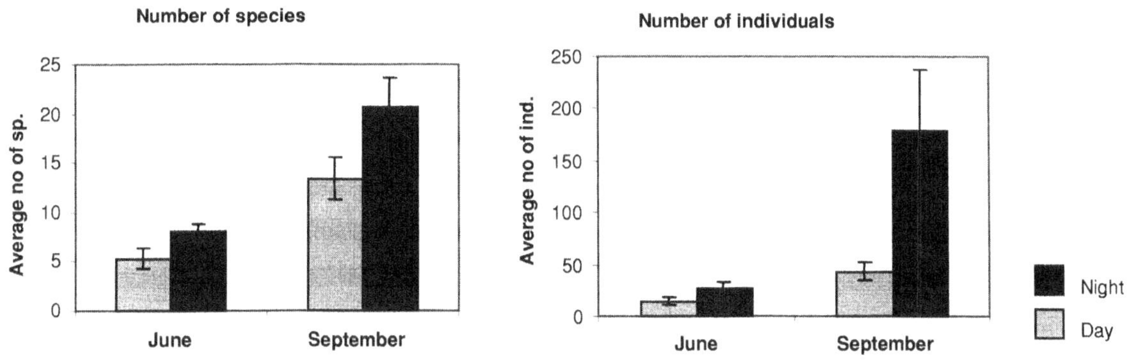

Figure 2. Colonization by kelp fauna of artificial substratum exposed by day and night in June and September, presented as mean number of species and mean number of individuals per artificial substratum, with standard error bars.

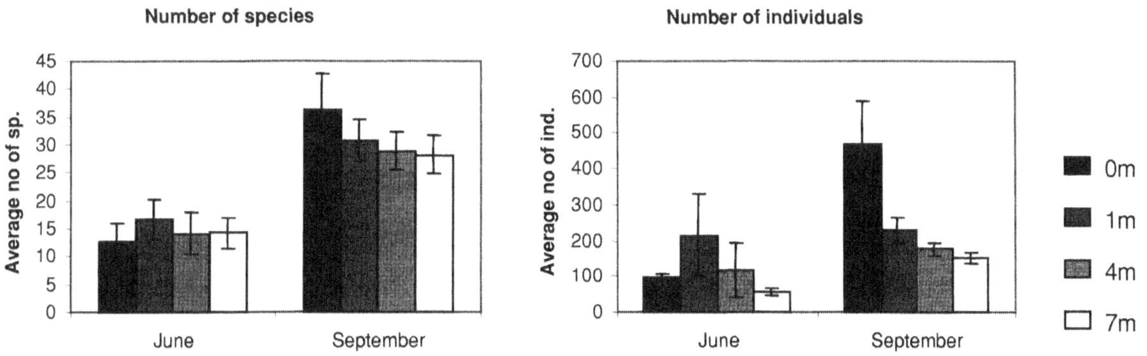

Figure 3. Colonization by kelp fauna of artificial substratum exposed for 4 days inside a kelp forest (0 m) and at different distances (1, 4 and 7 m) from the kelp on a sandy bottom in June and September, presented as mean number of species and mean number of individuals per artificial substratum, with standard error bars.

Ordination plot of species distribution (Fig. 5) showed that samples from each of the three main vertical parts of the kelp plants were grouped together, whereas samples from the vertical experiment mimics fell into two groups. In the case of the samples from the natural kelp plants, the holdfast, the three parts of the stipes, and the lamina were clearly separated, whereas only the 8 samples from the lowest level of the artificial substrata were separated from samples from the other four levels (32 samples). Samples from the four upper levels were very similar. Like the holdfast samples, samples from the lowest level of the artificial substrata appeared towards the bottom of the plot, while species distribution in all the other artificial substrata from 40 cm above the bottom to 40 cm above the kelp canopy (those not encircled) was more similar to that in the samples from the stipes. The plot also shows that the composition of the fauna was similar at the two sites investigated.

Discussion

Our results show that the kelp forest fauna colonizes artificial substrata in high numbers within a short time, indicating that the animals are highly mobile or have high dispersal rates. These results clearly support the conclusion of Norderhaug et al. (2002) that the kelp fauna is highly mobile and frequently moves between plants in the kelp forest. However, this study has given further information on diurnal and spatial patterns of dispersal of the kelp-associated fauna. One experiment showed that the animals dispersed both during the day and at night, but were more active at night. Another two experiments showed that large numbers of animals dispersed out of the kelp forest, both horizontally along a benthic gradient and vertically along a pelagic gradient. Although the results that emerged from each experiment were very consistent and clear, it is important to note that the validity of the conclusions is limited by the way the study was performed. Experiments were conducted with a limited number of

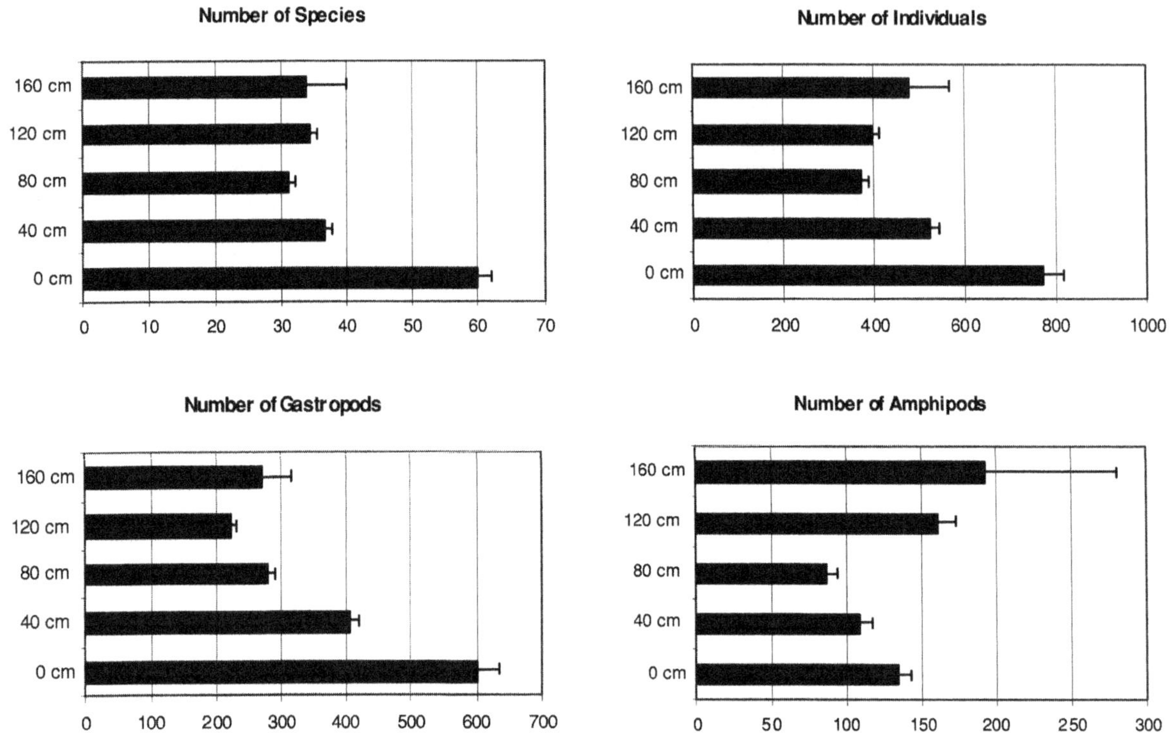

Figure 4. Colonization by kelp fauna of artificial substratum exposed at five different vertical levels inside and above the kelp forest. Total number of species, total number of individuals, number of gastropods and number of amphipods are presented as mean number per artificial substratum, with standard error bars. The figure shows data from both sites (n = 8); the kelp canopy was 120 cm above the bottom level at each site.

replicates, at either two seasons or two sites in the same year and within the same area. It is, therefore, difficult to be sure that the data are fully representative. On the other hand, the dominant species in this study have also been found to be abundant in kelp forests in other parts of Norway (Christie et al., 2003) and in the North Sea (Jones, 1971; Moore, 1973, 1986; Schulze et al., 1990).

Similar patterns of horizontal faunal dispersal (Virnstein & Curran, 1986; Waage-Nielsen et al., 2003), diurnal dispersal (Watkin, 1941; Colman & Segrove, 1955; Edgar, 1983; Leber, 1985; Howard, 1987) and nocturnal vertical migration by epifauna (Alldredge & King, 1977, 1985; Stretch, 1985; Sanchez-Jerez et al., 1999) have been found in other habitats. This suggests that frequent dispersal is favourable for animals in certain benthic habitats in contrast to other benthic habitats where faunal dispersal occurs mainly as pelagic larval drift during a limited reproduction period. The kelp forest is a very dynamic habitat, where limited disturbance occurs on several different time scales, i.e. the duration of the kelp habitat (years), seasonal fluctuations of epiphytes (months, Whittic, 1983), and disturbance by storms and grazers (hours–days). Generally, mobility may be favourable for faunal community stability in unstable habitats (Kuno, 1981). Norderhaug et al. (2002) also suggested that frequent dispersal might be favorable for exploitation of the habitat if resources like food (cf. Charnov, 1976) or space (c.f. Winder, 1990) were limited. The high densities of fauna found in both natural kelp and artificial substrata in this study support the suggestion that any one of these resources may be limited.

Colonization of the artificial substrata was higher in September than in June. This probably reflects higher fauna densities rather than higher dispersal rates. The most abundant colonizers were epiphyte-associated species (see Norderhaug et al., 2002), which were found mainly in the upper parts of the kelp plants. Figure 5 and Table 1 indicates that holdfast species may prefer to crawl more close to the bottom, while the epiphyte fauna was able to disperse in several ways and at all vertical levels. According to Norderhaug et al. (2002) *Lembos websterii* (Bate), *Corophium bonelli* (G. O. Sars) and *Turtonia*

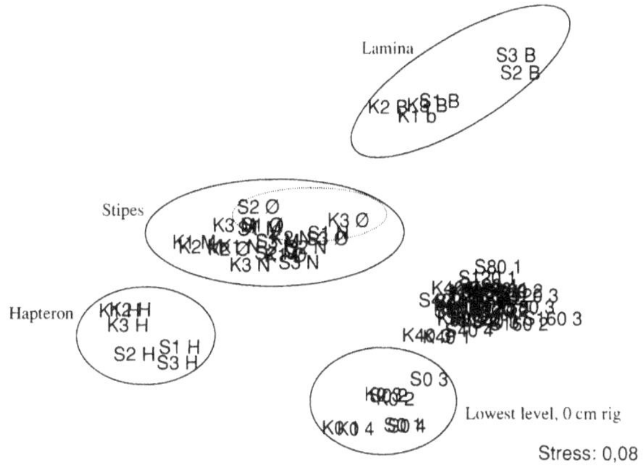

Figure 5. MDS plot of species distribution on different parts of natural kelp plants and the artificial substrata at the different vertical levels of the rig. The replicate samples from the main parts of the kelp plants are circled, as well as the replicated samples from the bottom level of the rigs.

minuta (Fabricius) belong to the holdfast fauna, while the other species in Table 1 are epiphyte-associated species or ubiquitous kelp species.

As both the mimics outside the kelp forest and those above the kelp canopy were colonized, swimming or drifting in the water must be an important method of dispersal. Amphipods and isopods are able to swim, but drifting with tidal currents and wave action is an alternative method and may be more relevant for juvenile amphipods. The amphipod *Jassa falcata* is able to swim upwards and then spread out its antennae and 'paraglide' in the water column (own observations). The most abundant gastropod species in the kelp forest are also reported to drift, using a mucous thread to increase buoyancy (Vahl, 1983; Martel & Diffenbach, 1993; own observations). This may explain the rapid dispersal of gastropods out of the kelp forest. Personal observations of gastropods drifting within and above the kelp forest support this. However, the vertical distribution of these fauna groups (Fig. 4) indicates that amphipods swim upwards to a greater extent while gastropods tend to sink towards the bottom.

However, a high dispersal rate involves risks for the fauna. Dispersal increases the risk of being discovered by common predators like fish that feed in macrophyte habitats (Nelson, 1979; Kennelly, 1983, 1991; Holmlund et al., 1990). A high dispersal rate at night, such as we found in this study (Fig. 2), may reduce this risk (Zaret & Suffern, 1976; Howard, 1987). The results of this study also indicate that part of the fauna is continuously being exported (emigrating) from the kelp forest. It has been shown earlier that kelp production is exported out of the kelp forest (Vetter, 1995), and that 1–2% of the biomass of amphipods and isopods emigrates from a seaweed community every day (Christie & Kraufvelin, 2003). Our results suggest that the kelp forest in the studied area may be a source of secondary production (invertebrate fauna) exported to other food webs (pelagic systems and other benthic systems adjacent to the kelp forest) at the coast.

Acknowledgements

This study was funded by the Research Council of Norway. The authors wish to thank Kjell Magnus Norderhaug for advice on statistical and numerical analysis and for correcting the manuscript, and Stein Fredriksen for help during the field work.

References

Alldredge, A. L. & J. M. King, 1977. Distribution, abundance and substrate preferences of demersal reef zooplankton at Lizard Island Lagoon, Great Barrier Reef. Mar. Biol. 41: 317–333.

Alldredge, A. L. & J. M. King, 1985. The distance demersal zooplankton migrates above the benthos: implications for predation. Mar. Biol. 84: 253–260.

Bray, J. R. & J. T. Curtis, 1957. An ordination of the upland forest communities of southern Wisconsin. Ecol. Monogr. 27: 325–349.

Bustenes, J. O., H. Christie & S. H. Lorentsen, 1997. Seabirds, kelp beds and kelp trawling: a summary of knowledge. NINA Oppdragsmelding 472: 1–43 (in Norwegian with English summary).

Charnov, E. L., 1976. Optimal foraging theory: the marginal value theorem. Theoret. Popul. Biol. 9: 129–136.

Christie, H., S. Fredriksen & E. Rinde, 1998. Regrowth of kelp and colonization of epiphyte and faunal community after kelp trawling at the coast of Norway. Hydrobiologia 375/376: 49–58.

Christie, H. & P. Kraufvelin, 2003. Mechanisms regulating amphipod population density within macro algal communities with restricted predator impact. Scientia Marina, EMBS36 special issue: Biological Oceanography at the turn of the Millenium. (In press).

Christie, H., N. M. Jørgensen, K. M. Norderhaug & E. Waage-Nielsen, 2003. Species distribution and habitat exploitation of fauna associated with kelp, *Laminaria hyperborea*, along the Norwegian coast. J. mar. biol. Ass. U.K. 83: 687–699.

Colman, J. S. & F. Segrove, 1955. The tidal plankton over Stoupe Beck Sands, Robin Hoods Bay (Yorkshire, North Riding). J. anim. Ecol. 24: 445–462.

Cruz-Riveira, E. & M. E. Hay, 2000. The effect of diet mixing on consumer fitness: macro algae, epiphytes, and animal matter as food for marine amphipods. Oecologia 123: 252–264.

Duggins, D. O., C. A. Simenstad & J. S. Estes, 1989. Magnification of secondary production by kelp detritus in coastal marine ecosystems. Science 245: 170–173.

Duggins, D. O. & J. E. Eckman, 1994. The role of kelp detritus in the growth of benthic suspension feeders in an understory kelp forest. J. exp. mar. Biol. Ecol. 176: 53–68.

Edgar, G. J., 1983. The ecology of South-East Tasmanian phytal animal communities, 4. Factors affecting the distribution of amphithoid amphipods among algae. J. exp. mar. Biol. Ecol. 70: 205–225.

Hagermann, L., 1966. The macro- and microfauna associated with *Fucus serratus* with some ecological remarks. Ophelia 3: 1–43.

Holmlund, M. B., C. H. Peterson & M. E. Hay, 1990. Does algal morphology affect amphipod susceptibility to fish predation? J. exp. mar. Biol. Ecol. 139: 65–83.

Howard, R. K., 1987. Diel variation in the abundance of epifauna associated with sea grasses of the Indian River, Florida, U.S.A. Mar. Biol. 96: 137–142.

Jensen, A., 1999. The seaweed resources of Norway. In Critchley, A. T. & M. Ohno (eds), Seaweed Resources of the World. Japan International Cooperation Agency: 200–209.

Jones, D. J., 1971. Ecological studies on macro invertebrate populations associated with polluted kelp forests in the North Sea. Helgol. wiss. Meereresunters: 22: 417–441.

Kain, J. M., 1971. The biology of *Laminaria hyperborea*. Some Norwegian populations. J. mar. biol. Ass. U.K. 51: 387–408.

Kennelly, S. J., 1983. An experimental approach to the study of factors affecting algal colonization in a sublittoral kelp forest. J. exp. mar. Biol. Ecol. 68: 257–276.

Kennelly, S. J., 1991. Caging experiment to examine the effect of fishes on understory species in a sublittoral kelp community. J. exp. mar. Biol. Ecol. 147: 207–230.

Kuno, E., 1981. Dispersal and the persistence of populations in unstable habitats: a theoretical note. Oecologia 49: 123–126.

Kruskal, J. B.,1964a. Multidimensional scaling by optimizing goodness-of-fit to a nonmetric hypothesis. Psychomterica 29: 1–28.

Kruskal, J.B.,1964b. Nonmetric multidimensional scaling: a numerical method. Psychomterica 29: 115–129.

Leber, K. M., 1985. The influence of predatory decapods, refuge and microhabitat selection on sea grass communities. Ecology 1985: 1951–1964.

Mann, K. H., 1972. Ecological energetics of the seaweed zone in a marine bay on the Atlantic coast of Canada. II. Productivity of the seaweeds. Mar. Biol. 14: 199–209.

Martel, A. & T. Diefenbach, 1993. Effect on body size, water current and microhabitats on the mucous-thread drifting in post-metamorphic gastropods *Lacuna* spp. Mar. Ecol. Prog. Ser. 99: 215–220.

Moore, P. G., 1973. The kelp fauna of Northeast Britain. II. Multivariate classification: Turbidity as an ecological factor. J. exp. mar. Biol. Ecol. 13: 127–163.

Moore, P. G., 1986. Levels of heterogeneity and the amphipod fauna of kelp holdfasts. In Moore, P. G. & R. Seed (eds), The Ecology of Rocky Coasts. Colombia University Press, New York: 274–290.

Nelson, W. G., 1979. Experimental studies of selective predation on amphipods: Consequences for amphipod distribution and abundance. J. exp. mar. Biol. Ecol. 38: 225–245.

Norderhaug, K. M., H. Christie & E. Rinde, 2002. Colonization of epiphyte and holdfast fauna; a study of mobility patterns. Mar. Biol. 141: 965–973.

Sanchez-Jerez, P., C. B. Cebrian & A. R. Espla, 1999. Daily vertical migrations in the epifauna associated with Posidoniaoceanica meadows. J mar. biol. Ass. UK 79: 971–977.

Shultze, K., K. Janke, A. Krüß & W. Weidemann, 1990. The macrofauna and macroflora associated with *Laminaria digitata* and *L. hyperborea* at the island of Helgoland (German bight, North sea). Helgoländer Meeresunters. 44: 39–51.

Shepard, R. N., 1962. The analysis of proximities: multidimensional scaling with an unknown distance function. II. Psychometrica 27: 219–246.

Sjøtun, K., S. Fredriksen, T. E. Lein, J. Rueness & K. Sivertsen, 1993. Population studies of *Laminaria hyperborea* from its northern range of distribution in Norway. Hydrobiologia 260/261: 215–221.

Stretch, J. J., 1985. Quantitative sampling of demersal zooplankton: re-entry and airlift dredge sample comparisons. J. exp. mar. Biol. Ecol. 91: 125–136.

Vahl, O., 1983. Mucus drifting in the limpet *Helcion* (= *Patina*) *pellucida* (Prosobranchia, Patellidae). Sarsia 68: 209–211.

Vetter, E. W., 1995. Detritus-based patches of high secondary production in the near shore benthos. Mar. Ecol. Prog. Ser. 120: 251–262.

Virnstein, R. V. & M. C. Curran, 1986. Colonization of artificial sea grass versus time and distance from source. Mar. Ecol. Prog. Ser. 29: 279–286.

Waage-Nielsen, E., H. Christie & E. Rinde, 2003. Short-term dispersal of kelp fauna to cleared (kelp-harvested) areas. Hydrobiologia 503 (Dev. Hydrobiol. 174): 77–91.

Watkin, E. E., 1941. Observations on the night tidal migrant Crustaceans of Kanes Bay. J. mar. biol. Ass. U.K. 25: 81–96.

Whittick, A., 1983. Spatial and temporal distribution of dominant epiphytes on the stipes of *Laminaria hyperborea* (Gunn.) Fosl. (Phaeophyta: Laminariales) in S. E. Scotland. J. exp. mar. Biol. Ecol. 73: 1–10.

Winder, L., 1990. Predation of the cerial ahid Sitobion avenae by polyphagous predators on the ground. Ecol. Entomology 15: 105–110.

Zaret, T. M. & J. S. Suffern, 1976. Vertical migration in zooplankton as a predator-avoidance mechanism. Limnol. Oceanogr. 21: 804–813.

Hydrobiologia **503:** 77–91, 2003.
M.B. Jones, A. Ingólfsson, E. Ólafsson, G.V. Helgason, K. Gunnarsson &
J. Svavarsson (eds), Migrations and Dispersal of Marine Organisms.

Short-term dispersal of kelp fauna to cleared (kelp-harvested) areas

Elisabeth Waage-Nielsen, Hartvig Christie & Eli Rinde
Norwegian Institute for Nature Research, P.O. Box 736, Sentrum, 0105 Oslo, Norway
E-mail: hartvig.christie@nina.no

Key words: artificial substratum, dispersal, invertebrates, kelp trawling, recovery

Abstract

The kelp *Laminaria hyperborea* forms large forests and houses a numerous and diverse fauna, especially in the kelp holdfast and stipe epiphytes. Kelp harvesting creates cleared areas and fragmentizes the kelp forest. We investigated the dispersal ability of kelp fauna to cleared, harvested areas by studying their colonization pattern to artificial substrata (kelp mimics) exposed for a short (3 days) and longer time period (35 days) at different sites within the kelp forest (one site) and at a cleared area (two sites). Most of the kelp fauna (111 species) showed a rapid dispersal and colonized the artificial substrata within the cleared area. The similarity of the faunal community in the mimics with the natural kelp holdfast community increased with the length of the exposure period. During the experiments, 87% of the mobile species in the kelp plants were found in the kelp mimics, indicating good dispersal for slow-moving animals like gastropods, polychaetes and tube-building crustaceans. Relating the frequency of the different faunal groups in the untrawled kelp forest to their frequency in the kelp mimics, showed gastropods, amphipods and decapods to have relatively high dispersal rates, whereas isopods, bivalves, polychaetes and tanaids showed a lower dispersal rate than expected. Amphipods dispersed as juveniles and adults. No significant differences were found between the faunal composition and number of species in the mimics placed inside the kelp forest and in the cleared area. Remaining holdfasts and pebbles were identified as refuges/alternative habitats in the harvested area, and may together with the nearest kelp vegetation, serve as sources for colonization to new substrata. The high dispersal ability of most of the kelp fauna provides maintenance of the faunal composition of disturbed habitats and ensures colonization of recovering algal habitats regardless of reproduction strategy.

Introduction

Kelps form large forests along the north-eastern Atlantic coastline, where *Laminaria hyperborea* (Gunn.) Foslie is the dominant species. Kelp forests are estimated to extend between 5000 and 10 000 km^2 along the Norwegian coast (Jensen, 1999). The kelp forest is highly productive (Mann, 1972) and houses many macrofaunal species both in the holdfasts (hapterons) and on the epiphytes attached to the stipes; this has been reported in Great Britain (Jones, 1971; Moore, 1972, 1973), Ireland (Edwards, 1980), Germany (Schultze et al., 1990), and Norway (Norderhaug et al., 2002; Christie et al., 2003). The kelp forest is also very important as a habitat and feeding ground for several fish species (Nelson, 1979; Kennelly, 1983; Holmlund et al., 1990; Nordeide & Fosså, 1992) and for seabirds (Bustnes et al., 1997).

Norwegian kelp forests are affected by both biological and physical disturbances. Grazing by sea urchins clears large areas (Hagen, 1983; Skadsheim et al., 1995), whereas storms and harvesting (kelp trawling) often clear smaller areas due to variations in topography and depth, leaving a fragmented kelp forest landscape rather than large cleared areas (Rinde et al., 1992; Christie et al., 1994). Kelp harvesting takes place in all seasons on the Norwegian west coast (from 58° to 64° N), and approximately 170 000 tons of kelp are harvested every year. Generally, a kelp trawl removes all canopy-forming kelp plants in a 4-m wide track. Depending on the composition of vegetation in the understory, the track is either barren or vegetated by small understory algae, including kelp sporelings. All that remains of the canopy plants is a few holdfasts. From harvesting statistics, it has been estimated that, on average, 6–13% of the kelp population is harvested within the trawled areas. However, in areas particularly suitable for kelp trawling, such as large expanses of smooth rock, the harvest efficiency is higher, and canopy-forming kelp plants may be cleared totally from larger areas (Christie et al., 1994).

Besides the more destructive disturbances mentioned above, kelp fauna lives in a habitat that changes more or less continuously. Individual *L. hyperborea* plants live for about 5–15 y (Kain, 1963; Sjøtun et al., 1993). The quality of the habitat changes as a plant becomes older; its size increases and structural changes take place in the stipe, especially with the establishment of an increasing number of epiphytic red algae (Christie et al., 1998). Even the epiphytes change as they undergo seasonal fluctuations in both size and composition (Whittick, 1983). The kelp fauna has to withstand these changes, and a well-developed dispersal ability is probably advantageous for reaching new habitats and escaping from habitats that have deteriorated.

In a study of the regrowth of kelps and the accompanying recovery of kelp fauna and epiphytic algae in trawled tracks of different ages, Christie et al. (1998) found that kelp plants reached the normal canopy height within 3–5 y. Within the same period, large numbers of organisms colonized the recovering kelp plants, but the diversity of the holdfast fauna and stipe epiphytes did not reach the same level as in the natural kelp forest. Animal groups almost recovered to their previous abundance, except for amphipods and isopods (Christie et al., 1998). Their lack of recovery may be explained by low dispersal ability, related either to behaviour (e.g. tube building, crawling instead of swimming or drifting) or to reproduction strategy (brooding instead of having planktonic larvae). However, another explanation may be that the habitat had not fully recovered. Christie et al. (1998) found that the regrowing kelp forest was more homogenous in structure than natural kelp forests, and contained fewer and less voluminous stipe epiphytes (mainly red algae).

Although the most abundant animals in kelp forests are usually slow-moving gastropods and tube-building crustaceans (Foster-Smith & Shillaker, 1977; Moore, 1986; Shillaker & Moore, 1987; Ulrich et al., 1995), Norderhaug et al. (2002) found that these groups exhibited high mobility between kelp plants in a dense forest. Mobility was particularly high amongst fauna associated with the epiphytes, and this was explained as being advantageous for species living in a habitat of limited duration (Norderhaug et al., 2002). Less mobile components, such as the holdfast fauna, are likely to recover more slowly.

The strictly sessile fauna colonise new habitats only by their pelagic larvae, mainly in a short settlement season. Mobile animals associated with kelp may reach new habitats in several ways. They may disperse as adults or juveniles by swimming, crawling, or drifting. In addition to active mobility among kelp fauna (Norderhaug et al., 2002; Jørgensen & Christie, 2003), drifting has been observed among the most abundant gastropods (Southgate, 1982; Vahl, 1983; Martel & Diefenbach, 1993; Jørgensen & Christie, 2003) and post-settlement drifting among bivalves (Bayne, 1964; Norderhaug et al., 2002). In tube-building amphipods and slow-moving tanaids (Johnson & Attramadal, 1982), which reproduce by brooding, ability to disperse as adults or juveniles is particularly important for reaching new habitats. An alternative to colonise new habitats in cleared areas is to retain part of the population in other habitats (refugia/alternative sources) that may be nearer than the kelp itself, and to spread to new kelp habitats from these.

The aim of this study was to investigate the short-term dispersal of kelp fauna to areas recently cleared by kelp trawlers. We wanted to see whether dispersal, measured as the colonization rate on artificial substrata at realistic distances from undisturbed kelp forests, was a significant factor in the slow recovery of the original kelp fauna community in such disturbed areas. We wanted to determine whether the fauna dispersed only from the undisturbed kelp forest or whether there were alternative habitats (refugia) within the cleared areas from which dispersal could take place. We also investigated the relative mobility of adult and juvenile amphipods (brooders). Contrary to most species with a planktonic larval stage, amphipods reproduce several times a year (Tully & O'Ceidigh, 1986; Costa & Costa, 1999). As the main season for planctonic larval settlement is in July (Thorson, 1946), juvenile amphipods occur over longer time periods than the planktonic stages. As we wanted to study less well-known dispersal abilities in mobile animals, the study took place after larval settlement to avoid settlement interfering with active juvenile postlarval or adult dispersal.

Table 1. The results of the ANOVA tests comparing numbers of individuals on the artificial substrata at different sites in the long-term term experiment. The most numerous species or taxa are shown. Except for *Lacuna vincta* and *Rissoa parva* (square-root transformed), the data are not transformed. The degrees of freedom are 8. Significant differences are marked with an asterisk *

Species/taxa	*F*	*p*
Mytilus edulis	7.397	0.024*
Musculus discors	1.401	0.317
Turtonia minuta	3.478	0.099
Rissoa parva	3.189	0.080
Lacuna vincta	1.831	0.239
Bittum reticulatum	2.791	0.139
Margarites spp.	1.415	0.314
Stenothoidae	0.090	0.915
Apherusa ssp.	10.251	0.012*
Caprellidea	5.290	0.047*
Corophium bonnellii	0.877	0.463
Aoridae	5.739	0.040*
Ischyroceridae	1.567	0.284

Materials and methods

Study area

A small bay at about 60° N in the Norwegian west coast archipelago (Langøyvågen outside Bergen) was chosen as the study area. The bay is exposed, facing towards the North Sea. Before the manipulation, the entire bay had a dense natural kelp forest with a canopy height of about 125 cm. With the exception of the sloping shorelines and some large rocks in one area of the bay, the bottom consisted of a smooth rocky surface at a depth of 7–9 m. At the start of this study (22 and 23 September 1997), the northern half of the bay was cleared of kelp by harvesting (kelp trawling). The trawled part measured approximately 4800 m^2 and was about 40 m wide.

Artificial substrata

Untwisted hemp rope was used to mimic kelp habitats (holdfasts and epiphytic algae). Studies have demonstrated that hemp rope bundles are useful in studies of colonization by fauna from red, green and brown algae (Kraufvelin et al., 2002), and fauna from *L. hypereborea* holdfasts and epiphytes (Norderhaug et al., 2002). Because the mimics were not stable in the exposed bay, the design was changed during the experiment. In the original design used in September, the hemp was filled inside a plastic funnel and stabilised by weights placed on top of the funnel (see Norderhaug et al. (2002) for further description). These mimics were placed on the bottom. As the weather conditions became harsher, we changed the design of the artificial substratum to make the units more stable. In November, the funnel was replaced with a plastic bottle anchored by weights fastened to the bottom of the bottle. For the longer-term experiment, the bundle of hemp rope was attached (tied) to weights placed on the bottom and secured with guy ropes. Slits were cut in the funnel and the bottle to allow animals to pass freely in and out. The displacement volume of the rope was 450 cm^3 for the mimics used in September (funnel) and in the long-term exposure experiment, and 300 cm^3 in the November experiment (bottle).

Study design

Three experiments were performed. Two short-term (3-day) experiments and one longer-term (35-d) were carried out to detect differences in dispersal rates. The first 3-day experiment was performed in late summer (23–26 September 1997); the second 3-day experiment from 3 to 6 November and was intended to illustrate early winter conditions; and the longer-term exposure experiment was performed from 28 September to 3 November.

In each experiment, artificial substrata were placed at three different sites. Site 1 was located inside the kelp forest a few metres from the cleared part of the bay. Site 2 was in the cleared area, 10 m from the kelp forest. The bottom here was dominated by large rocks and a few kelps were left after trawling. Site 3 was also in the cleared area, 25 m out from the untrawled kelp area. It was completely cleared of macroscopic kelp plants, but the nearest kelp vegetation (left after trawling) was within 10 m. The size of the cleared area in this study was comparable with what has been observed in other trawled areas, resulting in a fragmented kelp forest rather than large cleared areas (Christie et al., 1998).

At each site, five replicate kelp mimics were exposed for the short-term experiments and three replicates were used for the longer-term experiment. The distance between replicates was 2 m. At each sampling, the artificial substrata were sealed in separate plastic bags by SCUBA divers. Substrata were washed and sieved (250 μm) to retain invertebrates, which were preserved and later identified and counted. Each individual was identified to species or higher taxonomic level (see Appendix 1). Brooders (amphipods) were classified as adults or juveniles (not developed full species-specific characters). Meiofauna (copepods, ostracods, nematods) and colony-forming sessile species (hydroids, bryozoans) were abundant but not counted; sessile species were not included as they do not have the ability to disperse actively as adults.

To assess the species composition and abundance of the natural kelp forest fauna, three kelp plants were sampled in the area to be cleared just before harvesting (September). In addition, three kelp plants were sampled from the untrawled area at the start of the experiment (September) and at the end of the experiments (November). A cylindrical fine-mesh net was closed around each stipe after the lamina was removed. The holdfast was loosened carefully with a knife and sealed separately in a bag. Fauna associated with the kelp plants was treated as described for the artificial substrata. Fauna of the kelp plants from this study is included in Christie et al. (2003).

For the study of alternative habitats (refugia/sources) of kelp fauna, three replicate samples were taken from gravel (pebbles, about the same volume as a kelp holdfast) close to site 2, and of old remaining holdfasts within the trawled area, close to site 3. This sampling was repeated in September and November.

Statistical analysis

Differences in colonization of the faunal components between the three sites were tested using one way ANOVA. Prior to analysis, all data were checked for normal distribution and transformed in the few instances where the distribution was not normal. Tukey's test was used to identify the samples that differed when the ANOVA test was significant. To identify patterns of similarity between samples, SIMPER and MDS (Multi Dimensional Scaling) (Shepard, 1962; Kruskal, 1964 a,b) ordinations, based on similarity matrices, were calculated using the PRIMER (Plymouth Routines In Multivariate Ecological Research, 5.2.1) computer package. The Bray-Curtis similarity index was used (Bray & Curtis, 1957) and all data were square-root transformed.

As differences in colonization rate between different taxonomical groups may be due to differences in occurrence in the natural kelp plants, and not caused by real differences in active dispersal abilities, we tested whether observed numbers of the various species or faunal groups in the artificial substrata differed from the number expected on the basis of their relative abundance in the natural kelp plants using Monte Carlo simulations to achieve 95% confidence levels of the expected numbers in each group. The confidence intervals were based on simulations replicated 200 times, and the number of individuals in each replicate equalled the average total number of observed individuals in the artificial substrata. Each individual was put randomly into a group according to the probability given by the relative frequency of the groups. When the observed number of a group was higher (alternatively lower) than the upper 95% confidence level, we concluded that the group was a faster (alternatively slower) coloniser than expected. The observed number divided by the expected number was used as a measure of the relative mobility (rel.m) or activity of the group.

Results

A total of 50 199 individuals from 183 species/taxa was identified. Of these, 17 species were sessile (mainly Cnidaria, the polychaete family Serpulidae, Bryozoa, and Ascidiacea). Appendix 1 shows the 166 mobile taxa (in total 49 864 individuals) and their numerical occurrence (with exception of four meiofauna groups) in the kelps and the different substrata (the artificial kelps, the remaining holdfasts and the pebble substratum). In total, 128 mobile taxa were found in the kelp plants and 111 (86.7%) of these colonised the artificial substrata during the three experiments. Between 79 and 96% of the species/taxa found in each experiment or substratum sample were species associated with kelp (Appendix 1). The occurrence of 82% of the species found in kelp plants in the long-term experiment compared to the 69% in the remaining holdfasts, further underlined how good the rope mimics were as supplements for the natural kelp plants with regard to the mobile part of the kelp fauna. On all types of substratum, amphipods and gastropods dominated in terms of number of species and number of individuals.

Short-term experiments

The artificial substrata were colonized rapidly by large numbers of species and individuals; 106 species and

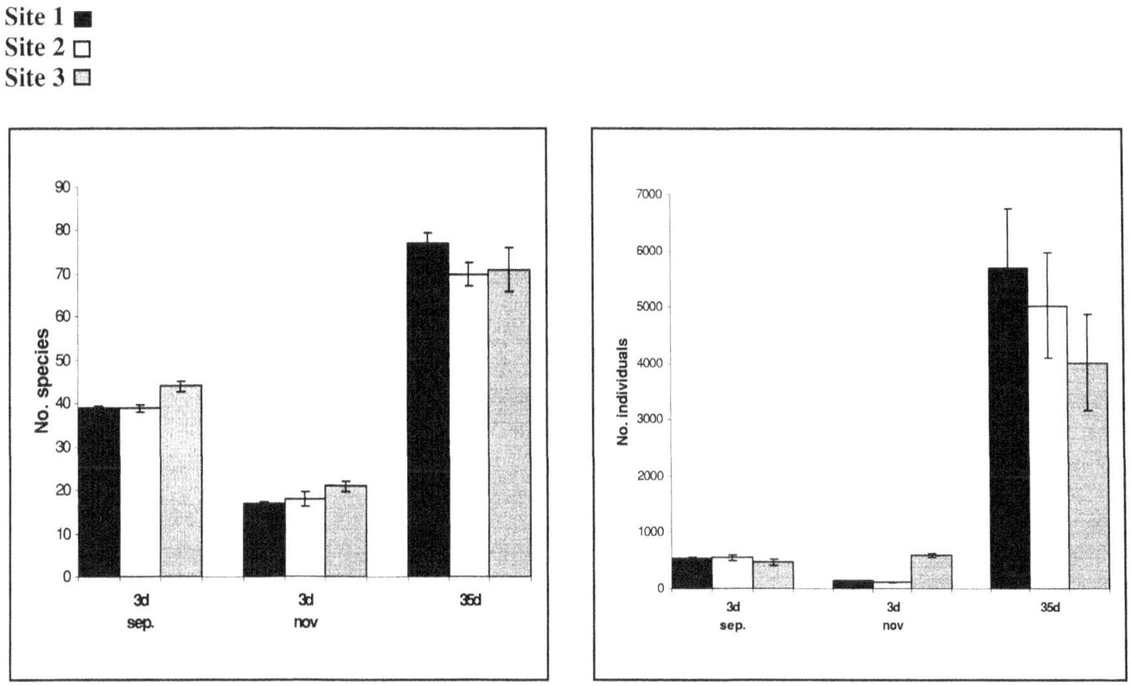

Figure 1. Colonization of artificial holdfasts in undisturbed kelp forest (Site 1, black), and at sites 2 (white) and 3 (grey) in the kelp trawled area, measured as number of individuals and number of species. Standard error bars are shown.

9801 individuals (a mean of 653 individuals per artificial substratum) were identified after 3 days' exposure in September, and 52 species and 5047 individuals (a mean of 336 individuals per artificial substratum) in November. Amphipods and gastropods were the dominant groups (Appendix 1). The genus *Apherusa* accounted for 80% of the total number of amphipods in September and 79% in November. *Athylus swammerdami* accounted for 1% of the total in September and 16% in November. *Rissoa parva* was the dominant gastropod and accounted for 83 and 90% of the total number of gastropods in September and November, respectively. Mussels (primarily *Turtonia minuta, Musculus discors* and juvenile *Mytilus edulis*), isopods and decapods were also abundant on the artificial substrata. Small numbers of polychaetes and tanaids colonized the 3-day mimics.

No reduction in the rate of colonization of the mimics, either in number of species or in number of individuals, could be seen from site 1 (within the kelp forest) to sites 2 and 3 at increasing distances from the edge of the trawled area (Fig. 1). In September, no significant differences were found between sites in the number of species and individuals (F=1.32, p=0.303; F=1.318, p=0.304, respectively), however, in November, the number of individuals and species were higher at site 3 than at sites 1 and 2 (ANOVA F=110.884, p=0.0001, F=155.128, p=0.0001, followed by Tukey's test). The difference in the number of individuals was explained mainly by the large number of amphipods, primarily *Athylus swammerdami*, found only at site 3.

Long-term experiment

Numbers of individuals and species increased with the length of exposure period (Fig. 1, Appendix 1). Almost 2500 individuals per substratum, and a total of 21 933 individuals and 128 species/taxa, were found on the artificial substrata after 35 days. There were no significant differences between sites in number of species (p=0.346) or total abundance (p=0.494). The longer exposure period led to a closer resemblance to the fauna found on natural kelp holdfast, as demonstrated in the ordination plot (Fig. 3). The longer period of exposure also resulted in higher diversity. Seven species accounted for 48% of the total number of individuals after 3 days exposure in September, whereas 13 species accounted for 50% after 35 days exposure, and 14 species accounted for 49% of the total in the natural holdfasts.

Table 2. Confidence intervals (95% c.i.) for expected number of individuals in the various taxa on the basis of their relative abundance in the natural kelp plants, and for three different exposure periods (3 days (September and November) and 35 days). Relative mobility (rel. mob) >1 indicates that the group colonises faster than expected on the basis of its relative frequency in the natural kelp plants

Taxa	Site 1			Site 2			Site 3		
	95% c.i.	Obs.	Rel. Mob.	95% c.i.	Obs.	Rel. Mob.	95% c.i.	Obs.	Rel. Mob.
3 days' exposure, September									
Gastropoda	185–224	187.6	0.92	189–227	290.4	1.4	159–191	165.2	0.94
Amphipoda	105–134	291.6	2.4	105–137	170	1.4	89–120	271.8	2.6
Bivalvia	133–169	20.8	0.14	135–167	31.4	0.21	113–143	18.2	0.14
Polychaeta	34–55	0.2	0.005	35–57	0.2	0.005	29–46	2	0.053
Decapoda	2–9	30.2	5.5	2–10	25.2	4.5	1–8	6.6	1.4
Isopoda	2–9	1.4	0.25	2–10	3.6	0.64	1–9	0.8	0.17
Tanaidacea	2–10	0	0	2–9	0.2	0.031	2–8	0	0
Others	10–24	19.2	1.2	10–25	40.6	2.4	9–20	8	0.56
(sum)		(551)			(561)			(473)	
3 days' exposure, November									
Gastropoda	37–56	31	0.69	30–46	26.8	0.72	171–212	61	0.32
Amphipoda	31–50	102.2	2.6	26–42	80.6	2.5	150–185	530.8	3.2
Bivalvia	27–45	2.8	0.076	24–37	4	0.13	138–173	3.4	0.022
Polychaeta	4–13	0	0	3–12	0.2	0.028	28–48	0.2	0.006
Decapoda	0–3	1.8	1.3	0–3	0.4	0.34	2–10	0	0
Isopoda	0–3	0.2	0.14	0–3	0.8	0.68	3–11	2.8	0.47
Tanaidacea	1–7	0	0	1–6	0	0	11–26	0	0
Others	1–7	3	0.71	1–7	3.8	1.1	12–25	1.4	0.078
(sum)		(141)			(117)			(600)	
35 days' exposure									
Gastropoda	974–1067	1322	1.3	870–950	1277.7	1.4	689–760	535.3	0.74
Amphipoda	645–723	1174.3	1.7	566-630	1004	1.7	454–512	1253.3	2.6
Bivalvia	708–782	181	0.24	617–619	124.7	0.19	490–556	97	0.19
Polychaeta	177–221	33.3	0.17	155–197	23.6	0.13	123–159	17.7	0.13
Decapoda	19–37	15	0.53	18–33	25	0.99	12–27	12.7	0.63
Isopoda	21–37	18.7	0.66	18–32	12	0.48	13–18	37.3	1.9
Tanaidacea	44–71	7.33	0.13	39–62	4	0.08	30–50	5	0.13
Others	71–103	101	1.2	61–90	46	0.61	48–72	53.7	0.89
(sum)		(2845)			(2513)			(2007)	

There were no significant differences between the sites in number of gastropods (F=3.868, p=0.083) or amphipods (F=0.377, p=0.701). Furthermore, there were no significant differences between the sites for the four most abundant gastropod species (Table 1). *Rissoa parva* was the dominant gastropod and accounted for 90% of the total number of gastropods after 35 days. Amongst the amphipods, *Apherusa* spp. and Stenothoidae accounted for 31 and 28% of the total number, respectively. Of the most abundant amphipods, *Apherusa* spp. occurred in significant higher density at site 3, whereas the families Caprelidae and Aorida showed a significant decline in numbers from the site within the kelp forest and with increasing distance into the cleared area (Table 1). *Jassa falcata*, which accounted for more than 10% of the amphipods found on the natural kelp plants, accounted for less than 1% in the artificial substrata after 35 days. Mussels were found in large numbers on the natural kelp plant, but only in small numbers after 35 days. The dominant bivalves were *Turtonia minuta, Musculus discors* and juvenile *Mytilus edulis*. The latter showed lower density in the cleared area than inside kelp (Table 1).

Relative mobility

Although colonization was rapid and a diverse community was found on the artificial substrata after a short exposure period, differences in mobility between the different fauna groups emerged (Table 2). The confidence intervals achieved by the Monte Carlo simulations showed that, in general, gastropods and amphipods colonised the artificial substrata faster than or as fast as expected in both seasons, regardless of the time period of exposure (except in November for gastropods). Mussels, polychaetes, isopods and tanaids were slower colonisers than expected in all experiments. Decapods showed a more variable pattern and were found to be faster colonisers than expected in September, slower than expected in November and as expected in the 35-day experiment.

There was a higher proportion of adult than juvenile amphipods both on the natural kelp plants and on the artificial substrata, except in November at site 2 (Fig. 2). However, the Monte Carlo simulations indicated that relative mobility was higher for the juveniles than for the adults (Table 3).

Alternative habitats/refugia

Faunal species common on kelp were found in both pebble substrata and remaining holdfasts collected from the trawled area. Fauna in the remaining holdfasts included a larger number of species and had more species in common with the natural kelp plants than the fauna found in the pebbles (91 and 83%, respectively). These similarities are illustrated in the MDS plot (Fig. 3), where the samples of remaining holdfasts and holdfasts from undamaged kelp plants are grouped together, whereas samples of the pebbles are grouped further away from the natural kelp holdfasts. SIMPER analysis supports this, and shows that the remaining holdfasts were more similar (63% in September, 60% in November) to the undisturbed kelp holdfasts than the fauna collected in the pebbles (39% in both September and November). There were fewer crustaceans and polychaetes, and larger numbers of gastropods in the pebbles than in the remaining and undisturbed holdfasts.

Discussion

The fauna associated with *L. hyperborea* is highly mobile and showed a high dispersal rate to both sites in the trawled area. This was demonstrated by the large number of both individuals and species in the artificial substrata exposed in the trawled area for only 3 days, and underlined by the results that substrata in the cleared area were colonised as fast as those placed within the kelp forest. The rapid colonization of the artificial holdfasts within the kelp forest agrees with the results of another investigation from a Norwegian kelp forest (Norderhaug et al., 2002). However, a major difference between this study and Norderhaug et al. (2002) is the greater distance to the nearest kelp plant in this study (about 10 m compared to less than 0.3 m in Norderhaug et al., 2002). The present study shows that most species of the kelp fauna are able to disperse rapidly across the trawled tracks made in harvested areas. Dispersal of fauna from macrophytes has also been reported for kelp (Jørgensen & Christie, 2003) and other seaweed species and sea grasses (Virnstein & Curran, 1986; Edgar, 1991).

Despite the high dispersal rate of most species, dispersal ability varied from one taxonomic group to another. This was demonstrated by: (1) significant increase in numbers of both species and individuals after an increase in exposure time from 3 to 35 days, (2) differences in relative mobility revealed by the Monte Carlo simulations, and (3) the fact that species composition in the artificial holdfasts became more similar to that in the natural kelp holdfasts over time. Most fauna groups were present after 35 days, and new substrata may be colonised regardless of season and reproductive strategy.

Amphipods and gastropods were the dominant faunal groups, as has been found in natural kelp forest (Moore, 1972, 1973; Schultze et al., 1990; Christie et al., 1998, 2003; Norderhaug et al., 2002). In addition, both these taxa colonised the artificial substrata more quickly than expected on the basis of their relative abundance in the natural kelp plants. The high mobility of amphipods has been demonstrated earlier (Edgar & Aoki, 1993; Norderhaug et al., 2002), but differences in mobility were also found between the various amphipod species. *Apherusa* spp., characterised as 'epiphyte species' (Norderhaug et al., 2002), and hence as a fast coloniser, dominated after 3 days of exposure. In contrast, a 'holdfast species' such as *Corophium bonnellii* was characterised as slow moving and increased in abundance with increasing time

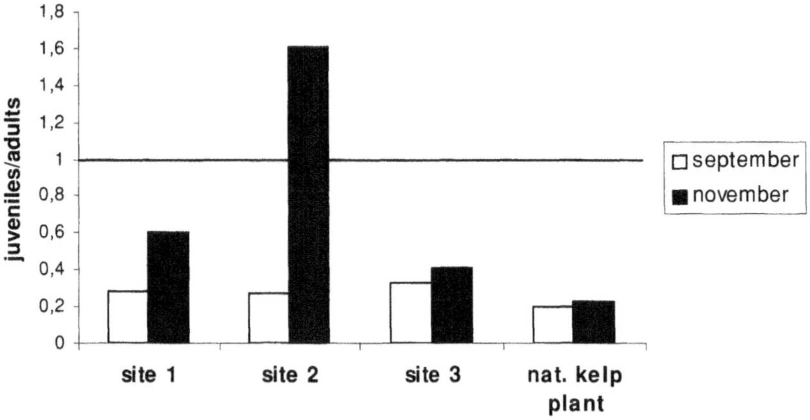

Figure 2. The proportions of juvenile and adult amphipods on the artificial substrata and on natural kelp plants in September and November. Values >1 indicate the dominance of juvenile amphipods, and values <1 indicates the dominance of adult amphipods.

Table 3. Confidence intervals, (95% c.i.) for expected number of amphipods in the various taxa on the basis of their relative abundance in the natural kelp plants, and for two different exposure periods (3 days September and November). Relative mobility (rel. mob) >1 indicate that the group colonises faster than expected on the basis of its relative frequency in the natural kelp plants

3 days' exposure	Site 1			Site 2			Site 3		
	95% c.i.	Obs.	Rel. Mob.	95% c.i.	Obs.	Rel. Mob.	95% c.i.	Obs.	Rel. Mob.
September									
Adults	233–252	229.6	0.96	133–149	135.8	0.97	214–235	207.2	0.93
Juveniles	40–59	62	1.2	21–37	34.2	1.1	37–58	65	1.3
(sum)		(292)			(170)			(272)	
November									
Adults	80–92	66.6	0.79	63–72	31.2	0.46	431–458	376.6	0.81
Juveniles	10–22	35.6	2.1	9–18	49.4	3.6	73–100	154.2	1.7
(sum)		(102)			(81)			(531)	

of exposure. Similar differences in mobility of amphipod species have also been demonstrated previously (Virnstein & Curran, 1986; Pavia et al., 1999). *Athylus swammerdami* appeared on the artificial substrata in large numbers, especially in November, and was found in small numbers on the natural kelp plants. This species is more often found on sandy bottom (Lincoln, 1979), rather than being closely associated with living kelp plants. The present study suggests that amphipods and isopods may, on the whole, be less vulnerable to plant destruction than earlier presumed (Lessios et al., 1994; Christie et al., 1998). Mussels, polychaetes, isopods and tanaids were slower colonisers than expected, but even the tanaids colonised the mimics after 35 days, despite a very low expected dispersal rate (Johnson & Attramadal, 1982). The high dispersal rate of the gastropods in this study was also observed by Norderhaug et al. (2002), and is, probably, a consequence of their ability to drift in the water column both as juveniles and as adults (Southgate, 1982; Vahl, 1983; Martel & Chia, 1991; Martel & Diefenbach, 1993; Jørgensen & Christie, 2003).

Observed differences in mobility may be more important for species with direct development than for species with pelagic larvae. The latter have the advantage of being able to colonise new or cleared areas in each settlement period (mainly around July, Thorson, 1946), whereas species with direct development are dependent on the dispersal abilities of juveniles and adults (Christie et al., 1998). Lessios et al. (1994) concluded that animals with direct development have a limited ability to spread. This present study indicates that a large part of the kelp fauna (the amphipods) spread as both juveniles and adults.

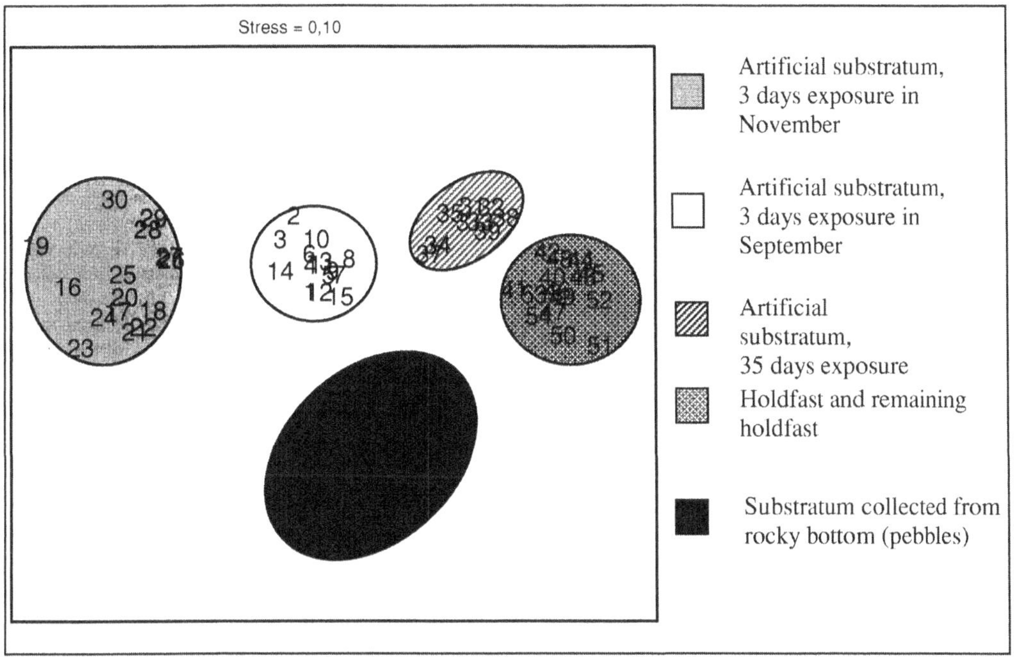

Figure 3. MDS ordination based on similarity in the species composition of the artificial substrata (short- and long-term exposure), undisturbed and remaining kelp holdfasts, and pebble substratum collected from the rocky bottom.

Hence, they probably have the opportunity to colonise available substrata throughout the year.

Furthermore, the present study has demonstrated that the kelp fauna has several alternative habitats or refuges other than the kelp plants themselves, and can spread from these to the recovering kelp forest in trawled areas. The relative importance of these sources will depend probably on factors such as topography and degree of exposure to wave action. In areas that are topographically complex, more holdfasts are likely to be left in the trawled tracks than in areas where the bottom is smooth. The remaining holdfasts were the best refugia or alternative habitat in this study, as they contained a fauna very similar in composition to the fauna associated with the natural kelp plants. The fauna in the holdfasts included many species with low mobility (Norderhaug et al., 2002). On the basis of the species found in the remaining holdfasts and in the pebble samples, amphipods would be most likely to spread from the remaining holdfasts and gastropods from the pebbles. Colonization of cleared areas with such species-rich refuges will not be dependent solely on recruitment from the nearest kelp plant. The occurrence of such refuges will be especially important when large areas have been cleared of kelp and the distance to the nearest kelp plant will be less important. However, the remaining holdfasts only persist for a limited time (Vasquez, 1993) and the pebbles contained neither all that many individuals nor species associated with kelp.

The diversity in the kelp mimics increased with time, as found by Dean & Connell (1987). This may be partly an effect of randomness, since the probability of a new species encountering a mimic will increase as time goes by. However, increase in diversity may also be due to the observed differences in dispersal ability (mobility) of the fauna. Another explanation is that the variety of niches in artificial substrata increases with time as microalgae grow on them (Edgar, 1991), and as the supply of nutrients and POM increases. Natural holdfasts act as sediment traps (Moore, 1972) and the artificial holdfast may do so too. Microalgae and POM are important food resources for invertebrate kelp fauna (Duggins & Eckman, 1994, 1997). Although a large proportion of the species, and all species groups of the fauna associated with the natural kelp plants, were found on the artificial substrata in this relatively short-term experiment, some species associated with kelp were missing. The missing species were among those that were less abundant on the natural kelp plants. Their absence on the mimics might be explained by a lack of certain habitat properties in the artificial substrata (compared to mature kelp plants) or low dispersal ability; however, the low numbers of these species may mean that they have less chance encountering a mimic.

After pulse disturbances such as kelp trawling, the kelp forest faunal community shows a certain ability to recover quickly. Thus the fauna shows little vulnerability to such disturbances. In addition to the dispersal ability (mobility) found in this study, many species have pelagic larvae that colonise new habitats efficiently. The incomplete recovery of many species associated with the kelp plants, as found for many crustaceans by Christie et al. (1998), seems not to be explained by low dispersal ability. A more likely explanation is incomplete recovery of the habitat. Christie et al. (1998) found that kelp plants less than 4 years old had smaller holdfasts and the stipe epiphytes were not fully developed. The most crucial factor for the establishment of an ecologically mature kelp forest community is probably sufficient time to allow kelp size to recover completely and a diverse epiphyte community to become established.

Acknowledgements

We are grateful to an unknown kelp trawler, FMC Biopolymer, and the Norwegian Institute of Marine Research for inviting us to take part in this field manipulation and for their support during the field work. We would also like to thank Kjell Magnus Norderhaug for his help with the numerical analysis.

References

Bayne, B. L., 1964. Primary and secondary settlement in Mytilus edulis L (Mollusca). J. anim. Ecol. 33: 513–523.

Bray, J. R. & J. T. Curtis, 1957. An ordination of the upland forest communities of southern Wisconsin. Ecol. Monogr. 27: 325–349.

Bustnes, J. O., H. Christie & S. H. Lorentsen, 1997. Seabirds, kelp beds and kelp trawling: a summary of knowledge. NINA Oppdragsmelding 472: 1–43 (in Norwegian with English summary).

Christie, H., E. Rinde, S. Freriksen & A. Skadsheim, 1994. Ecological consequences of kelp trawling: reestablishment of kelp forest, epiphytes and holdfast fauna after kelp trawling at the Rogaland coast. NINA Oppdragsmelding 295: 1–29 (in Norwegian with English summary).

Christie, H., S. Fredriksen & E. Rinde, 1998. Regrowth of kelp and colonization of epiphyte and fauna community after kelp trawling at the coast of Norway. Hydrobiologia 375/376: 49–58.

Christie, H., N. M., Jørgensen, K. M. Norderhaug & E. Waage-Nielsen, 2003. Species distribution and habitat exploitation of fauna associated to kelp (*Laminaria hyperborea*) at the Norwegian coast. J. mar. biol. Ass. U.K. 83:687–699.

Costa, F. O. & M. H. Costa, 1999. Life history of the amphipod *Gammarus locusta* in the Sado estuary (Portugal). Acta Oecologica 20: 305–314.

Dean, R. L. & J. H. Connell, 1987. Marine invertebrates in algal succession. I. Variations in abundance and diversity with succession. J. exp. mar. biol. Ecol. 109: 196–215.

Duggins, O. D. & J. E. Eckman, 1994. The role of kelp detritus in the growth of benthic suspension feeders in an understory kelp forest. J. exp. mar. biol. Ecol. 176: 53–68.

Duggins, D. O. & J. E. Eckman, 1997. Is kelp detritus a good food for suspension feeders? Effects of kelp species, age and secondary metabolites. Mar. Biol. 128: 489–495.

Edgar, G. J., 1991. Artificial algae as habitats for mobile epifauna: factors effecting colonization in japanese Sargassum bed. Hydrobiologia 226: 111–118.

Edgar, G. J. & M. Aoki, 1993. Resource limitation and fish predation: their importance to mobile epifauna associated with Japanese Sargassum. Oecologia 95: 122–133.

Edwards, A., 1980. Ecological studies of kelp, *Laminaria hyperborea*, and its associated fauna in south-west Ireland. Ophelia 19: 47–60.

Foster-Smith, R. L. & R. O. Shillaker, 1977. Tube irrigation by *Lembos websteri* Bate and *Corophium bonelli* Milne Edwards (Crustacea: Amphipoda). J. exp. mar. biol. Ecol. 26: 289–296.

Hagen, N. T., 1983. Destructive grazing of kelp beds by sea urchins in Vestfjorden, northern Norway. Sarsia 68: 177–190.

Holmlund, M. B., C. H. Peterson & M. E. Hay, 1990. Does algal morphology affect amphipod susceptibility to fish predation? J. exp. mar. biol. Ecol. 139: 65–83.

Jensen. A., 1999. The seaweed resources of Norway. In Critchley, A. T. & M. Ohno (eds), Seaweed Resources of the World. Japan International Cooperation Agency: 200–209.

Johnson, B. J. & Y. G. Attramadal, 1982. A functional-morphological model of *Tanais cavolinii* Milne-Edwards (Crustacea, Tanaidacea) adapted to a tubicolous life-strategy. Sarsia 67: 29–42.

Jones, D. J., 1971. Ecological studies on marine invertebrate populations associated with polluted kelp forests in the North Sea. Helgoländer wiss. Meeresunters. 22: 417–441.

Jørgensen, N. M. & H. Christie, 2003. Diurnal, horizontal and vertical dispersal of kelp-associated fauna. Hydrobiologia 503 (Dev. Hydrobiol. 174): 69–76.

Kain, J. M., 1963. Aspects of the biology of *Laminaria hyperborea*. II. Age, weight and length. J. mar. biol. ass. U.K. 43: 129–151.

Kennelly, S. J., 1983. An experimental approach to the study of factors affecting algal colonization in a sublittoral kelp forest. J. exp. mar. Biol. Ecol. 68: 257–276.

Kraufvelin, P., H. Christie & M. Olsen, 2002. Macrofauna (secondary) responses to experimental nutrient addition to rocky shore mesocosms and a coastal lagoon. Hydrobiologia 484: 149–166.

Kruskal, J. B., 1964a. Multidimensional scaling by optimising goodness-of-fit to a nonmetric hypothesis. Psychometrica 29: 1–28.

Kruskal, J. B., 1964b. Nonmetric multidimensional scaling: a numerical method. Psychometrica 29: 115–129.

Lessios, H. A., J. R. Weinberg & V. R. Starczak, 1994. Temporal variation in the populations of the marine isopod *Excirolana* – how stable are gene frequencies and morphology? Evolution 48: 549–563.

Lincoln, R. J., 1979. British marine Amphipoda: Gammaridea. Brittish Museum (Natural History), London: 1–657.

Mann, K. H., 1972. Ecological energetics of the seaweed zone in a marine bay on the Atlantic coast of Canada. II. Productivity of the seaweeds. Mar. Biol. 14: 199–209.

Martel, A. & F. S. Chia, 1991. Drifting and dispersal of small bivalves and gastropods with direct development. J. exp. mar. biol. Ecol. 150: 131–147.
Martel, A. & T. Diefenbach, 1993. Effect of body size, water current and microhabiata on the mucous-thread drifting in post-metamorphic gastropods *Lacuna* spp. Mar. Ecol. Prog. Ser. 99: 215–220.
Moore, P. G., 1972. The kelp fauna of Northeast Britain. I. Introduction and the physical environment. J. exp. mar. biol. Ecol. 13: 97–125.
Moore, P. G., 1973. The kelp fauna of Northeast Britain. II. Multivariate classification: Turbidity as an ecological factor. J. exp. mar. biol. Ecol. 13: 127–163.
Moore, P. G., 1986. Levels of heterogeneity and the amphipod fauna of kelp holdfast. In Moore, P. G. & R. Seed (eds), The Ecology of Rocky Coasts. Colombia University Press, New York: 274–290.
Nelson, W. G., 1979. Experimental studies of selective predation on amphipods: Consequences for amphipod distribution and abundance. J. exp. mar. biol. Ecol. 38: 225–245.
Nordeide, J. T. & J. H. Fosså, 1992. Diet overlap between to subsequent year-classes of juvenile coastal cod (*Gadus morhua* L.) and the wild and reared cod. Sarsia 77: 111–117.
Norderhaug, K. M., H. Christie & E. Rinde, 2002. Colonization of kelp imitations by epiphyte and holdfast fauna; a study of mobility patterns. Mar. Biol. 141: 965–973.
Pavia, H., H. Carr & P. Åberg, 1999. Habitat and feeding preferences of crustacean mesoherbivores inhabiting the brown seaweed *Ascophyllum nodosum* (L.) Le Jol. and its epiphytic macroalgae. J. exp. mar. biol. Ecol. 236: 15–32.
Rinde, E., H. Christie, S. Fredriksen & A. Sivertsen, 1992. Ecological consequences of kelp trawling: Importance of the structure of the kelp forest for abundance of fauna in the kelp holdfasts, benthic fauna and epiphytes. NINA Oppdragsmelding 127: 1–37 (in Norwegian with English summary).
Shepard, R. N., 1962. The analysis of proximities: multidimensional scaling with an unknown distance function 2. Psychometrica 27: 219–246.
Shillaker, R. O. & P. G. Moore, 1987. The feeding habits of the amphipods *Lembos websteri* Bate and *Corophium bonelli* Milne Edwards. J. exp. mar. biol. Ecol. 110: 93–112.
Schultze, K., K. Janke, A. Krüß & W. Weidemann, 1990. The macrofauna and macroflora associated with *Laminaria digitata* and *L. hyperborea* at the island of Helgoland (German bight, North sea). Helgoländer wiss Meeresunters. 44: 39–51.
Sjøtun, K., S. Fredriksen, T. E. Lein, J. Rueness & K. Sivertsen, 1993. Population studies of *Laminaria hyperborea* from its northern range of distribution in Norway. Hydrobiologia 261: 215–221.
Skadsheim, A., H. Christie & H. P. Leinaas, 1995. Population reductions of *Strongylocentrotus droebachienis* (Echinodermata) in Norway and the distribution of its endoparasite *Echinomermella matsi* (Nematoda). Mar. Ecol. Prog. Ser. 119: 199–209.
Southgate, T., 1982. Studies on intertidal population of *Rissoa parva* in Southwest Ireland. J. Nat. Hist. 16: 183–194.
Thorson, G., 1946. Reproduction and larval development of Danish marine bottom invertebrates. Meddelelser fra kommisjonen for Danmarks fiskeri- og havundersøgelser. 4: 1–523.
Tully O. & P. O'Ceidigh, 1986. The ecology of *Idotea* species (Isopoda) and *Gammarus locusta* (Amphipoda) on surface driftweed in Galway bay (west of Ireland). J. mar. biol. Ass. U.K. 66: 931–942.
Ulrich, I., K. Anger & U. Schöttler, 1995. Tube-building in two epifaunal amphipods species, *Corophium insidiosum* and *Jassa falcata*. Helgoländer wiss Meeresunters. 49: 393–398.
Vahl, O., 1983. Mucus drifting in the limpet *Helcion* (=*Patina*) *pellucida* (Prosobranchia, Patellidae). Sarsia 68: 209–211.
Vasquez, J. A., 1993. Effects on the animal community of dislodgement of holdfasts of *Macrocystis pyrifera*. Pacific Sci. 47: 180–184.
Virnstein, R. W. & M. C. Curran, 1986. Colonization of artificial seagrass versus time and distance from source. Mar. Ecol. Prog. Ser. 29: 279–288.
Whittick, A., 1983. Spatial and temporal distribution of dominant epiphytes on the stipes of *Laminaria hyperborea* (Gunn.) Fosl. (Phaeophyta: Laminariales) in S. E. Scotland. J. exp. mar. biol. Ecol. 73: 1–10.

Appendix 1. Number of individuals recorded (or present, x) in the different substrata of mobile species/taxa

Species/taxa	Kelp plants *n*=9	Artificial kelp 3 days September n=15	Artificial kelp 3 days November n=15	Artificial kelp 35 days November n=9	Remaining holdfast *n*=6	Pebble substrata *n*=6
Foraminifera indet	x	x	x	x	x	x
Plathyhelmintes indet	7	1	0	0	0	0
Nematoda indet	x	x	x	x	x	x
Nemertea indet	13	0	0	0	4	2
Oligochaetae indet	70	0	0	0	0	1
Enchytraidae, Cernosvitoviella	21	0	0	6	45	1
Polynoidae indet	45	1	0	36	20	6
Hesionidae indet	29	0	0	2	23	32
Neriomyra punctata (O F Muller)	9	0	0	0	0	0
Kefersteinia cirrata (Keferstein)	61	0	0	0	1	0
Nereidae indet	1	0	0	2	0	0
Nereis pelagica (L)	7	0	0	4	2	0
Platyneris dumerili (Audouin & Milne-Edwards)	2	0	0	0	2	0
Neanthes irroruta (Malmgren)	10	0	0	0	2	0
Lumbrinereidae indet	0	0	0	4	0	0
Phyllodoce macutta (L)	21	0	0	28	29	5
Phyllodocidae indet	47	1	0	18	4	1
Spionidae indet	0	1	0	0	0	0
Eunicidae indet	28	0	0	0	8	0
Terebellidae indet	2	0	0	0	0	0
Eusyllis sp.	3	0	0	0	0	0
Typosyllis spp.	65	0	0	28	34	5
Syllidae indet	138	3	1	3	15	1
Capitellidae indet	0	0	0	6	0	0
Fabricia stellaris (Blainville)	72	1	0	13	27	0
Polychaeta indet juv	223	4	1	77	62	5
Polyplacophora indet	2	0	0	0	3	0
Littorina saxatilis (Olivi)	0	3	1	3	0	1
Littorina obtusata (L)	13	0	0	8	0	15
Littorina mariae (Sacchi & Rastelli)	2	4	0	0	0	0
Littorina littorea (L)	2	1	0	24	0	5
Margarites helicinus (Fabricius)	21	14	1	96	6	1
Margarites groenlandicus (Gurelin)	0	0	0	0	0	1
Margarites striata	0	0	0	1	0	0
Margarites indet juv	72	172	15	156	16	11
Gibbula cineraria (L)	1	8	3	5	2	12
Cingula sp.	0	14	0	0	0	0
Lacuna parva (daCosta)	3	4	0	1	1	0
Lacuna vincta (Montagu)	390	217	51	70	13	29
Lacuna pallidula (daCosta)	0	1	0	0	0	0
Onoba semicostrata (Montagu)	412	20	3	135	178	75
Rissoa parva (daCosta)	1606	4095	884	8450	152	162
Rissoidae indet	5	24	2	11	0	1
Iodioluna phaseolina	0	8	0	0	0	0
Epitonium sp.	0	0	0	0	0	1
Alvania punctata (Montagu)	24	0	0	4	0	2

Continued on p.89

Appendix 1. Continued

Species/taxa	Kelp plants *n*=9	Artificial kelp 3 days September n=15	Artificial kelp 3 days November n=15	Artificial kelp 35 days November n=9	Remaining holdfast *n*=6	Pebble substrata *n*=6
Alvania beani (Hanley in Thorpe)	0	1	0	4	1	0
Skeneopsis planorbis (Fabricius)	1	6	0	11	0	14
Omalogyra sp	9	29	2	53	3	10
Velutina velutina (O F Muller)	0	0	0	1	0	1
Vitreolina philipii (Reureval & Ponzi)	49	13	0	32	33	5
Rhaphithoma leufroy (Michaud)	46	0	0	0	0	0
Bittium reticulatum (daCosta)	125	45	2	228	71	29
Crysalidia spiralis (Montagu)	3	9	0	4	1	1
Marshallora adversa (Montagu)	64	3	0	3	1	2
Hinia incrasata (Strom)	43	6	0	0	0	2
Helicon pellucidum (L)	4	3	0	0	0	1
Buccinum undatum (L)	3	1	0	1	0	2
Prosobranchia indet juv	2	10	2	24	3	1
Retusa obtusa (Montagu)	22	0	0	1	1	0
Retusa truncatula (Bruguriere)	17	0	1	6	12	0
Odostomia spp.	3	0	0	3	3	0
Onchidorididae indet	362	4	0	13	46	4
Aplysia punctata (Cuvier)	297	161	2	32	36	3
Nudibranchia indet	160	0	2	17	15	11
Mytilus edulis (L)	687	66	13	469	88	21
Modiolus modiolus (L)	60	17	5	77	30	1
Pectinidae indet	2	0	0	0	0	0
Cardidae indet	5	1	0	19	1	1
Musculus discors (L)	1512	107	14	457	291	7
Hiatella arctica (L)	376	1	0	8	111	6
Turtonia minuta (Fabricius)	98	159	24	156	18	3
Bivalvia indet juv	7	3	1	17	1	4
Nymphon sp.	2	1	0	0	0	0
Pycnogonidae indet	1	0	0	0	0	0
Copepoda harpacticoida	x	x	x	x	x	x
Ostracoda indet	x	x	x	x	x	x
Eualus pusiolus (Kroyer)	0	1	0	11	0	0
Hippolyte varians (Leach)	2	22	1	16	0	0
Athaens nitchens (Leach)	0	1	0	19	0	3
Thoralus cranchii (Leach)	2	223	8	46	0	1
Cardion steveni (Lebour)	0	0	0	2	0	0
Natantia indet	1	4	0	8	0	0
Galathea nexa (Embleton)	7	0	0	5	3	2
Galathea strigosa (L)	2	1	1	6	1	4
Galathea sp. juv.	11	15	0	11	6	3
Eyronome spinosa (Hailstone)	15	0	0	3	0	0
Primela denticilata (Montagu)	1	3	0	4	4	0
Macropodia rostrata (L)	2	0	0	2	0	0
Pisidia longicornis (L)	5	0	0	2	0	0
Pilumnus hirtellus (L)	3	0	0	3	3	0
Macropipus arcuatus (Leach)	2	6	0	13	1	3

Continued on p.90

Appendix 1. Continued

Species/taxa	Kelp plants *n*=9	Artificial kelp 3 days September n=15	Artificial kelp 3 days November n=15	Artificial kelp 35 days November n=9	Remaining holdfast *n*=6	Pebble substrata *n*=6
Paguridae indet	1	19	1	5	0	9
Decapoda indet juv	2	17	0	1	1	0
Praunus inermis (Rathke)	6	67	3	5	0	2
Praunus neglectus (G O Sars)	0	5	0	1	0	0
Mycidacea indet	1	11	0	2	0	0
Cumacea indet	38	77	10	168	11	6
Pseudoparatanais batei (G O Sars)	194	1	0	48	115	3
Tanais dulongi (Audouin)	0	0	0	1	0	0
Idotea baltica(Pallas)	1	0	0	44	0	0
Idotea pelagica (Leach)	1	4	21	73	0	1
Idotea granulosa (Rathke)	0	1	2	5	0	0
Idotea viridis (Slabber)	0	0	0	6	0	0
Janira maculosa (Leach)	25	0	0	9	6	0
Janiropsis breviremis (G O Sars)	2	0	0	6	1	0
Munna kroeyeri (Goodsir)	25	27	0	55	7	1
Isopoda indet juv	2	1	1	6	1	0
Tryphosa nana (G O Sars)	60	241	22	57	4	19
Tryphosa hoeringi (Boeck)	35	6	0	24	0	0
Lyseannasidae indet	86	4	0	0	0	0
Iphimedia minuta (G O Sars)	0	0	0	2	0	0
Iphimedia obesa (Boeck)	4	0	0	9	0	0
Amphilocus manudens (Bate)	5	3	0	1	0	0
Stenothoidae indet	351	87	30	2877	20	5
Calliopius laeviusculus (Kroyer)	7	0	6	57	0	0
Gammarella fuciola (Leach)	0	0	0	0	1	2
Cheirocratus sundevallii (Rathke)	0	0	0	9	8	16
Apherusa bispinosa (Bate)	152	1667	1240	2352	11	7
Apherusa jurinei (Milne-Edwards)	18	774	511	289	0	2
Apherusa juv	26	672	1106	541	3	2
Gammarellus homari (Fabricius)	4	15	1	2	2	0
Calliopiidae indet	0	1	0	2	0	0
Melitidae indet	0	0	0	4	0	0
Trieta gibbosa (Bate)	9	0	0	0	0	0
Dexamine spinosa (Montagu)	0	1	0	0	0	0
Dexamine thea (Boeck)	290	66	19	609	37	4
Atylus vedlomensis (Bate & Westwood)	0	0	0	11	0	0
Atylus swammerdami (Milne-Edwards)	4	5	900	16	0	0
Atylus sp.	0	15	6	2	0	0
Ampithoe rubricata (Montagu)	54	5	0	210	30	5
Aora typica (Kroyer)	10	2	0	82	2	0
Lembos websteri (Bate)	40	2	0	58	4	3
Microdeutopus sp.	9	0	0	0	3	1
Aoridae juv	33	65	5	385	3	4
Leptocheirus sp.	18	0	0	0	52	0
Corophium bonnellii (G O Sars)	377	1	0	273	109	6
Jassa falcata (Montagu)	265	15	1	73	14	2

Continued on p.91

Appendix 1. Continued

Species/taxa	Kelp plants *n*=9	Artificial kelp 3 days September n=15	Artificial kelp 3 days November n=15	Artificial kelp 35 days November n=9	Remaining holdfast *n*=6	Pebble substrata *n*=6
Ischyrocerus megalops (G O Sars)	0	2	0	0	0	0
Ischyrocerus anguipes (Kroyer)	69	2	1	62	11	0
Parajassa pelagica (Leach)	0	1	0	0	1	0
Ischyrceridae juv	40	14	2	179	6	3
Hyale nilssoni (Rathke)	0	2	0	21	0	0
Gammarus locusta (L)	17	0	0	50	5	0
Gammaridae juv	7	1	0	127	5	2
Stenopleustes sp.	0	0	0	1	0	0
Phthisica marina (Slabber)	31	22	0	12	8	0
Caprella septentrionalis (Kroyer)	6	1	0	23	1	0
Caprella linearis (L)	0	11	0	23	0	0
Caprella acantifera (Leach)	69	64	2	466	20	0
Caprella punctata (Boeck)	0	1	0	1	0	0
Aegina echinata (Boeck)	0	0	0	1	0	0
Caprellidae juv	47	1	0	366	20	1
Amphipoda indet juv	313	101	88	823	96	9
Chironomidae indet	4	0	0	3	1	0
Halacaridae indet	2	0	0	4	0	0
Henricia sanguinolenta (O F Muller)	2	0	0	0	1	0
Asterias rubens (L)	49	0	0	32	21	1
Martasterias glacialis (L)	0	1	0	0	0	0
Ophiopolis aquleata (L)	9	1	0	0	4	0
Amphipholis squamata (DelleChiaje)	57	25	1	219	17	31
Echinoidea juv	1	0	0	7	2	0
Spadella sp.	1	152	28	155	0	0
Liparis montagui (Donovan)	0	1	0	0	0	0
Centrolabrus rupestris (L)	0	1	0	0	0	0
Pholis gunnellus (L)	0	1	0	0	0	0
Total no. of individuals	10 319	9801	5047	21 933	2097	667
Total no. of species	128	106	52	128	92	80
No. species common with kelp		84	49	105	88	73

Hydrobiologia **503:** 93–102, 2003.
M.B. Jones, A. Ingólfsson, E. Ólafsson, G.V. Helgason, K. Gunnarsson & J. Svavarsson (eds), *Migrations and Dispersal of Marine Organisms*.

Regulation of species richness by advection and richness-dependent processes in a coastal fish community

Kyrre Lekve[1], Nils Chr. Stenseth[1,2] & Jakob Gjøsæter[2]
[1]*Department of Biology, Division of Zoology, University of Oslo, P.O. Box 1050, Blindern, N-0316 Oslo, Norway*
[2]*Institute of Marine Research, Flødevigen Marine Research Station, N-4817 His, Norway*
E-mail: n.c.stenseth@bio.uio.no

Key words: biodiversity, dispersal, NAO, Skagerrak, community ecology, top-down control, bottom-up control

Abstract

In coastal fish communities, both top-down and bottom-up processes influence the dynamics of individual species. Here, we demonstrate how advective processes in the coastal zone of the Skagerrak basin affect the main predator, the cod (*Gadus morhua*), as a bottom-up process, while gobies are influenced by top-down processes such as predation and competition. A model for the regulation of fish community dynamics, incorporating both bottom-up processes, corresponding to disturbance, and top-down processes are outlined. Wind stress may be seen as a proxy variable for wind-driven currents that results in dispersal of eggs and larvae and advection of prey species into coastal areas. About half of the fjords along the southern Norwegian Skagerrak coast are well fitted when wind stress is included as an environmental forcing factor in the period 1957–94. Along with richness-dependent regulation, the dynamics of fish-species richness are well predicted with the simple model presented.

Introduction

A recurrent challenge facing ecologists is to understand and predict the factors that determine structure in natural and managed communities (Menge et al., 1997). Due to strong variability (Hjort, 1914), this challenge is particularly demanding for marine ecologists. A pioneering approach towards improved understanding of marine communities was taken in the 1960s by Paine (1966), who demonstrated the effects of a keystone predator on sessile organisms on rocky shores, and by Connell (1961, 1978) who demonstrated the influence of interspecific competition. The experiment of Paine is a classic example of top-down control – where top-down control refers to situations where the structure (abundance, distribution, and/or diversity) of lower trophic levels depends directly or indirectly on activities of higher trophic levels (Menge, 1992). Along with the related concept of 'trophic cascades' (Polis et al., 2000), top-down control has been demonstrated for many systems (see Menge, 2000 for a review).

A different approach to the study of regulation of marine communities was taken by the Danish marine biologist, Gunnar Thorson, who observed that species with part of their life cycle as free-floating larvae display large fluctuations in abundance compared with otherwise similar one-phase species (Thorson, 1950: cited in Roughgarden et al., 1988). Thorson hypothesised that the larval phase was the most influential in determining marine population dynamics. This is an example of bottom-up control, which refers to direct or indirect dependence of community structure on factors producing variation at lower trophic levels or in their resources (Menge, 1992). As the larval stage is part of the plankton, we here interpret the larval stage as a lower trophic level in relation to dynamics of populations within the fjords. Often referred to as 'supply-side ecology' (Lewin, 1986), this approach to the study of regulation of communities gained increased attention in the 1980s, with theoretical developments (Roughgarden et al., 1988) and empirical demonstrations (see Menge, 2000 for examples). Top-down and bottom-up regulation may not be mutually exclusive; the relevant question is under what conditions bottom-up or top-down factors are important and what is the relative strength of each factor? (Menge, 1992; Sinclair et al., 2000). For instance, fluctuation in the quantity of larvae reaching shore will have little impact on the abundance of adults living along the shore if the shoreline habitat is already saturated with organisms. Roughgarden et al. (1988) identified three sources of variation influencing the abundance of adults in a coastal community:

(1) supply of larvae (when the supply of larvae is limited, oceanographic variability will lead to variable abundance of settling organisms),
(2) intraspecific competition (through density-dependent interaction among adults abundance will be regulated by competition and cannibalism), and
(3) interspecific competition (predation and competition among species will influence the community ecology in the coastal zone).

The purpose of this paper is to provide a synthesis of investigations of single fish species and fish species communities in a top-down/bottom-up framework, and to provide a modelling approach to species richness dynamics including both processes. First, we will establish how oceanographic features of the Skagerrak influence the coastal zone of the southern coast of Norway. Then, we will explain how cod (*Gadus morhua*) exhibits a pattern of variability being consistent with a bottom-up influence, while demersal gobies (Black goby *Gobius niger*, Sand goby *Pomatoschistus minutus*, and Two-spot goby *Gobiusculus flavescens*) exhibit a pattern of variability that might be consistent with top-down regulation. We synthesise existing investigations and present new data. Finally, we outline a general model for the regulation of fish communities in the coastal zone and report how data are consistent with such a model.

Oceanographic features of the Skagerrak

The dominant currents in the Skagerrak basin flow counter clockwise, flowing westward along the Norwegian coast as the 'Norwegian Coastal Current' (NCC; see e.g., Danielssen et al., 1996; Danielssen et al., 1997). NCC dominates over other currents along the Skagerrak coast, flowing westward while low saline water originating in the Baltic Sea is mixed with sub-surface North Sea water of higher salinity. Thus, the NCC accumulates increasing saline content as it flows along the Norwegian coast (Fig. 1). In the Northern Hemisphere, a pull from the wind will divert the currents to the right due to the Coriolis force. Thus, wind blowing from the northeast along the Norwegian Skagerrak coast will create currents towards the shore and accumulate water along the coast by Ekman transport (Mann & Lazier, 1991). As the dominant current along the Skagerrak coast is westward, both wind from the east and onshore wind stress will create onshore transport of water, as demonstrated by Johannessen et al. (1995) for the western coast of Norway.

The wind regimes may represent the forcing factor for bottom-up control of fish populations in (at least) two ways. Firstly, in the dispersal phase, favourable wind-driven currents will bring larvae to suitable areas before metamorphosis and settlement. Secondly, in the post-metamorphosis phase, wind-driven currents supply young fish with zooplankton nauplii [see, e.g., (Aksnes et al., 1989; Kaartvedt, 1991); the most important species being *Calanus finmarchius* (Sysoeva & Degtereva, 1965; Hop et al., 1992)]. Wind- and tidal-driven currents cause localized variation in zooplankton abundance (Lindahl & Perissinotto, 1987; Lindahl & Hernroth, 1988; Lindahl et al., 1998), and are a major source of variability in the spatio-temporal distribution of zooplankton (Bagøien, 1999). A link between wind, currents, zooplankton and the distribution of larvae of herring (*Clupea harengus* L.) has been demonstrated by Johannessen et al. (1995) for the coastal waters of western Norway.

Results

Data sets

Since 1919, the near-shore coastal communities of the Norwegian Skagerrak coast have been sampled by beach seine each autumn [more than 250 stations; (Stenseth et al., 1999)]. The sampling resolution of the monitoring program has expanded by more stations and quantitative information has been compiled for an increasing number of fish species ('the Flødevigen data'). For this study, three subsets of the 'Flødevigen data' have been compiled as described in the following to contain the maximum information available for the purpose of the studies: (1) to analyse juvenile cod length-frequency distributions, Lekve et al. (2002a) utilised a total of 38 stations from west of Kristiansand to the Kragerø area (as these run continuously since 1919; Fig. 1). As homogenous variability

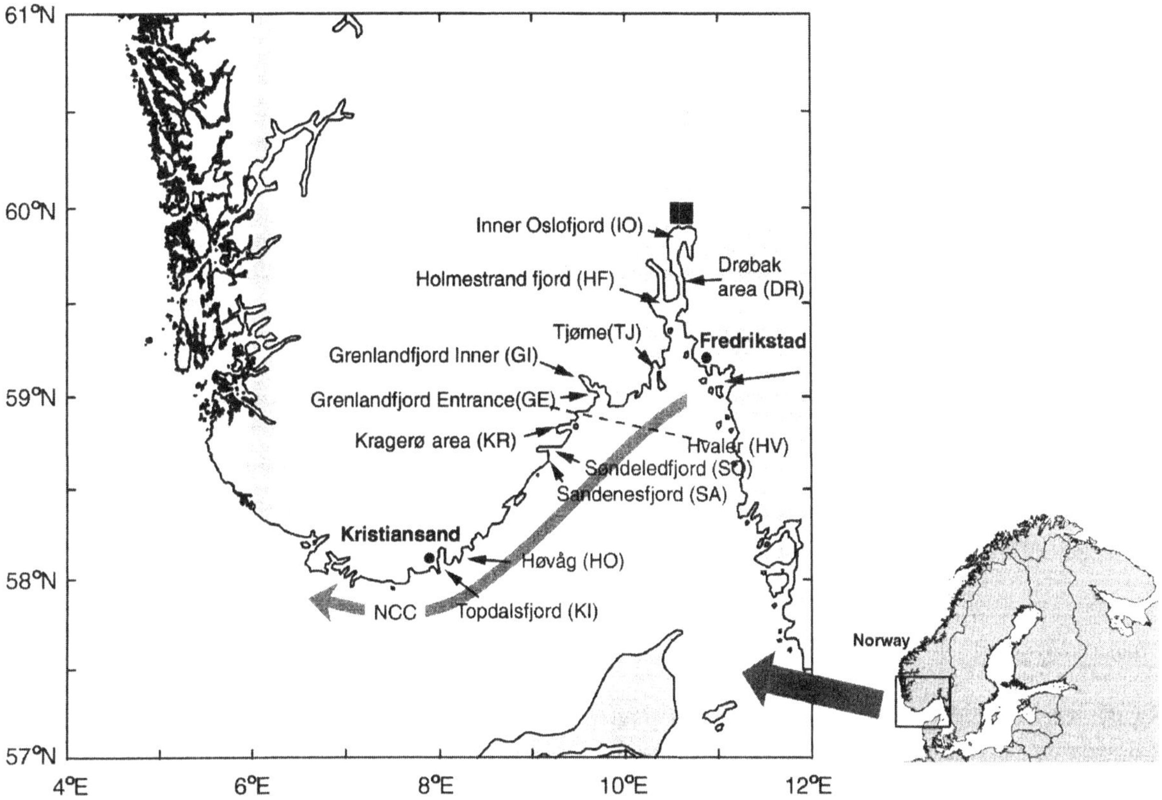

Figure 1. General location of the study areas along the Norwegian Skagerrak coast (lower right-hand side). Coastline of the Norwegian Skagerrak coast (main) indicating the location of the 12 fjords/areas of the analysis (abbreviations as in Table 1). The stippled line indicates the eastern boundary for the analysis of cod length–frequency distributions. The Norwegian Coastal Current (NCC; see text) is indicated along the coast.

could be demonstrated for the cod populations of these stations, samples from individual stations were pooled for analyses of length-frequency distributions. Properties of these distributions of juvenile cod were related to environmental data (available from 1957; see below). (2) The relationship between gobies and gadoids were studied in the same area, however, only in the latter part of the period (1978–97). At that time, 52 stations that run continuously were available. (3) A model for dynamics of species richness was developed for the whole area of the 'Flødevigen data' (from Kristiansand to Fredrikstad; Fig. 1). To account for variability in detection probability of a given fish species among species (Burnham & Overton, 1979), a fjord-based estimate of species richness was developed (Lekve et al., 2002b). To account for non-detectability and sampling variability, we constructed a time series of estimated species richness for 12 fjords/areas based on four stations within each location (see Lekve et al., 2002b for more details).

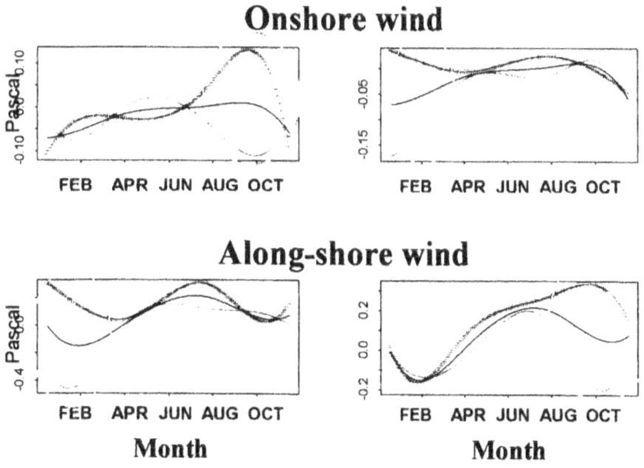

Figure 2. 'Functional Prinicpal Components Analysis' (F-PCA; Ramsey & Silverman, 1997) of onshore (upper) and along-shore (lower) wind stress (see Ottersen & Sundby, 1995 and Lekve et al., 2002a for decomposition of wind stress) measured three times a day 1957–94 at Arendal. The data are smoothed by B-splines. Left: main (Principal) temporal pattern ± 2 standard deviations. Right: Second component of temporal variability ± 2 standard deviations (not correlated with the principal component; see e.g., Legendre & Legendre, 1998).

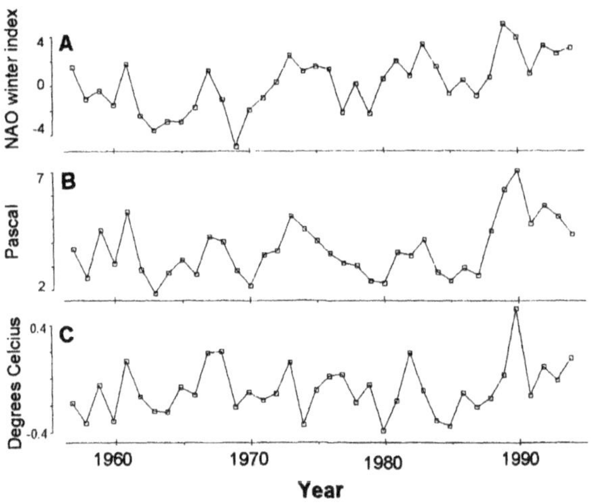

Figure 3. Environmental forcing variables. (A) The North Atlantic Oscillation winter Index, NAO (Hurrell, 1995), (B) Spring wind stress and (C) Spring sea surface temperature (Lekve et al., 2002a).

Climatic variables

For the Norwegian Skagerrak coast, there are no data that measure directly the availability of planktonic food in the coastal zone. Thus, we use wind data as an indicator of advective transport of water (and thus plankton; see above). Data on wind and sea surface temperature were obtained from the Norwegian Meteorological Institute (for details, see Fromentin et al., 1998). Temperature and wind were measured three times a day from 1957 to 1994. Wind direction and speed were decomposed into directional wind stress, and alongshore wind stress was used as an environmental forcing variable (Ottersen & Sundby, 1995; Lekve et al., 2002a). The annual variability of wind stress is dominated strongly by winds in the early spring (January–March) and during the fall (found by Principal Component Analysis; Fig. 2; Lekve et al., 2002a). However, as most fish species along the Norwegian Skagerrak coast spawn in winter, spring or summer (Bøhle, 1993), we used wind stress in the spring as a forcing factor. To match the spawning season of cod, wind and temperature were averaged from March to April (Ottersen & Sundby, 1995; Lommond et al., 1998; Lekve et al., 2002a).

The influence of environmental conditions on fish species in a given year is probably not limited to conditions in that year. However, the indirect influence of environmental conditions the previous year(s) will most likely be highly diluted. A surrogate measure of climate (the NAO index) was used to represent past and residual environmental variability (Hurrell, 1996). The NAO reflects the difference in surface level air pressure between the high-pressure zone concentrated around the Azores and the low-pressure zone over Iceland. The NAO winter index is strongly correlated with the local environment (Hurrell & van Loon, 1997). In the Skagerrak area, temperature, wind and precipitation (and thus freshwater run-off) are correlated with the NAO winter index (Fromentin et al., 1997). The normalised NAO index for December–March (Hurrell, 1995) was obtained from the web site www.cgd.ucar.edu:80/ jhurrell/nao.html. For our study area, the effect of weather on species richness thus comprises of wind stress and temperature in the present year, and the North Atlantic Oscillation in the present and previous year as possible influential factors (Fig. 3).

Bottom-up control of cod and top-down control of gobies

Bottom-up control of cod

Along the Skagerrak coast, cod (*Gadus morhua*) spawn in the fjords in early winter (sometimes under the ice). Each female spawns several millions eggs (Kjesbu et al., 1996, 1998) that drift for about two weeks before hatching, and then for another 3–4 weeks prior to settlement in March–April (Stenseth et al., 1999). Juvenile and young cod are strongly site associated (Tupper & Boutilier, 1995) until maturation [for the Skagerrak at about 3 years of age; (Kjesbu et al., 1996)], after which some may undertake spawning migrations. During the initial stationary period, the main food item of cod is copepods, such as *Calanus finmarchius;* (Sysoeva & Degtereva, 1965; Hop et al., 1992). In late summer, the young cod switch to other prey, e.g. small fish (Hop et al., 1992; Lommond et al., 1998).

Lekve et al. (2002a) investigated life-history traits (length, density, skewness of frequency distributions of length) of juvenile cod and found a significant correlation between the skewness of the length–frequency distribution and alongshore wind stress (correlation of −0.72; Fig. 4), but not between alongshore wind stress and density. They suggested that the conditions for juvenile cod in the spring were affected by advection (as measured by alongshore wind stress), and that winds setting up off-coast transport increased competition among juvenile cod, leading to an increased

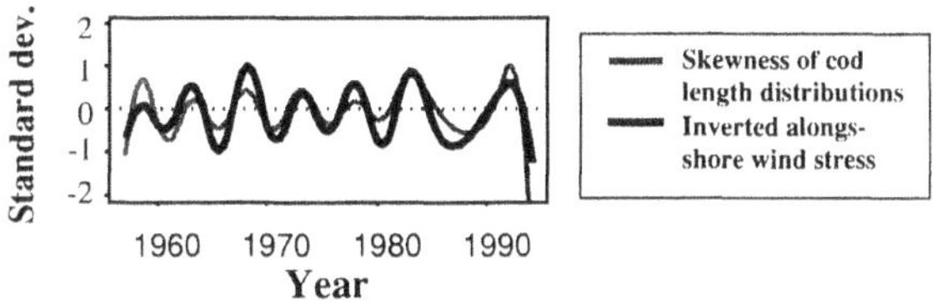

Figure 4. Relationship between skewness of the distribution of length frequencies of cod (solid line) and spring along-shore wind stress *inverted* (stippled line) 1957–1994. Displayed are the derivatives of patterns through time (functions created and data smoothed by B-splines in the 'Functional Data Analysis' approach; FDA; Ramsey & Silverman, 1997).

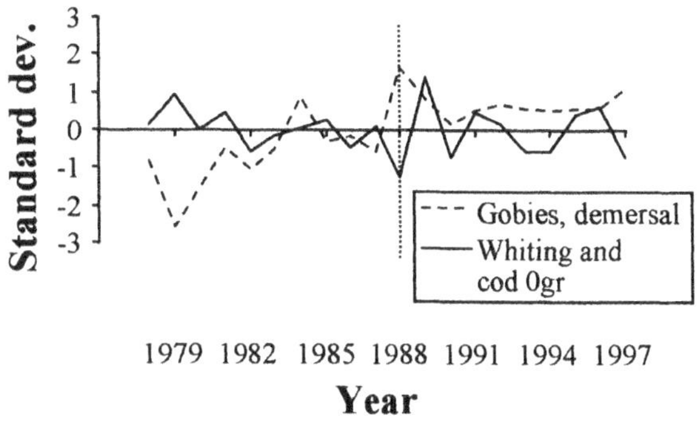

Figure 5. Relationship between demersal gobies (family Gobiidae; Black goby *Gobius niger*, Sand goby *Pomatoschistus minutus*, and Two-spot goby *Gobiusculus flavescens*) and gadoids (family Gadidae; Cod *Gadus morhua* and Whiting *Merlangius merlangus*). The data are derived from the 'Flødevigen Data Set', beach seine monitoring data sampled each fall since 1919 at more than 250 stations along the Southern Norwegian Skagerrak Coast. For the 0 group of cod and whiting the average abundance per station for 52 stations between Kristiandsand and Kragerø (cf. Fig. 1) are normalised by the mean and standard deviation for the study period. For the demersal gobies an index of abundance is normalised in the same way.

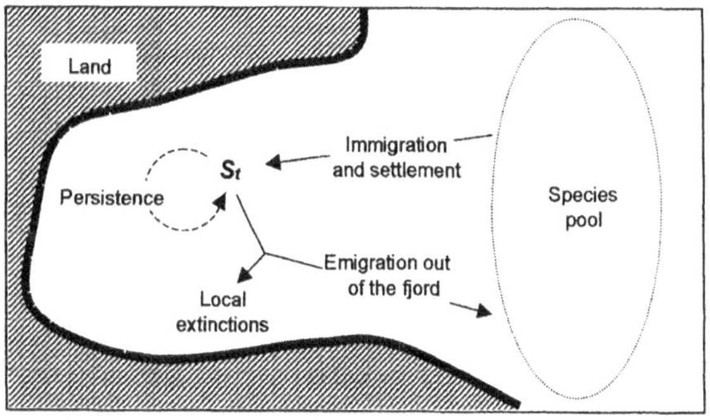

Figure 6. Conceptual framework. Schematic overview of some assumptions and processes taking place in fish communities in the coastal zone along the Norwegian Skagerrak coast. S_t is the number of species at time t.

skewness of the length–frequency distributions (increased asynchrony of the distributions; Lekve et al., 2002a). As the density was not correlated with wind stress, they suggested that it is not the dispersal of cod larvae *per se* that is affected by wind, but that the subsequent availability of copepod food items is affected by wind-driven currents.

If wind can be interpreted correctly as a measure of advection, the relationship between wind and skewness can be seen as an indicator of bottom-up control of juvenile cod in the coastal zone. The skewness can be interpreted as a result of increased competition between juvenile cod, which is caused also by decreased supply of advected food items.

Top-down control of gobies

The demersal gobies (family Gobiidae, e.g., Black goby, Sand goby, and Two-spot goby) have a life history very different from cod. Gobies spend their entire life in the coastal zone and spawn in May-August in simple nests guarded by males. During this period, there is no variability in wind (Fig. 2) and thus no variability in advective transport (consequently no relationship has been found between gobies and wind).

Although individual gadoids (family Gadidae) feed on gobies (Hop et al., 1992), this relationship has been difficult to quantify at the *population* level as several goby species can serve as food for gadoids. However, in May 1988, a bloom of the toxic alga *Chrysochromulina polylepis* killed an estimated 60% of juvenile cod (Chan et al., 2003) and comparable amounts of whiting *Merlangius merlangus* (Gjøsæter et al., 2000). This 'natural experiment' provided some insight into the relationship between gobies and gadoids. As the gobies spawned after the effect of the toxic bloom had vanished (Julliard et al., 2001), we hypothesised that they were released from predation pressure from juvenile cod and whiting. Indeed, the demersal gobies increased their abundance dramatically in 1988, as the gadoids decreased dramatically (Fig. 5; correlation 1988–1997 of -0.3, $p < 0.1$), however, as the abundance of gadoids rebounded the following year, the abundance of gobies decreased.

Gobies of the Southern Norwegian Skagerrak coast display a pattern compatible with top-down control. As they experience little variability in wind, advective processes such as dispersal to suitable habitats or inflow of food items seems to be of less importance to gobies. Biotic interactions, e.g., predation and competition, may be important factors regulating population fluctuations of demersal gobies.

A model of community regulation

We have demonstrated how dispersal of prey by advection affects one important species (cod is the most abundant fish species in the coastal zone) while other species are not influenced by advection. We now turn to developing a model incorporating both top-down (mostly biotic) and bottom-up effects.

If we assume that free-floating eggs and larvae are available all along the coast from a pool of recruits in the open ocean (e.g., due to offshore transport of eggs; Johannessen et al., 1995), the dynamics of fish communities within a fjord will be determined by the dispersal success modified by advective processes and the subsequent persistence of a species in the habitat (as illustrated in Fig. 6). Following Brown et al. (2001), a model describing regulation of coastal fish community may thus consist of three elements: (1) available niches in the habitat will change 'slowly' (i.e., compared to the generation time of fish). The growth of brown algae is an example of a slowly changing abiotic factor (providing shelter for many fish species). If the community is close to some overall equilibrium this can be modelled as a 'baseline species settlement potential' which is not time dependent, but will vary among fjords (expressed as $R_{0,f}$). This baseline settlement potential is furthermore modified by the number of other species already present in the fjord (which we have expressed as $S_{f,t-1}$, with f being the fjord and t being year). These species may be important as prey species, or may be competitors, forming the expression $S_{f,t-1}R_{0,f}$, analogous to the constant term in a linear model. (2) Ability of a fish species to persist is dependent upon interaction with other species. With reference to population ecology (Royama, 1992), we model this as an exponential relationship. The settlement potential is thus further modified by $\exp\{-a_f s_{f,t-1}\}$, where $s_{f,t-1}$ is the natural logarithm of the number of species present in the fjord. This 'richness-dependent regulation' of the fish community in the fjord is quantified by the coefficient a_f, which is unique for each fjord, but not time dependent. (3) Dispersal by advective processes and effects of weather is expressed by the term $\exp\{\underline{b}_f^T \underline{w}_t\}$, where $\underline{w}_t$ is a matrix of the environmental variables (wind, temperature, NAO) while $\underline{b}_f^T$ is a vector of coefficients determining the strength of the relationships.

Altogether, the different elements of the model may be put together in a multiplicative manner as:

$$S_{f,t} = S_{f,t-1}R_{0,f}\exp\{-a_f s_{f,t-1}\}\exp\{\underline{b}_f^T \underline{w}_t\} = S_{f,t-1}R_{0,f}\exp\{-a_f s_{f,t-1} + \underline{b}_f^T \underline{w}_t\}. \quad (1)$$

Model fitting to the data

To estimate the coefficients of the model, the negative log likelihood of the Poisson distributions (McCullagh & Nelder, 1989) for the derived model of the number of fish species [Equation (1)] were minimised by the Splus routine 'nlminb' (Venables & Ripley, 1997) for each of the fjords independently. Different combinations of wind and temperature at time t and NAO at time t and t-1 entered the models as environmental forcing variables. Corrected Akaike Information Criterion (AIC_C; Brockwell & Davies, 1991) was applied for model selection. The AIC_C is a model selection criterion that measures the fit of the model (the log likelihood) and penalises the number of parameters in the model. Models with a difference in AIC_C value of less than 2 are generally seen as indistinguishable in terms of model selection criteria. When confronted with indistinguishable models, we selected parameters to end up with the same parameter sets for several fjords/areas, making biological interpretation and comparison between fjords possible. The chosen 'biological models' were those containing factors making biological interpretations possible and, at the same time, falling within an acceptable AIC_C-level. To assess the model fit and significance of parameter values, we resampled the residuals (Davison & Hinkley, 1997) and optimised the likelihood of the Poisson distributions (McCullagh & Nelder, 1989) 100 times (see Lekve et al., 2003 for details).

Results for community regulation

Lekve et al. (2003) performed a complete analysis of the fit of this model (see also Stenseth et al., 2002). Here, the focus is on fjords displaying both bottom-up and top-down control (i.e., the fjords/areas for which wind enter the model for regulation of fish-species richness).

For six of the fjords, wind was included as one of the variables by the model selection procedure (Table 1). The rather simple model outlined above was able to capture important features of the dynamics of species richness in the fjords/areas studied along the Norwegian Skagerrak coast (Fig. 7). For several of the fjords, the model estimates closely follow the observed number of species (see e.g., Søndeledfjord $R^2 = 0.19$,

Table 1. Environmental forcing variables entering models of fish species richness fitted separately for 12 fjords/areas along the Norwegian Skagerrak coast, chosen by the selection tool AIC_C (Corrected Akaike Information Criterion; Brockwell & Davies, 1991) at year t and year t-1. $AIC_{C,best}$ $-AIC_{C,model}$ is the difference between the AIC_C value of the model with the lowest value and the chosen 'biological model' for each of the fjords

	Forcing at time t	Delayed forcing (time t-1)	$AIC_{C,best}$ − $AIC_{C,model}$
Kristiansand (Topdalsfjord)– KI	–	NAO_{t-1}	2.0
Høvåg (Steindalsfjord) – HO	–	NAO_{t-1}	0
Sandnesfjord, Risør – SA	–	NAO_{t-1}	1.3
Søndeledfjord, Risør – SO	$Wind_t$, $Temp_t$	NAO_{t-1}	1.3
Kragerø area – KR	$Wind_t$, $Temp_t$	NAO_{t-1}	1.1
Grenlandfjord Entrance – GE	Wind	NAO_{t-1}	1.3
Grenlandfjord Inner – GI	$Wind_t$, $Temp_t$	NAO_{t-1}	0
Tjøme – TJ	$Wind_t$, $Temp_t$	NAO_{t-1}	0.6
Holmestrand fjord – HF	–	NAO_{t-1}	0.2
Hvaler – HV	NAO_t	–	1.6
Drøbak area – DR	NAO_t	–	1.0
Inner Oslofjord – IO	$Wind_t$	NAO_{t-1}	1.6

Kragerø $R^2 = 0.27$, Grenland Entrance $R^2 = 0.19$, Grenlandfjord Inner $R^2 = 0.18$, and Tjøme $R^2 = 0.50$ in Fig. 7). For four of these fjords, the wind entered the model significantly (Fig. 8). Estimates and interpretation of variables can be found in Lekve et al. (2003).

Discussion

The dynamics of coastal fish communities may be modelled in various ways. We have chosen an empirical approach, explicitly including processes known to be important for the involved fish species. Even with our approach, several assumptions had to be made. The modelling approach outlined in this work is a first approach and needs to be discussed and tested further. The focus here will be on conceptual issues to clarify the ideas underpinning the model described.

Fish communities in the coastal zone are composed of species with widely differing life histories. As the larval stage is critical in determining year class strength of fish (Hjort, 1914; Cushing, 1966; Myers & Cadigan, 1993), it is important to include factors affecting the larval stage when modelling community dynamics in the coastal zone. Due to differences in e.g., onset of spawning, fish with different life histories will not respond to changes in external forcing in the same way. In this work, we have demonstrated that the cod is sensitive to advective processes (and we suggest that the advection of prey is the causal mechanism) and, therefore, bottom-up processes affect cod. The same is probably true for other species that spawn in winter (Nielsen et al., 1998 have demonstrated wind-induced dispersal of Plaice *Pleuronectes platessa* in the Skagerrak-Kattegat). Such species are likely to respond to changes in the environmental variables determining dispersal. In this work, we have shown that wind stress is a variable that may properly describe changes in dispersal and advection (temperature is another important environmental variable, however, beyond the scope of this study). Wind is a relevant variable affecting the community dynamics of several of the fjords along the Skagerrak coast. The fact that the sign of the estimated coefficient changes along the coast is probably due to the interplay between the orientations of the fjords in relation to the coastal current.

Some species appear to be unaffected by advective processes (e.g., the gobies). However, all species in the coastal zone are involved in biotic interactions. The variability due to advection induced on gadoids will subsequently affect interspecific competition in the habitats by means of competition or predation. Furthermore, many species experience density-dependent regulation (see, e.g., Stenseth et al., 1999; Fromentin et al., 2001). These basic properties of fish species result in differential ability of fish to occupy available niches and to compete with other species, and constitute the mechanisms determining the persistence

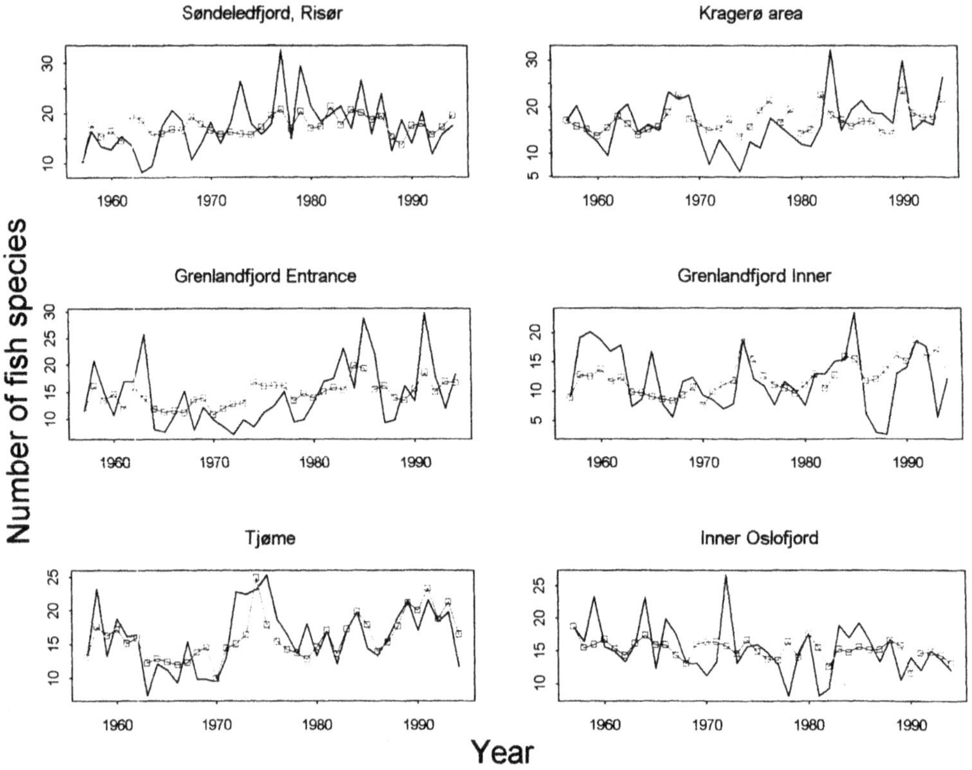

Figure 7. Model predictions of fish species richness-regulation of six of the 12 fjords analysed along the Norwegian Skagerrak coast (cf. Fig. 1). Displayed are the observed number of species (e.g., estimated from repeated sampling of four stations; Lekve et al., 2002b; solid line) and the one-step prediction of the model using only the environmental data as forcing variables (see main text; stippled line with boxes).

of a fish species in the community. By analogy to the 'density-dependent regulation' of abundance in population dynamics, this may be seen as 'richness-dependent regulation' (Brown et al., 2001) of fish species richness. The analogy is obviously not complete – the most important difference from a modelling point of view is that stochasticity in the population dynamics of each species weakens the biotic signals in community dynamics.

The concepts presented here are similar to the disturbance-productivity concepts (Kondoh, 2001) and the 'inter-mediate disturbance hypothesis' (Connell, 1978). By including three elements of the community [the quality of the niche, environmental variability (corresponding to disturbance) and richness-dependent regulation], we have suggested a feasible way of modelling fish species richness.

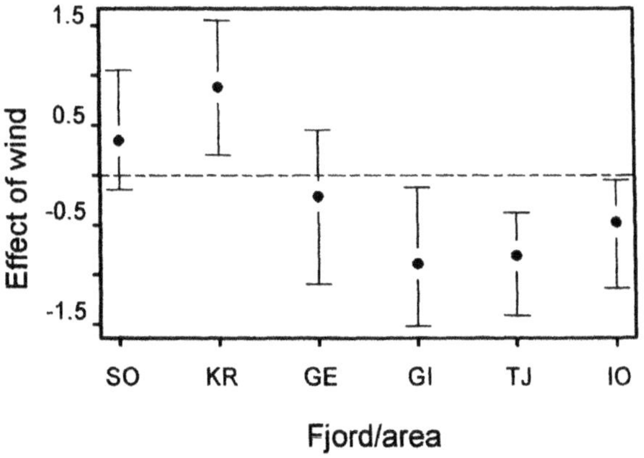

Figure 8. Effect of wind stress on species richness. The credibility intervals are found from the frequency distribution of the estimated coefficient from 100 reoptimisations of the likelihood of the Poisson distributions for each fjord (McCullagh & Nelder, 1989) after resampling the residuals of the model predictions (Table 1). See main text and Lekve et al. (2003) for further details. Abbreviations as in Table 1.

Acknowledgements

We thank present and previous staff at Flødevigen Research Station for carrying out the sampling resulting in the 'Flødevigen data set'. We thank the Norwegian Meteorological Institute for data series on wind and sea-surface temperature. Thanks to Rune Johannessen and Ole Chr. Lingjærde for assistance in the development of the model. Financial support from the Norwegian Science Council (NFR; KL and NCS) made the analyses possible.

References

Aksnes, D. L., J. Aure, S. Kaartvedt, T. Magnesen & J. Richard, 1989. Significance of advection for the carrying capacities of fjord populations. Mar. Ecol. Prog. Ser. 50: 263–274.

Bagøien, E., 1999. Predatory impact of invertebrate and fish on overwintering Calanus. Biology. University of Oslo, Oslo.

Brockwell, P. J. & R. A. Davies, 1991. Time Series: Theory and Methods. Springer-Verlag Inc., New York. 577 pp.

Brown, J. H., S. K. M. Ernest, J. M. Parody & J. P. Haskell, 2001. Regulation of diversity: maintenance of species richness in changing environments. Oecologia 126: 321–332.

Burnham, K. P. & W. S. Overton, 1979. Robust estimation of population size when capture probabilities vary among animals. Ecology 60: 927–936.

Bøhle, B., 1993. Sårbare fiskeressurser i Skagerrak. Rapport fra forskningsstasjonen Flødevigen 1993.

Chan, K.-S., N. C. Stenseth, K. Lekve & J. Gjøsæter, 2003. Modeling pulse disturbance impact on cod population dynamics: the 1988 algal bloom of Skagerrak, Norway. Ecol. Monogr. 73: 151–171.

Connell, J. H., 1961. The influence of interspecific competition and other factors on the distribution of the barnacle *Chtamalus stellatus*. Ecology 42: 710–723.

Connell, J. H., 1978. Diversity in tropical rain forests and coral reefs. Science 199: 1302–1310.

Cushing, D. H., 1966. Biological and hydrographic changes in British Seas during the last thirty years. Biol. Rev. 41: 221–258.

Danielssen, D. S., E. Svendsen & M. Ostrowski, 1996. Long-term hydrographic variation in the Skagerrak based on the section Torungen-Hirtshals. ICES J. mar. Sci. 53: 917–925.

Danielssen, D. S., L. Edler, S. Fonselius, L. Hernroth, M. Ostrowski, E. Svendsen & L. Talpsepp, 1997. Oceanographic variability in the Skagerrak and northern Kattegat, May–June, 1990. ICES J. mar. Sci. 54: 753–773.

Davison, A. C. & D. V. Hinkley, 1997. Bootstrap Methods and their Application. Cambridge University Press, Cambridge.

Fromentin, J.-M., N. C. Stenseth, J. Gjøsæter, T. Johannessen & B. Planque, 1998. Long-term fluctuations in cod and pollack along the Norwegian Skagerrak coast. Mar. Ecol. Prog. Ser. 162: 265–278.

Fromentin, J.-M., N. C. Stenseth, J. Gjøsæter, O. N. Bjørnstad, W. Falck & T. Johannessen, 1997. Spatial patterns of the temporal dynamics of three gadoid species along the Norwegian Skagerrak coast. Mar. Ecol. Prog. Ser. 155: 209–222.

Fromentin, J.-M., R. A. Myers, O. N. Bjørnstad, N. C. Stenseth, J. Gjøsæter & H. Christie, 2001. Effects of density-dependent and stochastic processes on the regulation of cod populations. Ecology 82: 567–579.

Gjøsæter, J., K. Lekve, N. C. Stenseth, H. P. Leinaas, H. Christie, E. Dahl, D. D. Danielssen, B. Edvardsen, F. Olsgaard, E. Oug & E. Paasche, 2000. A long-term perspective on the *Chrysochromulina* bloom on the Norwegian Skagerrak coast 1988: a catastrophe or an innocent incident? Mar. Ecol. Prog. Ser. 207: 201–218.

Hjort, J., 1914. Fluctuations in the great fisheries of northern Europe viewed in the light of biological research. Cons. int. Explor. Mer. 20: 1–228.

Hop, H., J. Gjøsæter & D. S. Danielssen, 1992. Seasonal feeding ecology of cod (*Gadus morhua* L.) on the Norwegian Skagerrak coast. ICES J. mar. Sci. 49: 453–461.

Hurrell, J. W., 1995. Decadal trends in the North Atlantic Oscillation: regional temperatures and precipitation. Science 269: 676–679.

Hurrell, J. W., 1996. Influence of variations in extratropical wintertime teleconnections on Northern Hemisphere temperature. Geophys. Res. Lett. 23: 665–668.

Hurrell, J. W. & H. van Loon, 1997. Decadal variations in climate associated with the north Atlantic oscillation. Clim. Change 36: 301–326.

Johannessen, A., G. Blom, A. Folkvord & H. Svensen, 1995. The effect of local wind on the distribution of Norwegian spring spawning herring (*Clupea harengus* L.) larvae. In Skjoldal, H. R., C. Hopkins, K. E. Erikstad & H. P. Leinaas (eds), Ecology of Fjords and Coastal Waters: Proceedings of the Mare Nor Symposium on the Ecology of Fjords and Coastal Waters, Tromsø, Norway, 5–9 December, 1994. Elsevier, Amsterdam: 365–384.

Julliard, R., N. C. Stenseth, J. Gjøsæter, K. Lekve, J.-M. Fromentin & D. S. Danielssen, 2001. Natural mortality and fishing morality in a coastal cod population: a release-recapture experiment. Ecol. Appl. 11: 540–558.

Kaartvedt, S., 1991. Plankton as potential prey in fjords. ICES mar. Sci. Symp. 192: 156–160.

Kjesbu, O. S., P. Solemdal, P. Bratland & M. Fonn, 1996. Variation in annual egg production in individual captive Atlantic cod (*Gadus morhua*). Can. J. Fish. aquat. Sci. 53: 610–620.

Kjesbu, O. S., P. R. Witthames, P. Solemdal & M. G. Walker, 1998. Temporal variations in the fecundity of Arcto-Norwegian cod (*Gadus morhua*) in response to natural changes in food and temperature. J. Sea. Res. 40: 303–321.

Kondoh, M., 2001. Unifying the relationships of species richness to productivity and disturbance. Proc. r. Soc. Lond. Ser. B-Biol. Sci. 268: 269–271.

Legendre, P. & L. Legendre, 1998. Numerical Ecology. Elsevier, Amsterdam.

Lekve, K., G. Ottersen, N. C. Stenseth & J. Gjøsæter, 2002a. Length dynamics in juvenile coastal Skagerrak cods: effects of biotic and abiotic processes. Ecology 83: 1676–1688.

Lekve, K., N. C. Stenseth, R. Johansen, O. C. Lingjærde & J. Gjøsæter, 2003. Richness-dependent and climatic factors regulating the dynamics of coastal fish diversity. Ecol. Lett. 6: 428–439.

Lekve, K., T. Boulinier, N. C. Stenseth, J. Gjøsæter, J.-M. Fromentin, J. E. Hines & J. D. Nichols, 2002b. Spatio-temporal dynamics of species richness in coastal fish communities. Proc. r. Soc. Lond. B 269: 1781-1789.

Lewin, R., 1986. Supply-side ecology. Science 234: 25–27.

Lindahl, O. & R. Perissinotto, 1987. Short-term variations in the zooplankton community related to water exchange processes in the Gullmar fjord, Sweden. J Plankton Res 9: 1113–1132.

Lindahl, O. & L. Hernroth, 1988. Large-scale and long-term variation in the zooplankton community of the Gullmar fjord, Sweden, in relation to advective processes. Mar. Ecol. Prog. Ser. 43: 161–171.

Lindahl, O., A. Belgrano, L. Davidsson & B. Hernroth, 1998. Primary production, climatic oscillations, and physico-chemical processes: the Gullmar Fjord time-series data set (1985–1996). ICES J. mar. Sci. 55: 723–729.

Lommond, T. M., D. C. Schneider & D. A. Methven, 1998. Transition from pelagic to benthic prey for age group 0 – I Atlantic cod, *Gadus morhua*. Fish. Bull. 96: 908–911.

Mann, K. H. & J. R. N. Lazier, 1991. Dynamics of marine ecosystems. Blackwell, Cambridge.

McCullagh, P. & J. A. Nelder, 1989. Generalized linear models. Chapman and Hall, London.

Menge, B. A., 1992. Community regulation: under what conditions are bottom-up factors important on rocky shores. Ecology 73: 755–765.

Menge, B. A., 2000. Top-down and bottom-up community regulation in marine rocky intertidal habitats. J. exp. mar. Biol. Ecol. 250: 257–289.

Menge, B. A., B. A. Daley, P. A. Wheeler, E. Dahlhoff, E. Sanford & P. T. Strub, 1997. Benthic-pelagic links and rocky intertidal communities: Bottom-up effects on top-down control? Proc. natn. Acad. Sci. 94: 14530–14535.

Myers, R. A. & N. G. Cadigan, 1993. Density-dependent juvenile mortality in marine demersal fish. Can. J. Fish. aquat. Sci. 50: 1576–1590.

Nielsen, E., O. Bagge & B. R. MacKenzie, 1998. Wind-induced transport of plaice (*Pleuronectes platessa*) early life-history stages in the Skagerrak-Kattegat. J. Sea. Res. 39: 11–28.

Ottersen, G. & S. Sundby, 1995. Effects of temperature, wind and spawning stock biomass on recruitment of Arcto-Norwegian cod. Fish. Oceanogr. 4: 278–292.

Paine, R. T., 1966. Food web complexity and species diversity. Am. Nat. 100: 65–75.

Polis, G. A., A. L. W. Sears, G. R. Huxel, D. R. Strong & J. Maron, 2000. When is a trophic cascade a trophic cascade? Trends Ecol. Evol. 15: 473–475.

Ramsey, J. O. & B. W. Silverman, 1997. Functional data analysis. Springer-Verlag New York, Inc., New York.

Roughgarden, J., S. Gaines & H. Possingham, 1988. Recruitment dynamics in complex life cycles. Science 241: 1460–1466.

Royama, T., 1992. Analytic population dynamics. Chapman & Hall, London. Sinclair, A. R. E., C. J. Krebs, J. M. Fryxell, R. Turkington, S. Boutin, R. Boonstra, P. Seccombe-Hett, P. Lundberg & L. Oksanen, 2000. Testing hypotheses of trophic level interactions: a boreal forest ecosystem. Oikos 89: 313–328.

Stenseth, N. C., K. Lekve & J. Gjøsæter, 2002. Modeling species richness controlled by community-intrinsic and community-extrinsic processes: coastal fish communities as an example. Population Ecol. 44: 165–178.

Stenseth, N. C., O. N. Bjørnstad, W. Falck, J.-M. Fromentin, J. Gjøsæter & J. S. Gray, 1999. Dynamics of coastal cod populations: intra- and intercohort density dependence and stochastic processes. Proc. r. Soc. Lond. B 266: 1645–1654.

Sysoeva, T. K. & A. A. Degtereva, 1965. The relation between feeding of cod larvae and pelagic fry and the distribution and abundance of their principal food organisms. ICNAF, Spec. publ. 6: 411–416.

Tupper, M. & R. G. Boutilier, 1995. Effects of habitat on settlement, growth, and postsettlement survival of Atlantic cod (*Gadus morhua*). Can. J. Fish. aquat. Sci. 52: 1834–1841.

Venables, W. N. & B. D. Ripley, 1997. Modern Applied Statistics with S-PLUS. Springer, New York.

Hydrobiologia **503:** 103–116, 2003.
M.B. Jones, A. Ingólfsson, E. Ólafsson, G.V. Helgason, K. Gunnarsson & J. Svavarsson (eds), Migrations and Dispersal of Marine Organisms.

Secondary settlement of cockles *Cerastoderma edule* as a function of current velocity and substratum: a flume study with benthic juveniles

X. de Montaudouin[1], G. Bachelet[1] & P.-G. Sauriau[2]
[1]*Laboratoire d'Océanographie Biologique, UMR 5805 Université Bordeaux 1 - CNRS, 2 Rue du Professeur Jolyet, F-33120 Arcachon, France*
[2]*Centre de Recherche sur les Ecosystèmes Marins et Aquacoles de l'Houmeau, UMR 10 CNRS – IFREMER, Place du Séminaire, B.P. 5, F-17137 L'Houmeau, France*
E-mail: x.de-montaudouin@epoc.u-bordeaux1.fr

Key words: Cerastoderma edule, resuspension, flume experiment, sediment, juveniles

Abstract

Some newly-settled bivalve molluscs can experience a second dispersal stage in the water column and colonize areas distant from initial settlement zone ('secondary settlement'). To document mechanisms involved in such a process, experiments were conducted in a recirculating 13-m flume, using juvenile cockles *Cerastoderma edule* (shell length between 0.8 and 5.7 mm). After 4 h and under current surface velocities of 10, 20 and 24 cm s^{-1}, all juveniles left the plexiglass substratum (site of initial introduction) and 42.4, 58.6 and 76.2% of juveniles, respectively, were retrieved from a downstream sand area (which only represented 7.7% of the total flume surface). Naked-eye observations showed that smaller individuals were borne within the water column, whereas larger individuals tended to roll or slip on the substratum. Byssus threads produced by the juveniles were often seen, sometimes covered in sand grains. These observations were confirmed by finding larger juveniles in the upstream part of the sand area. When sand was replaced by mud, a reduced proportion of cockles was retrieved in the experimental substratum (18.4, 20.0 and 16.4%, respectively). Observations showed that juveniles rarely succeeded in anchoring themselves in mud. When initially introduced on a favourable substratum (medium sand), more than 87% of juveniles were retrieved from that sand array at all flow velocities. This study shows that secondary settlement occurs for juvenile cockles up to 5.7 mm in shell length and depends not only on flow velocity but on substratum type.

Introduction

In marine benthic invertebrates with bentho-pelagic life cycles, settlement of planktonic larvae is determined by a combination of behaviour and hydrodynamic processes (Butman & Grassle, 1992; Grégoire et al., 1996; Harvey & Bourget, 1997; Snelgrove et al., 1999). Planktonic larvae do not simply fall passively onto the bottom, but rather have an explorative behaviour (Butman, 1987, 1989; Jonsson et al., 1991). They drift passively in the same direction as the water current but control their buoyancy with byssus threads (Sigurdsson et al., 1976; Jonsson et al., 1991; de Montaudouin, 1997), mucus secretions (Lambert, 1991; Abelson & Denny, 1997) or by swimming (Grant, 1980; Butman & Grassle, 1992; Grassle et al., 1992b). These mechanisms provide the larvae with an opportunity to choose or reject the substrata tested during the drifting period (Butman et al., 1988; Butman, 1989; Mullineaux & Butman, 1990; Jonsson et al., 1991; André et al., 1993).

Although the ability of settled organisms to resuspend themselves in the water column is likely to decrease as development proceeds, little data exist describing the stage, or size, at which individuals of a given species become definitively benthic. Three sets of results relative to invertebrate juveniles suggest, that for numerous species, drifting migration does not necessarily cease after the initial settlement: (1) samples collected by suprabenthic sleds and plankton nets often contain juveniles of benthic invertebrates (Butman, 1987; Beukema & de Vlas, 1989; Armonies, 1992, 1994a, b; Jaklin & Günther, 1996; Olivier et al., 1996b; Thiébaut et al., 1996); (2) juveniles of numerous molluscs possess adaptative structures, i.e. mucus or byssus threads, enhancing buoyancy and allowing resuspension (Sigurdsson et al., 1976; Lane et al., 1982, 1984; Vahl, 1983; Prezant & Chalermwat, 1984; Young, 1985; Yankson, 1986; Martel & Chia, 1991; Beaumont & Barnes, 1992; Cummings et al., 1993; Martel & Dieffenbach, 1993; de Montaudouin, 1997); and (3) the occurrence of large juveniles on a site where no recruitment had been initially observed suggests a drifting migration by juveniles (Beukema & de Vlas, 1989).

In most field studies, it is often difficult to distinguish whether substratum properties, flow characteristics, or a combination of both are responsible for the observed settlement or recruitment patterns (Nowell & Jumars, 1984; Wu & Shin, 1997). Flume experiments, in which sediments can be manipulated and boundary layer flows controlled, offer an alternative approach for recruitment studies. Indeed, much effort has been devoted through flume experiments to understand the precise mechanisms involved during larval settlement (Woodin, 1986, 1991). Few flume studies, however, have dealt with the behaviour of juvenile (de Montaudouin & Bachelet, 1996; Olivier et al., 1996a; de Montaudouin, 1997) or adult (Grant, 1980) infaunal marine invertebrates. The aim of the present work was to investigate how bivalve molluscan juveniles behave in different hydrodynamic and sediment conditions. Flume experiments were conducted in turbulent flow using juveniles of the edible cockle *Cerastoderma edule* (L.), a dominant suspension-feeding bivalve in intertidal and shallow soft sediments along the European Atlantic coast. Cockle larvae settle at a shell length of about 280 μm, after 2–3 weeks of planktonic life (Creek, 1960). The purpose of the present work was to answer the following questions: (1) will juvenile cockles 0.5–5.0 mm long leave an unsuitable substratum, and how? (2) will they able to settle secondarily on a more suitable substratum? (3) when settled on a suitable substratum, will they resuspend themselves in the water column? (4) what are the relative contributions of flow velocities and sediment type? and (5) does the behaviour of juvenile cockles differ from the motion of passive particles?

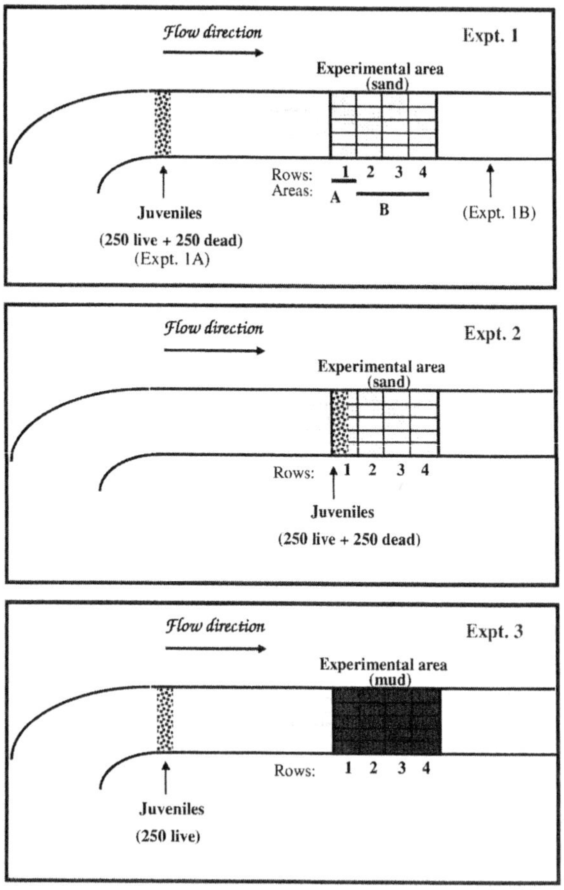

Figure 1. Experimental overview. Cockle juveniles were placed upstream (Experiments 1A & 3) or into the sediment array (Experiment 2). The five sediment boxes were filled either with sand (Experiments 1A & 2) or mud (Experiment 3). At the end of experiments, each box was sampled following four rows, numbered from 1 (upstream) to 4 (downstream). In Experiment 1B, juveniles were placed as in Experiment 1A but downstream of the sediment array.

Materials and methods

Flume characteristics

Sediment selection and resuspension experiments were conducted in the 'Canal Hycobenthos' (Lambert, 1991), a disc-wheel, non-toxic plexiglass flume located at the Laboratoire du Muséum National d'Histoire Naturelle at Dinard (North Brittany). The flume channel is shaped as a recirculating racetrack with a total length of 13 m and two straight sections of 3.5 m. The flume channel width is 0.5 m and the water depth was maintained during experiments at 0.15 m. Water is recirculated by friction by means of an 80-cm diameter wheel made of 14 vertical discs located in the middle of one of the two straight lines of the flume. The working section (i.e., sediment array) is situated on the other straight line. This section measures 0.5 × 1.0 m and consists of five 0.08 × 1.0 m boxes containing 5-cm deep sediment (Fig. 1). Sediment boxes were filled with either sand (median grain size Md = 240 μm) or mud (Md = 60 μm) to a level where the sediment was flush with the flume bottom. The upstream edge of the working section was situated 6.0 m from the wheel. At this distance, the thickness δ of the boundary layer should be well developed. In turbulent flow (Vogel, 1989),

$$\delta = 0.376x(Ux/\nu)^{-1/5},$$

where δ is the boundary layer thickness (cm), x is the flow length (600 cm), U is the free-stream velocity (10, 20 and 24 cm s^{-1}) and ν is the kinematic viscosity of sea water (0.01 cm^2 s^{-1}).

In our experiments, δ was 15, 14 and 13 cm for free-stream velocity of 10, 20 and 24 cm s^{-1}, respectively. Consequently, the processes of resuspension and juvenile behaviour took place within the boundary layer.

Flow speed was measured in the centre of the working section using an ultrasonic current meter (Minilab SD-12, Sensor Data, Bergen, Norway). During experiments, juvenile cockles explored the whole channel. Consequently, velocity profiles should have been performed on most of the flume surface, but this was not the case in the present work. We are aware that profiles measured above the sediment area did not represent the flow regime of the whole flume but only that restricted area where settlement and resuspension were investigated. Shear velocities (u_*) were calculated from the logarithmic section of the velocity profiles by fitting the velocity measurements to the function described by the 'law of the wall' (Jumars & Nowell, 1984):

$$U(z) = u_*/\kappa\ Ln(z/z_0),$$

where $U(z)$ is the mean downstream velocity at height z above the bottom, z_0 is the boundary roughness and κ is the von Kármán's constant (0.41). Estimates of u_* were then used to calculate the roughness Reynolds number, Re_*:

$$Re_* = u_* k_s/\nu,$$

where k_s is the bed roughness scale (assimilated to the medium grain size of the sediment) and ν is the kinematic viscosity of sea water (0.01 cm^2 s^{-1}).

In our experiments, mean surface velocities were based on current speeds and shear velocities calculated by different authors in natural soft-sediment sites (Table 1). Surface current velocities were 10, 20 and 24 cm s^{-1}, corresponding to shear velocities ranging from 0.51 to 0.99 cm s^{-1} (Table 2). Roughness Reynolds numbers indicated a smooth turbulent flow (Re_* <3.5) (Nowell & Jumars, 1984). In the sediment array, at 600 cm from the wheel, the local Reynolds number Re_x, defined as:

$$Re_x = Ux/\nu$$

ranges between 600 000 (U = 10 cm s^{-1}) and 1 560 000 (U = 24 cm s^{-1}). These values (>500 000) characterize turbulent flow (Vogel, 1989).

Experimental design

Juvenile cockles were collected from Arcachon Bay one day prior to a laboratory flume experiment, at the time of maximum field occurrence (May 1992, May and June 1993, May 1994). The top centimetre of sediment was collected with a shovel at two sites where recruitment was known to be high (Bachelet et al., 1992b). Sand was sieved gently (500-μm mesh) and the contents poured into an isothermal box with sea water. In the laboratory, water was changed and cockles were fed algal paste (diatoms *Skeletonema costatum* produced by SATMAR, Barfleur, France) every day. Experiments did not last more than 10 d to work with juveniles in 'good shape', i.e. still active and with no mortality. At the beginning of each experiment, 500 juveniles (shell lengths from 0.6 to 5.8 mm, mostly between 1 and 2 mm) were sorted from the sediment under a binocular microscope. Half of these juveniles were used for the experiment. The other half was fixed in 70% ethanol, stained with Rose Bengal and used to compare the behaviour of living animals with the motion of passive particles (dead stained cockles). Dead juveniles were soaked again a couple of hours before experiment in sea water to rinse the ethanol which would decrease their density. We hypothesized that the shell contributed to most of the density of juveniles which was consequently not affected by ethanol. Sediment (sand or mud) was defaunated by soaking in fresh water for at least 48 h before being immersed again in sea water. The five compartments of the experimental area were filled with sediment until flush with the surrounding bottom. When the experimental sediment was mud, it was prepared at least 12 h before an experiment to obtain a

Table 1. Examples of surface flow velocities (U) and shear stress velocities (u_*) measured in natural sites

U (cm.s^{-1})	u_* (cm s^{-1})	Site	Study	Reference
–	0.52	Intertidal area of	Recruitment	(Eckman, 1983)
–	1.08	Skagit Bay (U.S.A.)		
–	1.67	in 5.3 and 1.75-cm depth		
65	2.64–3.17	Skagit Bay Channel (U.S.A.)	Calculation of the boundary layer	(Gross & Nowell, 1983)
4.6	0.20	Buzzards Bay	Larval settlement in relation	(Buttman, 1986)
9.8	0.40	(U.S.A.)	to hydrodynamics	
15.3	0.60			
15.3	0.98			
15.3	0.82			
–	≤0.60	Tidal current	Substratum selection by larvae	(Butman et al., 1988)
–	≤3.00	Storm		
5.2–15.0	0.52–0.78	St-Lawrence estuary (Canada)	Mussel feeding	(Fréchette et al., 1989)
-	0.10–0.50	Extreme habitat	Settlement in the boundary layer	(Eckman et al., 1994)
6–66	0.31–2.34	Subtidal area (7 m) of Baie de Seine (France)	Resuspension of post-larvae	(Oliver et al., 1996b)
7–48	0.26–1.65	Subtidal area (7 m) of Baie de Seine (France)	Polychaete post-larvae dispersal	(Thiébaut et al., 1996)
0–13.5	–	Subtidal area (22 m) of Baie des Chaleurs (Québec)	Settlement and Substratum structure	(Harvey & Bourget, 1997)
0–100	0–5	Cordova Channel (British Columbia)	Calculation of the boundary layer in a tidal channel	(Lueck & Lu, 1997)

superficial oxidized layer. The flume was filled with sea water of a salinity close to that of the collection site in Arcachon Bay (≈30–32). Salinity and water temperature were measured at the beginning and at the end of each experiment. Juveniles were introduced in the flume in different positions and with various flow velocities according to the experiment (Table 3, Fig. 1).

Experiment 1A

To investigate the ability of juvenile cockles to leave an unsuitable substratum (plexiglass) and to assess their ability to 'select' a suitable sediment, 250 live juveniles and 250 dead juveniles were placed together in a 2-cm wide lane on plexiglass surface, 1.2 m upstream of the sand area. This abundance (≈25 000 ind. m^{-2}) was similar to natural conditions (Bachelet

Table 2. Hydrodynamic characteristics of the flow in the centre of the experimental array of the flume for two sediments, calculated from the logarithmic section of the velocity profiles by fitting the velocity measurements to the velocity function described by the 'law of the wall' (Jumars & Nowell 1984). The number (*N*) of measurement points in the logarithmic section was always higher than 6 as recommended by Gross & Nowell (1983) to reduce the error of u_* estimates. Different parameters were calculated in relation to mean surface flow speed (*U*) for a 15-cm water depth and a 0.01-cm^2 s^{-1} kinematic viscosity. k_s is the bed roughness scale (cm), a and b the parameters of the equation $u = a\text{Ln}(z) + b$ (where u is the flow velocity in cm s^{-1} at z, height in cm from the bottom), r^2 the squared coefficient of the correlation, u_* the shear velocity (cm s^{-1}) with its confident interval (c.i. 95%) and Re_* the roughness Reynolds number

Sediment	U (cm s^{-1})	N	k_s (cm)	a	b	r^2	u_* (cm s^{-1})	c.i. (95%)	Re_*
Sand	10	6	0.024	1.24	7.90	0.997	0.51	0.04	1.22
Sand	20	7	0.024	2.30	13.93	0.999	0.94	0.03	2.67
Sand	24	9	0.024	2.42	16.84	0.998	0.99	0.04	2.38
Mud	10	10	0.006	1.34	8.50	0.998	0.55	0.02	0.33
Mud	20	10	0.006	2.35	14.68	0.999	0.96	0.03	0.58
Mud	24	11	0.006	2.16	18.73	0.998	0.88	0.03	0.53

et al., 1992b). Three surface flow velocities were tested (10, 20, 24 cm s^{-1}) and each manipulation was duplicated (two 'runs').

Experiment 1B

To verify whether or not the initial position of cockles, in relation to the working array and the wheel, could have any serious effect on the final distribution of juveniles in the sediment array, a supplementary run was performed with the cockles (250 live + 250 dead), initially placed 0.4 m downstream of the sand area. Two flow velocities were tested (10 and 24 cm s^{-1}) with no replicate.

Experiment 2

This experiment was designed to investigate the behaviour of juvenile cockles on a suitable substratum (sand). Two hundred and fifty live juveniles and 250 dead juveniles were placed in the experimental array filled with sand, about 10 cm from the upstream boundary of the array. Still water (0 cm s^{-1}) and three flow velocities were tested (10, 20, 24 cm s^{-1}), and each manipulation was duplicated (except 0 cm s^{-1}).

Experiment 3

To compare the 'sediment selection' ability in relation to sediment type and flow velocity, Experiment 1 was repeated and duplicated (two 'runs') with mud in the experimental array. To test whether juvenile cockles were able to burrow into this sediment, a small experiment was conducted concurrently. Six Petri dishes were filled with 1 cm of the same mud and placed in an aquarium filled with sea water. Ten 0.5–2.5-mm length cockles were placed on the sediment of three Petri dishes and ten 2.5–5.0-mm length cockles were placed on the sediment of the other three dishes. After 4 h in this still environment, cockles which had not burrowed (i.e., with umbo out of sediment) were counted.

In all experiments, flow conditions within the flume were maintained for 4 h, simulating half a tide for an intertidal species such as *C. edule*. After the flume was gently drained, localisation of dead stained cockles was recorded, and they were picked out and counted. The top centimetre of the sediment area was sampled in four successive 8×25-cm rectangles in each of the 5 boxes. Each sample was collected in a plastic jar, fixed with 70% ethanol, and stained with Rose Bengal. After at least 24 h, the 20 samples were sieved (200-μm mesh) and cockles were sorted under a stereo microscope. Cockle lengths were measured under a stereo microscope using a micrometer scale (except for the second run of Experiment 2). Finally, the flume was cleaned for the next run.

Statistical procedure

In Experiments 1A and 3, the proportion of individuals (= response variable) found within the different rows (1, 2, 3 and 4) of the experimental array, in different types of sediment (sand and mud), and under three flow velocities (10, 20 and 24 cm s^{-1}), was compared by means of a two-way main effects split-plot ANOVA

Table 3. Percentage (± SD) of live and dead *Cerastoderma edule* juveniles retrieved in the experimental array in relation to sediment type and flow velocity. Position of juveniles at the beginning of experiments is given (see also Fig. 1): 1.2 m upstream of the sediment array (upstr.), first row of the experimental area (row 1) or 0.36 m downstream of the sediment array (downstr.). U is the mean surface current speed (cm s^{-1}), u_* the shear velocity (cm s^{-1}), Re_* the roughness Reynolds number, S the salinity and T the temperature range (°C) during the experiments (Expt.)

Expt.	Sediment	Initial position	U	u_*	Re_*	S	T	% live	% dead
1A	sand	upstr.	10	0.51	1.22	29.7	15.8–16.4	42.4 (20.9)	6.4 (9.1)
	"	"	20	0.94	2.67	"	14.5–16.3	58.6 (0.8)	26.6 (6.5)
	"	"	24	0.99	2.38	"	14.4–16.5	76.2 (11.6)	17.0 (14.4)
1B	"	downstr.	10	0.51	1.22	32.0	14.1–14.5	40.4	0.0
	"	"	24	0.99	2.38	32.0	16.0–16.5	84.0	0.0
2	sand	row 1	0	-	-	29.7	14.5	93.2	100
	"	"	10	0.51	1.22	"	16.5–16.7	87.0 (4.2)	97.2
	"	"	20	0.94	2.67	"	14.5–16.7	96.6 (0.3)	53.6
	"	"	24	0.99	2.38	"	15.8–17.3	94.0 (3.4)	32.8
3	mud	upstr.	10	0.55	0.33	29.7	15.2–17.1	18.4 (4.0)	
	"	"	20	0.96	0.58	"	14.1–18.0	20.0 (1.7)	
	"	"	24	0.88	0.53	"	13.6–18.0	16.4 (3.4)	

(Potvin, 1993). This analysis involved two sediment types and three flow velocities as main plot factors, four row treatments as a subplot factor, and flume runs as main plot. Runs were considered as random factor.

In Experiment 2, a split-plot ANOVA was performed (Potvin, 1993) to compare the proportion of individuals (= response variable) found within the different rows (1, 2, 3 and 4) of the experimental array and under three flow velocities (10, 20 and 24 cm s^{-1}). This design involved both a between-main plot factor and a within-main plot treatment factor. The three flow velocities were considered a main plot factor, the four rows treatment a subplot factor and the two flume runs a main plot. Runs were then considered a random factor nested under flow velocities.

Prior to both analyses, data were arcsin-transformed to provide homoscedasticity. Normality of residuals and the other assumptions were graphically verified after each analysis. The ANOVAs were followed by *a posteriori* comparisons on the significant results using Tukey's procedure (Winer, 1971).

To determine whether the distribution of individuals in the experimental array was homogeneous in terms of size distribution, shell length of cockles was compared for the three experimental flow speeds and between the first 25 cm of the sediment array (row 1) and the next 75 cm (rows 2 + 3 + 4) (Fig. 1). Rows 2, 3 and 4 were combined to obtain sufficient number of cockles and to compare with row 1. Size distribution did not fit a Gaussian curve due to the sampling method (truncation by sieving). Therefore, non-parametric tests were conducted: Mann & Whitney test when two samples were compared, and Kruskal-Wallis test with more than two samples (Sokal & Rohlf, 1981).

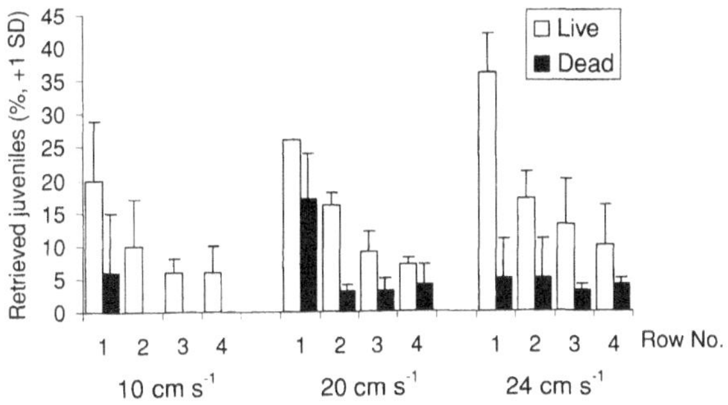

Figure 2. Experiment 1A. Percentage of live and dead juvenile cockles (± 1 SD) retrieved in the experimental array filled with sand, after 4 h of flow. Experiments were conducted at three surface flow velocities (10, 20 and 24 cm s^{-1}) and were duplicated. Individuals were sampled in the experimental array according to four rows numbered 1–4 from the upstream to the downstream part of the array (see Fig. 1).

Table 4. Parameters of the two-way main effects split-plot ANOVA comparing the proportion of juvenile cockles retrieved in the 4 different rows at the end of experiments 1 and 3, for two sediment types and three surface flow velocities (10, 20 and 24 cm s^{-1})

Source of variation	df	SS	MS	*F*
Sediment	1	1130	1130	48.30***
Velocity	2	83	41	1.77 ns
Sediment × Velocity	2	113	56	2.42 ns
Run (Sediment × Position)	6	140	23	2.76*
Position	3	490	163	19.26***
Sediment × Position	3	415	138	16.31***
Velocity × Position	6	58	10	1.13 ns
Sediment × Velocity × Position	6	107	18	2.11 ns
Error	18	153	8	
Total	47	2688		

ns: $p>0.05$; *: $p<0.05$; ***: $p<0.001$.

Results

Quantitative results

Experiment 1A

This experiment documented the behaviour of juvenile cockles settled on an unsuitable substratum (plexiglass) and their ability to settle secondarily on a suitable sediment (sand). After 4 h of unidirectional smooth-turbulent flow, different proportions of individuals were retrieved in the sediment as a function of the surface velocity (but $p>0.05$, due to high variability between the two runs at 10 cm s^{-1}) (Table 4): 42.4%, 58.6%, and 76.2% of live juveniles for surface velocities of 10, 20, and 24 cm s^{-1}, respectively (Fig. 2, Table 3). Abundance of individuals was significantly higher (Tukey test, $p<0.001$) in the first upstream row (Area A: 20–36% of initial abundance) than in the three other rows (Area B: 6 – 10%) at all flow speeds. The 'velocity × position' interaction was not significant ($p>0.05$) (Table 4).

In May 1992, cockle mean shell lengths in the experiments were relatively small: 1.4–1.9 mm (Table 5). The size distribution in the experimental array was homogeneous for the three tested flow velocities, without any significant difference between the upstream area and the rest of the sediment surface ($p>0.05$). When the whole surface was considered (Areas A+B), the length of retrieved cockles was higher ($p<0.001$) at the lowest velocity (10 cm s^{-1}).

In May 1994, cockle mean sizes were larger than two years before: 1.9–2.4 mm (Table 5). The mean shell length of retrieved individuals was larger in the first row of the experimental array than in the other three rows for 10 cm s^{-1} (but $p=0.06$) and 20 cm s^{-1} ($p<0.05$) flow velocities, whereas it was the opposite trend ($p<0.05$) for 24 cm s^{-1}. The flow velocity had no significant effect ($p>0.05$) on the mean length of juveniles retrieved from the sediment (Table 5). Although it was difficult to find juveniles out of the sedimentary array, live individuals were always retrieved (1–9% of initial abundance) between the post-experimental area and the site of initial introduction, suggesting they were able to migrate at least for a whole flume lap.

The position of dead cockles was different from the distribution of live cockles (Fig. 2). At 10 cm s^{-1}, only 6.4% of dead cockles were able to reach the experimental array (vs 42.4% for live juveniles) (Table 3). At 20 cm s^{-1}, dead cockles had spread more widely in the flume with only 26.6% of individuals in the experimental array (vs. 58.6% for live juveniles). At 24 cm s^{-1}, most dead juveniles had passed through the experimental array (where only 17.0% individuals were retrieved, vs 76.2% live juveniles) and accumulated in a 'retention area' close to the wheel.

Experiment 1B

This experiment was performed to verify whether the initial position of cockles in the flume affected the results of Experiment 1. For surface velocities of 10 and 24 cm s^{-1}, 40.4 and 84.0%, respectively, of the juveniles were retrieved in the sediment array (Fig. 3, Table 3). The proportion of cockles was homogeneous in all rows at 10 cm s^{-1} but decreased from the upstream (39%) to the downstream row (13%) at

Table 5. Experiment 1A: mean shell length (mm ± SD) and shell length range (mm) of juvenile cockles sampled in Row 1 (A, see Fig. 1) and in the other three rows (B) of the experimental array after 4 h of flow (see Expt. 1A in the text). The dates of experiment, the surface mean velocities (U, in cm s^{-1}) and the number of measured individuals (N) are included. Shell length in areas A and B were compared with a Mann & Whitney test for which U_S observed value and probability (p) are given. The effect of flow velocity on shell length of all retrieved cockles was tested with a Kruskal–Wallis procedure for which H' observed value and probability (p) are given

Date	U	Area	Mean shell length (±SD)	Shell length range	N	$U_S(p)$	Total shell length (±SD)	H' (p)
05-15-92	10	A	1.8 (±0.6)	0.8–3.9	67	2147.5 (0.77 ns)		
		B	1.9 (±0.8)	0.8–4.2	68		1.8 (±0.7)	
05-19-93	20	A	1.6 (±0.6)	1.0–4.3	66	2951.0 (0.17 ns)		22.66
		B	1.4 (±0.3)	0.8–4.3	79		1.5 (±0.5)	(0.00***)
05-20-93	24	A	1.4 (±0.4)	0.8–2.6	77	4465.5 (0.14 ns)		
		B	1.6 (±0.6)	0.8–4.2	132		1.5 (±0.6)	
05-13-94	10	A	2.4 (±1.1)	0.8–5.2	36	772.5 (0.06 ns)		
		B	1.9 (±1.0)	0.8–4.3	34		2.1 (±1.1)	
05-14-94	20	A	2.4 (±1.1)	0.8–5.0	65	3328.0 (0.02 *)		2.04
		B	1.9 (±1.0)	0.8–4.3	79		2.1 (±1.1)	(0.36 ns)
05-15-94	24	A	1.9 (±1.0)	0.8–4.8	97	2975.5 (0.04 *)		
		B	2.1 (±1.1)	0.8–4.2	72		2.0 (± 1.1)	

ns = $p > 0.05$; * = $p < 0.05$; *** = $p < 0.001$.

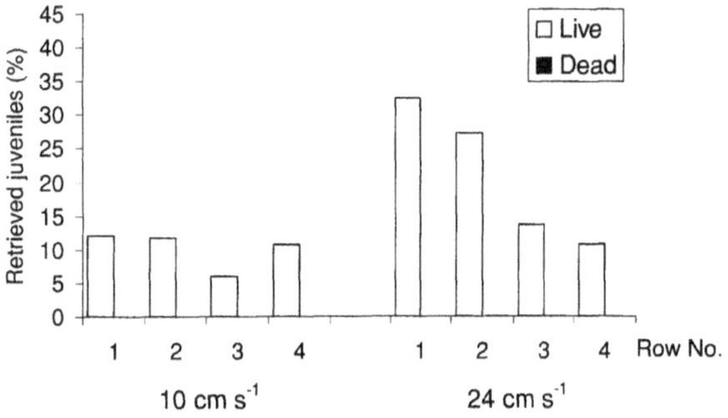

Figure 3. Experiment 1B. See legend of Figure 2, except that experiments were conducted at two surface flow velocities (10 and 24 cm s^{-1}).

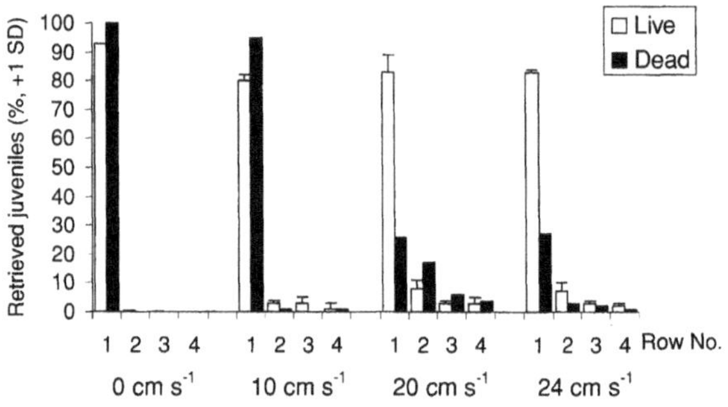

Figure 4. Experiment 2. See legend of Figure 2, except that experiments were conducted in still water and at three surface flow velocities (10, 20 and 24 cm s^{-1}).

24 cm s^{-1}. This latter result was not statistically verified due to the lack of replication in this experiment. There was no size-dependent distribution of juveniles in the sediment array at both velocities but bigger individuals reached the sediment array at higher flow velocity (Table 6). None of the dead cockles reached the sediment array.

Experiment 2

This experiment documented the behaviour of juvenile cockles settled on a suitable sediment (sand). At the end of each run, at least 87.0% of the live juveniles were retrieved in the sediment array (Table 3), most of them at the same place they were set initially (Fig. 4). The proportions of individuals found in each area did not differ with flow velocity ($p>0.05$) (Table 7). The mean shell length of retrieved individuals did not differ ($p>0.05$) in the first row and in the three following rows (Table 8) at all flow velocities. Flow velocity had no significant effect ($p>0.05$) on the mean shell length of juveniles retrieved from the sediment. The experiment in still water (Fig. 4, Table 3) demonstrated that juvenile migration without current was negligible over 4 h.

Table 6. Experiment 1B. See legend of Table 5

Date	U	Area	Mean shell length (±SD)	Shell length range	N	U_S (p)	Total mean shell length (±SD)	H' (p)
05-22-93	10	A	1.4 (±0.4)	0.8–3.1	30	435.5 (0.99 ns)		
		B	1.4 (±0.5)	0.8–3.0	30		1.4 (±0.4)	3833.0
05-24-93	24	A	1.5 (±0.3)	1.0–2.7	36	1323.5 (0.76 ns)		(0.02 *)
		B	1.5 (±0.3)	1.0–2.4	72		1.5 (±0.3)	

ns = $p > 0.05$; * = $p < 0.05$.

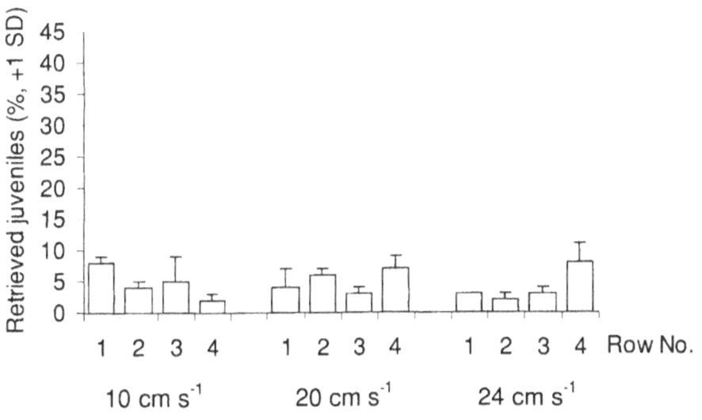

Figure 5. Experiment 3. See legend of Figure 2, except that the experiment area of the flume was filled with mud.

As in Experiment 1A, the motion of dead juveniles was different from that of live juveniles (Fig. 4). As expected, most individuals (97.2%) remained in the sediment area for a 10-cm s^{-1} flow velocity (Table 3). With increasing velocity (20 and 24 cm s^{-1}), dead cockles were carried out of this area and the retrieved densities were 53.6 and 32.8%, respectively, and many individuals were found close to the wheel as in the first experiment.

Experiment 3

This experiment was similar to Experiment 1A except that the sediment boxes were filled with mud instead of sand. After 4 h of flow, ≈ 18% of live juveniles were retrieved from the sediment array (Fig. 5, Table 3) with no significant effect of the flow velocity ($p>0.05$) (Table 4). This proportion was two to three times less than when the experiment area contained sand ($p<0.001$). With mud, distribution of cockles in the sediment array was more homogeneous (Tukey test, $p>0.05$) than with sand in Experiment 1A with no decrease of abundance between positions 1 and 4 (significant interaction Sediment × Position, Table 4) (Fig. 5).

The mean shell length of retrieved individuals was the same ($p>0.05$) in the first row compared to the three following rows (Table 9) at all flow velocities, except twice when the length in the first row was larger: in May 1994 for 24 cm s^{-1} ($p<0.05$) and in June 1994 for 10 cm s^{-1} ($p<0.01$). The flow velocity had no significant effect ($p>0.05$) on the mean length of juveniles retrieved from the sediment (Table 9). The small concurrent experiment in still water demonstrated that most juvenile cockles were perfectly able to burrow in mud (80%, standard deviation = 11%) with no effect of shell length (in the range 0.5-5.0 mm).

Behavioural observations

In our experimental flow velocity range, dead juvenile cockles were never resuspended in the water column. They drifted or rolled on the bottom, with their speed of transport increasing as flow velocity increased. At the lowest flow velocity (10 cm s^{-1}), most dead juveniles did not move at all from the area where they were placed initially. Some dead individuals, and some live juveniles, were always retrieved from a retention area, a few centimetres downstream of the disc-wheel.

At all flow velocities, no live individual remained at its initial place. Three methods of displacement have been observed. Firstly some juveniles simply drifted on the bottom, either on the plexiglass or on the sediment. Most of the time, cockles were observed dragging ≈1 cm long byssal threads with sand grains stuck on them. In this case, the umbo was directed downstream. It was less common to observe live juveniles simply rolling at the bottom surface, without obvious use of byssal threads. When passing through the experimental array, juveniles sometimes stopped 'instantaneously' as if their byssal threads had anchored in the sediment, stood up with the umbo downstream and the foot buried in the sediment to be

Table 7. Parameters of the split-plot ANOVA comparing the proportion of juvenile cockles retrieved in the 4 different rows at the end of Experiment 2, and for three surface flow velocities (10, 20 and 24 cm s^{-1})

Source of variation	df	SS	MS	F
Velocity	2	51	25	2.50 ns
Run(Velocity)	3	30	10	1.07 ns
Position	3	13520	4507	472.76 ***
Velocity × Position	6	43	7	0.75 ns
Error	9	86	10	

ns: $p>0.05$; ***: $p<0.001$.

in profile against the flow direction. Then, the cockles burrowed rapidly when sediment was sand but had much more difficulties when sediment was mud. In the latter case, juveniles often failed in settling and resumed their drifting, especially at higher flow velocities. Secondly some juveniles were observed drifting in the water column, at heights between 1 and 10 cm above the bottom, but never floating at the surface. Few individuals were followed visually during a full lap and these usually drifted with the foot upwards. As juveniles were approaching the bottom, the foot was directed upwards as well as downwards. When the foot touched the sediment, cockles stopped drifting instantaneously, as described in the first case. Another behaviour was observed twice when sand was used in the experimental area: juveniles (1–2 mm length) were drifting horizontally at 2–3 cm above the bottom, when they suddenly fell vertically onto the sediment and burrowed within a few seconds. The third means of locomotion was a combination of the two previous ones. Juveniles were observed drifting at the sediment surface over a few centimetres or even metres and eventually resuspended. Sometimes 'large' individuals (≈5 mm length) carrying long byssal threads with entangled sand grains seemed very heavy, however, these individuals managed to resuspend themselves with their byssal load. In all situations, juvenile cockles were remarkable for the intense activity of their foot, which they extruded out of the shell and agitated constantly.

Discussion

Effect of flow velocity

In Experiment 1A, the motion of live juveniles was completely different from the motion of dead cockles at all experimental flow velocities. Initially placed on an unsuitable substratum (plexiglass), the ratio 'number of live juveniles retrieved in the sandy area / total number of live juveniles' was 5–10 times greater than the ratio 'surface of the sandy experimental array / total surface of the flume bottom'. These results suggest that juveniles actively moved off the unsuitable plexiglass and selected suitable sediment (Fig. 6). For example, a 10 cm s^{-1}current was not sufficient to remove most dead juveniles from their initial position on plexiglass, while at this current velocity no live individuals remained on this flume section. The fact that half the retrieved cockles were found in the first row (Fig. 2) suggests that some juveniles rolled and crawled to reach this place. This was confirmed by observations. Logically, the smallest juvenile cockles resuspended in the water column more easily than the largest ones. A previous experiment (de Montaudouin, 1997) showed that 70.5% of cockle juveniles with shell length <2.5 mm migrated by resuspension, vs. 21.5% of individuals with shell length comprised between 2.5 and 4.8 mm. In the first data set (1992), there was no size-dependent distribution in the experimental array (Table 5). However, in the second run (1994), a size-dependent distribution was found (Table 5). At 10 and 20 cm s^{-1}, some of the bigger individuals rolled and accumulated in the first row; at 24 cm s^{-1}, these bigger juveniles had difficulties resisting the current and spent more time trying to resettle and burrow. When flow velocity increased, the mean length of retrieved juveniles at the end of most experiments decreased, as if bigger individuals had more difficulties stopping on the sediment array (Tables 5, 8 and 9).

In Experiment 1A, the number of retrieved cockles in relation to flow velocity was difficult to interpret. Indeed, higher numbers of retrieved juveniles under high velocity may be an experimental artifact: during a 4 h experiment, floating juveniles had more opportunities to settle on the experimental array when flow velocity was 24 cm s^{-1} than when it was 10 cm s^{-1}. However, the correlation between flow velocity and percentage of retrieved cockles was neither linear nor significant (Table 4): this percentage only increased from 42 to 59% while surface water velocity doubled from 10 to 20 cm s^{-1}; the percentage was 76 when surface velocity was 24 cm s^{-1} (Table 3). Pawlik & Butman (1993) showed that larvae of the annelid *Phragmatopoma lapidosa* achieved a better sediment selection at higher current speeds. Another hypothesis, supported by our observations, is that a current threshold existed

Table 8. Experiment 2. See legend for Table 5

Date	U	Area	Mean shell length (±SD)	Shell length range	N	U_S (p)	Mean shell length (±SD)	H' (p)
06-17-93	10	A	1.3 (±0.4)	0.8–2.9	69	729.0 (0.78 ns)		
		B	1.3 (±0.4)	0.8–2.1	22		1.3 (±0.7)	
06-18-93	20	A	1.2 (±0.3)	0.8–2.7	92	1289.5 (0.35 ns)		4.15
		B	1.1 (±0.2)	0.8–1.8	25		1.2 (±0.5)	(0.13 ns)
06-19-93	24	A	1.2 (±0.3)	0.8–2.1	81	960.0 (0.20 ns)		
		B	1.1 (±0.2)	0.8–1.7	20		1.2 (±0.6)	

ns = $p > 0.05$.

Table 9. Experiment 3. See legend for Table 5

Date	U	Area	Mean shell length (±SD)	Shell length range	N	U_S(p)	Mean shell length (±SD)	H' (p)
06-20-93	10	A	1.6 (±0.6)	0.8–2.8	21	280.5 (0.01 **)		
		B	1.2 (±0.2)	0.8–1.8	18		1.4 (±0.5)	
06-21-93	20	A	1.2 (±0.3)	0.8–2.1	17	313.5 (0.89 ns)		2.55
		B	1.2 (±0.3)	0.9–2.0	36		1.2 (±0.3)	(0.28 ns)
06-22-93	24	A	1.3 (±0.3)	0.9–2.0	9	156.5 (0.78 ns)		
		B	1.3 (±0.5)	0.8–3.9	43		1.3 (±0.5)	
05-16-94	10	A	1.9 (±1.0)	0.8–4.0	18	314.5 (0.99 ns)		
		B	2.0 (±1.2)	0.8–4.8	35		2.0 (±1.1)	
05-17-94	20	A	1.6 (±1.0)	1.0–3.4	6	122.0 (0.50 ns)		0.24
		B	2.0 (±1.3)	0.8–5.7	25		2.0 (±1.2)	(0.89 ns)
05-18-94	24	A	2.0 (±0.9)	1.0–3.5	9	163.5 (0.05 *)		
		B	1.5 (±0.7)	0.9–3.5	25		1.6 (± 0.7)	

ns = $p > 0.05$; * = $p < 0.05$; ** = $p < 0.01$.

between 20 and 24 cm s^{-1}, beyond which juveniles had more difficulties sticking to the plexiglass substratum and were resuspended in the water column at higher numbers.

The complementary experiment (Experiment 1B) confirmed the previous results and observations. The fact that no dead cockles could reach the experimental area when 40–84% of live juveniles were retrieved, demonstrates the active participation of cockles in their migration and suggests that most of them resuspend in the water column. This resuspension of juveniles is demonstrated at 10 cm s^{-1} by the homogeneous distribution of cockles in the experimental array (Fig. 3), with no size-dependent pattern (Table 6). At 24 cm s^{-1}, current strength probably allowed more juveniles to reach the experimental array by crawling and rolling. As a consequence, more juveniles were retrieved in the first rows (Fig. 3) and their mean length was significantly higher than at 10 cm s^{-1} (Table 6). The total percentages of retrieved cockles (40.4 and 84.0% at 10 and 24 cm s^{-1}, respectively) in Experiment 1B were similar to those observed in Experiment 1A (42.4 and 76.2%, respectively), suggesting that the initial position of live juveniles did not modify the results.

Bivalve juveniles that initially settle in soft sediments can be resuspended due to tidal currents and waves (Emerson & Grant, 1991; Armonies, 1994a). The intensity of this resuspension is due mainly to the strength of currents, effects of which can be magnified by processes such as bioturbation (Davis, 1993; Olivier et al., 1996b; Blanchard et al., 1997), species behaviour (Palmer, 1986), and nycthemeral rhythm (Grant, 1980). However, no correlation between resuspension and current strength was found by Thiébaut et al. (1996). In a non-erosive system like Experiment

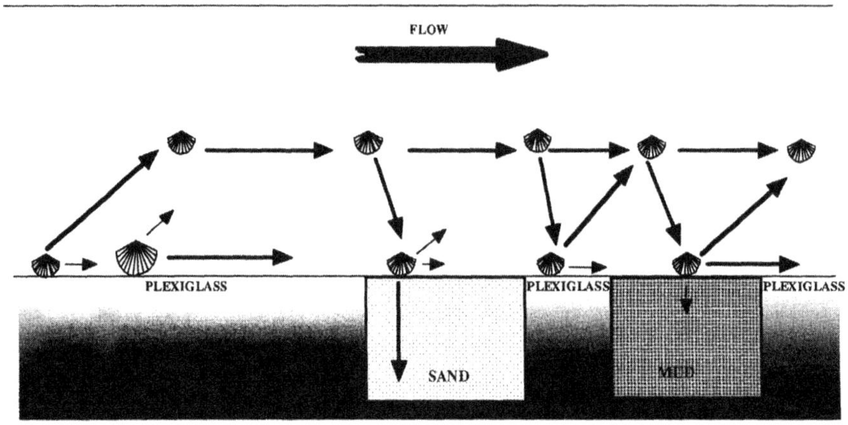

Figure 6. Side-view of summarized options of juvenile cockle motions in flow conditions, with three substrata: plexiglass, sand and mud. Size and direction of arrows indicate the chosen option (drifting, rolling, resuspension, landing, burrowing). Scales and disposition of substrata are arbitrary.

2, very few cockles were retrieved out of their initial position. Most burrowed rapidly and remained out of reach of shearing force of the boundary layer.

Effect of the interaction between flow velocity and sediment type

With mud in the experimental array, the percentage of retrieved cockles was two to four times lower than when the experimental array was filled with sand. The type of sediment affected significantly the number of retrieved juveniles from the experimental array, with no effect of flow velocity nor interaction (Sediment × Velocity, Table 3). Selection of sediment could be achieved in relation to diet. For example, larvae of the deposit-feeding polychaete *Capitella* sp. prefer natural mud to inorganic sediment (Butman & Grassle, 1992; Grassle et al., 1992a; Snelgrove et al., 1993). Conversely, for suspension feeders (that usually select high flow conditions), larval settlement of the tube worm *Phragmatopoma lapidosa californica* was less intense in still water (Pawlik et al., 1991; Pawlik & Butman, 1993), and mussel *Mytilus edulis* juveniles adhered better to a filamenteous substratum in a turbulent environment than in still water (Eyster & Pechenik, 1987). However, the effect of hydrodynamics on settlement depends primarily on the species concerned (Mullineaux & Garland, 1993) and the smallest modification of the boundary layer may alter settlement (Mullineaux & Butman, 1990).Thus, although the bivalve *Mercenaria mercenaria* is a suspension feeder, experiments with larvae in still water showed no significant habitat selection between organic rich mud and abiotic glass beads (Bachelet et al., 1992a). In our experiments, the hypothesis that a 'diet determinism' explains the preferential secondarily settlement of cockles on sand rather than on mud was not investigated but is doubtful. From visual observations, it seems that the habitat selection displayed by juvenile cockles has a mechanical basis. Mud is smoother and more cohesive than sand, and juveniles often failed to stop on this 'slippery' substratum. Their intense efforts to interrupt their drifting on the mud surface suggest that there were no repellent chemicals. Moreover, it is possible that part of the 20% juveniles retrieved in the mud took advantage of the slight heterogeneity of the bottom surface at the border of sediment boxes to anchor themselves. Experiments in still water demonstrated that juvenile cockles had no difficulty burrowing in this muddy sediment.

The ecological implications of the hydrodynamic transport of recently-metamorphosed, surface-dwelling bivalves are important. There may be selective advantages in a decoupled recruitment pattern consisting of an initial settlement period followed by a mobile juvenile phase (Roegner et al., 1995). Initial settlement into an area occupied by dense concentrations of adults has been theorized to induce high settler mortality (Bayne, 1964; André & Rosenberg, 1991; de Montaudouin & Bachelet, 1996). In contrast, secondary settlement, as demonstrated in the present study with cockles <5.8 mm in shell length, could reduce juvenile mortality (Beukema & de Vlas, 1989). In terms of population dynamics, these results imply that the disappearance of juvenile cohorts from a given site could be the result of mortality as well as of a migration with secondary settlement (Butman, 1987; Beaumont & Barnes, 1992; Armonies, 1996). Further more, the observed spatial high variability in cockle abundance often intends to suggest competition and predation as possible processes (Lindegarth et al., 1995). Post-settlement dispersal over distances at least >13 m (Experiment 1B), but certainly larger in natural environment, may also contribute to explain this variability.

Acknowledgements

This research was supported partly by CNRS and IFREMER through the French Programme on Recruitment Studies (PNDR) and Coastal Environment (PNEC). We thank C. Retière for providing research facilities at the Laboratoire Maritime du Muséum at Dinard, F. Olivier for his contribution to flow velocity measurements, and K. Hudson and R. Carnegie for language corrections of the manuscript. The authors are grateful to two anonymous referees who significantly helped them in improving the manuscript.

References

Abelson, A. & D. Denny, 1997. Settlement of marine organisms in flow. Ann. Rev. Ecol. Syst. 28: 317–339.

André, C., P. R. Jonsson & M. Lindegarth, 1993. Predation on settling bivalve larvae by benthic suspension feeders: the role of hydrodynamics and larval behaviour. Mar. Ecol. Prog. Ser. 97: 183–192.

André, C. & R. Rosenberg, 1991. Adult-larval interactions in the suspension-feeding bivalves *Cerastoderma edule* and *Mya arenaria*. Mar. Ecol. Prog. Ser. 71: 227–234.

Armonies, W., 1992. Migratory rhythms of drifting juvenile molluscs in tidal waters of the Wadden Sea. Mar. Ecol. Prog. Ser. 83: 197–206.

Armonies, W., 1994a. Turnover of postlarval bivalves in sediment of tidal flats in Königshafen (German Wadden Sea). Helgolander wiss. Meeresunters. 48: 291–297.

Armonies, W., 1994b. Drifting meio- and macrobenthic invertebrates on tidal flats in Königshafen: a review. Helgolander wiss. Meeresunters. 48: 299–320.

Armonies, W., 1996. Changes in distribution patterns of 0-group bivalves in the Wadden sea: byssus-drifting releases juveniles from the constraints of hydrography. J. Sea Res. 35: 323–334.

Bachelet, G., C. A. Butman, C. M. Webb, V. R. Starczak & P. V. R. Snelgrove, 1992a. Non-selective settlement of *Mercenaria mercenaria* (L.) larvae in short-term, still-water, laboratory experiments. J. exp. mar. Biol. Ecol. 161: 241–280.

Bachelet, G., J. Guillou & P.-J. Labourg, 1992b. Adult-larval and juvenile interactions in the suspension-feeding bivalve, *Cerastoderma edule*: field observations and experiments. In Colombo, G., I. Ferrari, V. Ceccherelli & R. Rossi (eds), Marine Eutrophication and Population Dynamics. Olsen & Olsen, Fredensborg: 175–182.

Bayne, B. L., 1964. Primary and secondary settlement in *Mytilus edulis* L. (Mollusca). J. anim. Ecol. 33: 513–523.

Beaumont, A. R. & D. A. Barnes, 1992. Aspects of veliger larval growth and byssus drifting of the spat of *Pecten maximus* and *Aequipecten (Chlamys) opercularis*. ICES Mar. Sci. Symp. 49: 417–423.

Beukema, J. J. & J. de Vlas, 1989. Tidal-current transport of thread-drifting postlarval juveniles of the bivalve *Macoma balthica* from the Wadden Sea to the North Sea. Mar. Ecol. Prog. Ser. 52: 193–200.

Blanchard, G. F., P.-G. Sauriau, V. Cariou-Le Gall, D. Gouleau, M.-J. Garet & F. Olivier, 1997. Kinetics of tidal resuspension of microbiota: testing the effects of sediment cohesiveness and bioturbation using flume experiments. Mar. Ecol. Prog. Ser. 151: 17–25.

Butman, C. A., 1986. Sediment trap biases in turbulent flows: results from a laboratory flume study. J. mar. Res. 44: 645–693.

Butman, C. A., 1987. Larval settlement of soft-sediment invertebrates: the spatial scales of patterns explained by active habitat selection and the emerging role of hydrodynamical processes. Oceanogr. mar. biol. ann. Rev. 25: 113–165.

Butman, C. A., 1989. Sediment-trap experiments on the importance of hydrodynamical processes in distributing settling invertebrate larvae in near-bottom waters. J. exp. mar. Biol. Ecol. 134: 37–88.

Butman, C. A. & J. P. Grassle, 1992. Active habitat selection by *Capitella* sp. I larvae. I. Two-choice experiments in still water and flume flows. J. mar. Res. 50: 669–715.

Butman, C. A., J. P. Grassle & C. M. Webb, 1988. Substrate choices made by marine larvae settling in still water and in a flume flow. Nature 333: 771–773.

Creek, G. A., 1960. The development of the Lamellibranch *Cardium edule* L. Proc. Zool. Soc. Lond. 135: 243–260.

Cummings, V. J., R. D. Pridmore, S. F. Thrush & J. E. Hewitt, 1993. Emergence and floating behaviours of post-settlement juveniles of *Macomona liliana* (Bivalvia: Tellinacea). Mar. Behav. Physiol. 24: 25–32.

Davis, W. H., 1993. The role of bioturbation in sediment resuspension and its interaction with physical shearing. J. exp. mar. Biol. Ecol. 171: 187–200.

de Montaudouin, X., 1997. Potential of bivalves' secondary settlement differs with species: a comparison between cockle (*Cerastoderma edule*) and clam (*Ruditapes philippinarum*) juvenile resuspension. Mar. Biol. 128: 639–648.

de Montaudouin, X. & G. Bachelet, 1996. Experimental evidence of complex interactions between biotic and abiotic factors in the dynamics of an intertidal population of the bivalve *Cerastoderma edule*. Oceanol. Acta 19: 449–463.

Eckman, J. E., 1983. Hydrodynamic processes affecting benthic recruitment. Limnol. Oceanogr. 28: 241–257.

Eckman, J. E., F. E. Werner & T. F. Gross, 1994. Modelling some effects of behavior on larval settlement in a turbulent boundary layer. Deep Sea Res. II, 41: 185–208.

Emerson, C. W. & J. Grant, 1991. The control of soft-shell clam (*Mya arenaria*) recruitment on intertidal sandflats by bedload sediment transport. Limnol. Oceanogr. 28: 1288–1300.

Eyster, L. S. & J. A. Pechenik, 1987. Attachment of *Mytilus edulis* L. larvae on algal and byssal filaments is enhanced by water agitation. J. exp. mar. Biol. Ecol. 114: 99–110.

Fréchette, M., C. A. Butman & W. R. Geyer, 1989. The importance of boundary-layer flows in supplying phytoplankton to the benthic suspension feeder, *Mytilus edulis* L. Limnol. Oceanogr. 34: 19–36.

Grant, J., 1980. A flume study of drift in marine infaunal amphipods (Haustoriidae). Mar. Biol. 56: 79–84.

Grassle, J. P., C. A. Butman & S. W. Mills, 1992a. Active habitat selection by *Capitella* sp. I larvae. II. Multiple-choice experiments in still water and flume flows. J. mar. Res. 50: 717–743.

Grassle, J. P., P. V. R. Snelgrove & C. A. Butman, 1992b. Larval habitat choice in still water and flume flows by the opportunistic bivalve *Mulinia lateralis*. Neth. J. Sea Res. 30: 33–44.

Grégoire, Y., E. Bourget & J.-L. Verrette, 1996. Deposition of mimics of planktonic invertebrate larvae on simple and complex substrata in flume flows. Mar. Ecol. Prog. Ser. 135: 89–100.

Gross, T. F. & A. R. M. Nowell, 1983. Mean flow and turbulence scaling in a tidal boundary layer. Cont. Shelf Res. 2: 109–126.

Harvey, M. & E. Bourget, 1997. Recruitment of marine invertebrates onto arborescent epibenthic structure: active and passive processes acting at different spatial scales. Mar. Ecol. Prog. Ser. 153: 203–215.

Jaklin, S. & C.-P. Günther, 1996. Macrobenthic driftfauna of the Gröninger Plate. Senckenbergiana Marit. 26: 127–134.

Jonsson, P. R., C. André & M. Lindegarth, 1991. Swimming behaviour of marine bivalve larvae in a flume boundary-layer flow: evidence for near-bottom confinement. Mar. Ecol. Prog. Ser. 79: 67–76.

Jumars, P. A. & A. R. M. Nowell, 1984. Fluid and sediment dynamic effects on marine benthic community structure. Amer. Zool. 24: 45–55.

Lambert, R., 1991. Recrutement d'espèces benthiques à larves pélagiques en régime mégatidal. Cas de *Pectinaria koreni* (Malmgren), annélide polychètes. PhD University Rennes 1.

Lane, D. J. W., A. R. Beaumont & J. R. Hunter, 1984. Byssus drifting and the drifting threads of the young post-larval mussel *Mytilus edulis*. Mar. Biol. 84: 301–308.

Lane, D. J. W., J. A. Nott & D. J. Crisp, 1982. Enlarged stem glands in the foot of the post-larval mussel, *Mytilus edulis*: adaptation for bysso-pelagic migration. J. Mar. Biol. Ass. U.K. 62: 809–818.

Lindegarth, M., C. André & P. R. Jonsson, 1995. Analysis of the spatial variability in abundance and age structure of two infaunal bivalves, *Cerastoderma edule* and *C. lamarcki*, using hierarchical sampling programs. Mar. Ecol. Prog. Ser. 116: 85–97.

Lueck, R. G. & Y. Lu, 1997. The logarithmic layer in a tidal channel. Cont. Shelf Res. 17: 1785–1801.

Martel, A. & F.-S. Chia, 1991. Drifting and dispersal of small bivalves and gastropods with direct development. J. exp. mar. Biol. Ecol. 150: 131–147.

Martel, A. & T. Dieffenbach, 1993. Effects of body size, water current and microhabitat on mucous-thread drifting in post-metamorphic gastropods *Lacuna* spp. Mar. Ecol. Prog. Ser. 99: 215–220.

Mullineaux, L. S. & C. A. Butman, 1990. Recruitment of encrusting benthic invertebrates in boundary-layer flows: a deep-water experiment on Gross Seamount. Limnol. Oceanogr. 35: 409–423.

Mullineaux, L. S. & E. D. Garland, 1993. Larval recruitment in response to manipulated field flow. Mar. Biol. 116: 667–683.

Nowell, A. R. M. & P. A. Jumars, 1984. Flow environments of aquatic benthos. Ann. Rev. Ecol. Syst. 15: 303–328.

Olivier, F., N. Desroy & C. Retière, 1996a. Habitat selection and adult-recruit interactions in *Pectinaria koreni* (Malmgren) (Annelida: Polychaeta) post-larval populations: results of flume experiments. J. Sea Res. 36: 217–226.

Olivier, F., C. Vallet, J.-C. Dauvin & C. Retière, 1996b. Drifting in post-larvae and juveniles in an *Abra alba* (Wood) community of the eastern part of the Bay of Seine (English Channel). J. exp. mar. Biol. Ecol. 199: 89–109.

Palmer, M. A., 1986. Hydrodynamics and structure: interactive effects on meiofauna dispersal. J. exp. mar. Biol. Ecol. 104: 53–68.

Pawlik, J. R. & C. A. Butman, 1993. Settlement of a marine tube worm as a function of current velocity: interacting effects of hydrodynamics and behavior. Limnol. Oceanogr. 38: 1730–1740.

Pawlik, J. R., C. A. Butman & V. R. Starczak, 1991. Hydrodynamic facilitation of gregarious settlement of a reef-building tube worm. Science 251: 421–424.

Potvin, C., 1993. ANOVA: experiments in controlled environments. In Scheiner, M. & J. Gurewitch (eds), Design and Analysis of Ecological Experiments. Chapman & Hall, New York: 46–68.

Prezant, R. S. & K. Chalermwat, 1984. Flotation of the Bivalve *Corbicula fluminea* as a means of dispersal. Science 225: 1491–1493.

Roegner, C., C. André, M. Lindegarth, J. E. Eckman & J. Grant, 1995. Transport of recently settled soft-shell clams (*Mya arenaria* L.) in laboratory flume flow. J. exp. mar. Biol. Ecol. 187: 13–26.

Sigurdsson, J. B., C. W. Titman & P. A. Davies, 1976. The dispersal of young post-larval bivalve molluscs by byssus threads. Nature 262: 386–387.

Snelgrove, P. V. R., C. A. Butman & J. P. Grassle, 1993. Hydrodynamic enhancement of larval settlement in the bivalve *Mulinia lateralis* (Say) and the polychaete *Capitella* sp I in microdepositional environments. J. exp. mar. Biol. Ecol. 168: 71–109.

Snelgrove, P. V. R., J. P. Grassle, J. F. Grassle, R. F. Petrecca & H. Ma, 1999. *In situ* habitat selection by settling larvae of marine soft-sediment invertebrates. Limnol. Oceanogr. 44: 1341–1347.

Sokal, R. & F. Rohlf, 1981. Biometry. WH Freeman, New York. 859 pp.

Thiébaut, E., J.-C. Dauvin & Z. Wang, 1996. Tidal transport of *Pectinaria koreni* postlarvae (Annelida: Polychaeta) in the Bay of Seine (eastern English Channel). Mar. Ecol. Prog. Ser. 138: 63–70.

Vahl, O., 1983. Mucus drifrting in the limpet *Helcion* (= *Patina*) *pellucidus* (Prosobranchia, Patellidae). Sarsia 68: 209–211.

Vogel, S., 1989. Life in moving fluid. The Physical Biology of Flow. Princeton University Press, Princeton. 335 pp.

Winer, B. J., 1971. Statistical Principles in Experimental Design. McGraw-Hill, New York. 907 pp.

Woodin, S. A., 1986. Settlement of infauna: larval choice? Bull. mar. Sci. 39: 401–407.

Woodin, S. A., 1991. Recruitment of infauna: positive or negative cues. Amer. Zool. 31: 797–807.

Wu, R. S. S. & P. K. S. Shin, 1997. Sediment characteristics and colonization of soft-bottom benthos: a field manipulation experiment. Mar. Biol. 128: 475–487.

Yankson, K., 1986. Observations on byssus systems in the spat of *Cerastoderma glaucum* and*C. edule*. J. mar. biol. Ass. U.K. 66: 277–292.

Young, G. A., 1985. Byssus-thread formation by the mussel *Mytilus edulis*: effects of environmental factors. Mar. Ecol. Prog. Ser. 24: 261–271.

Hydrobiologia **503:** 117–120, 2003.
M.B. Jones, A. Ingólfsson, E. Ólafsson, G.V. Helgason, K. Gunnarsson &
J. Svavarsson (eds), Migrations and Dispersal of Marine Organisms.

Anchovy egg and larval distribution in relation to biological and physical oceanography in the Strait of Sicily

A. Cuttitta[1], V. Carini[1], B. Patti[1], A. Bonanno[1], G. Basilone[1], S. Mazzola[1], J. García Lafuente[2], A. García[3], G. Buscaino[4], L. Aguzzi[5], L. Rollandi[1], G. Morizzo[1] & C. Cavalcante[1]
[1]*Istituto di ricerche sulle Risorse Marine e l'Ambiente, Consiglio Nazionale delle Ricerche – Via L.Vaccara 61, 91026 Mazara del Vallo (TP), Italy*
[2]*Department of Applied Physics, University of Malaga, Campus de Teatinos, E.T.S.I. Telecomunicacion, Malaga, Spain*
[3]*Instituto Español de Oceanografia (IEO), Fuengirola, Spain*
[4]*Dipartimento Scienze Ambientali, Università di Venezia "Ca Foscari", Calle Santa Marta, Italy*
[5]*Agenzia Regionale per la Protezione dell'Ambiente Lazio – Via Garibaldi 114, 02100 Rieti, Italy*
E-mail: angela@irma.pa.cnr.it

Key words: Engraulis encrasicolus, anchovy, zooplankton, eggs, larvae, Strait of Sicily

Abstract

The European Anchovy (*Engraulis encrasicolus*, Linnaeus, 1758) represents one of the most important fishery resources in some areas of the Mediterranean. This short-lived, small pelagic fish is characterized by large interannual fluctuations, probably as a result of environmental variability. As part of the European Project Med 98-070, the main aim of which was the study of the anchovy population in the Strait of Sicily, icthyoplankton surveys were carried out between 1999 and 2001, during the peak spawning season for anchovy. Present work reports the relationship between meso-zooplankton biomass and the abundance of anchovy eggs and larvae in the Strait of Sicily. Data on anchovy egg abundance showed that the main spawning area was located in the north-western region of the study area. The branch of the Atlantic Ionian Stream, running parallel to the southern Sicilian coast, acts as a transport mechanism for anchovy eggs and larvae towards the southernmost end of the island, off Cape Passero. Observed distributions were largely consistent with local hydrographic features, which allow larvae to be retained in areas providing the necessary feeding conditions for recruitment success.

Introduction

The European anchovy (*Engraulis encrasicolus*, Linnaeus, 1758) represents one of the most important fishery resources in some areas of the Mediterranean (Fig. 1). Interannual variability in biomass of such small pelagic fish species may be due to fishing effort and/or recruitment success (Borja et al., 1996). The latter is dependent on several biological and environmental parameters, amongst which the most important limiting factor is the presence and availability of food (Casavola et al., 1998; Tikhonova et al., 2000; James et al., 2003). The dynamics of the biomass of the anchovy population in the Strait of Sicily were addressed by two European projects (Med 96-052 and Med 98-070). Within the framework of these projects, relationships between meso-zooplankton biomass and anchovy egg and larval abundance and distribution in the Strait of Sicily were examined in relation to possible transport mechanisms and interannual variations of their distribution. Distribution of anchovy early stages is highly dependent on surface water dynamics (García Lafuente et al., 2002). The surface circulation in the study area is controlled mainly by the Modified Atlantic Water (MAW) motion, the so-called Atlantic-Ionian Stream (AIS, Fig. 1) (Robinson et al., 1999). According to Robinson et al. (1999), the current enters the Strait of Sicily from the west, describing a large cyclonic meander which embraces the Adventure Bank, then approaches the shore near

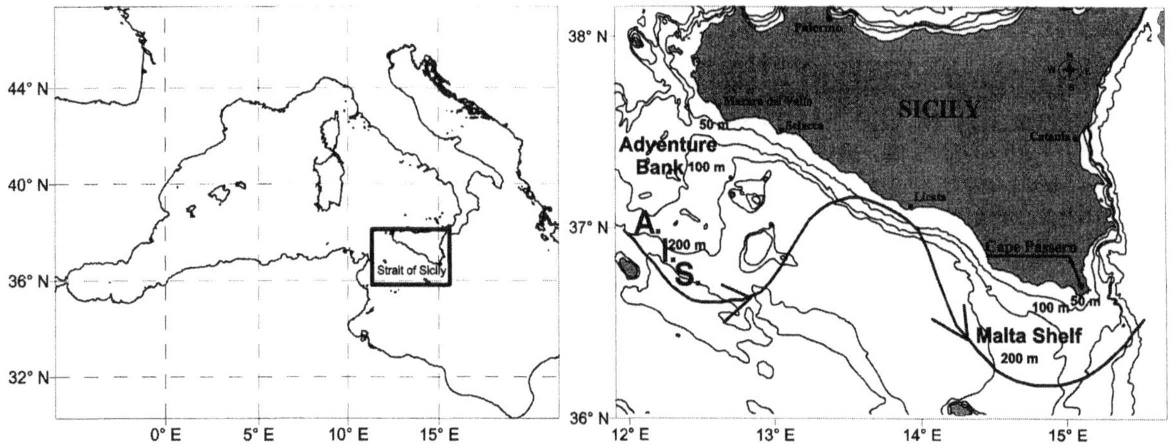

Figure 1. Western and Central Mediterranean, with the Strait of Sicily shown within the box (left panel), and a detailed map of the study area shown in the right panel with the average path of Atlantic Ionian Stream (A.I.S.) superimposed.

the middle of the southern coast of Sicily and separates again when it encounters the shelf of Malta. In doing so, the AIS forms two large cyclonic vortices, the first, already mentioned, and the second around the Malta shelf, off Cape Passero. The path of AIS and its year-to-year variability has consequences for other predominant hydrological phenomena occurring in the region, such as the extension of upwelling and the formation of frontal structures.

To achieve the aims of this study, hydrological and ichthyoplanktonic surveys were conducted from 1999 to 2001, during the peak of the anchovy spawning period (Giráldez & Abad, 1995; Cuttitta et al., 1999).

Materials and methods

Annual summer cruises were carried out from 1999 to 2001 using the O/V "Urania". In 1999, the cruise was carried out from 6 to 17th June; in 2000, from 23rd June to 13th July; and in 2001, from 5th to 26th July. Plankton samples, collected on the continental shelf of the southern Sicilian coast from stations based on a 4 n/mile step grid, were obtained using Bongo 40 oblique hauls carried out at a speed of 2 knots using a 200 μm mesh size net for both sides of the frame. Hauls were towed from the bottom to the surface or from 100 m to the surface where depth was more than 100 m. Ichthyoplankton and meso-zooplankton samples were collected simultaneously. A total of 345 stations was carried out: 70 stations in 1999, 131 stations in 2000, and 144 stations in 2001. Hydrological data were acquired during the same cruises by means of a CTD probe model SBE 911. CTD casts were analyzed using Ocean Data View software and the path of AIS was depicted by the minimum salinity technique. Biological samples were preserved in 4% buffered formaldehyde and seawater solution. Sorting of ichthyoplankton was performed in the laboratory by means of a stereo binocular at $\times 12$ magnification. Fish eggs and larvae were extracted, and anchovy eggs and larvae counted. Zooplanktonic biomass, as dry weight (mg/m^3), was obtained by drying half the sample in an oven at 60 °C for 48 h, after removing the fish eggs and larvae. Data interpolation (kriging method) and drawing of surface distribution maps was done by means of Surfer Golden software.

Results

Presence of eggs in the middle and southern part of the continental shelf (Fig. 2 a, b, c) was interpreted as indicating the spawning areas. Development from egg to larval stage is temperature dependent and, for this species in this area, takes about 36 h (Holden & Raitt, 1975). The 1999 and 2000 distributions, in particular, indicated high interannual variability in egg distribution. The 2001 data identified a main spawning ground in the northern area, with additional spawning noted in the south area of Cape Passero. Larval distribution maps (Fig. 3 a, b, c), particularly for 2000 and 2001, indicated a more regular pattern than that for eggs, with a clear are of concentration on the southernmost Maltese shelf. Offshore larval transport was indicated, probably due to local circulation features

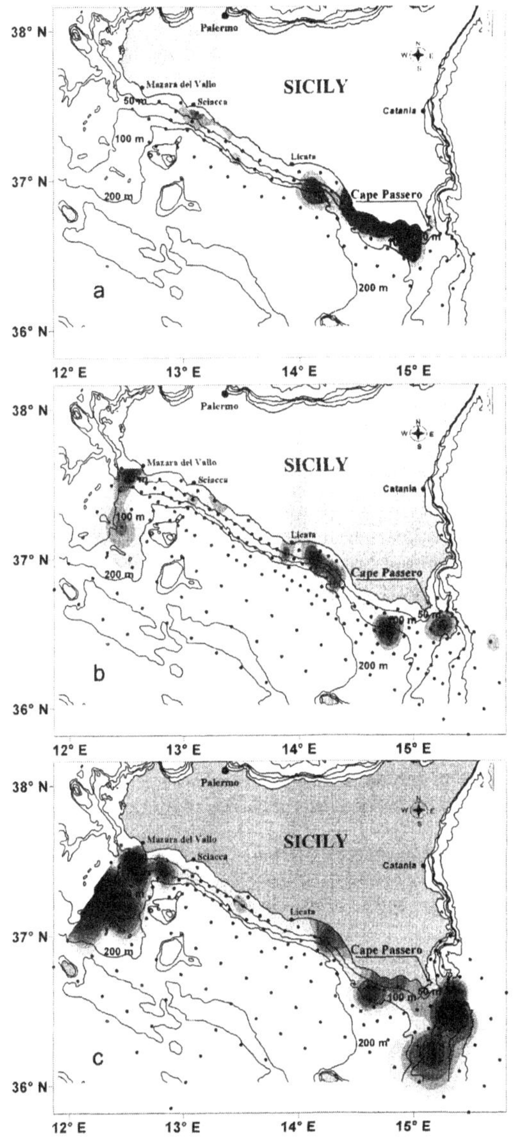

Figure 2. Distribution of anchovy eggs (counts) during surveys carried out along the Sicilian coast in 1999 (a), 2000 (b) and 2001 (c).

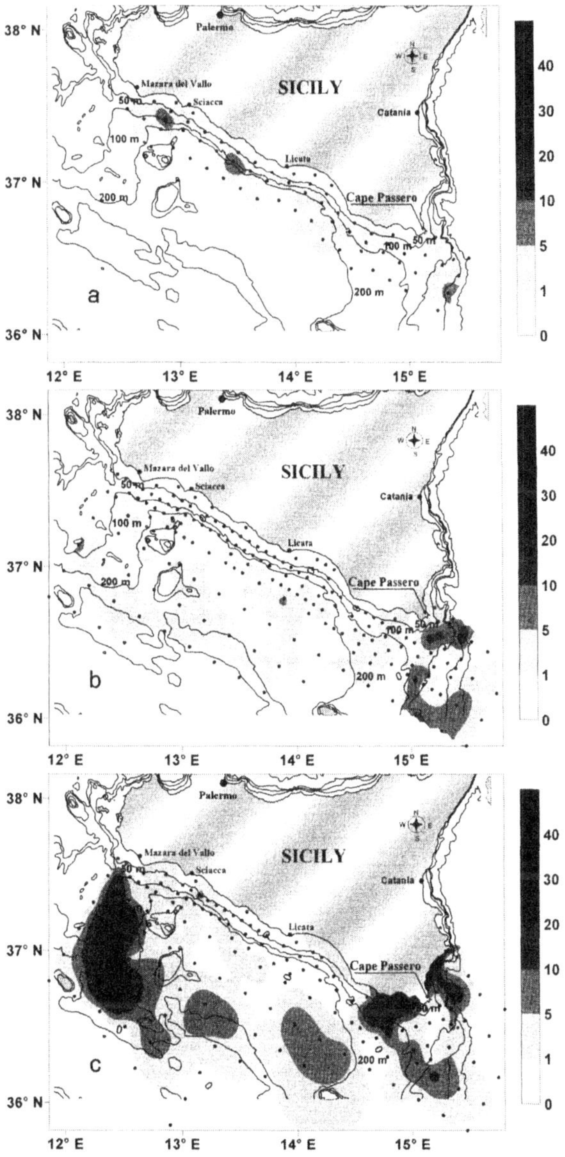

Figure 3. Distribution of anchovy larvae (counts) during surveys carried out along the Sicilian coast in 1999 (a), 2000 (b) and 2001 (c).

(García Lafuente et al., 2002). Zooplankton distribution (Fig. 4 a, b, c) reflected the same trends shown by eggs and larvae (i.e. concentrated mainly in the north and off the southern tip of the Sicilian coast); highest densities were found in Mazara del Vallo and the Cape Passero sea areas. Analysis of CTD casts showed that the AIS path deviated from the 'mean' path shown in Figure 1, especially in 1999 and 2000, when the AIS flow was much more distant from the shore, and only closed in to the shores of Sicily in the southern region. As a consequence, coastal waters in the northern and central regions were relatively colder. In 2001, there was an indication of a warmer and fresher vein of Atlantic water in the northern region.

Discussion

The hydrographic features of the circulation in the Strait of Sicily indicate that the main feature responsible for the distribution of anchovy early life stages is

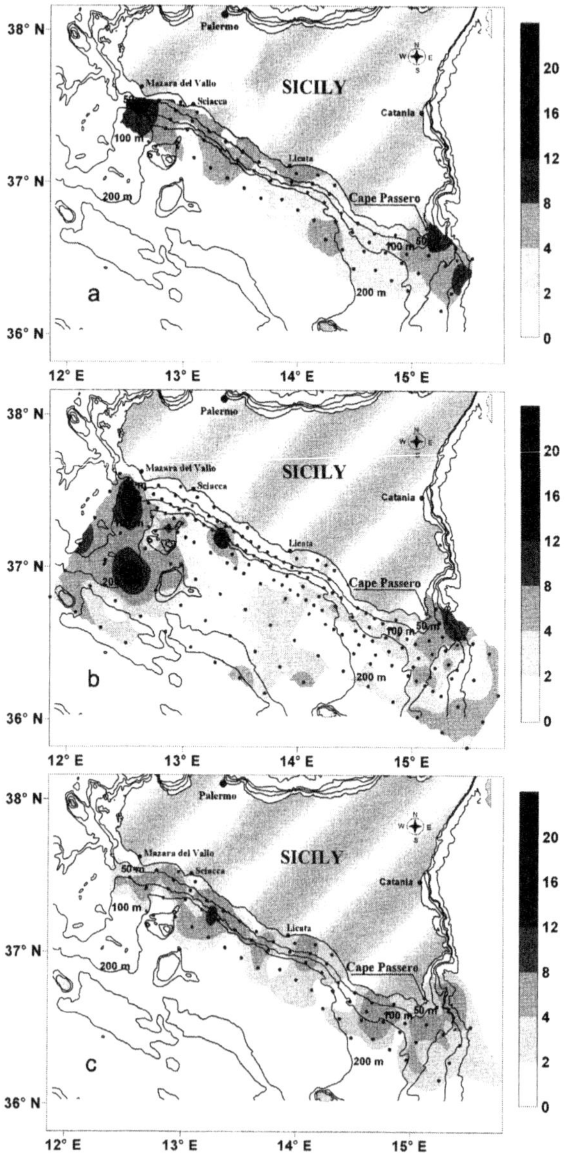

Figure 4. Distribution of meso-zooplankton (dry weight, mg/m^3) during surveys carried out along the Sicilian coast in 1999 (a), 2000 (b) and 2001 (c).

the Atlantic Ionian Stream (AIS). García-Lafuente et al. (2002) proposed a conceptual model for anchovy spawning strategy, whereby, the AIS is a transport mechanism for eggs and larvae towards the southernmost region of the Sicilian coastline. In this area, the presence of a cyclonic gyre aids retention of the early life stages of anchovy, and the convergence of different water masses forms a front that concentrates the food (zooplankton) required by the larvae. Present data on anchovy eggs and larvae and zooplankton are consisted with such a model, which explains the contemporary presence of fish eggs and larvae, and food (zooplankton). However, the AIS path was quite variable from year-to-year, and such variability may cause changes in temperature and extent of upwelling in the northern part of the area. In turn, this may have consequences for the distributional pattern of the anchovy spawning grounds, as low temperature can inhibit the spawning process (see Figs 2 a, b and 3 a, b). Conversely, the relatively high egg and larval densities, characterizing the northern region during the 2001 survey (Figs 2 c and 3c), appear to be related to the warmer surface temperature conditions induced by the intrusion of modified Atlantic water. Our findings support the hypothesis that the distributional patterns of anchovy eggs and larvae are linked to changes in the AIS path and distance from the shoreline.

References

Borja, A., A. Uriarte, V. Valencia, L. Motos & A. Uriarte, 1996. Relationships between anchovy (*Engraulis encrasicolus,* L.) recruitment and the environment in the Bay of Biscay. Sci. mar. 60: 179–192.

Casavola, N., E. Hajderi & G. Marano, 1998. Relationship between the density of *Engraulis encrasicolus* eggs and zooplanktonic biomass in the southern adriatic sea. Biol. mar. Mediterr. 5: 56–62.

Cuttitta, A., G. Basilone, B. Patti, A. Bonanno, S. Mazzola & B. Giusto, 1999. Trends of anchovy (*Engraulis encrasicolus*) condition factor and gonadosomatic index in the Sicilian Channel (in Italian). Biol. mar. Mediterr. 6: 566–568.

García-Lafuente, J., A. García, S. Mazzola, L. Quintanilla, A. Delgado, A. Cuttita & B. Patti, 2002. Hydrographic phenomena influencing early life stages of the Sicilian Channel anchovy. Fish. Oceanogr. 11: 31–44.

Giráldez, A. & R. Abad, 1995. Aspects on the reproductive biology of the Western Mediterranean anchovy from the coast of Málaga (Alboran Sea). Sci. mar. 59: 15–23.

Holden, M. J. & D. F. S. Raitt (eds), 1975. Manual de Ciencia Pesquera. Parte 2 - Métodos Para Investigar los Recursos y su Aplicaciòn. Doc. Tec. FAO Pesca, (115) Rev. 1. 211 pp.

James, A., J. W. Pitchford & J. Brindley, 2003. The relationship between plankton blooms, the hatching of fish larvae, and recruitment. Ecol. Model. 160: 77–90.

Robinson, A. R., J. Sellschopp, A. Warn-Varnas, W. G. Leslie, C. J. Lozano, P. J. Haley, L. A. Anderson & P. F. J. Lermusiaux, 1999. The Atlantic Ionian Stream. J. mar. Syst. 20: 113–128.

Tikhonova, I., O. Arino, G. Ivanitskii, H. Malchow & A. Medvinskii, 2000. The dependence of fish school movement and plankton spatial distributions on the phytoplankton growth rate. Biofizika 45: 352–359.

Hydrobiologia **503:** 121–130, 2003.
M.B. Jones, A. Ingólfsson, E. Ólafsson, G.V. Helgason, K. Gunnarsson & J. Svavarsson (eds), Migrations and Dispersal of Marine Organisms.

Juveniles stick to adults: recruitment of the tube-dwelling polychaete *Lanice conchilega* (Pallas, 1766)

Ruth Callaway
(nee R. Zühlke) School of Biological Sciences, University of Wales Swansea, Singleton Park, Swansea, SA2 8PP, U.K.
E-mail: rm.callaway@swansea.ac.uk

Key words: attached juveniles, *Lanice conchilega*, polychaetes, recruitment, tubes, Wales

Abstract

Juveniles of the tube-dwelling polychaete *L. conchilega* were found attached to adult tubes at an exposed beach in South Wales (U.K.). They attached to the upper part of tubes protruding above the sediment. The aim of this study was to determine the timing of juvenile settlement on to adult tubes, their abundance and size distribution, and the time of detachment and densities in the sediment after re-settlement. Furthermore, to establish whether the sediment composition of juvenile tubes differed from those of adults and the surrounding sediment. In 1998, 1999 and 2000, tubes of adult *L. conchilega* were sampled and numbers of attached juveniles were quantified and their sizes measured. Densities of solitary *L. conchilega* in the sediment were estimated, as well as the size distribution of solitary tubes. On average, 5–13 juveniles were attached to single adult tubes in spring, but numbers varied greatly with up to 72 juveniles on a single tube in April 1998. Availability of a hard substratum and chemical cues are discussed as potential factors triggering the attachment of juveniles to adult tubes. Juveniles started to detach and re-settle in the sediment 1 month after attachment. Once settled in the sediment, juvenile numbers decreased steeply, but the size distribution of solitary *L. conchilega* suggested that some juveniles survived in the population. However, the extent to which numbers of juvenile recruits contribute to the density of adults remained unclear. Juveniles used a smaller mean particle size for tube building than adults, but both actively selected particles.

Introduction

Lanice conchilega (Pallas, 1766) is a tube-building terebellid polychaete with an amphiboreal distribution, found on all coasts of Europe except arctic waters (Holthe, 1978). *Lanice conchilega* colonises intertidal and subtidal sediments down to about 100 m and reaches densities of several thousand individuals per m^2 (Hertweck, 1995; Ropert & Dauvin, 2000); Buhr & Winter (1977) reported up to 20 000 individuals per m^2 from shallow subtidal areas off the German coast. The tubes of adult *L. conchilega* have a diameter of about 0.5 cm and are up to 65 cm long (Ziegelmeier, 1952). They consist of sand particles attached to an inner thin organic layer. The particles are of a wide range of different material including silica grains, parts of bivalve and gastropod shells, parts of sea urchins, foraminiferans and ostracods (Ziegelmeier, 1952). The anterior end of the tube protrudes above the sediment surface by 1–4 cm and is crowned with a sand-fringe.

The planktonic larva of *L. conchilega* was termed 'aulophore' by Kessler (1963) and has a transparent tube, which is used as a floating device. A detailed description of embryogenesis from spiral cleavage to the larval stage ready for metamorphosis was given by Heimler (1981). The pelagic phase is relatively long, up to 8 weeks. Morphologically, aulophore larvae represent juveniles endowed with larval characteristics which persist until they settle and metamorphose (Bhaud, 2000). Tentacles play an important role during the settlement of aulophore larvae. They glue the anterior end of the larval tube to the substratum and extend the width and length of the tube by gathering sediment particles (Heimler, 1981). Information about

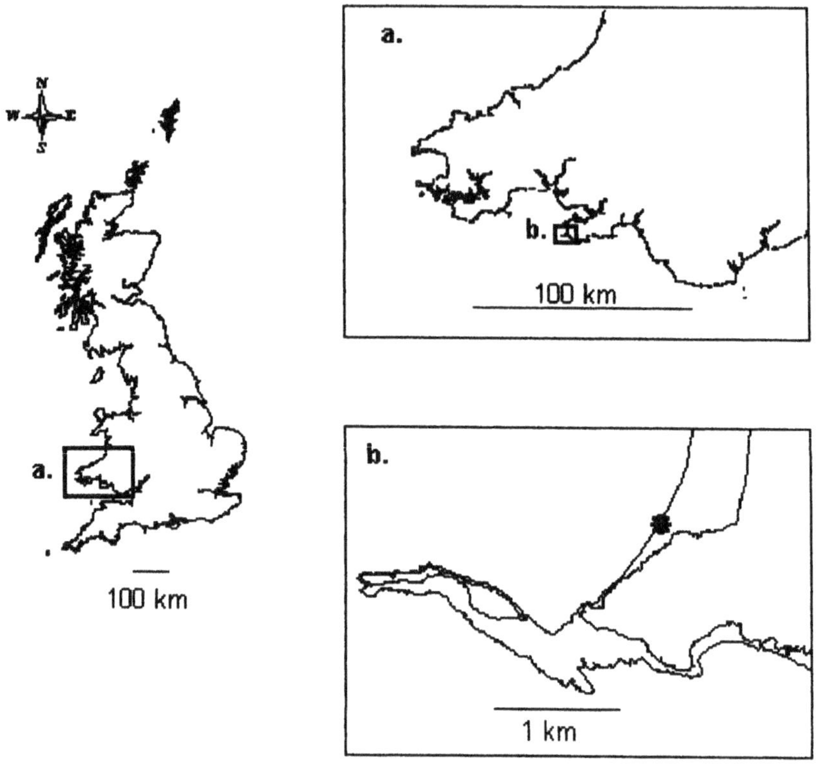

Figure 1. Sampling location; (A) South Wales, (B) sampling position at Rhossili Bay (★).

substratum-preferences of aulophore larvae is contradictory, as laboratory and field observations suggest they settle in a variety of sediment types (Heimler, 1981; Bhaud, 2000). Juveniles may attach to different hard substrata in intertidal areas (Heuers & Jaklin, 1999; Strasser & Pieloth, 2001) but most settle on tubes of adult *L. conchilega* (Carey, 1987; Heuers et al., 1998). Many marine invertebrate species possess larvae which settle on or close to adult conspecifics, including members of Hydrozoa, Bivalvia, Gastropoda, Polychaeta and Echinodermata (Burke, 1986).

Here, I report the density and size of attached juveniles, timing of settlement, eventual detachment and re-settlement in the substratum and differences in the sediment composition of tubes of adults and juveniles from a *L. conchilega* population of an exposed sandy beach in South Wales (U.K.).

Materials and methods

Description of site

The *L. conchilega* population was located at Rhossili Bay (4° 17.88′ W, 51° 34.08′ N), a sandy beach at the western tip of the Gower Peninsula, South Wales (U.K.) (Fig. 1). The bay is renowned for its exposure to waves and tidal currents are strong with differences between low and high tide of up to 11 m (Environmental Agency, Wales, unpublished data). The *L. conchilega* population was emersed during spring tides, usually for 2–3 d each fortnight. However, due to strong onshore winds, which decreased the tidal range, the *L. conchilega* population remained submerged during several spring tides over the duration of the study.

Attached juvenile L. conchilega

In 1998, 1999 and 2000 juvenile *L. conchilega* were found attached to adult tubes. On each sampling occasion, at least 20 adult tubes were collected randomly. Tubes were preserved in 70% ethanol and examined

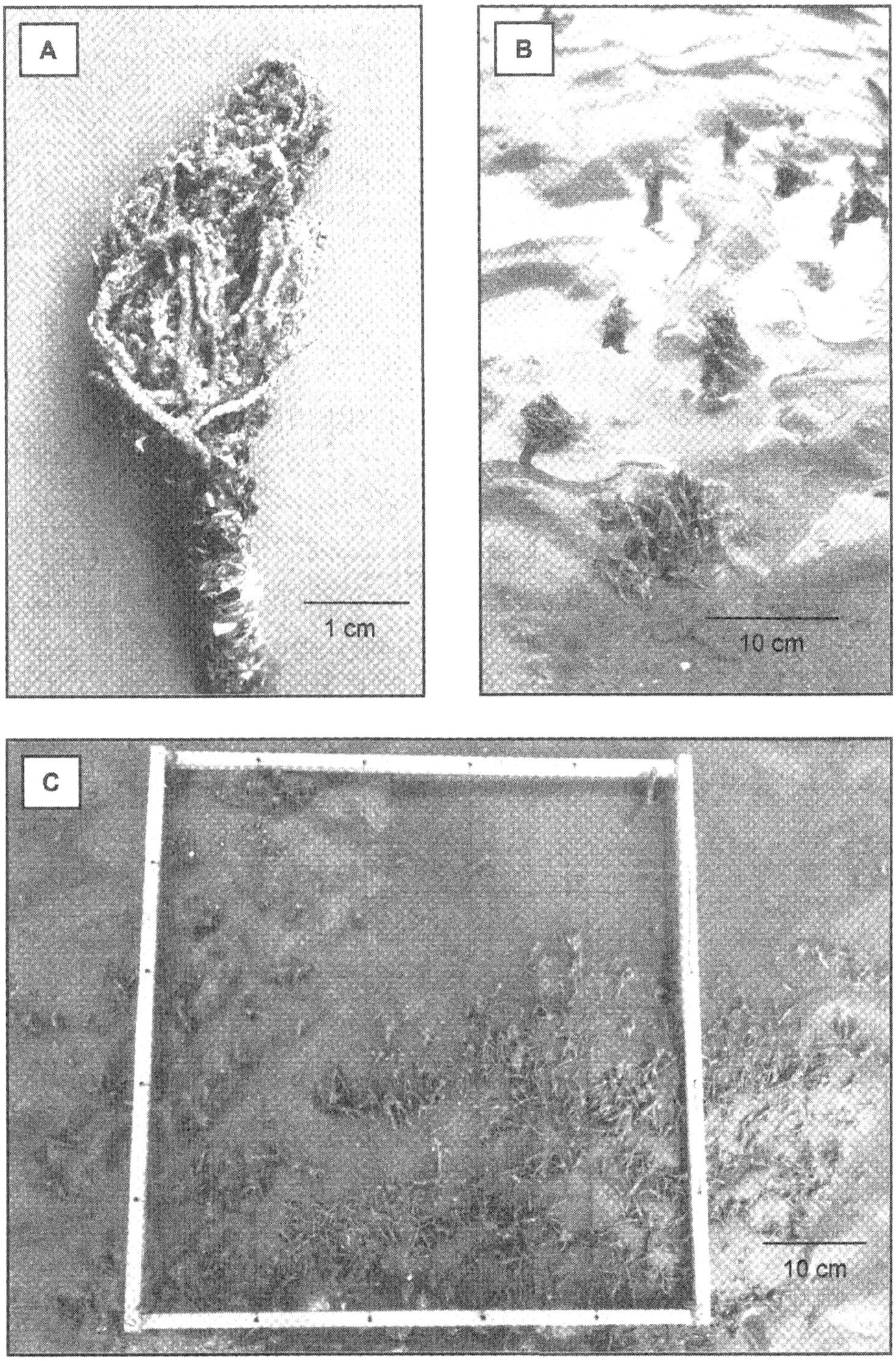

Figure 2. (A) Tubes of juvenile *Lanice conchilega* attached to the upper part of an adult tube. (B) Adult *L. conchilega* with attached juveniles in sandflat at Rhossili Bay. (C) Detached juveniles in sediment.

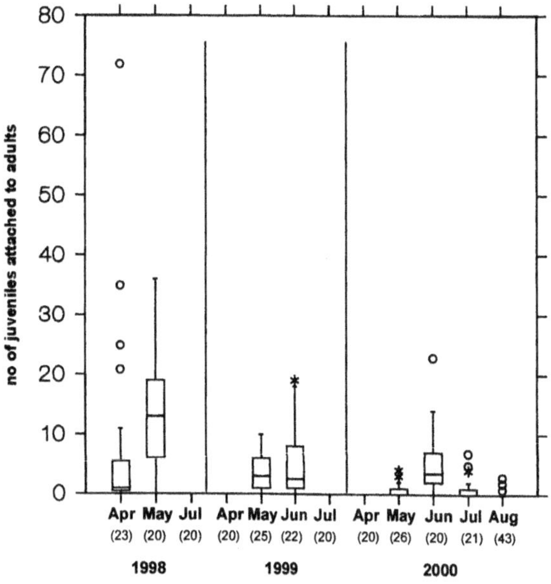

Figure 3. Number of attached juvenile *Lanice conchilega* on individual tubes of adults between April 1998 and August 2000 (median, standard error and range). Numbers of adult tubes analysed (n) are given in parentheses.

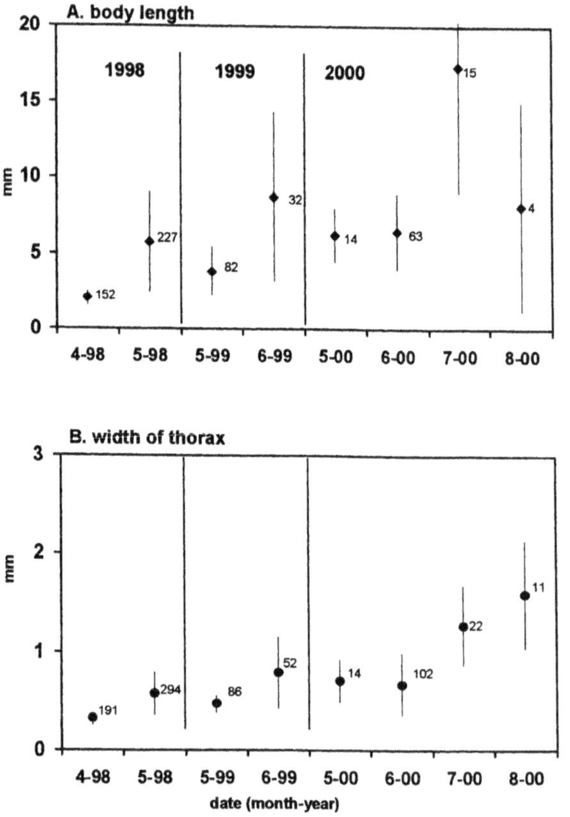

Figure 4. Mean width of thorax and body length (±SD) of juvenile *Lanice conchilega* attached to tubes of adults. The number of measurements (n) are given with each mean.

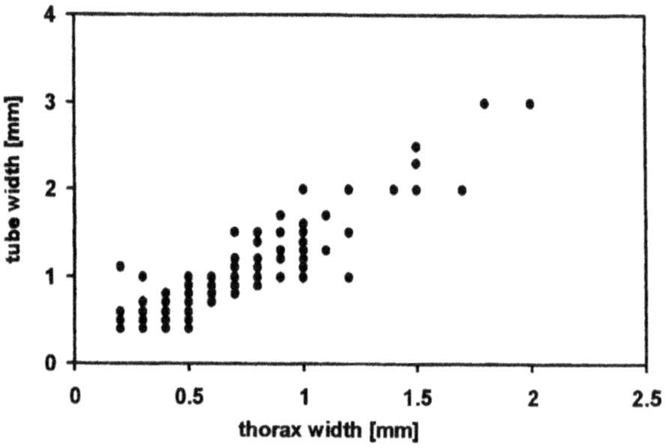

Figure 5. Correlation of the width of the thorax and the sand tube of juvenile *Lanice conchilega* attached to adult tubes. Product moment correlation coefficient $r = 0.91$, $n = 348$.

in the laboratory. All juveniles attached to adult tubes were removed and counted. Differences between median numbers of attached juveniles per adult each month were tested for significance with the Kruskal–Wallis test ($p < 0.05$). Pairwise comparisons were made with Dunn's multiple comparison test ($p < 0.05$) (Daniel, 1978). Juveniles were removed from their tubes; the width of the tubes, thorax width and body length of each worm were measured under a microscope. In many cases, only the anterior part of the body could be removed from the tubes and the length could not be measured. Differences in mean size between months were tested with ANOVA ($p < 0.05$) and the Tukey-test was used for pairwise comparisons ($p < 0.05$). Significance of the relationship between the width of the thorax and tube width of attached juveniles was tested by calculating the product moment correlation coefficient.

Solitary L. conchilega

From April 1998 to January 2001, the density of *L. conchilega* at Rhossili Bay was estimated at least six times each year. Densities of the polychaete were determined by counting tube fringes in a randomly placed 50×50 cm frame ($n = 10$). Estimating the abundance of *L. conchilega* using corers was not practical as these worms retreat rapidly deep into their tubes. Counting tube tops with intact sand fringes has been used previously to estimate densities of *L. conchilega* (Ropert & Dauvin, 2000). Strasser & Pieloth (2001) tested the reliability of this method and found a strong linear correlation between the number of tube top counts and living worms, although the counts underestimated slightly the number of animals present. However, counting tube tops gives a reliable reflection of the densities of *L. conchilega*. Juveniles and adults were separated; tubes <3 mm width were defined as juvenile, tubes >3 mm were defined as adult. Differences in density between months were tested with Kruskal–Wallis, and Dunn's test was used for multiple comparisons ($p < 0.05$). In 2000, the widths of all tubes were measured. Differences between the distribution of tube widths each month were tested with Kolmogorov–Smirnov ($p < 0.05$).

Sediment composition of L. conchilega *tubes*

In April 1998, tubes of adult and juvenile *L. conchilega* and sediment samples were collected at Rhossili Bay. Tubes were dried at 55 °C for 3 days and carefully disaggregated under a microscope using forceps. Disaggregated tube sand was sprinkled onto a slide and the maximum length of each particle was measured to the nearest 0.1 mm. All grains were sorted into calcareous or silica particles. Between 400 and 500 grains were measured per sample. Grain size distributions were described as frequencies per size class. Furthermore, following Gruet (1984), granulometric analysis based on weights was employed. The sum of all lengths of particles in a size class was calculated, which corresponds to the weight of the grains per size class. Mean and median grain size for juvenile and adult tubes, as well as sorting of the sediments was calculated (Tucker, 1988).

Results

Attached juvenile L. conchilega

In April or May juvenile *L. conchilega* started to attach to the upper part of the adult tube protruding above the sediment, giving the appearance of miniature birch-brooms stuck in the sediment (Figs 2 and 3). Most juveniles were found among the sand fringes. The peak in juvenile attachment was delayed for about 1 month each year. In 1998, attached juveniles were found in April and May (no data available for June), in 1999, in May and June, and in 2000, numbers peaked in June with a few attached juveniles found in August (Fig. 3). Highest average numbers of juvenile *L. conchilega* attached to adults were found in May 1998 (13.0±0.3, mean±SD). They were significantly higher than in May 2000, but no significant difference was found to numbers in May 1999 or June 2000 (Dunn's test, $p < 0.05$). A maximum of 72 juvenile *L. conchilega* was found attached to a single adult tube in April 1998.

The smallest juveniles were recorded in April 1998 with an average thorax width of 0.3±0.1 mm (mean±SD) and a length of 2.0±0.5 mm (mean±SD) (Fig. 4). One month later, their average size had doubled. Significant seasonal growth in body size was found each year (Tukey test, $p < 0.05$). The thorax and tube width of attached juveniles were significantly correlated ($n = 348$, $r = 0.91$, $p < 0.01$) (Fig. 5). Sand tubes were, on average, 0.3 mm wider than the thorax.

Solitary L. conchilega

Solitary juvenile tubes, detached from adults, were found in the sediment 1 month after the first juveniles had attached to adults (Fig. 6). Highest numbers of solitary juveniles were found in May 1998, coinciding with high numbers of attached juveniles that year (Figs 3 and 6). With the exception of 2000, densities of solitary juveniles decreased sharply after they were initially recorded in the sediment. In 2000, relatively few detached juveniles had settled in the sediment 1 month after the juveniles had attached to adults. However, solitary juveniles were recorded until October as in 1998 (no autumn values were available for 1999).

Although average densities of adult *L. conchilega* were generally lower in 2000 than in the preceeding years, few significant interannual differences were found. Densities were significantly lower in April and July 2000 than in the same months in 1998, and in March 2000 than in March 1999 (Dunn's test, $p < 0.05$).

The size distribution of all solitary *L. conchilega* tubes reflected the settlement of detached juveniles in the sediment (Fig. 7). The distribution was unimodal from February until May 2000, but a distinct cohort of juveniles with a tube width of 2–3 mm led to a bimodal distribution in July (data for June were not available). The distribution was significantly different between July and the preceeding month (Kolmogorov–Smirnov, $p < 0.05$) but not between July and August (Kolmogorov–Smirnov, $p > 0.05$). In October, the size distribution differed significantly from July and August due to high numbers of tubes 3 and 4 mm wide. Although the settlement of ju-

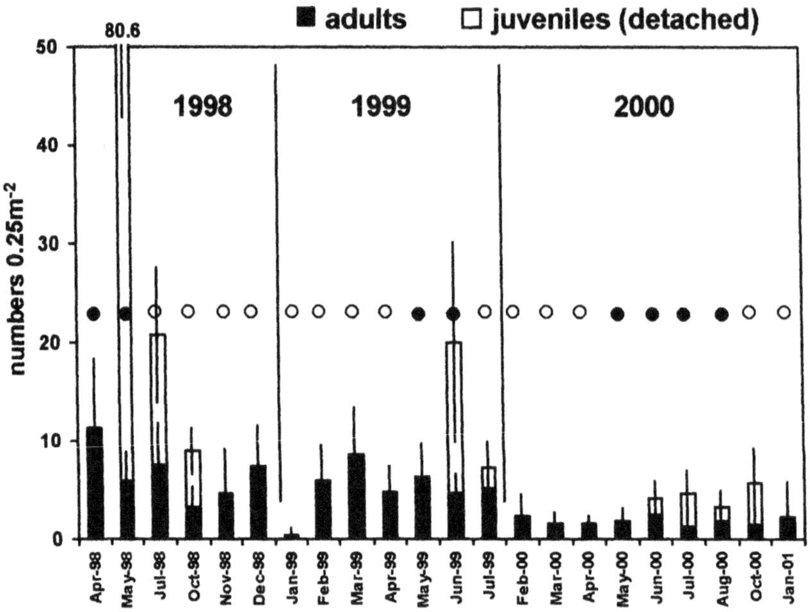

Figure 6. Density of *Lanice conchilega* from 1998 to 2000 at Rhossili Bay (mean±SD, $n = 10$). Juveniles (white bars): detached *L. conchilega* tubes <3 mm wide, adults (black bars): tubes >3 mm wide; ● juveniles attached to adults present (for abundance see Fig. 3), ○ no attached juveniles present.

Table 1. Sediment parameters of tubes of adult and juvenile *Lanice conchilega* (April 1998).

	Surrounding sediment	Adults	Juveniles
Mean particle size (phi)	2.31	1.03	1.43
Mean particle size (mm)	0.20	0.49	0.37
Median (phi)	2.20	1.07	1.40
Median (mm)	0.22	0.48	0.38
Sorting	0.51*	0.91**	0.71**
Calcareous particles (%)	9.0	26.7	38.5
Silicate (%)	91.0	73.3	61.5

* Moderately well sorted, ** moderately sorted

veniles was the most obvious reason for differences in the distribution of size classes, significant differences were also found between February and March as well as March and April 2000 (Kolmogorov–Smirnov, $p < 0.05$).

Sediment composition of tubes

The mean and median particle sizes were smaller for tubes of juvenile *L. conchilega* than for adults, but showed the same sorting (Table 1, Fig. 8). Sediment from the surrounding area was considerably finer and better sorted than that of *L. conchilega* tubes. In fact, larger sized particles (>0.4 mm) regularly used by juvenile and adult *L. conchilega* were not recorded in the sample of the surrounding sand. The proportion of calcareous particles was higher in the tubes than in the surrounding sediment (Table 1).

Discussion

At Rhossili Bay, Wales (U.K.), juvenile *L. conchilega* were found attached to the sand tubes of adult conspecifics. This phenomenon has been observed before, albeit to a much lesser extent (Heuers & Jaklin, 1999; Strasser, pers. comm.). Heuers & Jaklin (1999) found up to four juveniles attached per adult in the Wadden Sea, while up to 72 juveniles were counted on a single tube at Rhossili Bay. Generally, settling in areas occupied by adult conspecifics increases the probability of adequate survival conditions in terms of food supply, predation pressure and abiotic environmental conditions (Pawlik, 1992). The main advantage for juvenile *L. conchilega* attaching to adults is probably secure anchorage in the sand. Settling directly in the sediment of exposed sites would subject juveniles to a high risk of passive dislodgment. Polychaetes are

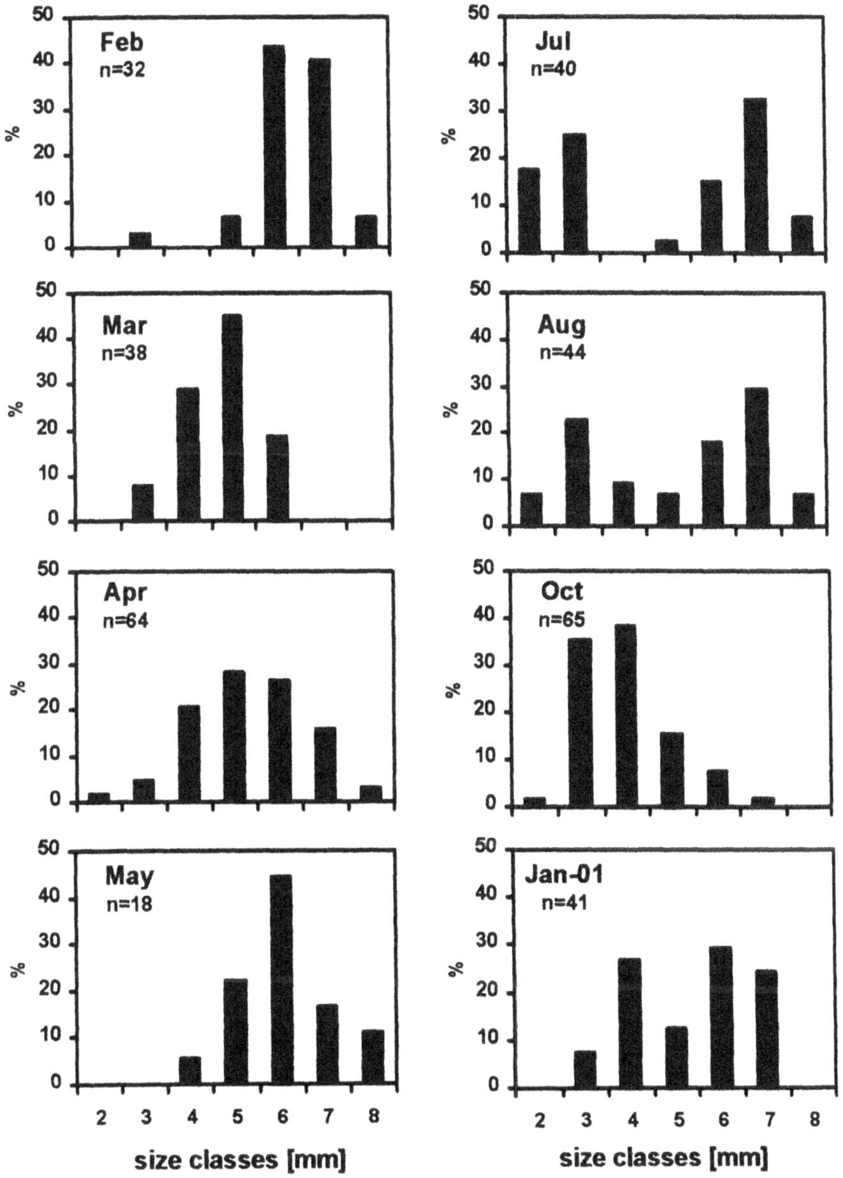

Figure 7. Size distribution of solitary *Lanice conchilega* in the year 2000 and January 2001. Width of sand tubes is shown.

prone to dislodgement by waves or tidal currents if they live close to the sediment–water interface (Tamaki, 1987). Bhaud (2000) reported that larvae of *Eupolymnia nebulosa* (Montagu) settling in areas unoccupied by adults were dislodged by hydrodynamic action once they reached the tubiculous stage. Hydrodynamic forces may have been responsible for the rapidly decreasing numbers of juvenile *L. conchilega* once they detached from adult tubes and re-settled in the sediment. Most were probably passively relocated, although predation by fish or birds may also have been of importance (Petersen & Exo, 1999). The disadvantage of clusters of juveniles attaching to adults may be intraspecific competition for food, which might also affect the adult host. This may be counterbalanced by the increased size and more complex structure of the tube top, due to the attached juveniles, trapping more food particles.

Two causal factors are generally discussed as triggers for polychaete larvae settling on or close to conspecific adults: (a) the availability of habitat structures and effects of these on the local hydrodynamic regime, and (b) chemical cues. Heuers & Jaklin (1999) found that *L. conchilega* larvae settled on artificial tubes in similar numbers as on natural tubes and also found juvenile *L. conchilega* attached to dead cockle shells, while sediment without epibenthic structure was not colonised. They concluded that hard epibenthic structures are a prerequisite for settlement of juvenile *L. conchilega* in exposed areas (Heuers & Jaklin, 1999). Degree of exposure could explain the different numbers of attaching juveniles at different sites. Intertidal sandflats of the Wadden Sea are less exposed to currents and waves than Rhossili Bay in South Wales. If small-scale turbulences around the adult tubes trigger settlement of *L. conchilega* larvae, this would be more pronounced at an exposed than a sheltered site.

Alternatively, settlement of *L. conchilega* larvae may be induced by chemical cues. Several tube-building polychaetes produce chemical cues, which attract larvae to settle on or near adult conspecifics (Pawlik, 1992; Bryan et al., 1997). Larvae respond to water-soluble chemical cues associated with living conspecific worms as well as to biofilmed surfaces (Unabia & Hadfield, 1999; Toonen & Pawlik, 2001). Tubes themselves, notably those built from cemented sediment particles, can provide chemical cues. Jensen & Morse (1984) showed that the presence of the anterior part of tubes of *Phragmatopoma californica* (Fewkes) initiated settlement and metamorphosis of conspecific larvae. Tubes of adults, their remains, or tubes of recently-settled juveniles also provided a strong stimulus to settle for larvae of *Sabellaria alveolata* (Linnaeus) (Wilson, 1968); cement secreted by young or adult *S. alveolata* in constructing their tubes was quickly recognised by larvae making contact with the substance. If larval *L. conchilega* respond to the cement used in constructing the tube, this could explain observations by Carey (1987), who reported that areas with empty tubes of *L. conchilega* became sites of dense larval settlement.

However, attachment of juvenile *L. conchilega* to adults or hard substratum is not obligatory. Strasser & Pieloth (2001) showed that juveniles recolonised a sandflat that had experienced a collapse of the *L. conchilega* population and was devoid of adult tubes. Other studies made no mention of juvenile *L. conchilega* attaching to adults and showed that high densities of *L. conchilega* were maintained even in the absence of juvenile recruitment (Marcano & Bhaud, 1995; Ropert & Dauvin, 2000). Ropert & Dauvin (2000) concluded that population renewal occurred by transfer of adults from offshore areas towards the coast.

Interannual differences of numbers of attaching juveniles and the timing of their settlement were probably related to numbers of larvae in the water column, as Strasser & Pieloth (2001) reported for the Wadden Sea. These in turn may be dependent on the larval output of the supply site and environmental factors such as water temperature, currents and storms. However, seasonal variations in water temperatures were similar over the 3 years with temperatures ranging from 7 °C in winter to 18 °C in summer (Environmental Agency, unpublished data).

Further clarification is required of the process of juvenile detachment from the adult tube. Juveniles grew in size whilst attached to adults. Solitary juveniles were detached about 1 month after settlement and re-settled in close proximity to the adults. However, the exact timing, triggers starting the process and the actual mechanisms, e.g. whether they remain or leave their tubes for the migration, are unknown. Body size might play a role or juveniles may respond to favourable hydrodynamic conditions, allowing them to settle in close vicinity to adults without being washed away.

Although the size distribution of solitary *L. conchilega* for 2000 suggested that some juveniles recruited to the population, this study did not indicate clearly the extent to which numbers of recruits affected the density of the adult population at Rhossili Bay. Assuming that juveniles of one year contribute to the adult population of the subsequent year, decreasing number of juveniles settling in the sediment from 1998 to 1999 may have resulted in the lower densities of adults in 2000 than in 1999. Although mean densities of adults were lower in 2000 than the previous year, the difference was only significant in March. Further studies are needed to clarify the issue.

Sediment composition of L. conchilega *tubes*

The average size of particles used by juvenile *L. conchilega* to build tubes was smaller than those of adults. This coincides with results of Féral (1989) who investigated granulometric properties of a wide range of size classes of *L. conchilega* tubes from the French coast. He found increasing mean grain sizes with increasing tube width. Similar relationships were found for the tube-building polychaetes *S. alveolata* and *Owenia fusiformis* Chiaje (Fager, 1964; Gruet, 1984). For *S. alveolata*, Gruet (1984) showed that size of the sand grains used in tube construction correlated

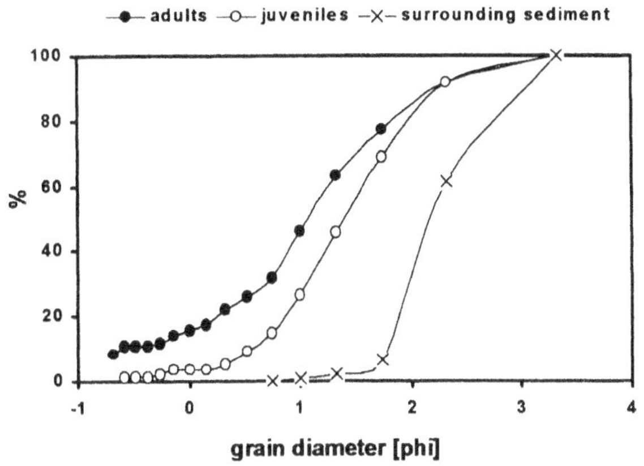

Figure 8. Grain size composition of tubes of adult and juvenile *Lanice conchilega* and the surrounding sediment. Cumulative frequency distribution is shown.

significantly with the height of the 'building organ' – a specially formed lip, which was in turn correlated with the anterior diameter of the animal. For *L. conchilega*, Féral (1989) found a significant correlation between the diameter of the tube and the width of the upper lip, while this study showed that the widths of tube and thorax are significantly correlated.

Comparing particle size frequency of surrounding sediment with that of sediment used in tube construction supported the hypothesis that juvenile and adult *L. conchilega* select particles for their tubes. The fact that they used size classes of particles which were not even present in the sample of the surrounding sediment, and considerably higher quantities of calcareous material than present in the vicinity of the tubes, indicates that *L. conchilega* may take every opportunity to pick up rare, large particles. Féral (1989) found similar sediment compositions for *L. conchilega* tubes from different areas, with either coarse or relatively fine substratum, showing that sediment selection in *L. conchilega* was independent of the sediment composition of its surroundings. Jones & Jago (1993) suggested that uptake of shell fragments by *L. conchilega* could lead to carbonate depletion of the surrounding area, resulting in an inverse correlation between *L. conchilega* density and carbonate content of the study site.

It appears that polychaetes are capable of selecting particles of desirable size and shape if they construct tubes using building organs, mostly referred to as lips (Ziegelmeier, 1952; Gruet, 1984; Féral, 1989), which cement particles to the tube with mucus secreted from a gland in the mouth region. In *L. conchilega*, this ability appears to be present immediately after the transition from pelagic to benthic life.

Acknowledgements

I thank my supervisor Peter Hayward, Martin Thiel and two anonymous reviewers for their valuable comments on the first draft of the manuscript.

References

Bhaud, M. R., 2000. Two contradictory elements determine invertebrate recruitment: dispersion of larvae and spatial restrictions on adults. Oceanol. Acta 23: 409–422.

Bryan, P. J., P. Y. Qian, J. L. Kreider & F. S. Chia, 1997. Induction of larval settlement and metamorphosis by pharmacological and conspecific associated compounds in the serpulid polychaete *Hydroides elegans*. Mar. Ecol. Prog. Ser. 146: 81–90.

Buhr, K. J. & J. E. Winter, 1977. Distribution and maintenance of a *Lanice conchilega* association in the Weser estuary (FRG), with special reference to the suspension-feeding behaviour of *Lanice conchilega*. In Keegan, B. F., P. O. Ceidigh & P. J. S. Boaden (eds), Biology of Benthic Organisms. 11th European Symposium on Marine Biology, Gallaway, Oct. 1976. Pergamon Press, Oxford and New York: 101–113.

Burke, R. D., 1986. Pheromones and the gregarious settlement of marine invertebrate larvae. Bull. mar. Sci. 39: 323–331.

Carey, D. A., 1987. Sedimentological effects and paleoecological implications of the tube-building polychaete *Lanice conchilega* Pallas. Sedimentology 34: 49–66.

Daniel, W. W, 1978. Applied Nonparametric Statistics. Houghton Mifflin Company.

Fager, E. W., 1964. Marine sediments: Effects of a tube-building polychaete. Science 143: 356–358.

Féral, P., 1989. Influence des populations de *Lanice conchilega* (Pallas) (Annelida, Polycheta) sur la sedimetation sableuse intertidale de deux plages bas-normandes (France). Bull. soc. geol. France 8: 1193–1200.

Gruet, Y., 1984. Granulometric evolution of the sand tube in relation to growth of the polychaete annelid *Sabellaria alveolata* (Linne) (Sabellariidae). Ophelia 23: 181–193.

Heimler, W., 1981. Untersuchunger zur Lavalentwicklung von *Lanice conchilega* (Pallas) 1766 (Polychaeta, Terebellomorpha) Tiel 1: Entwicklungsablauf. Zool. Jb. Anat. 106: 12–45.

Hertweck, G., 1995. Verteilung charakteristischer Sedimentkörper und der Benthossiedlungen im Rückseitenwatt der Insel Spiekeroog, südliche Nordsee. 1. Wattkartierung 1988–1992. Senckenbergiana marit. 26: 81–94.

Heuers, J. & S. Jaklin, 1999. Initial settlement of *Lanice conchilega*. Senckenbergiana marit. 29 (Suppl.): 67–69.

Heuers, J., S. Jaklin, R. Zühlke, S. Dittmann, C. P. Günther, H. Hildebrandt & V. Grimm, 1998. A model on the distribution and abundance of the tube-building polychaete *Lanice conchilega* (Pallas, 1766) in the intertidal of the Wadden Sea. Verhandl. Ges. Ökol. 28: 207–215.

Holthe, T., 1978. The Zoogeography of the Terebellomorpha (Polychaeta) of the Northern European Waters. Sarsia 63: 191–198.

Jensen, R. A. & D. E. Morse, 1984. Intraspecific facilitation of larval recruitment: gregarious settlement of the polychaete *Phrag-*

matopoma californica (Fewkes). J. exp. mar. Biol. Ecol. 83: 107–126.

Jones, S. E. & C. F. Jago, 1993. Insitu assessment of modification of sediment properties by burrowing invertebrates. Mar. Biol. 115: 133–142.

Kessler, M., 1963. Die Entwicklung von *Lanice conchilega* (Pallas) mit besonderer Berücksichtigung der Lebensweise. Helgol. wiss. Meeresunters. 8: 425–476.

Marcano, G. & M. R. Bhaud, 1995. New observations on the terebellid (Polychaeta) aulophore larvae on the French coasts. Ophelia 43: 229–244.

Pawlik, J. R., 1992. Chemical ecology of the settlement of benthic marine invertebrates. Oceanogr. mar. biol. Ann. Rev. 30: 273–335.

Petersen, B. & K. M. Exo, 1999. Predation of waders and gulls on *Lanice conchilega* tidal flats in the Wadden Sea. Mar. Ecol. Prog. Ser. 178: 229–240.

Ropert, M. & J. C. Dauvin, 2000. Renewal and accumulation of a *Lanice conchilega* (Pallas) population in the bale des Veys, western Bay of Seine. Oceanol. Acta 23: 529–546.

Strasser, M. & U. Pieloth, 2001. Recolonization pattern of the polychaete *Lanice conchilega* on an intertidal sand flat following the severe winter of 1995/1996. Helgol. mar. Res. 55: 176–181.

Tamaki, A., 1987. Comparison of resistivity to transport by wave action in several polychaete species on an intertidal sand flat. Mar. Ecol. Prog. Ser. 37: 181–189.

Toonen, R. J. & J. R. Pawlik, 2001. Settlement of the gregarious tube worm *Hydroides dianthus* (Polychaeta: Serpulidae). I. Gregarious and nongregarious settlement. Mar. Ecol. Prog. Ser. 224: 103–114.

Tucker, M., 1988. Techniques in Sedimentology. Blackwell Scientific Publications. 394 pp.

Unabia, C. R. C. & M. G. Hadfield, 1999. Role of bacteria in larval settlement and metamorphosis of the polychaete *Hydroides elegans*. Mar. Biol. 133: 55–64.

Wilson, D. P., 1968. The settlement behaviour of the larvae of *Sabellaria alveolata* (L.). J. mar. biol. Ass. U.K. 48: 387–435.

Ziegelmeier, E., 1952. Beobachtungen ueber den Röhrenbau von *Lanice conchilega* (Pallas) im Experiment und am natürlichen Standort. Helgol. wiss. Meeresunters. 4: 107–129.

Hydrobiologia **503:** 131–141, 2003.
M.B. Jones, A. Ingólfsson, E. Ólafsson, G.V. Helgason, K. Gunnarsson & J. Svavarsson (eds), Migrations and Dispersal of Marine Organisms.

Settlement of bivalve spat on artificial collectors in Eyjafjordur, North Iceland

Elena Guijarro Garcia, Gudrun G. Thorarinsdottir & Stefan A. Ragnarsson
Marine Research Institute, Skulagata 4, 101 Reykjavik, Iceland
E-mail: elena@hafro.is

Key words: artificial collectors, bivalve settlement, byssal drifting, Iceland, secondary settlement, spat

Abstract

The seasonal pattern of bivalve spat settlement in Eyjafjordur, North Iceland, was investigated using artificial collectors of monofilament netting over 14 months (March 1998–January 2000) at 5, 10 and 15 m depth. SCUBA divers replaced the collectors at 4-weekly intervals. Twelve bivalve species settled on the collectors but only *Mytilus edulis* and *Hiatella arctica* were present throughout the year; they were the most abundant bivalve taxa. Of the remaining species, only *Chlamys islandica*, *Heteranomia* spp., *Arctica islandica*, *Serripes groenlandicus* and *Mya* spp. were sufficiently abundant to enable statistical analysis. All settled in late summer and autumn. Peak settlement of *M. edulis*, in September, consisted mainly of primary settlers (0.25–0.5 mm) although secondary settlers (>0.5 mm) were present in all samples. *Mytilus edulis* settled mostly at 5 m depth, especially larger individuals, possibly reflecting stronger currents at shallower depth and the proximity of mussel beds in the intertidal zone. Primary (<1 mm) and secondary *H. arctica* settlers (>1 mm) were present in most months, with the former being most numerous in September, 1999; settlement was equally abundant at 5 and 10 m depth. Primary settlement of *C. islandica* and *S. groenlandicus* occurred in autumn (mainly in September), and secondary settlers were very scarce and only seen in winter. *Arctica islandica, Heteranomia* spp. and *Mya* spp. settled mainly in September 1999 at 10 m depth, except for *A. islandica*, which was more numerous in August.

Introduction

Many biotic and abiotic factors affect larval abundance in the water column and influence the temporal variability of settlement. Amongst the most important are the reproductive cycles of adults, patterns of winds (Hawkins & Hartnoll, 1982), hydrography of the area (Gaines & Bertness, 1992) and changes in rates of larval mortality due to predation (Thorson, 1950; Day & McEdward, 1984) or to a prolonged planktonic stage (Day & McEdward, 1984). Alterations in abundance and spatial distribution of a population during early stages of benthic life can also arise from secondary settlement (Armonies, 1994). Following settlement, postlarvae of many bivalve species can secrete a byssal thread, resuspend in the water column and drift away to resettle somewhere else (Sigurdsson et al., 1976; Lane et al., 1985; Armonies, 1994). The process can occur several times, as long as postlarvae do not grow above a threshold size and become too heavy. Byssus drifting has been observed for several species of cockels and clams (Mountaudouin, 1997), mytilids (Sigurdsson et al., 1976), tellinids (Sörlin, 1988; Armonies & Hellwig-Armonies, 1992; Cummings et al., 1996; Baker & Mann, 1997) and scallops (Beaumont & Barnes, 1992). Each species has its own maximum size at which this kind of dispersal can be used (Beaumont & Barnes, 1992).

Artificial collectors of monofilament netting have been used successfully to collect many different species of drifting bivalve spat for use in aquaculture (Paul et al., 1981), estimation of recruitment in wild bivalve stock fisheries (Sause et al., 1987; King et al., 1990), monitoring settlement, early growth and mortality of bivalves (Knuckey, 1995; Chauvaud et al., 1996), and establishing patterns of distribution of benthic epifauna (Ardisson & Bourget, 1992). Sampling of bivalve spat on artificial collectors was

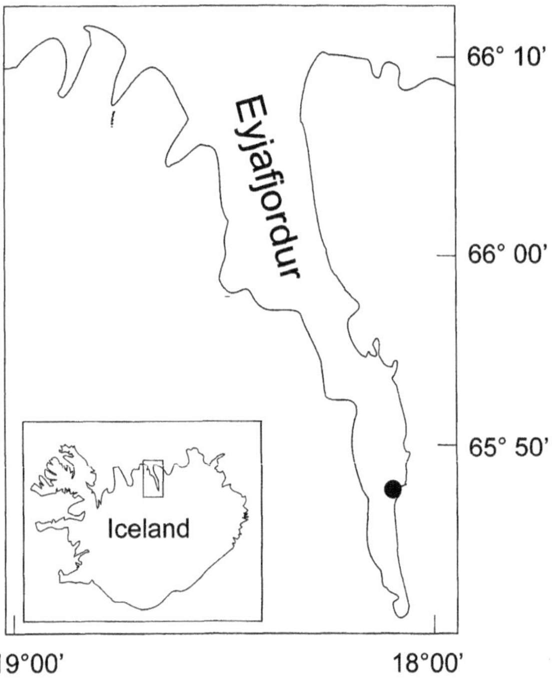

Figure 1. Map of Eyjafjordur, showing its location in Iceland. The sampling site is marked with a dot.

also used previously in Icelandic waters to determine the time of maximum settlement and optimal depth of settlement in two commercially important species, *Chlamys islandica* (Müller, 1776) and *Mytilus edulis* (L., 1750) (Thorarinsdottir, 1991, 1996).

In this paper, we present results from a study focusing on settlement patterns of bivalve larvae in shallow waters in relation to season and water depth. Some species recorded during our study are very rare or absent in the scientific literature for Iceland.

Materials and methods

Study area and experimental setup

The study was carried out in Eyjafjordur, North Iceland (Fig. 1). The fjord deepens from the head to the entrance, reaching a maximum depth of 200 m (Jonsson, 1996). Tides are semi diurnal, the amplitude of mean spring tide is 1.3 m and tidal currents flow at a few cm s^{-1} (Jonsson & Gudmundsson, 1994). Information on salinity, available for 1992 and 1993 (Jonsson, 1996), indicated a range from 30.0 to 34.3, with the greatest variability at 5 m depth (30.0–34.3). At 10 m depth, salinity ranged from 32.0 to 34.5 and at 15 m it was similar throughout the year (34.0–34.5). Highest salinity measurements were taken from January to April at 15 m depth, and the lowest from June and July at 5 m depth (Jonsson, 1996). Temperature was not recorded at the sampling site; however, data from a nearby location (65° 49.4′ N, 18° 08.0′ W) were available (MRI, unpubl. res.) (Fig. 2).

Spat collectors, deployed at the head of the fjord (65° 50.0′ N, 18° 10.0′ W) at 5, 10 and 15 m depth on a flat bottom composed of sand of medium grain size, consisted of plastic scourers made of a tubular monofilament net rolled over onto itself. The net had a mesh size of 5 mm and a volume of roughly 250 cc. Each scourer was attached to a stainless steel rod, and three of these were fixed onto a concrete block measuring 30×30×6 cm. Two blocks were attached together by a rope and weighted with stones to maintain stability on the bottom. At each depth (5, 10 and 15 m), a pair of blocks (with six spat collectors in total) was submerged and a buoy marked their location.

Sampling

SCUBA divers replaced the spat collectors at 4-weekly intervals from March 1998 to January 2000. The collectors were kept in separate plastic bags before taking them to the surface to minimize loss of animals. Once on deck, each collector was kept in a jar and filled with 10% formalin. No samples were collected from September to November, 1998 and from February to July, 1999. In the laboratory, spat collectors were washed thoroughly on a 63 μm sieve to collect the animals. The entire sample was analyzed, except for the months of heavy settlement, in which the sample was split. Bivalves were identified to species, counted and their shell length measured by means of a micrometer.

Data analysis

Tests of significance on the abundance of the most common species were carried out using two-way ANOVA (α=0.05), with depth and month as fixed factors after log $(x+1)$ transformation of the data (Zar, 1996). Lack of replication at depth, due to the occasional absence of bivalves, excluded the use of a two-way ANOVA to examine interaction effects. Consequently, the log $(x + 1)$ transformed shell length of spat (Zar, 1996) at month and depth was compared using two separate one-way ANOVAs.

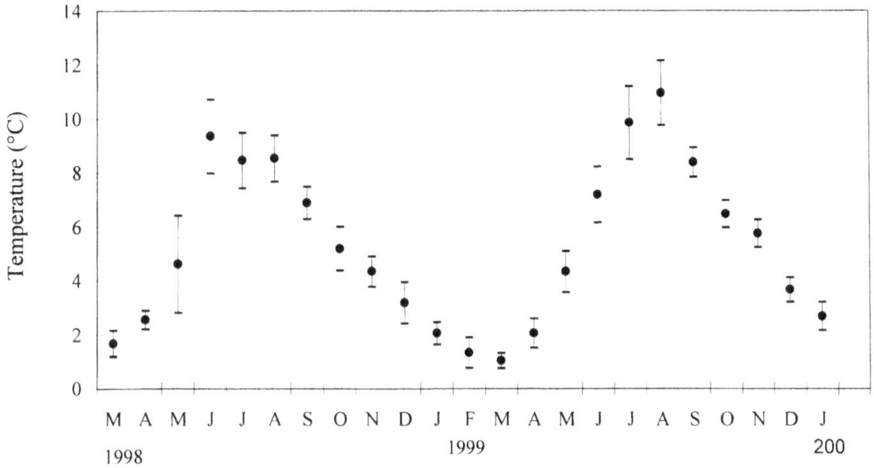

Figure 2. Mean temperature ± SD at month from March 1998 to January 2000. Note that no sampling was conducted in the time periods September–November 1998 and February–July 1999.

Table 1. Total number of individuals of the most common taxa collected each month and at each depth. Myt=Mytilidae, Hiat=*Hiatella arctica*, Cisl=*Chlamys islandica*, Heter=*Heteranomia* spp., Aisl=*Arctica islandica*, Sgro=*Serripes groenlandicus*, Mya=*Mya* spp

Month	Myt	Hiat	Cisl	Heter	Aisl	Sgro	Mya
March 98	23	1	0	0	0	0	0
April 98	31	0	0	0	0	0	0
May 98	142	1	0	0	0	0	0
Jun 98	264	0	0	0	0	3	0
Jul 98	248	34	0	0	0	1	0
Aug 98	423	853	0	0	1	0	0
Sep–Nov 98	*	*	*	*	*	*	*
Dec 98	107	60	0	0	0	0	0
Jan 99	219	83	0	0	0	0	0
Feb–Jul 99	*	*	*	*	*	*	*
Aug 99	1631	1027	13	0	129	26	30
Sep 99	165 398	43 084	1989	276	26	65	31
Oct 99	14 398	4201	594	38	29	4	0
Nov 99	686	251	29	3	1	1	6
Dec 99	208	56	1	0	0	1	0
Jan 00	577	78	1	0	0	0	0
Depth							
5 m	175 767	21 454	117	34	31	52	
10 m	6302	26 568	1537	280	149	56	30
15 m	2286	1707	973	1	6	40	35
All groups	184 355	49 729	2627	317	185	101	67

* No data were collected in these months.

Table 2. Results of the two-way ANOVA test (α=0.05) for number of individuals of the most abundant bivalve species recorded in the spat collectors in relation to months (m) and depth (d), and the interactions between both variables (m×d)

Taxa	Variable	Df	*F*	*P*-level
Mytilus edulis	m	13	77.89	0
	d	2	375.90	0
	m×d	26	7.17	<0.001
H. arctica	m	11	136.75	0
	d	2	156.29	0
	m×d	22	12.22	<0.001
C. islandica	m	5	185.55	0
	d	2	54.11	<0.001
	m×d	10	56.09	0
Heteranomia spp.	m	2	23.36	<0.001
	d	2	37.21	<0.001
	m×d	4	28.51	<0.001
A. islandica	m	4	13.51	<0.001
	d	2	8.00	<0.001
	m×d	8	5.31	<0.001
S. groenlandicus	m	6	18.17	<0.001
	d	2	11.61	<0.001
	m×d	12	5.97	<0.001
Mya spp.	m	2	1.65	0.202
	d	2	7.19	0.001
	m×d	4	1.40	0.246

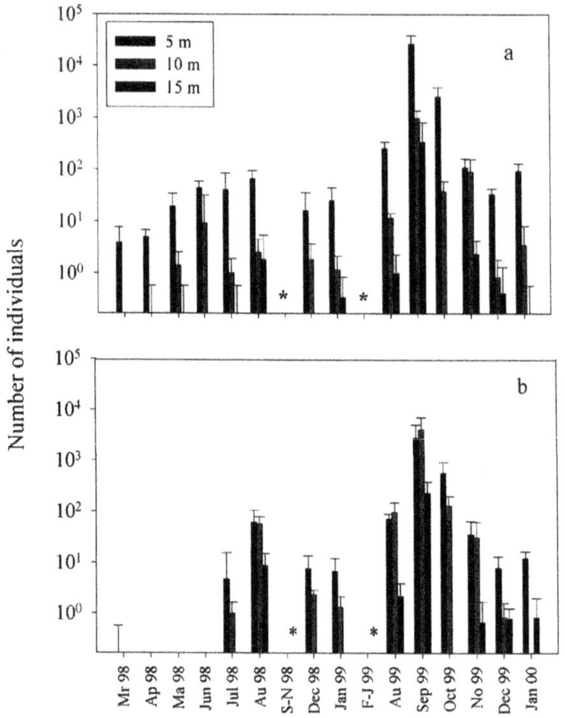

Figure 3. Logarithm of the average number (bars) of (a) *Mytilus edulis*, (b) *Hiatella arctica*, found each month and at each depth in the spat collectors. The whiskers indicate the standard deviation. The asterisks (*) indicate that no sampling was conducted in those months. S–N stands for the time period September–November 1998 and F–J for February–July 1999.

Results

Twelve bivalve taxa settled on the collectors during the sampling period, but five, *Thyasira* spp., *Thracia myopsis* (Möller), *Turtonia* spp., *Modiolaria discors* (L.) and *Astarte borealis* (Schumacher, 1817), were collected in too low number to carry out any analysis. The remaining species, *Mytilus edulis*, *Hiatella arctica* (L., 1767), *Chlamys islandica*, *Heteranomia* spp., *Arctica islandica* (L., 1767), *Serripes groenlandicus* (Bruguière, 1798) and *Mya* spp., showed marked seasonal changes in abundance and length, settling predominately from August to October (Table 1). Two-way ANOVA identified significant month and depth effects on abundance for each species except *S. groenlandicus* and *Mya* spp. (Table 2). Furthermore, significant month×depth interaction was also found for each species except *S. groenlandicus* and *Mya* spp., indicating that differences between depths and sampling months were not consistent. Similarly, significant month and depth effects were found for most species based upon shell length (one-way ANOVA). Findings indicate that even though there were marked seasonal differences, they were not concomitant at each depth.

Mytilus edulis (blue mussel) was the only species present in each sampling month and was the most abundant, constituting 78% of the total number of bivalves collected. From March to August 1998, there was a steady increase in *M. edulis* numbers. Most settled in September and October 1999, with over 165 000 and 14 000 mussels, respectively. A sharp decline followed from November onwards (Table 1). The majority of blue mussels (95%) settled at 5 m depth, with only 3.7 and 1.3% settling at 10 and 15 m, respectively (Fig. 3). The monthly mean shell length of mussels from April to August 1998 ranged from 1.10

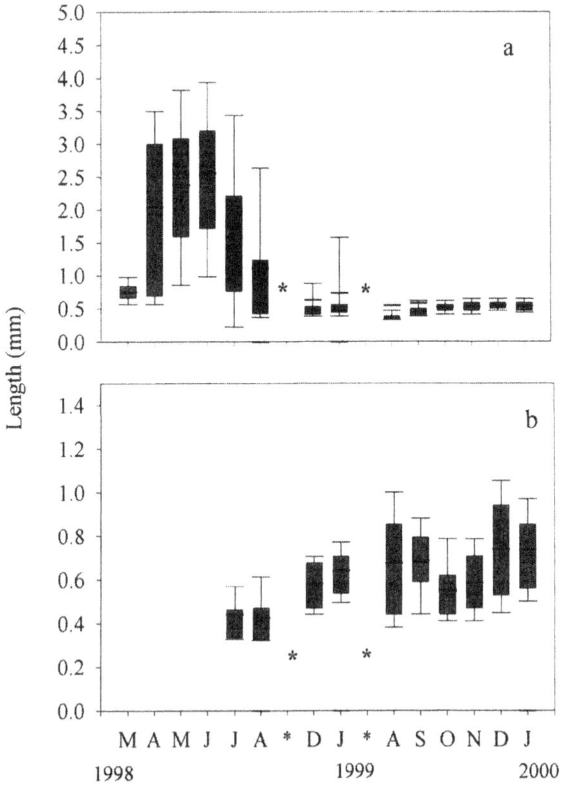

Figure 4. Mean length (bold line) and median length (regular line) at month of (a) *Mytilus edulis*, (b) *Hiatella arctica* recorded in the collectors. The 10th and 90th and 25 and 75 percentiles are represented by whiskers and boxes, respectively. The asterisks (*) indicate that no sampling was conducted in those months. The first asterisk in the tick labels stands for the time period September–November 1998 and the second for February–July 1999. Note the different scales in the *y*-axis of each graph.

Table 3. Results of the two one-way ANOVA (α=0.05) to test effect of the variables month and depth on length of the most common bivalve species

Taxa	Month			Depth		
	Df	*F*	*P*-level	Df	*F*	*P*-level
Mytilidae	13	303.64	<0.001	2	197.04	<0.001
H. arctica	9	206.95	<0.001	2	75.11	<0.001
C. islandica	2	14.00	<0.001	2	17.32	<0.001
Heteranomia spp.	2	0.63	0.533	1	5.63	0.019
A. islandica	2	23.27	<0.001	2	62.86	<0.001
S. groenlandicus	3	5.75	0.001	2	0.03	0.970
Mya spp.	2	11.51	0	2	3.89	0.027

to 2.60 mm. In contrast, the mean shell length was at minimum (0.50 mm) after the main settlement in September 1999, when most shells measured between 0.30 and 0.50 mm in length (Fig. 4a). The size range was from 0.25 to <5.00 mm in most sampling months, except for March 1998 and from October 1999 onwards, when individuals larger than 1.50 mm were rare or absent. Differences in the shell length of mussels between both months and depths was highly significant (Table 3). Mussels found at 5 m had an average length of 1.00 mm, whereas those collected at 10 and 15 m were 0.49 and 0.45 mm, respectively (Fig. 5a). These data indicate that most secondary settlers were collected at 5 m depth.

Spat of *H. arctica* settled in all but 2 months and amounted to 21% of the total number of bivalves collected. With the exception of August 1998, *H. arctica* settled scarcely from March 1998 to January 1999; predominant settlement being between August and October 1999 (Table 1). The 43 000 individuals that settled in September alone represented 87% of all the *Hiatella* collected during this study. The pattern of settlement at depth showed great variability between months; high numbers were generally at 5 m depth but, during August and September 1999, it was most abundant at 10 m depth (Fig. 3). Shell length of *H. arctica* ranged from 0.25 to 3.7 mm and rarely exceeded 2.00 mm; in August 1999 and January 2000; few individuals over 3.30 mm in shell length were found. The mean shell length of *H. arctica* varied significantly (Table 2) throughout the sampling period (0.43–0.74 mm), indicating that secondary settlers were scarce (Fig. 4b). The same conclusion was reached when comparing the mean shell length (0.57–0.66 mm) at all three depth levels (Table 3, Fig. 5b).

Chlamys islandica, *Heteranomia* spp., *A. islandica* and *S. groenlandicus* settled mainly between August and November 1999. All but *S. groenlandicus* were most abundant at 10 m depth, whereas *S. groenlandicus* was equally abundant at 10 and 15 m (Table 1). Two-way ANOVA showed significant month and depth effects in abundance for all species (Table 2). Despite the narrow range of shell length recorded for these species (Figs 5 and 6), variability in shell length at month and depth was also found to be significant for all but *Heteranomia* spp., for which no month effects were detected and *S. groenlandicus*, for which no depth effects were found (Table 3).

Mya spp. was most abundant in August and September 1999 (Table 1), but was scarce at 5 m depth and settled in similar numbers at 10 and 15 m

(Table 1). This difference between depths was significant (Table 2). Variability in shell length was small (Figs 5 and 6) but, nevertheless, differed significantly between months and depths (Table 3).

No correlation was found between monthly average water temperature and settlers abundance for any species ($R^2 = 0.137$).

Discussion

Of the seven bivalves, only *M. edulis* and *H. arctica* settled throughout the year, accounting for 98% of all bivalves collected. The remaining bivalve species were collected mostly in late summer or early autumn. The number of settlers of each species varied in the different months and depths. Unfortunately, lack of data from September and October 1998 increased greatly the variability between years, and precluded direct comparison of the settlement patterns between the two years. Nevertheless, all bivalve species showed significantly greater settlement from August to November 1999 than at other times, which might reflect the cycle of gonad development and spawning for these species. Secondary settlers of most species were present on the collectors during the study period. Although especially obvious for blue mussels and *H. arctica*, all species showed a decrease in main shell length during their main settlement.

Settlement patterns in time

The blue mussel, *M. edulis*, outnumbered all other taxa throughout the year, although both abundance and mean shell length varied greatly over the study period. Secondary settlement of *M. edulis* has been widely reported and it has been suggested that mussels over 1.0 mm in shell length are almost definitely secondary settlers (King et al., 1989; Snodden & Roberts, 1997). Unlike other studies of settlement of blue mussels on artificial collectors (King et al., 1990; Hunt & Scheibling, 1996), mussels with shell length over 5 mm settled during the period May–October 1998, although in very low number. It may be that the filamentous substrata, generally used in artificial collectors, may be more suitable for smaller, early postlarvae than for larger later stages. Blue mussels are thought to use byssus drifting as a means of transport to resettle until they are at least 2.5 mm (Sigurdsson, 1976; Seed & Suchanek, 1992), but mussels over 5 mm shell length may disperse by crawling, which could explain their presence on the collectors in this study, as they were deployed on the sea bottom. It is also possible that mussels grew in the collectors; Warren (1936), recorded 2–4 mm growth of *M. edulis* during 1 month in artificial collectors. Alternatively, larger-sized mussels may have been detached from their original substratum by currents or storms, and induced to secondary settlement (Cáceres-Martínez et al., 1994).

This study detected only one spawning period for blue mussels. The main spawning season in southwest Iceland has been observed to be from the middle of June to end of July, with primary settlement taking place from the middle of August to late September (Thorarinsdottir, 1996). However, settlement is delayed further north, and in western Iceland, the main spawning of *M. edulis* has been recorded in August (Thorarinsdottir & Gunnarsson, in press). Unlike other *M. edulis* populations (Ackefors & Haamer, 1987; King et al., 1989; Tammi et al., 1996), in Icelandic waters, blue mussels spawn only once a year, probably due to environmental restrictions. Short spawning periods have been described in other northern populations of *M. edulis* (Kautsky, 1982).

Settlement of *H. arctica* is poorly known from other areas. Spawning time and duration of the pelagic phase are unknown, but secondary settlement has been described (Sigurdsson et al., 1976) and a peak of settlement during July has been reported in France (Chauvaud et al., 1996). In our study, *H. arctica* was most abundant between August and October, but this could be related to the differences in sea temperature, which is known to be a trigger for spawning in bivalves (Bayne & Newell, 1983).

Harvey et al. (1995) obtained similar results for the Iceland scallop *C. islandica* as here when sampling in the Gulf of St. Lawrence. Data are in agreement with a previous study in which spawning was reported at the beginning of July in western Iceland, with settlement taking place in middle September at 0.3 mm in shell size (Thorarinsdottir, 1991). Byssus drifting and secondary settlement has been observed in other scallop species, such as *Pecten maximus* and *Aequipecten opercularis* (Beaumont & Barnes, 1992), and it is known that *C. islandica* spat can resettle when dislodged from the substrata (Harvey et al., 1993).

A previous study of *Heteranomia* spp., confirmed that settlement takes place in September in waters off Denmark (Jørgersen, 1936). Byssus drifting and secondary settlement have also been reported for *H. squamula* (Sigurdsson et al., 1976).

Absence of the ocean quahog, *A. islandica*, from the collectors during this study cannot be attributed directly to spawning time, as this species seems to spawn all year, with the main spawning from June to August (Mann, 1982; Thorarinsdottir, 2000). Spawning of *A. islandica* in northwestern Iceland is prolonged during the whole year, with the main spawning season in July (Thorarinsdottir, 2000). Larvae of *A. islandica* have been found in the plankton all months of the year in Denmark but settlement has only been observed from January to September, with a maximum in April (Muus, 1973). The shell length of ocean quahogs at metamorphosis is 0.23–0.3 mm, but smallest individuals collected in Denmark were 0.3–0.6 mm, a similar size to those measured during the present study. The mean shell growth during the first 6 months after settlement was less than 1 mm (Muus, 1973).

No investigations of settlement or growth of *Mya* spp. and *Serripes* spp. have been carried out in Icelandic waters, and published data on these genera are scarce in general. However, Muus (1973) recorded settlement of *Mya truncata* in Danish waters from January to July, with highest settlement in February. Shell length at metamorphosis was between 0.20 and 0.36 mm, and spat were first observed at 0.3–0.5 mm shell size. The mean monthly shell growth during the first 6 months after settlement was less than 0.1 mm (Muus, 1973), which may support the possibility of having secondary settlers of *Mya* spp. among the specimens collected in Eyjafjordur. Sigurdsson et al. (1976) reported byssal drifting and secondary settlement of *M. truncata*. Our data do not agree entirely with previous studies. *Mya arenaria* is known to spawn once a year off the east coast of North America and twice in Norway, where the main settlement has been recorded in September (Winther & Gray, 1985). Spawning behaviour of this species may also vary between years, switching from a large single spawning event to continuous small spawning events throughout the season (MacLean et al., 1999).

Nothing has been reported previously on spawning time or settlement for *S. groenlandicus*. In general, settlement may also have been conditioned by the material used in the artificial collectors, as it is known that bivalve spat shows preferences for certain types of substrata (Harvey et al., 1993, 1995; Cáceres-Martínez et al., 1994; Harvey & Bourget, 1995; Hunt & Scheibling, 1996).

Settlement patterns at depth

Most bivalve species settled in highest number at 10 m depth. Exceptions included *M. edulis* that was more abundant at 5 m, and *Mya* spp. that settled in very similar numbers at 10 and 15 m. However, 83.2% of the spat was collected at 5 m, but this was due to the high number of blue mussels. Excluding them from the analysis, 41% of all bivalves were collected at 5 m and 54% collected at 10 m. The significant month × depth interaction found for most species indicates that the proportion of settlers at the three different depths in each month varied throughout the period of study.

The relationship between depth and settlement has been investigated previously (Brand et al., 1980; Sause et al., 1987), using spat collectors attached to ropes anchored to the bottom and buoyed a few meters below the sea surface. In these studies, higher settlement was found in collectors in the middle of the water column. The authors suggested that settlement in near surface waters could be affected by fouling by algae or by the effects of wave action and turbulence; in collectors closest to the sea bed, silting could have inhibited settlement or caused high mortality among the settlers. However, settlement of many bivalve species on collectors appears to be greatest at a narrow depth near the sea bed, reflecting the behavior of the larvae at the time of settlement (Ventilla, 1977; Brand et al., 1980; Fraser & Mason, 1987). In the present study, all collectors were placed at the same distance from the sea bottom in contrast with the studies mentioned above. Changes in depth are accompanied by variations in several factors including temperature and food availability and quality. It might be that the bivalve distribution on the collectors at the three depths reflected better environmental conditions at 5 and 10 m than at 15 m (Mann, 1985). The significantly higher settlement of *M. edulis* at 5 m may be related directly to the fact that the spat comes from the intertidal zone and it is, therefore, more abundant in shallow waters.

Temperature and settlement

The average monthly temperature during 1998–1999 and 1987–1996 was similar with the exception of June 1998 and August and September 1999, which were considerably warmer (between 2 and 5 °C) compared to the same period in 1998 (Jonsson, 1999). August and September 1999 were the months of highest settlement; however, it is difficult to assess the influence of temperature alone on settlement processes. Some stud-

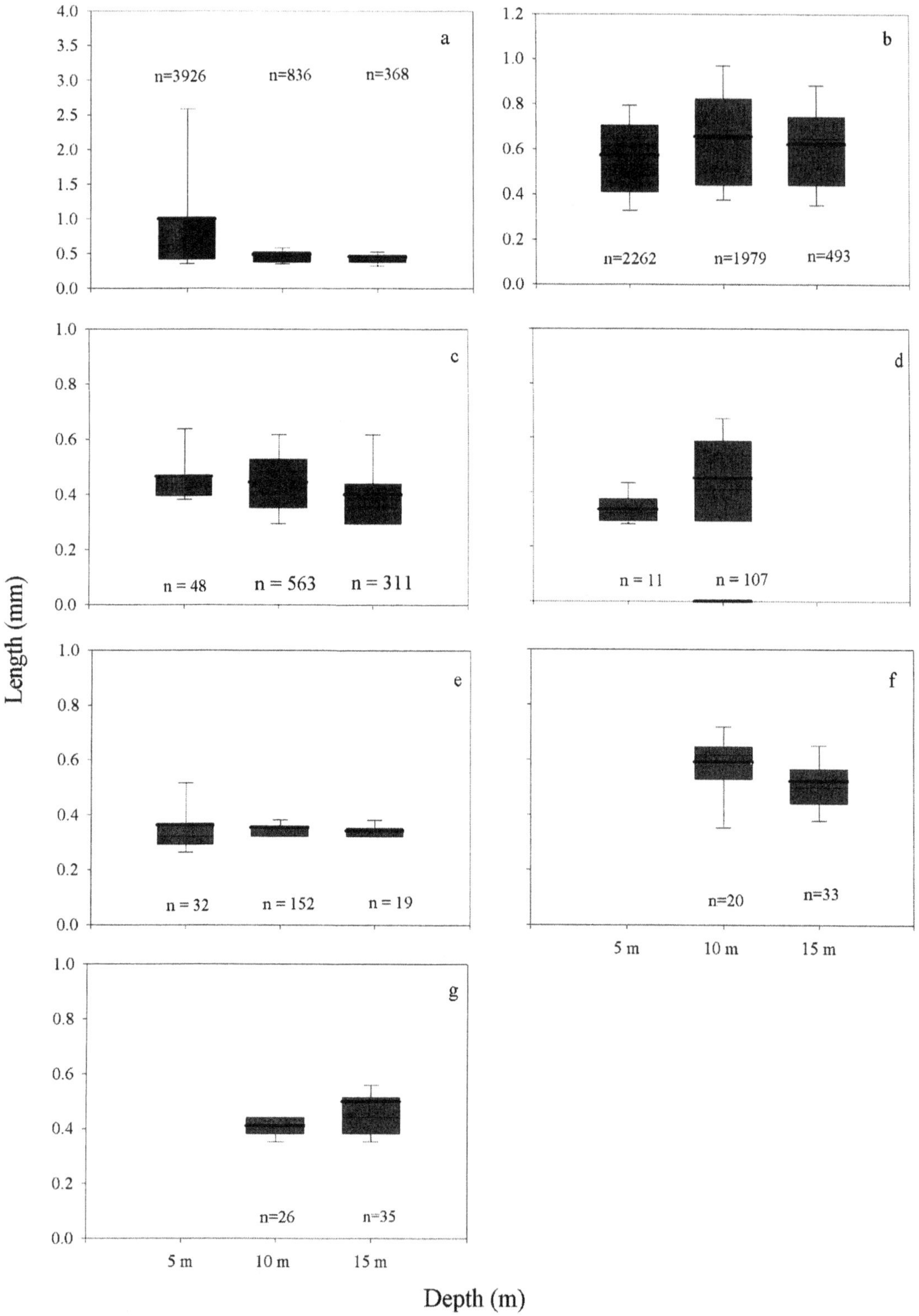

Figure 5. Mean length (bold line) and median length (regular line) at depth of (a) *Mytilus edulis*, (b) *Hiatella arctica*, (c) *Chlamys islandica*, (d) *Heteranomia* spp., (e) *Arctica islandica*, (f) *Serripes groenlandicus*, (g) *Mya* spp., recorded in the collectors. The 10th and 90th and 25 and 75 percentiles are represented by whiskers and boxes, respectively. The number of individuals measured (*n*) is given. Note the different scales in the *y*-axis for *M. edulis* and *H. arctica*.

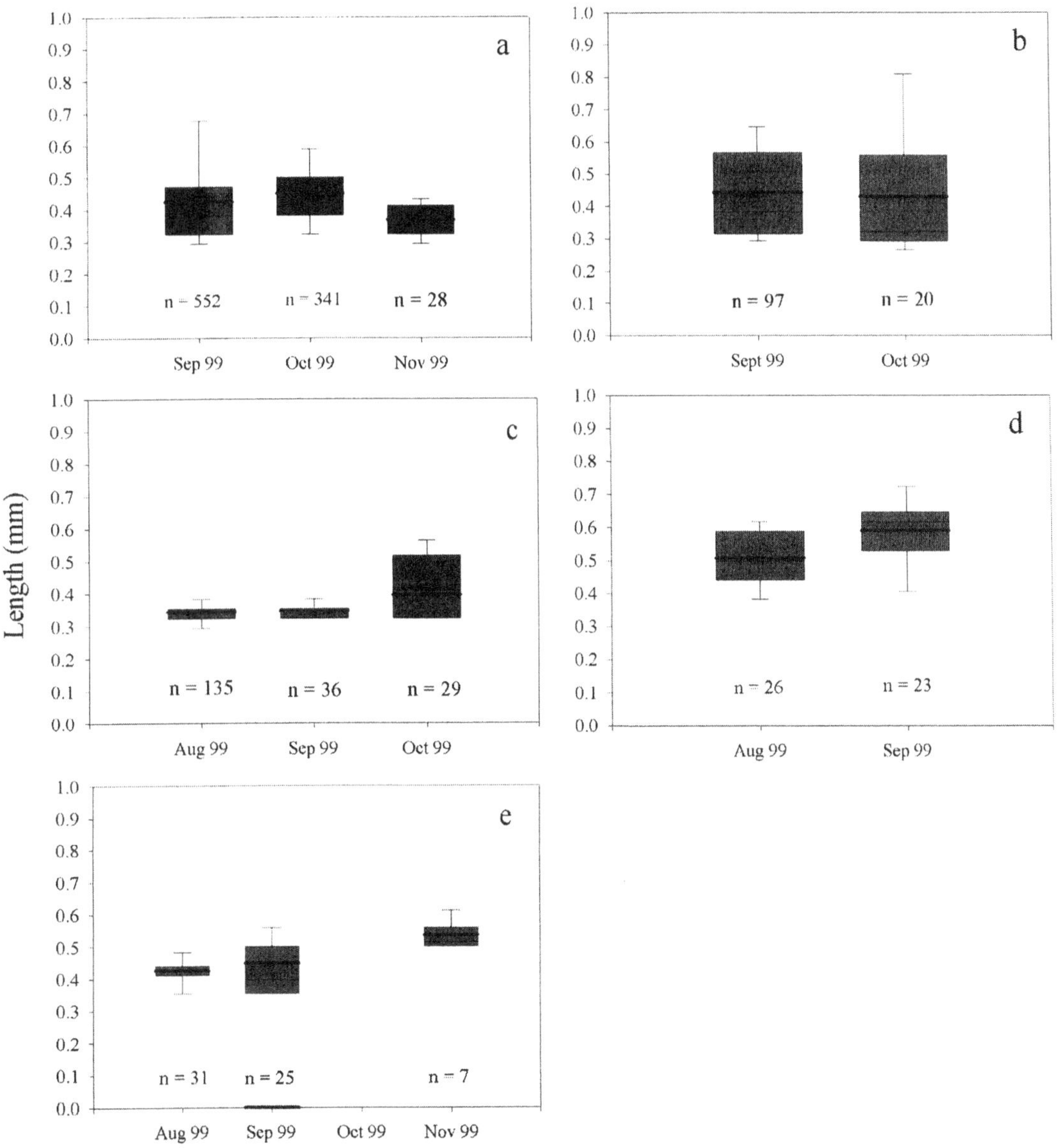

Figure 6. Mean length (bold line) and median length (regular line) at month of (a) *Chlamys islandica*, (b) *Heteranomia* spp., (c) *Arctica islandica*, (d) *Serripes groenlandicus*, (e) *Mya* spp. recorded in the collectors. The 10th and 90th and 25 and 75 percentiles are represented by whiskers and boxes, respectively. The number of individuals measured (n) is given. Note the different scales in the x-axis of each graph.

ies have demonstrated a positive correlation between water temperature and density of settlers (Robinson et al., 1991, 1999), whereas others have not found any correlation (Harvey et al., 1995). Nevertheless, it is widely regarded that settlement patterns are influenced predominately by the combined effects of temperature and food availability (Seed & Suchanek, 1992).

Acknowledgements

This study was funded by the Marine Research Institute of Reykjavik, Iceland. For that we thank the former director of the MRI, Prof. Jakob Jakobsson. Dr Lorna Taylor and Dr Karl Gunnarsson are thanked for comments on earlier versions of the manuscript, and Dr Roger Mann and an anonymous referee for their suggestions to improve this paper. Lina Gunnarsdottir, librarian at the MRI, is thanked for her valuable assistance in searching the literature, and Erlendur Bogason for the collection of samples.

References

Ackefors, H. & J. Haamer, 1987. A new Swedish technique for culturing blue mussel. ICES C.M.1987/K: 36.

Ardisson, P. L. & E. Bourget, 1992. Large-scale ecological patterns: discontinuous distribution of marine benthic epifauna. Mar. Ecol. Prog. Ser. 83: 15–34.

Armonies, W., 1994. Drifting meio- and marobenthic invertebrates on tidal flats in Königshafen: a review. Helgoländerwiss Meeresunters. 48: 299–320.

Armonies, W. & M. Hellwig-Armonies, 1992. Passive settlement of *Macoma balthica* spat on tidal flats of the Wadden Sea and subsequent migration of juveniles. Neth. J. Sea Res. 29: 371–378.

Bayne, B. L. & R. C. Newell, 1983. Physiological energetics of marine molluscs. In Saleuddin, A. S. M. & K. M. Wilbur (eds), The Mollusca 4. Academic Press, London: 407–515.

Baker, P. & R. Mann, 1997. The postlarval phase of bivalve molluscs: a review of functional ecology and new records of postlarval drifting of Chesapeake Bay bivalves. Bull. mar. Sci. 61: 409–430.

Beaumont, A. R. & D. A. Barnes, 1992. Aspects of veliger larval growth and byssus drifting of the spat of *Pecten maximus* and *Aequipecten (Chlamys) opercularis*. ICES J. mar. Sci. 49: 417–423.

Brand, A. R., J. D. Paul & J. N. Hoogesteger, 1980. Spat settlement of the scallops *Chlamys opercularis* (L.) and *Pecten maximus* (L.) on artificial collectors. J. mar. biol. Ass. U.K. 60: 379–390.

Cáceres-Martínez, J., J. A. F. Robledo & A. Figueras, 1994. Settlement and post-larvae behaviour of *Mytilus galloprovincialis*: field and laboratory experiments. Mar. Ecol. Prog. Ser. 112: 107–117.

Chauvaud, L., G. Thouzeau & J. Grall, 1996. Experimental collection of great scallop postlarvae and other benthic species in the Bay of Brest: settlement patterns in relation to spatio-temporal variability of environmental factors. Aquacult. Int. 4: 263–288.

Cummings, V. J., R. D. Pridmore & S. F. Thrush, 1996. Effect of the spionid polychaete *Boccardia syrtis* on the distribution and survival of juvenile *Macomona liliana* (Bivalvia: Tellinacea). Mar. Biol. 126: 91–98.

Day, R. & L. McEdward, 1984. Aspects of the physiology and ecology of pelagic larvae of marine benthic invertebrates. In Steidinger, K. A. & L. M. Walker (eds), Marine Plankton Life Cycle Strategies. CRC Press, Boca Ratón (FL): 93–120.

Fraser, D. I. & J. Mason, 1987. Pectinid spat studies in selected areas off the west coast of Scotland, 1982–1986. ICES C.M.1987/K: 4.

Gaines, S. D. & M. D. Bertness, 1992. Dispersal of juveniles and variable recruitment in sessile marine species. Nature 360: 579–580.

Harvey, M. & E. Bourget, 1995. Experimental evidence of passive accumulation of marine bivalve larvae on filamentous epibenthic structures. Limnol. Oceanogr. 40: 94–104.

Harvey, M., E. Bourget & G. Miron, 1993. Settlement of Iceland scallop *Chlamys islandica* spat in response to hydroids and filamentous red algae: field observations and laboratory experiments. Mar. Ecol. Prog. Ser. 99: 283–292.

Harvey, M., E. Bourget, C. Legault & R. G. Ingram, 1995. Short term variations in settlement and early spat mortality of Iceland scallop, *Chlamys islandica* (O. F. Müller). J. exp. mar. Biol. Ecol. 194: 167–187.

Hawkins, S. J. & R. G. Hartnoll, 1982. Settlement patterns of *Semibalanus balanoides* (L.) in the Isle of Man (1977–1981). J. exp. mar. Biol. Ecol. 62: 271–283.

Hunt, H. L. & R. E. Scheibling, 1996. Physical and biological factors influencing mussel (*Mytilus edulis*) settlement on a wave-exposed rocky shore. Mar. Ecol. Prog. Ser. 142: 135–145.

Jonsson, S. & K. Gudmundsson, 1994. An interdisciplinary study of Eyjafjordur, North Iceland. ICES C.M.1994/C: 6.

Jonsson, S., 1996. Ecology of Eyjafjordur Project: Physical parameters measured in Eyjafjordur in the period April 1992–August 1993. Hafrannsoknastofnun, Fjolrit 48 160 pp.

Jonsson, S., 1999. Temperature time series from Icelandic coastal stations. Rit Fiskid. 16: 59–68.

Jørgensen, C. B., 1946. Lamellibranchia. Meddl. Komm. Dan. Fisk. Havundersøg., Ser. Plankton 4: 277–311.

Kautsky, N., 1982. Growth and size structure in a Baltic *Mytilus edulis* population. Mar. Biol. 68: 117–133.

King, P. A., D. McGrath & E. M. Gosling, 1989. Reproduction and settlement of *Mytilus edulis* on an exposed rocky shore in Galway Bay, West Coast of Ireland. J. mar. biol. Ass. U.K. 69: 355–365.

King, P. A., D. McGrath & W. Britton, 1990. The use of artificial substrates in monitoring mussel (*Mytilus edulis* L.) settlement on an exposed rocky shore in the west of Ireland. J. mar. biol. Ass. U.K. 70: 371–380.

Knuckey, I. A., 1995. Settlement of *Pinctada maxima* (Jameson) and other bivalves on artificial collectors in the Timor Sea, Northern Australia. J. Shellfish Res. 14: 411–416.

Lane, D. J. W., A. R. Beaumont & J. R. Hunter, 1985. Byssus drifting and the drifting threads of the young post-larval mussel *Mytilus edulis*. Mar. Biol. 84: 301–308.

MacLean, L. A., N. G. MacNair, T. J. Davidson & G. G. Johnson, 1999. Two-year comparison of spawning patterns in soft shell clams (*Mya arenaria*). J. Shellfish Res. 18: 310.

Mann, R., 1982. The seasonal cycle of gonadal development in *Arctica islandica* from the southern New England shelf. Fish. Bull. 80: 315–326.

Mann, R., 1985. Seasonal changes in the depth-distribution of bivalve larvae on the southern New England shelf. J. Shellfish Res. 5: 57–64.

Montaudouin, X., 1997. Potential of bivalve's second settlement differs with species: a comparison between cockle (*Cerastoderma edule*) and clam (*Ruditapes philippinarum*) juvenile resuspension. Mar. Biol. 128: 639–648.

Muus, K., 1973. Settling, growth and mortality of young bivalves in the Øresund. Ophelia 12: 79–116.

Paul, J. D., A. R. Brand & J. N. Hoogester, 1981. Experimental cultivation of the scallops *Chlamys operculary s* (L.) and *Pecten maximus* (L.) using naturally produced spat. Aquaculture 24: 31–44.

Robinson, M. C. S., J. D. Martin, R. A. Chandler & G. J. Parsons, 1991. Spatial patterns of spat settlement in the sea scallop, *Placopecten magellanicus*, compared to hydrographic conditions in Passamaquoddy Bay, New Brunswick, Canada. J. Shellfish Res. 10: 274.

Robinson, M. C. S., J. D. Martin, R. A. Chandler & G. J. Parsons, 1999. An examination of the lingake between the early life history processes of the sea scallop and local hydrographic characteristics. J. Shellfish Res. 10: 272–273.

Sause, B. L., D. Gwyther & D. Burgess, 1987. Larval settlement, juvenile growth and the potential of use of spatfall indices to predict recruitment of the scallop *Pecten alba* Tate in Port Phillip Bay, Victoria, Australia. Fish. Res. 6: 81–92.

Seed, R. & T. H. Suchanek, 1992. Population and community ecology of *Mytilus*. In Gosling, E. G. (ed.), The Mussel *Mytilus*: Ecology, Physiology, Genetics and Culture. Dev. Aquacult. Fish. Sci. 25: 87–170.

Sigurdsson, J. B., C. W. Titman & P. A. Davies, 1976. The dispersal of young post-larval bivalve molluscs by byssus threads. Nature 262: 386–387.

Snodden, L. M. & D. Roberts, 1997. Reproductive patterns and tidal effects on spat settlement of *Mytilus edulis* populations in Dundrum Bay, Northern Ireland. J. mar. biol. Ass. U.K. 77: 229–243.

Sörlin, T., 1988. Floating behaviour in the tellinid bivalve *Macoma balthica* (L.). Oecologia 77: 273–277.

Tammi, K. A., W. H. Turner, M. Brumsted & M. A. Rice, 1996. Veliger vodoo and other witchcraft in the Wesport River: Bay scallop, *Argopecten irradians*, veliger abundance and the variability of spatfall recruitment to artificial spat collectors in the Wesport Estuary, Massachussetts. J. Shellfish Res. 15: 461–462.

Thorarinsdottir, G. G., 1991. The Iceland scallop, *Chlamys islandica* (O. F. Müller) in Breidafjordur, west Iceland. I. Spat collection and growth during the first year. Aquaculture 97: 13–23.

Thorarinsdottir, G. G., 1996. Gonad development, larval settlement and growth of *Mytilus edulis* L. in a suspended population in Hvalfjordur, south-west Iceland. Aquaculture Res. 27: 57–65.

Thorarinsdottir, G. G., 2000. Annual gametogenetic cycle in ocean quahog, *Arctica islandica* (Linnaeus, 1767), from north-western Iceland. J. mar. biol. Ass. U.K. 80: 661–666.

Thorson, G., 1950. Reproduction and larval ecology of marine bottom invertebrates. Biol. Rev. 25: 1–45.

Ventilla, R. F., 1977. A scallop spat collector trial off the Northern Ardnamurchan Coast. White Fish Authority, Field Report 485. 16 pp.

Warren, A. E., 1936. An ecological study of the sea mussel (*Mytilus edulis*). J. Biol. Bd. Can. 2: 89–94.

Winther, U. & J. S. Gray, 1985. The biology of *Mya arenaria* (Bivalvia) in the eutrophic inner Oslofjord. Sarsia 70: 1–9.

Zar, J. H., 1996. Biostatistical Analysis. Prentice Hall International Editions, New Jersey. 662 pp.

Hydrobiologia **503:** 143–151, 2003.
M.B. Jones, A. Ingólfsson, E. Ólafsson, G.V. Helgason, K. Gunnarsson & J. Svavarsson (eds), Migrations and Dispersal of Marine Organisms.

Barnacle larval supply to sheltered rocky shores: a limiting factor?

S.R. Jenkins & S.J. Hawkins
Marine Biological Association, The Laboratory, Citadel Hill, Plymouth, PL1 2PB, U.K.
E-mail: sjen@mba.ac.uk

Key words: larval dispersal, *Semibalanus balanoides*, settlement, supply, wave exposure

Abstract

In northwest Europe, sheltered rocky shores are dominated by fucoid canopy algae and barnacles are rare, although the latter are extremely abundant on exposed shores. The supply of the intertidal barnacle *Semibalanus balanoides* (L.) to sheltered, fucoid dominated rocky shores was investigated to determine the importance of larval supply in limiting the abundance of adults in shelter. Larval supply was measured at two spatial scales, at the scale of shore (100s of metres), by comparing larval concentrations at exposed and sheltered sites, and at a smaller spatial scale (m), by examining the role of fucoid canopies in limiting supply to the substratum. Replicate plankton trawls were carried out above the intertidal zone at high water at two sheltered sites and nearby exposed headlands. The concentration of *S. balanoides* cyprid larvae was significantly higher at the sheltered sites on two out of three sampling occasions with up to 14 times greater larvae on one occasion than the nearby exposed site. The effect of the macroalgal canopy on supply to the substratum was assessed in two ways: directly, by pumping water from the substratum in areas with and without *Ascophyllum nodosum* (L.) Le Jolis, and indirectly by measuring cyprid settlement in a canopy-manipulation experiment. Pumped plankton samples from mid tide level showed that the *A. nodosum* canopy did not form a barrier to larval supply and may have had a positive effect on larval concentrations at the substratum. Cyprid settlement was assessed in the mid shore *A. nodosum* and low shore *Fucus serratus* L. zones to areas with canopy algae (but protected from the sweeping effects of macroalgal fronds) and without canopy. Settlement over three consecutive 24-h periods showed a consistent pattern; settlement was consistently lower beneath the *F. serratus* canopy than in cleared areas, suggesting that this algal species forms a barrier, limiting supply of cyprid larvae to the substratum.

Introduction

The majority of marine benthic invertebrates exhibit a complex life cycle, that includes separate planktonic and bottom dwelling phases, linked by a settlement event (Thorson, 1950). Planktonic larvae undergo varying degrees of dispersal depending on the length of larval life, larval behaviour and the hydrographic regime which they encounter. Larval dispersal, or advection away from the spawning site, tends to limit competition between adults and offspring, prevents consumption of larvae by their parents and removes larvae from benthic predators. An obvious consequence of larval dispersal is that individuals are advected away from the habitat they will occupy as adults and return to this habitat will be dictated largely or wholly by physical transport processes. Several larval transport mechanisms have been studied including tidally forced internal waves (Shanks, 1986; Shanks & Wright, 1987), upwelling events (Roughgarden et al., 1988; Fairweather, 1991) and wind forcing (Hawkins & Hartnoll, 1982; McConnaughey et al., 1992; McQuaid & Phillips, 2000). However, we still know relatively little about patterns of dispersal and transport in benthic invertebrate larvae.

Variability in the level of dispersal of larvae from the spawning site and transport to an appropriate adult habitat will have a major effect on the level of larval supply. Studies over the last 15 years have shown convincingly that settlement rates can be dictated by the level of supply of larvae reaching a settlement site. This relationship operates over small (1–10 m, e.g. Gaines et al., 1985; Minchinton & Scheibling, 1991; Miron et al., 1995) and medium (100–1000 m, e.g. Eggleston & Armstrong, 1995; Bertness et al., 1996) scales. The level of larval settlement can have a major impact on adult population dynamics, although the way in which variability in settlement interacts with post-settlement processes to determine adult abundance is the subject of debate.

Acorn barnacles have been used extensively as a model organism to examine physical transport processes and the relationship between larval supply, settlement and adult population dynamics. They are extremely common world wide on hard substrata, and both planktonic and newly-settled larvae are easily identified and enumerated. In northwest Europe, acorn barnacles are major space occupiers at exposed intertidal sites and in the absence of mussels can dominate the entire mid shore area. Barnacle abundance declines markedly and fucoid canopy cover increases with a decline in wave exposure. At sheltered sites, barnacles often only occur in any numbers on the high shore, and on vertical surfaces and steep seaward facing slopes (Lewis, 1964). Lewis (1964) suggested a number of causative factors for the decline of barnacles in shelter including low larval supply, destruction of settling larvae and metamorphosed spat by the sweeping action of large algal fronds, the grazing activities of patellid limpets, and the lack of a suitable substratum. Interactions between settling barnacles and large canopy algae are clearly a major factor in limiting barnacle abundance on sheltered shores of northwest Europe (Jenkins et al., 1999a). It is not clear, however, to what extent supply of larvae to the substratum on sheltered shores is a contributory factor. The settlement rates of barnacles at shores with identical biota but differing in the degree of exposure have suggested that supply of larvae may be low in sheltered positions (Caffey, 1985; Bertness et al., 1992). On the other hand, there is evidence to suggest that shallow inlets and bays can be areas of larval retention through the formation of tidally driven eddy systems (e.g. Archambault et al., 1998).

One aspect of larval supply that is rarely considered is delivery of larvae to the substratum rather than merely to the general area. Hatton (1938) postulated that dense stands of canopy algae may form a barrier to the water-borne movement of plant and animal propagules and thus influence the establishment of benthic communities. For example, the presence of several algal canopy species has been shown to inhibit colonisation of *Sargassum muticum* (Yendo) Fensholt (Deysher & Norton, 1982). It was proposed that the algae formed a physical barrier between settling zygotes and the substratum. This phenomenon has also been suggested to inhibit the establishment of limpets on sheltered shores, although no tests of this hypothesis have been made (Fischer-Piette, 1948; Lewis & Bowman, 1975).

We aimed to examine the supply of cyprid larvae of *Semibalanus balanoides* to sheltered fucoid dominated shores to investigate the general model that low larval supply is a causative factor in limiting the abundance of adult *S. balanoides* on sheltered shores. Supply was examined at two spatial scales: firstly, at scales of 100s of metres between exposed and sheltered sites and, secondly, at the scale of metres by examining the role of fucoid canopies in limiting supply to the substratum. In so doing, we tested the hypotheses that larval concentration is lower at sheltered compared to exposed sites and that larval concentration at the substratum is negatively affected by the overlying canopy.

Materials and methods

Experimental sites

Sampling and experimental work was carried out at three locations in the south of the Isle of Man (Irish Sea) between 1993 and 2002 (Fig. 1). At all locations, sheltered sites where the intertidal community was dominated by fucoid canopy algae were chosen. Typically, these sites supported dense stands of fucoids, with the mid and low shore dominated by *Ascophyllum nodosum* and *Fucus serratus*, respectively. Descriptions of the communities in these fucoid assemblages on the Isle of Man are given by Jenkins et al. (1999b,c). At two of these locations (Fleshwick and Port St Mary), exposed shores, dominated by grazing patellid limpets and the acorn barnacle *Semibalanus balanoides*, were chosen in close proximity (200–400 m) to the sheltered bays (Fig. 1).

Supply of larvae to exposed and sheltered shores

The concentration of barnacle larvae in the water column above the intertidal zone at exposed and sheltered shores at Fleshwick and Port St Mary was assessed on two dates during the settlement period of *Semibalanus balanoides* in May 2002. Cyprid larvae were captured using a 0.5 m diameter 250 μm plankton net with attached flowmeter. This was towed from a small boat approximately 1 m below the sur-

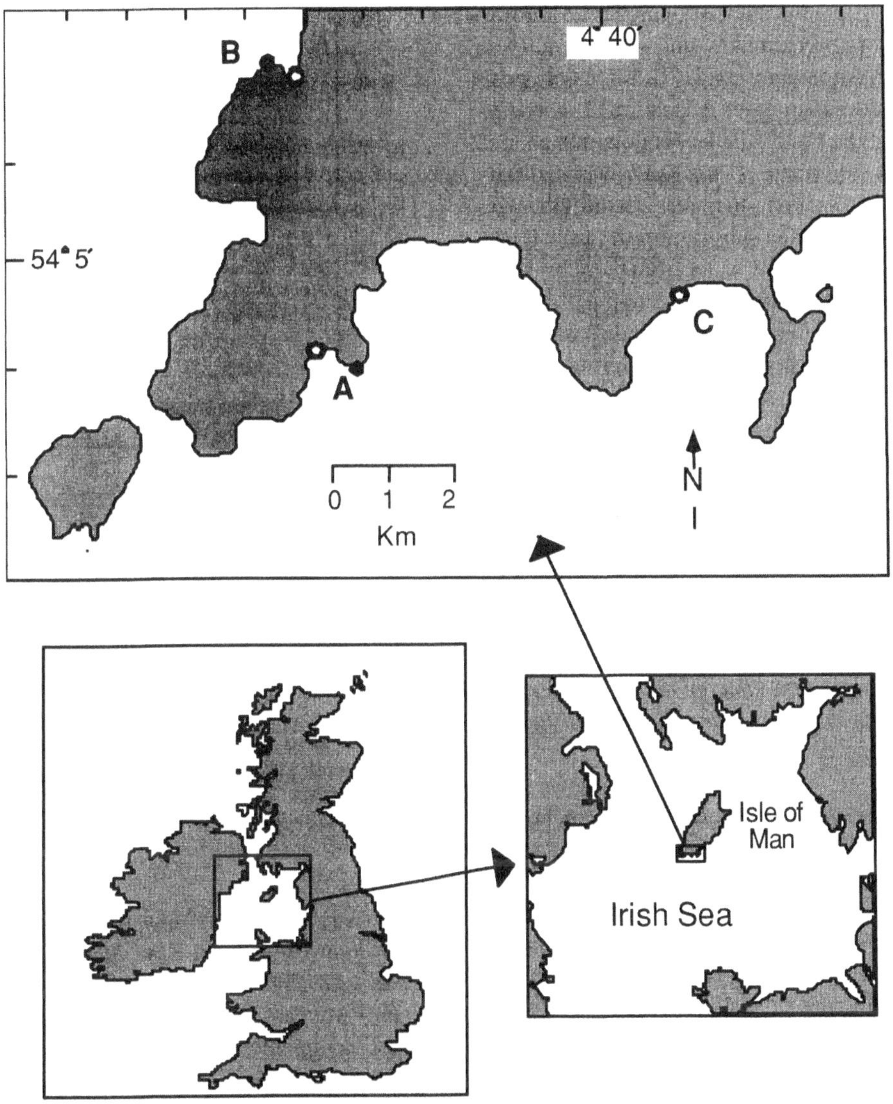

Figure 1. Location of sampling and experimental sites in the south of the Isle of Man, in the Irish Sea, British Isles. (A) Port St Mary location showing sheltered bay and exposed headland; (B) Fleshwick location showing sheltered bay and exposed headland; (C) Castletown Bay experimental site. Open circles represent sheltered sites, closed circles represent exposed headland sites.

face over the intertidal and shallow subtidal zones at a speed of 1–2 knots during the period of high water (HW ± 1.5 h). The water depth varied between 2.5 and 6 m. At each site on each sampling occasion, three tows, each of 10 min were made. The volume of water sampled in each tow ranged from 29 to 93 m^3, with a mean sampled volume of 56 m^3. The volume of water sampled varied owing to different flow conditions between exposed and sheltered sites and the varying degrees to which nets were clogged with phytoplankton. Plankton samples were preserved in 5% buffered formalin in seawater and the number of cyprid larvae of *S. balanoides* counted using a dissecting microscope.

Direct assessment of the effect of the Ascophyllum nodosum *canopy on larval supply to the substratum*

The concentration of *Semibalanus balanoides* larvae at the substratum in the presence and absence of a canopy of *Ascophyllum nodosum* was determined at a sheltered site in Castletown Bay during May 1993 using a petrol driven water pump (Honda WB10). The pump was located above the high water mark and connected via rigid tubing (internal diameter 25 mm) to two positions at approximately mid shore level (4 m above Lowest Astronomical Tide, L.A.T.). One position was located within a dense stand of *A. nodosum* and the other at the same tidal level but within an area at least 5 m wide, where *A. nodosum* was naturally absent. At each position, the inlet of the tube was fixed 10–20 cm above the substratum. In this way, the pump could be used to draw water from the substratum surface in areas with and without a canopy.

In May 1993, sampling was undertaken on 6 days during the barnacle settlement season, during a period 1 h either side of high water. At each date, three 10 min periods of pumping took place on each of the two tubes and water passed through a 100 μm plankton net. The pump rate for each of the two tubes on each date was calculated by determining the time taken to fill a container of known volume. The average pump rate over all sampling dates was 115 l per min.

Indirect assessment of the effect of the Ascophyllum nodosum *and* Fucus serratus *canopy on larval supply to the substratum*

Large macroalgae may affect barnacle settlement negatively through two processes: by forming a physical barrier limiting cyprid access to the substratum, or through a 'whiplash' or scouring effect of algal fronds at settlement. The potential barrier effect was examined in May 1997 by comparing settlement in two treatments, one in which the algal canopy was cleared (no barrier, no whiplash) and one in which the canopy was left intact but the substratum protected from algal whiplash by the use of cages (barrier but no whiplash). A third treatment, cleared areas with cages, was used to assess any artefacts associated with the use of cages. Each treatment was replicated six times at each of two shore levels, which corresponded with the middle of the *Fucus serratus* (2.8 m above L.A.T) and *Ascophyllum nodosum* (4.1 m above L.A.T.) zones.

Settlement tiles bearing adult conspecifics were used in all treatments to provide a cue to enhance settlement. Tiles were made from natural rock and had a minimum settling area of 12×12 cm with a patchy adult barnacle cover of between 33 and 50%. Where appropriate, tiles were caged. Cages were made from plastic-coated chicken wire (mesh size 13×13 mm) and were designed to form a complete cover over the whole tile with a 'roof' 3–4 cm above the settlement surface.

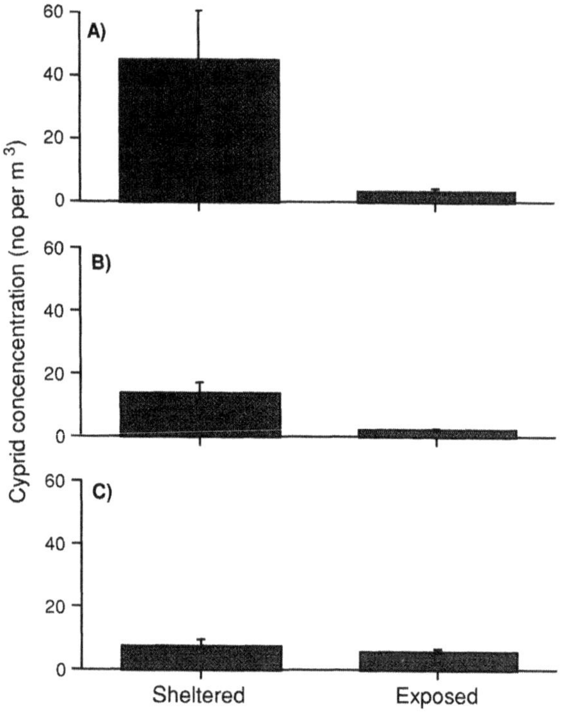

Figure 2. Mean concentration (±SE) of *Semibalanus balanoides* cyprid larvae above the intertidal zone at exposed and sheltered locations during May 2002. (A) Fleshwick – 27th May; (B) Port St Mary – 27th May; (C) Port St Mary – 31st May.

A stretch of shore approximately 100 m long was chosen at Castletown Bay. At each of the two shore levels, 18 plots were selected and experimental treatments were assigned at random to them. Canopy algae were completely cleared from a 2×2 m area in the appropriate plots. A single settlement tile was fixed in position using rapid setting cement at each plot. Settlement to experimental tiles was assessed over 24 h on three separate dates. At each date, all settled barnacles were cleared from the tile surfaces and the number of settled cyprids determined after 24 h. No examination of recruitment over longer time scales was made.

Statistical analysis

Analysis of variance (ANOVA) was used throughout to test for treatment effects. Cochran's test (Winer, 1971) was used to test for heterogeneity of variance and, where necessary, data were square root or log transformed. Multiple comparisons of levels within significant factors were made using Student Newman Keuls (SNK) tests.

Table 1. Two way mixed model ANOVA of *Semibalanus balanoides* cyprid concentration at the substratum with and without a canopy of *Ascophyllum nodosum*. Data Sqrt transformed. Cochran's test for homogeneity of variance: C= 0.285, $P > 0.05$

Source	df	MS	*F*	*P*	*F versus*
Canopy	1	6.4	1.1	>0.3	Ca×Da
Date	5	25.9	12.4	<0.001	Residual
Canopy×Date	5	5.9	2.8	<0.05	Residual
Residual	24	2.1			

SNK test on Canopy×Date Interaction:
Date 4: Canopy present>Canopy absent ($P < 0.05$).

Results

Supply of larvae to exposed and sheltered shores

The concentration of *Semibalanus balanoides* cyprid larvae above the intertidal zone at exposed and sheltered sites showed a similar trend on all three sampling occasions with higher larval numbers per unit volume of water in shelter (Fig. 2). This was very marked at Fleshwick Bay on 27th May where, on average, there were over 14 times higher numbers of cyprids in the sheltered bay compared to the exposed headland. There was, however, a high degree of variability among replicate hauls, indicating patchiness at a scale of 10s to 100s of metres. One way ANOVA showed a significantly higher concentration of larvae in shelter at only one of the three sampling occasions at a P value of 5%, but at two occasions with a P value of 10% (Fleshwick, 27th May – df 1,4, F 6.72, $P < 0.07$; Port St Mary 27th May – df 1,4, F 16.17, $P < 0.05$; Port St Mary 31st May – df 1,4, F 1.24, $P > 0.3$).

Direct assessment of the effect of the Ascophyllum nodosum *canopy on larval supply to the substratum*

The concentration of *Semibalanus balanoides* cyprid larvae at the substratum at the sheltered site in Castletown Bay was relatively high, but varied considerably among dates and among individual pumping periods within a date. Over all dates, the average cyprid concentration was 34 cyprids per m^3 in the presence of the canopy and 22 cyprids per m^3 in the absence of the canopy (Fig. 3). Data were analysed using a mixed model two way ANOVA with date considered a random factor. The effect of the *Ascophyllum nodosum* canopy on the concentration of cyprids differed according to the date as indicated by the significant Canopy×Date interaction (Table 1). Student Newman Keuls multiple comparison test on this interaction revealed only one date (May 27th) when there was a significant effect of the canopy. On this date, cyprid concentration was significantly greater in the presence of the canopy.

Indirect assessment of the effect of the Ascophyllum nodosum *and* Fucus serratus *canopy on larval supply to the substratum*

On all three sampling dates, the effect of the different treatments showed a similar trend. The effect of the three canopy treatments differed clearly between the *Fucus serratus* and *Ascophyllum nodosum* zones on all three dates, with highly significant interactions between treatment and shore height (Table 2, Fig. 4). In the *F. serratus* zone, settlement in the two cleared treatments was consistently higher than in the canopy treatment. Thus, despite being protected from the sweeping effects of *F. serratus* fronds, settlement beneath the canopy was apparently limited. Clearly, this was not an artefact of the cages since settlement in cleared areas was the same irrespective of the presence/absence of a cage. SNK tests confirmed these conclusions; tiles beneath the canopy showed significantly lower settlement than those in cleared areas on all three dates (Table 2, Fig. 4). In contrast, settlement in the *A. nodosum* zone was consistently highest in the canopy treatment. SNK tests showed settlement in this treatment was significantly higher than one or both canopy cleared treatments over the three dates (Table 2, Fig. 4).

Discussion

Assessment of larval supply to exposed and sheltered sites showed clearly that high larval concentrations of *Semibalanus balanoides* do occur at sheltered sites. Larval concentrations were, on average, greater in shelter than in exposed positions and, on one occasion, over 14 times higher. It is important to consider the relevance of measuring larval concentrations in the water column as an indicator of larval supply in areas of differing tidal flow. Larval concentrations such as are given by plankton tows or pumped water samples give a snapshot of larval supply, but do not integrate over time. Use of larval traps, which provide a time averaged signal of larval supply (e.g. Yund et al., 1991) or estimates of 'larval flux', which

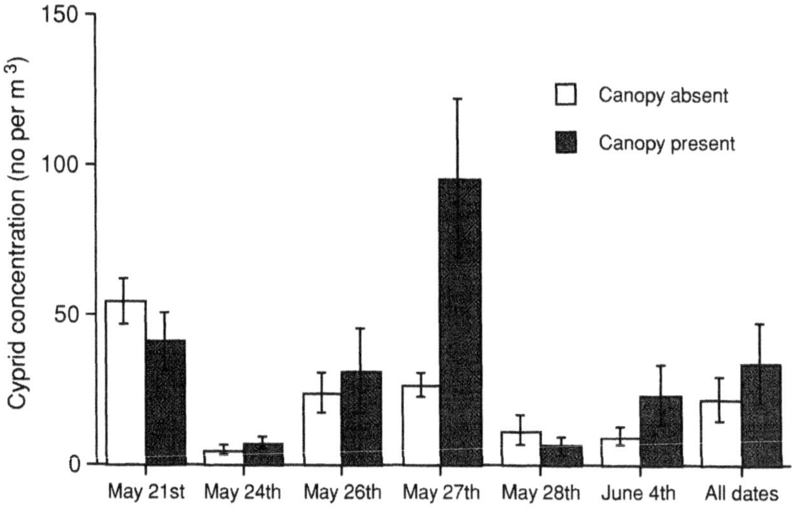

Figure 3. Mean concentration (±SE) of *Semibalanus balanoides* cyprid larvae at the substratum in two adjacent areas, one with a dense *Ascophyllum nodosum* canopy and one where canopy algae were naturally absent.

Table 2. Two way ANOVA of log transformed data testing for the effect of shore height and canopy treatments on the settlement of *Semibalanus balanoides* cyprids on three consecutive dates. C values give results of Cochran's test for homogeneity of variance

Source	df	Date 1 (C=0.59, $P < 0.01$)			Date 2 (C=0.27, $P > 0.05$)			Date 3 (C=0.28, $P > 0.05$)		
		MS	*F*	*P*	MS	*F*	*P*	MS	*F*	*P*
Treatment	2	2.1	8.6	<0.01	0.3	1.0	>0.3	0.6	2.4	>0.1
Height	1	66.2	274.5	<0.001	27.0	85.6	<0.001	0.2	0.9	>0.3
Tmt×Ht	2	5.8	24.1	<0.001	5.2	16.4	<0.001	4.5	17.0	<0.0001
Residual	30	0.2			0.3			0.3		

	Date 1		**Date 2**		**Date 3**	
	Mid	Low	Mid	Low	Mid	Low
Cleared (caged) v cleared (no cage)	S	NS	NS	NS	NS	NS
Cleared (caged) v canopy intact	NS	S	S	S	NS	S
Cleared (no cage) v canopy intact	S	S	S	S	S	S

SNK multiple comparisons of interaction. Comparisons between shore heights not shown (NS=$P > 0.05$; S=$P < 0.05$).

combine estimates of larval concentrations with measures of flow rates (Archambault & Bourget, 1999), may be more reliable indicators of true larval supply. Simple estimates of current velocity combined with larval concentration data have been used in previous studies to provide a better predictor of benthic recruitment than concentration alone (e.g. Archambault & Bourget, 1999). Although estimates of tidal flow are available for the exposed sites sampled, tidal flow inside the small sheltered bays was very low and data are unavailable. An alternative approach would be to use subtidal (Yund et al., 1991) or intertidal (Castilla & Varas, 1998) larval traps to estimate larval flux. Such traps, however, are likely to operate with varying efficiency in areas of differing tidal flow or wave exposure and, therefore, are probably not suitable for any comparison of sheltered and exposed sites.

Although it is likely, given the much higher tidal flow at exposed sites, that the number of larvae 'supplied' to exposed shores over the period of settlement is higher than to sheltered shores, we showed that the larval concentration at a point in time, and therefore the number of larvae available to settle can be high in shelter. Indeed, where conditions were suitable for settlement, with suitable settlement surfaces and without the sweeping or barrier effects of macroalgae, settlement was high. Jenkins (1995) showed that using identical settlement surfaces, recruitment of *S. balanoides* was higher on a fucoid dominated sheltered shore when the canopy was cleared than to an adjacent exposed site. It is clear, therefore, that patterns of larval dispersal do allow supply of larvae to sheltered shores. How accumulation of high larval concentrations in sheltered bays occurs is not clear, but retention of larvae in eddy systems generated by tidal flow is one candidate mechanism (Archambault & Bourget, 1999). An alternative is that local production of larvae by adult populations within sheltered bays undergoes limited dispersal. The pelagic larvae of benthic invertebrates are generally assumed to undergo extensive dispersal and large-scale export from the site of production. Hence populations are considered open, with the level of recruitment independent of local production. Recent work, however, has questioned this widely held view and stressed that local retention of pelagic larvae may be more prevalent than previously thought (e.g. Sponaugle et al., 2002; Swearer et al., 2002; Warner & Cowen, 2002). Sheltered embayments are undoubtedly areas where retention may be likely. This has obvious implications for local population dynamics and gene flow. Gaines & Bertness (1992) demonstrated high levels of barnacle recruitment (and hence probably larval retention) in a large bay when flushing times were long. The extent to which larval retention occurs in embayed areas on a much smaller scale, as in the present study, is not clear. Here retention will be related to the interaction of long shore currents with coastal geography and further work integrating measures of larval concentration and hydrography is required.

We examined supply of larvae not only at the scale of shore but also at a much smaller scale-supply to the substratum. Studies investigating larval 'supply' have examined larval concentration or larval flux at differing distances from the adult habitat from measures within centimetres (e.g. Minchinton & Scheibling, 1991) to measures over 50 m away (e.g. Bertness et al., 1996). Differential delivery of larvae from larval pools offshore of the adult habitat may disrupt any relationship between larval supply, measured 10s or 100s of metres from the shore, and settlement. On canopy-dominated sheltered shores, supply measured at the scale of shore may bear little relevance to supply to the substratum, owing to the potential of canopy algae to block larval movement to the substratum. Canopy algae may reduce supply of larvae to the substratum by forming a physical barrier to larval movement or reducing water circulation and mixing. For example, subtidal kelp forests have a significant impact on fluid

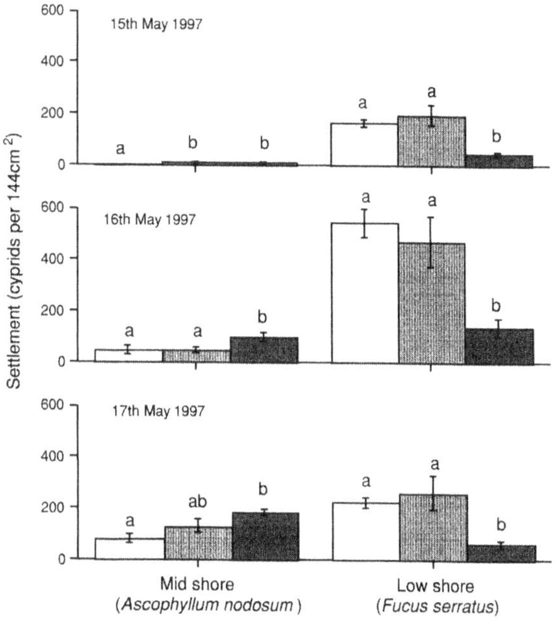

Figure 4. Mean level (±SE) of *Semibalanus balanoides* cyprid settlement over a 24-h period at three consecutive dates to experimental treatments in the mid and low shore of a sheltered site in Castletown Bay. For each shore level treatments sharing the same letter are not significantly different (from SNK multiple comparisons of the significant treatment×shore height interaction).

and particulate transport and can inhibit transport of suspended particles from the overlying water column to the bottom (Eckman et al., 1989). Alternatively, canopy algae may create a 'behavioural barrier'. Given the complex behaviour patterns and discriminatory processes seen in many planktonic larvae it is feasible that chemical or physical cues from canopy algae may discourage larvae from seeking the substratum.

A combination of direct and indirect measures of larval supply to the substratum in the mid shore *Ascophyllum nodosum* zone showed clearly that this canopy alga does not prevent cyprid access to the rock surface. In fact, evidence from both pumping and canopy manipulation suggests that supply may be higher where the canopy is present. This result is in agreement with observations of Leonard (1999) who found that the *A. nodosum* canopy in the northwest Atlantic created a depositional environment where larvae of *Semibalanus balanoides* accumulated. In contrast, indirect assessment of supply in the low shore *Fucus serratus* zone showed that this alga may form a barrier to larval supply. Settlement was considerably reduced where the canopy was left in place compared to cleared areas, even when settlers were protected from algal sweeping or whiplash. Clearly, measurement of settlement is not the same as measuring larval concentration or supply. However, given identical settlement surfaces, and the results showing no artefact of using cages, it seems likely that differences in settlement can be accepted as indicating differences in supply. The difference in result between the two shore zones may occur because of differences in the physical characteristics of the two dominant canopy algae. *Ascophyllum nodosum* stands erect in the water column, buoyed by numerous bladders, in contrast to the more bushy nature of *F. serratus*. When viewed from above during high water, *F. serratus* appears to present far more of a physical barrier to cyprids accessing the substratum.

In conclusion, we showed that larval concentrations on sheltered shores, where adult barnacles are scarce, can be very high. A more extensive sampling regime, incorporating estimates of larval flux over a number of years, is required to build on these preliminary observations to improve our understanding of how supply varies along indented coastlines. At a smaller scale, we found differing effects of two species of canopy algae on larval supply to the substratum. Larvae were not impeded by the mid-shore species *Ascophyllum nodosum* but, in the low-shore *Fucus serratus* zone, observations of settlement indicate that this fucoid species can limit larval access to settlement sites. Thus, at mid-shore level, biological interactions between settling larvae and macroalgae (both canopy and turf forming) are probably the main factors limiting barnacle abundance on sheltered shores, while in the low shore, supply to the substratum may be an additional limiting factor.

Acknowledgements

This work was supported by a PhD studentship from the Natural Environment Research Council to SRJ, the EUROROCK project (MAS3-CT95-0012) and fellowships funded by NERC grant in aid from the Marine Biological Association, UK to SRJ and SJH. Thanks to Trevor Norton and Richard Nash for providing facilities at Port Erin Marine Laboratory and Anna Markula for help with fieldwork. The manuscript was improved through the comments of two anonymous referees.

References

Archambault, P. & E. Bourget 1999. Influence of shoreline configuration on spatial variation of meroplanktonic larvae, recruitment and diversity of benthic subtidal communities. J. exp. mar. Biol. Ecol. 238: 161–184.

Archambault, P., J. C. Roff, E. Bourget, B. Bang & G. R. Ingram, 1998. Nearshore abundance of zooplankton in relation to shoreline configuration and mechanisms involved. J. Plankton Res. 20: 671–690.

Bertness, M. D., S. D. Gaines, E. G. Stephens & P. O. Yund, 1992. Components of recruitment in populations of the acorn barnacle *Semibalanus balanoides* (Linnaeus). J. exp. mar. Biol. Ecol. 156: 199–215.

Bertness, M. D., S. D. Gaines & R. A. Wahle, 1996. Wind-driven settlement patterns in the acorn barnacle *Semibalanus balanoides*. Mar. Ecol. Prog. Ser. 137: 103–110.

Caffey, H. M., 1985. Spatial and temporal variation in settlement and recruitment of intertidal barnacles. Ecol. Monogr. 55: 313–332.

Castilla, J. & M. Varas, 1998. A plankton trap for exposed rocky intertidal shores. Mar. Ecol. Prog. Ser. 175: 299–305.

Deysher, L. & T. A. Norton, 1982. Dispersal and colonisation in *Sargassum muticum* (Yendo) Fensholt. J. exp. mar. Biol. Ecol. 56: 179–195.

Eckman, J. E., D. O. Duggins & A. T. Sewell, 1989. Ecology of understorey kelp environments. 1. Effects of kelps on flow and particle transport near the bottom. J. exp. mar. Biol. Ecol. 129: 173–187.

Eggleston, D. & D. Armstrong, 1995. Presettlement and post-settlement determinants of estuarine Dungeness crab recruitment. Ecol. Monogr. 65: 193–216.

Fairweather, P. G., 1991. Implications of 'supply-side' ecology for environmental assessment and management. Trends Ecol. Evol. 6: 60–63.

Fischer-Piette, E., 1948. Sur les elements de prosperite des patelles et sur leur specificite. J. Conchyliologie 88: 45–96.

Gaines, S. D., S. Brown & J. Roughgarden, 1985. Spatial variation in larval concentration as a cause of spatial variation in settlement for the barnacle, *Balanus glandula*. Oecologia 67: 267–272.

Gaines, S. D. & M. D. Bertness, 1992. Dispersal of juveniles and variable recruitment in sessile marine species. Nature 360: 579–580.

Hatton, H., 1938. Essais de bionomie explicative sur quelques especes intercotidales d'algues et d'animaux. Annales de l'Institut Oceanographique 17: 241–348.

Hawkins, S. J. & R. G. Hartnoll. Settlement patterns of *Semibalanus balanoides* (L.) in the Isle of Man (1977–1981). J. exp. mar. Biol. Ecol. 62: 271–283.

Jenkins, S. R., 1995. The ecology of fucoid canopy dominated shores. PhD, Liverpool.

Jenkins, S. R., T. A. Norton & S. J. Hawkins, 1999a. Settlement and post-settlement interactions between *Semibalanus balanoides* (L.) (Crustacea: Cirripedia) and three species of fucoid canopy algae. J. exp. mar. Biol. Ecol. 236: 49–67.

Jenkins, S. R., S. J. Hawkins & T. A. Norton, 1999b. Direct and indirect effects of a macroalgal canopy and limpet grazing in structuring a sheltered inter-tidal community. Mar. Ecol. Prog. Ser. 188: 81–92.

Jenkins, S. R., S. J. Hawkins & T. A. Norton 1999c. Interaction between a fucoid canopy and limpet grazing in structuring a low shore intertidal community. J. exp. mar. Biol. Ecol. 233: 41–63.

Leonard, G. H., 1999. Positive and negative effects of intertidal algal canopies on recruitment and survival of barnacles. Mar. Ecol. Prog. Ser. 178: 241–249.

Lewis, J. R., 1964. The Ecology of Rocky Shores. English Universities Press, London.

Lewis, J. R. & R. S. Bowman, 1975. Local habitat induced variations in the population dynamics of *Patella vulgata* (L). J. exp. mar. Biol. Ecol. 17: 165–203.

McConnaughey, R., D. Armstrong, B. Hickey & D. Gunderson, 1992. Juvenile Dungeness crab (*Cancer magister*) recruitment variability and oceanic transport during the pelagic larval phase. Can. J. Fish. aquat. Sci. 49: 2028–2044.

McQuaid, C. D. & T. E. Phillips, 2000. Limited wind-driven dispersal of intertidal mussel larvae: in situ evidence from the plankton and the spread of the invasive species *Mytilus galloprovincialis* in South Africa. Mar. Ecol. Prog. Ser. 201: 211–220.

Minchinton, T. E. & R. E. Scheibling, 1991. The influence of larval supply and settlement on the population structure of barnacles. Ecology 72: 1867–1879.

Miron, G., B. Boudreau & E. Bourget, 1995. The use of larval supply in benthic ecology: testing correlations between larval supply and larval settlement. Mar. Ecol. Prog. Ser. 124: 301–305.

Roughgarden, J., S. Gaines & H. Possingham, 1988. Recruitment dynamics in complex life cycles. Science 241: 1460–1466.

Shanks, A. L., 1986. Tidal periodicity in the daily settlement of intertidal barnacle larvae and an hypothesised mechanism for the cross shelf transport of cyprids. Biol. Bull. 170: 429–440.

Shanks, A. L. & W. G. Wright, 1987. Internal-wave-mediated shoreward transport of cyprids, megalopae, and gammarids and correlated longshore differences in the settling rate of intertidal barnacles. J. exp. mar. Biol. Ecol. 114: 1–13.

Sponaugle, S., R. K. Cowen, A. Shanks, S. G. Morgan, J. M. Leis, J. S. Pineda, G. W. Boehlert, M. J. Kingsford, K. C. Lindeman, C. Grimes & J. L. Munro, 2002. Predicting self-recruitment in marine populations: biophysical correlates and mechanisms. Bull. mar. Sci. 70: 341–375.

Swearer, S. E., J. S. Shima, M. E. Hellberg, S. R. Thorrold, G. P. Jones, D. R. Robertson, S. G. Morgan, K. A. Selkoe, G. M. Ruiz & R.R. Warner, 2002. Evidence of self-recruitment in demersal marine populations. Bull. mar. Sci. 70: 251–271.

Thorson, G., 1950. Reproductive and larval ecology of marine bottom invertebrates. Biol. Rev. 25: 1–45.

Warner, R. R. & R. K. Cowen, 2002. Local retention of production in marine populations: Evidence, mechanisms, and consequences. Bull. mar. Sci. 70: 245–249.

Winer, B. J., 1971. Statistical Principles in Experimental Design. McGraw-Hill Kogakusha, Tokyo.

Yund, P. D., S. D. Gaines & M.D. Bertness, 1991. Cylindrical tube traps for larval sampling. Limnol. Oceanogr. 36: 1167–1177.

Hydrobiologia **503:** 153–161, 2003.
M.B. Jones, A. Ingólfsson, E. Ólafsson, G.V. Helgason, K. Gunnarsson & J. Svavarsson (eds), *Migrations and Dispersal of Marine Organisms.*

Go with the flow: tidal migration in marine animals

R.N. Gibson
Scottish Association for Marine Science, Dunstaffnage Marine Laboratory, Oban, Argyll, PA37 1QA Scotland
E-mail: robin.gibson@sams.ac.uk

Key words: intertidal migration, behaviour, selective tidal-stream transport, rhythms

Abstract

Many shallow-water organisms are subjected to regular, frequent and considerable changes in their environment imposed by the tides. Such changes range from complex alterations in local conditions, as in the intertidal zone, to simpler variations in current speed and direction offshore. To take advantage of the opportunities these changes provide, many animals undertake regular vertical and/or horizontal migratory movements in synchrony with the tidal cycle. Others use tidal currents selectively for transport to new habitats. Such movements vary in extent from a few mm to several km and may be modulated by the diel cycle of light intensity. This paper reviews the existence of tidal migration in marine animals from various taxa and habitats. Crustacean and fish examples illustrate the various functions of tidal migration, including transport, feeding, predator avoidance and reproduction. These functions are not necessarily mutually exclusive and may change throughout ontogeny. The mechanisms involved in migration are discussed in terms of direct reactions to environmental stimuli and the interaction of these responses with endogenous rhythms of activity and responsiveness.

Introduction

The common perception of the phenomenon of 'migration', both on land and in the sea, is one of seasonal movement of large numbers of animals over great distances. This perception is reflected in the many definitions of migration available; for example, 'the periodic or seasonal movement, typically of relatively long distance, from one area, stratum or climate to another' (Lincoln et al., 1982) or 'the mass directional movements of large numbers of a species from one location to another' (Begon et al., 1990). Such a perception is partially the result of the visibility of many of the large-bodied migrants on or over land (insects, birds, mammals) but also in the sea (fishes, turtles and whales). There are, nevertheless, other less visible, smaller scale movements, such as the diel vertical movements of plankton, that can be considered as migration with equal validity. On an even shorter scale of time and distance, many shallow-water animals undergo migratory movements at tidal frequencies. Such movements can be termed 'tidal migrations' and defined as 'the periodic movement of large numbers of a species from one location to another in synchrony with the tidal cycle'. As discussed below, tidal migrants also need to time and direct their movements correctly, and are subject to the same selective pressures as longer distance migrants. Consequently, the study of these 'migrations in miniature', which is considerably easier both in the laboratory and the field, has the potential to shed light on the adaptations available to long-distance migratory species (Smith, 1985). Previous selective accounts of tidal migration are available in the reviews of Brown & McLachlan (1990), Naylor & Rejeki (1996) and Gibson (1999).

This paper reviews the existence of tidal migration in marine animals from a variety of taxa and habitats. Examples selected from crustaceans and fishes are used to illustrate the various functions of tidal migration, including transport, feeding, predator avoidance and reproduction. The mechanisms involved in migration are discussed in terms of direct reactions to environmental stimuli (especially light and salinity) and the interaction of these responses with endogenous timing mechanisms. Inter- and intraspecific differences in migration patterns are also considered.

Types of tidal migration

Tidally-synchronised migrations can take several forms. The first, and probably the most common, intertidal migration, consists of the movement of animals into and out of the intertidal zone on each tide. Such migrations consist mostly of the movement of subtidal animals into the intertidal zone on the flood and their retreat on the ebb tide (upshore migrants). However, several birds and invertebrates enter the intertidal zone as the tide ebbs and return to higher levels as the tide floods (downshore migrants). Many resident intertidal species (e.g. limpets, crabs and fishes) make periodic, tidally-synchronised foraging excursions within the intertidal zone and in so doing may employ many of the mechanisms involved in migration over longer distances. Such movements are excluded from this review on the basis that they do not constitute a mass movement in the same direction.

In selective tidal-stream transport (STST), animals make use of strong tidal currents over one part of the tidal cycle (either ebb or flood) to transport them in the required direction (Forward & Tankersley, 2001). This behaviour has been studied mostly in estuaries and in offshore coastal waters where tidal streams are strong.

Other tidally-synchronised migrations consist of vertical movements through the sediment or on protruding structures. Some species also undertake foraging excursions to distant locations with a tidal frequency.

The occurrence of tidal migrations in marine animals

Intertidal migration

The existence of ancient intertidal fish trap remains is evidence that man has been aware of and exploited the movement of animals into the intertidal zone for millennia although published accounts of intertidal migration are relatively recent, one of the first being Watkin's (1941) account of sandbeach crustaceans. Subsequently, numerous examples of intertidal migration have been recorded in a wide range of species and habitats as shown in Table 1.

Selective tidal-stream transport (STST)

In a comprehensive review of STST, Forward & Tankersley (2001) recorded this form of behaviour in at least 25 species of decapod crustaceans from four infraorders together with unspecified examples from

Table 1. Selected examples of intertidal migration in different animal groups. References marked with an asterisk are review papers

Migration type	Group	Intertidal habitat	Reference
Downshore	Birds	Sedimentary shores	Evans & Peinkowski (1982)
	Amphipods	Sedimentary shores	Naylor & Rejeki (1996*)
	Mudskippers	Mangroves	Gibson (1999*)
Upshore	Molluscs	Sandy beaches	Brown & McLachlan (1990*)
	Decapods	Sandy beaches, rocky shores, sea grass beds	Brown & McLachlan (1990*) Naylor & Rejeki (1996*), Warman et al. (1993b) Emmerson (1987)
	Isopods	Sandy beaches	Brown & McLachlan (1990*) Naylor & Rejeki (1996*)
	Amphipods	Sandy beaches	Brown & McLachlan (1990*) Naylor & Rejeki (1996*)
	Mysids	Sandy beaches	Brown & McLachlan (1990*) Takahashi & Kawaguchi (1997)
	Fishes	Most habitats	Gibson (1992, 1997*)

Table 2. Occurrences of selective tidal-stream transport in different groups. Data from Forward & Tankersley (2001). *N* is the number of named species in each group. Percentages do not sum to 100 because STST may be shown in both young and adult stages

Group	Stage	Number	%
Decapod crustaceans	Larvae and juveniles	24	96
$N = 25$	Adults	2	8
Molluscs	Larvae and juveniles	2	100
$N = 2$	Adults	0	0
Polychaetes	Larvae and juveniles	1	100
$N = 1$	Adults	0	0
Fishes	Larvae and juveniles	29	81
$N = 36$	Adults	10	28

five families. Also listed were 36 fish species from five orders, one polychaete and two molluscs. In some species, more than one life-history stage performs STST but the majority of occurrences are in larval and juvenile stages (see Table 2).

Other tidally-synchronised migrations

Surprisingly, the first published account of tidally-synchronised movements is that of the intertidal flatworm *Convoluta roscoffensis* (Gamble & Keeble, 1904) rather than of larger and more visible species. This intertidal species migrates to the surface of the sediment at low tide and returns to deeper layers as the tide floods. Comparable small-scale vertical migrations are shown by the copepod *Porcellidium* on algal fronds (Gibbons, 1989).

On a larger spatial scale, kittiwakes (*Rissa tridactyla*) and guillemots (*Uria aalge*) make foraging excursions to distant feeding grounds that are synchronised with the tidal cycle from their roosting places on rocky cliffs (O'Connor, 1974; Slater, 1976).

Methods of studying tidal migration

Sequential capture

The most common method employed for studying tidal migration is the capture of individuals at different stages of the tide at the same or different locations. For example, the simplest evidence of the existence of upshore intertidal migration is the presence of individuals in the intertidal zone at high tide and their absence at low tide. This evidence can be obtained by fishing in the intertidal zone at high tide, by blocking off creeks with nets at high tide (Kneib, 1991), or setting traps (Hunter & Naylor, 1993) and fixed nets (Black & Miller, 1991) in the intertidal zone. More complex sampling schemes involve sequential sampling over the tidal cycle at several set positions in the intertidal and subtidal zones (Ansell & Gibson, 1990). Alternatively, continuous or sequential samples can be taken at a single site in, for example, an estuary (Jager, 1999) or offshore tidal streams (Arnold & Metcalfe, 1996). In all cases, the objective is to investigate the change in numbers of individuals over the tidal cycle.

Direct observation

Most examples of tidal migration recorded have been inferred from remote sampling procedures and relatively few from direct observations of behaviour. In many cases, observation of behaviour in the field is difficult or impossible because of the small size of the organisms and/or the turbidity and turbulence of the waters in which they live. Consequently, the behaviour of many larval stages is only known from laboratory observations. Such observations are essential for determining the likely mechanisms involved. Nevertheless, direct field observations have been made of the behaviour of larger animals in clear, calm waters from both above (e.g. Gibson, 1980; Colombini et al., 1995 on fishes) and below (e.g. Dare & Edwards, 1981; Warman et al., 1993b on *Carcinus maenas*) the surface. Direct observations can be used to validate net or trap catches as well as describing the behaviour involved in migration.

Underwater television (UWTV)

UWTV has been used on several occasions for long-term observation of the movements of larger animals such as crabs (Dare & Edwards, 1981), fishes (Tyler, 1971) or a mixture of migrating species (Burrows et al., 1994; Gibson et al., 1998). Combined with video-recording, it has the advantage of allowing long-term continuous observation and subsequent detailed analysis of behaviour.

Telemetry

Attachment of ultrasonic or recording tags allows individuals to be followed over long distances and for considerable periods. The size of tags means, however, that only larger individuals can carry tags and hence this method has been used principally on fishes (e.g. Greer Walker et al., 1978; Wirjoatmodjo & Pitcher, 1980; Szedlmayer & Able, 1993).

Behaviour involved in tidal migration

Although in the great majority of cases some form of locomotion is involved in tidal migration, in a few instances passive transport by tidal currents may be the only agent of transfer between two locations, as in juveniles of the bivalve *Macoma balthica*. Even here, though, transport is facilitated by the secretion of byssus threads (Beukema & de Vlas, 1989). Several species use the energy provided by waves to carry out their intertidal migrations. This behaviour, variously called 'surfing' (Brown & McLachlan, 1990) or 'swash riding' (Ellers, 1995a), involves using the swash for passive transport up and down the beach. For example, the bivalve mollusc *Donax variabilis* jumps out of the sand on the largest waves on a rising tide, extends its foot and siphons and is swept up the beach. It similarly enters the backwash on a falling tide but becomes progressively less active as low tide approaches (Ellers, 1995a). Comparable behaviour occurs in some species of the gastropod genus *Bullia* in which the foot is extended and acts as a 'sail' (Brown, 1982). The barrel-shaped bodies of the decapod crustaceans *Emerita* and *Hippa* enable them to be rolled up and down the beach by waves (Cubit, 1969; Brown & McLachlan, 1990).

In species with good locomotory abilities, migration is performed principally by active swimming in the water column or over the bottom, although shrimps and crabs may use a combination of walking and swimming just over the bottom. Direct observations (Gibson, 1980) of young plaice (*Pleuronectes platessa*) on sandy beaches showed that the migration of this species is an active process that consists of a series of short swimming movements interspersed with periods of rest on the bottom. Swimming is directed predominantly upshore on the flooding tide but downshore on the ebb. Similarly, direct observation of the mudskipper *Periophthalmus sobrinus* showed that this species follows closely the advancing and retreating waterline, and thereby moves up and down the banks of Kenyan mangrove creeks (Colombini et al., 1995). STST requires that individuals remain swimming in the water column at the appropriate stage of the tide but move to lower levels, or to the bottom, when the tidal direction changes. In negatively buoyant species, movement downwards may only require cessation of swimming.

Functions of tidal migration

Although few analyses of the costs and benefits of migration have been performed, it is tacitly assumed that migration confers a selective advantage on those individuals that migrate. The extent of this advantage, however, may vary with developmental stage such that migration may not be advantageous at some stages. Furthermore, different advantages may be conferred by the different types of migration.

Intertidal migration

Movement into and out of the intertidal zone can maximise food intake and, in some species, reproductive output, and minimise predation. In some cases, migration may provide more than one of these benefits. For example, movement into the intertidal zone to feed enables individuals to remain out of reach of predators in deeper water. These benefits may be counteracted by the energetic costs of migration, increased predation risk while actually migrating (Hiddinck et al., 2002), and the danger of stranding by the receding tide. Such behaviour can be modelled to explore the relative importance of the assumed benefits and shed light on the possible causative factors of migration. One such model is that developed for the intertidal movements of young plaice (*Pleuronectes platessa*) on sandy beaches (Burrows, 1994). In this model, the 'state' of individual fish is characterised by their vertical position on the shore, the fullness of their gut and the quantity of their energy reserves. Vertical position on the shore in relation to prey and predator abundance determines the probability of acquiring further energy (food) and the risk of predation. Assuming that the fish are visual predators and can only feed during the day, the model computes the optimum position for a fish to occupy at particular tidal states and times of day in relation to different combinations of predator and prey distributions. Although it takes no account of variation in physical factors such as gradients of light and temperature, the model still successfully predicts behaviour patterns observed under a range of natural conditions (Burrows et al., 1994; Gibson et al., 1998). Speirs et al. (2002) used a combination of modelling, field studies and laboratory experiments to examine the likelihood of several explanations for the intertidal migrations of the estuarine mysid *Neomysis integer*. Their results suggested that the observed movements were not the result of diffusive spread and that foraging was not the primary driving force. Behaviour involving movement towards a specific depth could not be ruled out but laboratory experiments indicated that depth-seeking was only weakly related to predator avoidance.

The intertidal zone can be a rich source of food and many intertidal migrants exploit this food source when it is available to them. For most species, foraging in the intertidal zone is only possible when it is submerged (upshore migrants) but others (downshore migrants, e.g. birds and talitrid amphipods) take advantage of the receding tide to feed on exposed benthic prey. Evidence for intertidal feeding may come from direct observation of feeding activity or analysis of gut contents at different states of the tide. Many fishes have been studied in this respect (Gibson, 1992, 1999) and the majority had fuller guts when leaving the intertidal zone than when entering it. Increased food intake can result in increased growth and a study of young plaice on Dutch intertidal sand- and mudflats indicated that migrant individuals grew at a faster rate than non-migrants (Berghahn, 1987).

A second function often assumed for intertidal migration is that of minimising predation risk. This assumption is based on several lines of evidence. First, some intertidal migrant species do not feed more intensively when in the intertidal zone and some additional function for their movement must be sought. Second, predators of migrating species are larger, more numerous and more varied subtidally (e.g. Ellis & Gibson, 1995); intertidal migration therefore reduces encounter time with these subtidal predators. Third, some predators may also move inshore with the tide. Prey species, therefore, minimise encounter with migrating predators by maintaining their position in shallow water as the tide rises. In many cases, young stages and adults of small-bodied species may not migrate at all but remain resident in the intertidal zone thereby obtaining refuge from predators over the low tide period (e.g. Kneib, 1987). Ontogenetic variation in migration strategies are discussed in greater detail below. The nocturnal modulation of some tidal migrations, in which individuals only migrate when the appropriate phase of the tide, occurs at night may also be an adaptation for reducing predation by diurnal predators.

Another means of reducing predation impact and thereby maximising reproductive success employed by several fish species is to lay eggs in the intertidal zone out of the reach of subtidal predators (DeMartini, 1999). During the reproductive period, mature fishes migrate into the intertidal zone on spring tides and lay eggs in sand (e.g. grunion *Leuresthes tenuis*), in mussel shells (e.g. killifish *Fundulus heteroclitus*) or among pebbles (e.g. fugu *Takifugu niphobles*). The eggs develop in the moist substrata over the ensuing 2 weeks and hatch when covered by the next set of spring tides (Gibson, 1999).

In upshore migrants, feeding, predator avoidance and spawning take place on the upshore phase of intertidal migration whereas the downshore phase is imposed by the receding tide. In upshore migrants, downshore movement is, therefore, a behaviour that results in the avoidance of adverse abiotic conditions or increased risk of predation from terrestrial or avian predators. Conversely, the upshore migratory phase of downshore migrants is necessitated by the submersion of intertidal areas by the flooding tide.

Selective tidal stream transport

The principal function of STST behaviour, as its name implies, is to transport individuals to their required destination, usually to spawning, feeding or nursery grounds. The distances covered in one tide can be in the order of several kilometres (Greer Walker et al., 1978; Jager & Mulder, 1999). Also implicit in STST behaviour is the reduction of energy expenditure, allowing individuals, particularly those with weak swimming abilities such as larval stages, to cover long distances by riding tidal currents. Larger animals, such as adult fishes, also use this mode of transport for the same reasons. However, tidal currents can also be used to prevent transport, i.e. to retain individuals in a required location and avoid displacement to less favourable areas. Behaviour that leads to retention, which mostly involves tidally-phased movement to levels in the water column where horizontal currents are weakest (Forward & Tankersley, 2001), is most common in estuaries and serves to retain larval stages in productive nursery areas. STST is also amenable to modelling studies, and the technique has been used to explore the transport possibilities open to organisms in different tidal and circulation regimes and the effects of diel migratory behaviour (Power, 1997, and summarised by Forward & Tankersley, 2001).

Mechanisms controlling tidal migration

To understand the behavioural mechanisms underlying tidal migration, it is necessary to know what cues trigger the beginning and end of activity and, once in motion, what clues are used to guide animals in the appropriate direction (Harden Jones, 1984). Clearly, synchronisation of activity with the tidal cycle is a prerequisite for all tidal migrants but the direction in which they migrate varies considerably. Although direct and indirect field observations may give some indication of which cues and clues could be used, field and laboratory experiments are needed to give definitive answers. Such experiments have indicated that, in most cases, responses to both exogenous and endogenous factors are involved. However, the relative contributions of exogenous and endogenous cues to the actual behaviour observed are difficult to determine. Nevertheless, if the presence of endogenous behaviour cannot be demonstrated, then migration patterns are presumably controlled solely by responses to exogenous factors.

One example of intertidal migration that is apparently a purely exogenous response is that of the sand crab *Emerita analoga*. This species inhabits the wash zones of sandy beaches and migrates with each tide in a partially passive fashion. On the upshore phase of the migration, crabs are washed or swim out of the area where the waves break and are carried up the beach by the swash. As the momentum of the swash slackens, they bury into the sand until the next breaking wave reaches them. On a falling tide, crabs swim out of areas of sand made thixotropic by water draining through it and are washed downshore (Cubit, 1969). The bivalve *Donax variabilis* also lives on sandy beaches and shows a similar wave-driven migration. Here, however, bivalves emerge from the sand in response to low frequency sound made by the largest breaking waves (Ellers, 1995a). Laboratory experiments demonstrated that this response was controlled by an endogenous rhythm because the emergence response to artificial sounds was maximal at subjective high tide and absent at subjective low tide (Ellers, 1995b).

A more complex example of a species responding principally to exogenous cues is provided by postlarvae of the blue crab *Callinectes sapidus* using STST to move upshore in estuaries. Although this species has an endogenous circadian locomotor rhythm, there is no comparable endogenous circatidal rhythm. Instead, the tidally-synchronised activity consists of a series of responses to changes in salinity and turbulence. The postlarvae are stimulated to swim into the water column by increased salinity on night-time flood tides. They remain swimming in response to high levels of turbulence but start to descend when turbulence decreases. On the ebb tide, the decrease in salinity inhibits swimming and the postlarvae descend to the bottom (Welch & Forward, 2001).

Endogenous circatidal rhythms in both locomotory activity and responsiveness to exogenous stimuli seem to be involved in the intertidal migration of the sand-beach isopod *Eurydice pulchra*. This species emerges from its low tide position in the sand on a rising tide and swims up into the water column to feed. At this stage of the tide, they are most abundant at the surface and are transported shoreward in upshore currents. In contrast, on the ebb tide, they congregate near the bottom and use the undertow to move back downshore where they rebury (Warman et al., 1991). The difference in the water levels occupied, and hence the direction of transport, is under the control of a circatidal rhythm of phototaxis during which the animals are photopositive on the flood tide and photonegative in the ebb (Warman et al., 1993a). The timing of activity is controlled by an endogenous circatidal rhythm of locomotion (Jones & Naylor, 1970). The intertidal migrations of the shore crab *Carcinus maenas* seem to be similarly timed and directed by endogenous rhythms of activity and responsiveness to light (Warman et al., 1993b). In this species, it has also been suggested (Hunter & Naylor, 1993) that there may also be a changing responsiveness to food because crabs are attracted to baited traps when moving upshore but not when retreating downshore. Endogenous tidal vertical migration rhythms of activity are present even in the early zoeal stages of this species and operate independently of the light regime (Zeng & Naylor, 1996), in contrast to the predominantly diel migrations of most zooplankton. Such rhythms of activity may enhance offshore dispersal of the larvae.

Although light, salinity and turbulence are important factors in timing and directing tidal movements in intertidal areas, estuaries and nearshore waters, many other environmental factors such as pressure (e.g. Morgan, 1965; Gibson, 1982), temperature, currents, and even electrical fields, may also be implicated (reviewed in relation to STST by Forward & Tankersley, 2001).

Variation in migration patterns

Many field studies of tidal migration have revealed that the pattern of migration differs both between closely related species and within the same species during its development. These differences in behaviour reflect differences in ecology between species and developmental stages, particularly in terms of their habitat and distribution. Closely related species in intertidal areas, for example, often have different depth distributions such that some species are predominantly subtidal and do not migrate, whereas others live in shallower water and do migrate. Instances of such interspecific differences in migratory behaviour can be found in mysids

(*Archaeomysis* and *Iliella,* Takahashi & Kawaguchi, 1997), crabs (*Liocarcinus holsatus* and *L. depurator*, Abello et al., 1991), isopods (*Eurydice* spp., Jones & Hobbins, 1970) and flatfishes (Ansell & Gibson, 1980; Wolff et al., 1981). Furthermore, there may be regional differences in behaviour in response to differences in tidal regimes or habitat structure. For example, newly-settled-plaice migrate intertidally on open shores (Gibson, 1973; Burrows et al., 1994) but not on the extensive sandflats of the Wadden Sea (van der Veer & Bergman, 1987); *Eurydice pulchra* modulates its intertidal migrations on a semilunar basis on open coasts but not in estuaries (Reid & Naylor, 1989).

The greatest changes in migration patterns, however, occur during development and are associated with the changing habitat requirements of each stage in the life history and, to some extent, increased locomotory capabilities. For example, adults may use STST to move from spawning to feeding grounds and vice versa; larvae use the same process to travel from the spawning sites to their nursery grounds. Once in the nursery areas, larvae may use STST for retention or transport into particularly favourable areas. Benthic species, following settlement, may begin a period of intertidal migration. In some species, the migratory behaviour that brought them to the nursery grounds may be reversed when, as juveniles or adults, they move into deeper water (Forward & Tankersley, 2001). The plaice can be used as an illustration of a species that changes its migratory behaviour over its life history (Gibson, 1997). In the southern North Sea, adults migrate southwards from their feeding grounds to their spawning grounds using STST. Once spawning is complete, they return northwards using the same process. Larvae drift towards the coast and enter the major nursery grounds on the Dutch coast also by STST. Once on the nursery grounds, juveniles undergo intertidal migration but migrate gradually into deeper water as they grow and mature. Where tidal currents are strong, juveniles may also leave nursery grounds using STST. Here again, though, regional differences are apparent. STST may not be used by the larvae and juveniles in other areas and in the virtually non-tidal Baltic, migrations in shallow water take place with diel rather than tidal frequency (Gibson et al., 1998).

In other cases, migratory behaviour may even cease temporarily at some point during development. As a consequence, animals at that stage of development may stay in one area until they grow to a larger size or different stage of development. The intertidal residency of juvenile *Crangon crangon* (Janssen & Kuipers, 1980; Naylor & Rejeki, 1996) and newly-settled individuals in some plaice populations (van der Veer & Bergman, 1987) on sandy beaches provide examples that contrast with the migrations of larger individuals. Comparable examples can be found in salt marshes where the juveniles of the shrimp *Palaemonetes pugio* and the killifish *Fundulus heteroclitus* also do not migrate but remain in the intertidal zone at low tide (Kneib, 1987). These, and other examples, can be regarded as mechanisms that reduce predation on the vulnerable juvenile stages.

A final question relating to variation in migration patterns is that of consistency of migration between and within individuals. What percentage of the migrating population migrates on each tidal cycle and, if it is <100%, are there migratory and non-migratory individuals or do particular individuals migrate on some tides and not others? Such variation between individuals, if it occurs, raises interesting questions about the functions and factors controlling migration – why do they not apply to all individuals all of the time?

The extent to which all individuals in a population migrate intertidally can be determined by fishing at different tidal levels over a tidal cycle. Three basic patterns might be expected. First, the shift of the whole population to a higher tidal level. Second, an upshore spreading of the population caused by differences in the distance moved by individuals, and third, a separation between migrant and non-migrant sections of the population on each tide. Studies of juvenile plaice have provided examples of each of these patterns: total movement to a higher tidal level (Kuipers, 1973), spreading of part of the population into the intertidal zone (Edwards & Steele, 1968, Ansell & Gibson, 1990), and separation of migrant and non-migrant individuals (Berghahn, 1987). The complex intertidal migrations of different sections of *Carcinus maenas* populations on rocky shores show similar patterns (Hunter & Naylor, 1993; Warman et al., 1993b).

Investigation of the variation within individual migrants requires that their behaviour be investigated over several tidal cycles. Data relating to this type of variation can be obtained by marking or tracking individuals and, in the few instances where such investigations have been made, it is apparent that individuals do vary their tide-to-tide migration patterns considerably. For example in a New Jersey tidal creek, an individual summer flounder (*Paralichthys dentatus*) differed in the distance it migrated on consecutive tidal cycles and did not migrate at all on some tides (Szedlmayer & Able, 1993).

Conclusions

Tidal migration takes a variety of forms and serves several functions that increase the fitness of individuals. Migratory behaviour is timed and directed by a range of mechanisms involving responses to both external and internal stimuli but further research is required for an understanding of the underlying sensory mechanisms. Study of individual spatial and temporal variation in migration patterns could provide greater insight into the functions of such behaviour.

References

Abello, P., D. G. Reid & E. Naylor, 1991. Comparative locomotor activity patterns in the portunid crabs *Liocarcinus holsatus* and *L. depurator*. J. mar. biol. Ass. U.K. 71: 1–10.

Ansell, A. D. & R. N. Gibson, 1990. Patterns of feeding and movement of juvenile flatfishes on an open sandy beach. In Barnes, M. & R. N. Gibson (eds), Trophic Relationships in the Marine Environment, Proc. 24th Europ. mar. biol. Symp., Aberdeen University Press, Aberdeen: 191–207.

Arnold, G. P. & J. D. Metcalfe, 1996. Seasonal migrations of plaice (*Pleuronectes platessa*) through the Dover Strait. Mar. Biol. 127: 151–160.

Begon, M., J. L. Harper & C. R. Townsend, 1990. Ecology. Individuals, Populations and Communities. Blackwell Scientific Publications, Oxford.

Berghahn, R., 1987. Effects of tidal migration on growth of 0-group plaice (*Pleuronectes platessa* L.) in the North Friesian Wadden Sea. Meeresforschungen 31: 209–226.

Beukema, J. J. & J. de Vlas, 1989. Tidal current transport of thread-drifting postlarval juveniles of the bivalve *Macoma balthica* from the Wadden Sea (Netherlands) to the North Sea. Mar. Ecol. Prog Ser. 52: 193–200.

Black, R. & R. J. Miller, 1991. Use of the intertidal zone by fish in Nova Scotia. Envir. Biol. Fish. 31: 109–121.

Brown, A. C., 1982. The biology of the sandy-beach whelks of the genus *Bullia* (Nassariidae). Oceanogr. mar. biol. Ann. Rev. 20: 309–361.

Brown, A. C. & A. McLachlan, 1990. Ecology of Sandy Shores. Elsevier, Amsterdam.

Burrows, M. T., 1994. An optimal foraging and migration model for juvenile plaice. Evol. Ecol. 8: 125–149.

Burrows, M. T., R. N. Gibson, L. Robb & C. Comely, 1994. Temporal patterns of movement in juvenile flatfishes and their predators: Underwater television observations. J. exp. mar. Biol. Ecol. 177: 251–268.

Colombini, I., R. Berti, A. Ercolini, A. Nocita & L. Chelazzi, 1995. Environmental factors influencing the zonation and activity patterns of a population of *Periophthalmus sobrinus* Eggert in a Kenyan mangrove. J. exp. mar. Biol. Ecol. 190: 135–149.

Cubit, J., 1969. The behaviour and physical factors causing migration and aggregation in the sand crab *Emerita analoga* (Stimson). Ecology 50: 118–123.

Dare, P. J. & D. B. Edwards, 1981. Underwater television observations on the intertidal movements of shore crabs, *Carcinus maenas*, across a mudflat. J. mar. biol. Ass. U.K. 61: 107–116.

DeMartini, E. E., 1999. Intertidal spawning. In Horn, M. H., K. L. M. Martin & M. A. Chotkowski (eds), Intertidal Fishes. Life in two Worlds. Academic Press, London: 143–164.

Edwards, R. R. C. & J. H. Steele, 1968. The ecology of 0-group plaice and common dabs at Lochewe. I. Population and food. J. exp. mar. Biol. Ecol. 2: 215–238.

Ellers, O., 1995a. Behavioral control of swash-riding in the clam *Donax variabilis*. Biol. Bull. mar. biol. Lab. Woods Hole 189: 120–127.

Ellers, O., 1995b. Discrimination among wave-generated sounds by a swash-riding clam. Biol. Bull. mar. biol. Lab. Woods Hole 189: 128–137.

Ellis, T. R. & R. N. Gibson, 1995. Size selective predation of 0-group flatfishes on a Scottish coastal nursery ground. Mar. Ecol. Prog. Ser. 127: 27–37.

Emmerson, W. D., 1987. Tidal migration and feeding of the shrimp *Palaemon pacificus* (Stimpson). S. Afr. J. Sci. 83: 413–416.

Evans, P. R. & M. W. Peinkowski, 1982. Behaviour of shelduck *Tadorna tadorna* in a winter flock: does regulation occur? J. anim. Ecol. 51: 241–262.

Forward, Jr, R. B. & R. A. Tankersley, 2001. Selective tidal-stream transport of marine animals. Oceanogr. mar. biol. Ann. Rev. 39: 305–353.

Gamble, F. W. & F. Keeble, 1904. The bionomics of *Convoluta roscoffensis*, with special reference to its green cells. Q. J. microsc. Sci. 47: 363–431.

Gibbons, M. J., 1989. Tidal migration of *Porcellidium* (Copepoda: Harpacticoida) on fronds of the rocky shore alga *Gigartina radula* (Esper) J. Agardh (Gigartinales: Rhodophyta). S. Afr. J. mar. Sci. 8: 3–7.

Gibson, R. N., 1973. The intertidal movements and distribution of young fish on a sandy beach with special reference to the plaice (*Pleuronectes platessa* L.). J. exp. mar. Biol. Ecol. 12: 79–102.

Gibson, R. N., 1980. A quantitative description of the behaviour of wild juvenile plaice (*Pleuronectes platessa*). Anim. Behav. 28: 1202–1216.

Gibson, R. N., 1982. The effect of hydrostatic pressure cycles on the activity of young plaice *Pleuronectes platessa*. J. mar. biol. Ass. U.K. 62: 621–635.

Gibson, R. N., 1992. Tidally-synchronised behaviour in marine fishes. In Ali, M.A. (ed.), Rhythms in Fishes. Plenum Press, New York: 63–81.

Gibson, R. N., 1997. Behaviour and the distribution of flatfishes. J. Sea Res. 37: 241–256.

Gibson, R. N., 1999. Movement and homing in intertidal fishes. In Horn, M. H., K. L. M. Martin & M. A. Chotkowski (eds), Intertidal Fishes. Life in Two Worlds. Academic Press, London: 97–125.

Gibson R. N., L. Pihl, M. T. Burrows, J. Modin, H. Wennhage & L. A. Nickell, 1998. Diel movements of juvenile plaice (*Pleuronectes platessa*) in relation to predators, competitors, food availability and abiotic factors on a microtidal nursery ground. Mar. Ecol. Prog. Ser. 165: 145–159.

Greer Walker, M., F. R. Harden Jones & G. P. Arnold, 1978. The movements of plaice (*Pleuronectes platessa*) tracked in the open sea. J. Cons. perm. int. Explor. Mer. 38: 58–86.

Harden Jones, F. R., 1984. A view from the ocean. In McCleave, J. D., G. P. Arnold, J. J. Dodson & W. H. Neill (eds), Mechanisms of Migration in Fishes. Plenum Press, New York and London: 1–26.

Hiddinck, J. G., R. P. Kock & W. J. Wolff, 2002. Active pelagic migrations of the bivalve *Macoma balthica* are dangerous. Mar. Biol. 140: 1149–1156.

Hunter, E. & E. Naylor, 1993. Intertidal migration by the shore crab *Carcinus maenas*. Mar. Ecol. Prog. Ser. 101: 131–138.

Jager, Z., 1999. Selective tidal stream transport of flounder larvae (*Platichthys flesus* L.) in the Dollard (Ems estuary). Estuar. coast. shelf Sci. 49: 347–362.

Jager, Z. & H. P. J. Mulder, 1999. Transport velocity of flounder larvae (*Platichthyus flesus* L.) in the Dollard (Ems estuary). Estuar. coast. shelf Sci. 49: 327–346.

Janssen, G. M. & B. R. Kuipers, 1980. On tidal migration in the shrimp *Crangon crangon*. Neth. J. Sea Res. 14: 339–348.

Jones, D. A. & C. S. T. C. Hobbins, 1970. The role of biological rhythms in some sandbeach cirolanid Isopoda. J. exp. mar. Biol. Ecol. 93: 45–59.

Jones, D. A. & E. Naylor, 1970. The swimming rhythm of the sandbeach isopod *Eurydice pulchra*. J. exp. mar. Biol. Ecol. 4: 188–199.

Kneib, R. T., 1987. Predation risk and use of intertidal habitats by young fishes and shrimp. Ecology 68: 379–386.

Kneib, R. T., 1991. Flume weir for quantitative collection of nekton from vegetated intertidal habitats. Mar. Ecol. Prog. Ser. 75: 29–38.

Kuipers, B., 1973. On the tidal migration of young plaice (*Pleuronectes platessa*) in the Wadden Sea. Neth. J. Sea Res. 6: 376–388.

Lincoln, R. J., G. A. Boxshall & P. F. Clark, 1982. A Dictionary of Ecology, Evolution and Systematics. Cambridge University Press, Cambridge.

Morgan, E., 1965. The activity rhythm of the amphipod *Corophium volutator* (Pallas) and its possible relationship to changes in hydrostatic pressure associated with the tides. J. anim. Ecol. 34: 731–746.

Naylor. E. & S. Rejeki, 1996. Tidal migrations and rhythmic behaviour of sandbeach Crustacea. Rev. Chil. Hist. nat. 69: 475–484.

O'Connor, R. J., 1974. Feeding behaviour of the kittiwake. Bird Study 21: 185–192.

Power, J.J., 1997. Time and tide wait for no animal: seasonal and regional opportunities for tidal stream transport or retention. Estuaries 20: 312–318.

Reid, D. G. & E. Naylor, 1989. Variations in rhythmic behaviour of an estuarine population of *Eurydice pulchra* (Leach). Proc. 21st Europ. mar. biol. Symp., Polish Academy of Sciences, Gdansk: 143–153.

Slater, P. J. B., 1976. Tidal rhythm in a seabird. Nature 264: 636–638.

Smith, R. J. F., 1985. The Control of Fish Migration. Springer-Verlag, Berlin.

Speirs, D., S. Lawrie, D. Raffaelli, W. Gurney & C.Eames, 2002. Why do shallow-water predators migrate? Strategic models and empirical evidence from an estuarine mysid. J. exp. mar. Biol. Ecol. 280: 13–31.

Szedlmayer, S. T. & K. W. Able, 1993. Ultrasonic telemetry of age-0 summer flounder, *Paralichthys dentatus*, movements in a southern New Jersey estuary. Copeia 1993: 728–736.

Takahashi, K. & K. Kawaguchi, 1997. Diel and tidal migrations of the sand-burrowing mysids, *Archaeomysis kokuboi*, *A. japonica* and *Iliella ohshimai*, in Otsuchi Bay, northeastern Japan. Mar. Ecol. Prog. Ser. 148: 95–107.

Tyler, A. V., 1971. Surges of winter flounder, *Pseudopleuronectes americanus*, into the intertidal zone. J. Fish. Res. Bd Can. 28; 1727–1732.

van der Veer, H. W. & M. J. N. Bergman, 1987. Development of tidally related behaviour of a newly settled 0-group plaice (*Pleuronectes platessa*) population in the western Wadden Sea. Mar. Ecol. Prog. Ser. 31: 121–129.

Warman, C. G., D. G. Reid & E. Naylor, 1993a. Circatidal variability in the behavioural responses of a sand-beach isopod *Eurydice pulchra* (Leach) to operational cues. J. exp. mar. Biol. Ecol. 168: 59–70.

Warman, C. G., D. G. Reid & E. Naylor, 1993b. Variation in the tidal migratory behaviour and rhythmic light-responsiveness in the shore crab *Carcinus maenas*. J. mar. biol. Ass. U.K. 73: 355–364.

Warman, C. G., T. J. O'Hare & E. Naylor, 1991. Vertical swimming in wave-induced currents as a control mechanism of intertidal migration by a sandy-beach isopod. Mar. Biol. 111: 49–54.

Watkin, E. E., 1941. Observations on the night tidal migrant Crustacea of Kames Bay. J. mar. biol. Ass. U.K. 25: 81–96.

Welch, J. M. & R. B. Forward, Jr, 2001. Flood tide transport of blue crab, *Callinectes sapidus*, postlarvae: behavioral responses to salinity and turbulence. Mar. Biol. 139: 911–918.

Wirjoatmodjo, S. & T. J. Pitcher, 1980. Flounders follow the tides to feed: Evidence from ultrasonic tracking in an estuary. Estuar. coast. shelf Sci. 19: 231–241.

Wolff, W. J., M. A. Manders & A. J. J. Sandee, 1981. Tidal migration of plaice and flounders as a feeding strategy. In Jones, N. V. & W. J. Wolff (eds), Feeding and Survival Strategies of Estuarine Organisms. Plenum Press, London: 159–171.

Zeng, C. & E. Naylor, 1996. Endogenous tidal rhythms of vertical migration in field collected zoea-1 larvae of the shore crab *Carcinus maenas*: implications for ebb tide offshore dispersal. Mar. Ecol. Prog. Ser. 132: 71–82.

Hydrobiologia **503:** 163–170, 2003.
M.B. Jones, A. Ingólfsson, E. Ólafsson, G.V. Helgason, K. Gunnarsson &
J. Svavarsson (eds), Migrations and Dispersal of Marine Organisms.

A review of the adaptive significance and ecosystem consequences of zooplankton diel vertical migrations

Graeme C. Hays
School of Biological Sciences, University of Wales Swansea, Singleton Park, Swansea SA2 8PP, U.K.
E-mail: g.hays@swan.ac.uk

Key words: predator evasion, biogeochemical cycling, deep scattering layer, leatherback turtles, penguin, seal, whale, diving, satellite tracking, TDR, isolume

Abstract

Diel vertical migration (DVM) by zooplankton is a universal feature in all the World's oceans, as well as being common in freshwater environments. The normal pattern involves movement from shallow depths at night to greater depths during the day. For many herbivorous and omnivorous mesozooplankton that feed predominantly near the surface on phytoplankton and microzooplankton, minimising the risk of predation from fish seems to be the ultimate factor behind DVM. These migrants appear to use deep water as a dark daytime refuge where their probability of being detected and eaten is lower than if they remained near the surface. Associated with these vertical movements of mesozooplankton, predators at higher trophic levels, including invertebrates, fish, marine mammals, birds and reptiles, may modify their behaviour to optimise the exploitation of their vertically migrating prey. Recent advances in biotelemetry promise to allow the interaction between migrating zooplankton and diving air-breathing vertebrates to be explored in far more detail than hitherto.

Introduction

Diel vertical migration (DVM) occurs in a wide range of both freshwater and marine zooplankton taxa and probably represents the biggest animal migration, in terms of biomass, on the planet. It is, consequently, not surprising that ever since the first descriptions of this behaviour a century or more ago, there has been extensive consideration of the adaptive significance and wider ecosystem consequences of this phenomenon (e.g. Forel, 1878; Hardy, 1936; Cushing, 1951; Pearre, 2003). The normal pattern (normal DVM) involves animals occurring deeper in the water column during the day and shallower at night, this being achieved by a dusk ascent and a dawn descent, with the difference between the day and night depths being termed the amplitude of DVM. A number of different aspects of DVM have been studied. For example, there has been considerable discussion of the proximate cues initiating vertical movements, with a strong role being implicated for the relative change in light intensity around dusk and dawn (Ringelberg, 1995, 1999). Over the last two decades, international programmes such as the Joint Global Ocean Flux Study (JGOFS) have increased awareness of the role of the oceans in global biogeochemical cycles and have led to consideration of the implications of DVM in terms of the net removal of carbon and nitrogen from the ocean surface. By feeding near the surface at night, and then fasting at depth during the day where they continue to defecate, respire and excrete, migrating zooplankton may remove carbon and nitrogen from the surface layers and release it at depth (Longhurst & Harrison, 1989; Hays et al., 1997; Schnetzer & Steinberg, 2002).

In this review, I will first concentrate on the ultimate reasons for DVM and, second, I will examine the wider ecosystem implications in terms of how upper trophic levels adapt their behaviour to exploit vertically migrating prey.

Ultimate reasons for DVM

Given the wide occurrence of DVM in different marine and freshwater environments and in different taxa, it is naive to hope to find a single ultimate reason to explain this behaviour. For example, DVM may serve to reduce the risk of damage to animals from ultraviolet radiation (Leech & Williamson, 2001), but avoidance of exposure to ultraviolet radiation certainly cannot explain many cases where the vertical movements have an amplitude of many tens or even hundreds of metres. However, the fact that DVM is so widespread and found within practically all taxonomic groups, suggests that, in many cases, there must be a common underlying ultimate driving force. Generally, when considering the ultimate reason for normal DVM, this question is couched in terms of mesozooplankton that feed either herbivorously or omnivorously on phytoplankton and microzooplankton that occur in maximal densities in near surface waters. For these mesozooplankton, there is a clear cost to DVM, in that during the day, animals will be separated spatially from their near-surface food and so cannot feed at high rates. In addition, cold sub-thermocline temperatures may reduce the reproductive fitness of migrants (Aksnes & Giske, 1990; Dawidowicz & Loose, 1994).

Counterbalancing this cost, it is axiomatic that there must be some benefit to DVM. Hypotheses for the ultimate reason for DVM have been divided into two broad categories by Lampert (1989). Firstly, it has been suggested that residing in cold water during the day and feeding in warmer water at night may provide a metabolic advantage for migrants (e.g. McLaren, 1963, 1974; Enright, 1977). However, empirical support of this group of hypotheses is scant and most evidence suggests that the reverse is true: vertical migration is energetically disadvantageous (Lampert et al., 1988; Lampert, 1989; Aksnes & Giske, 1990). Secondly, it has been suggested that normal DVM serves to reduce the risk of predation from visually orientating predators, the so-called predator evasion hypothesis. This hypothesis is elegantly simple and has intuitive appeal. The predator evasion hypothesis suggests that if mesozooplankton remained near the surface during the day, they would have a high chance of being seen by visually orientating predators (principally fish) and hence consumed. Consequently, migrants descend around dawn to use the ocean depths as a dark daytime refuge where the probability of being detected by a visually hunting predator, and hence eaten, is lower that if they remained near the surface (Zaret & Suffern, 1976). Essentially, the benefit of a reduced probability of predation is suggested to outweigh the cost of a reduced potential for daytime feeding, with the maxim being 'better hungry than dead' (Kremer & Kremer, 1988). There is considerable support for the predator evasion hypothesis.

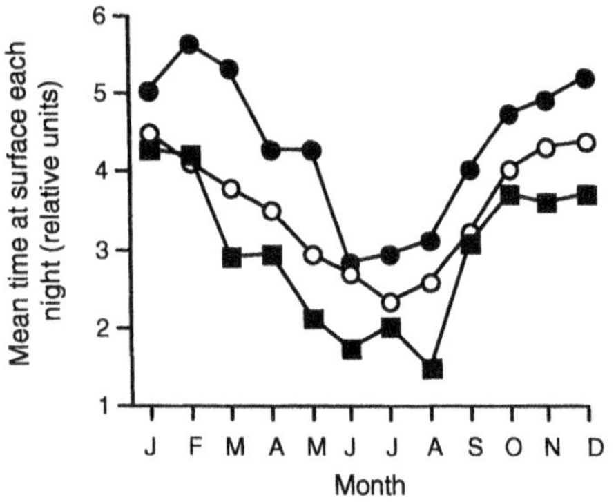

Figure 1. The mean daily length of near-surface occupation by different populations of copepods in the North Atlantic, using samples from the Continuous Plankton Recorder survey. Filled circles = juvenile *Metridia*, open circles = adult *M. lucens*, filled squares = adult *M. longa*. For adult *Metridia longa*, adult *M. lucens* and juvenile *Metridia*, there were seasonal changes in the length of near-surface occupation in line with seasonal changes in the length of the night. For example, in winter, when nights are longer, populations spent longer at the surface each night compared to during the summer. In addition to this seasonal pattern, the length of near-surface occupation varied with body size, being shortest for the largest group, adult *M. longa*, and longest for the smallest group, juvenile *Metridia*. Adapted from Hays (1995).

In freshwater environments, there has been particularly elegant work showing how DVM is a plastic behaviour modified in accord with the presence of planktivorous fish or fish kairomones: when fish are present, zooplankton show stronger DVM (Dawidowicz & Loose, 1992, 1994; Dodson et al., 1997; Van Gool E. & J. Ringelberg, 1998, 2002). Comparable evidence also exists for marine systems. For example, at times when planktivorous fish are more abundant, normal DVM may be more pronounced (Bollens & Frost, 1989; Frost & Bollens, 1992). Individual differences in DVM may also be under genetic control leading to long-term evolutionary changes in the DVM behaviour of populations associated with changes in the intensity of predation pressure from fish (Cousyn et al., 2001).

While these types of study show that, for the selected species, normal DVM is coupled to the abundance of planktivorous fish, they cannot be used to suggest that normal DVM serves an anti-predator function in all other cases. One approach to consider whether predator evasion can be used as a general explanation for the occurrence of normal DVM in a range of taxa, is to examine inter-specific patterns in this behaviour. Furthermore, we can use the predator evasion hypothesis to make specific *a priori* predictions about inter-specific patterns in DVM. Specifically, the predator evasion hypothesis predicts that DVM should be more pronounced in those species that are most susceptible to visually orientating predators. Since the visibility, and hence susceptibility, of mesozooplankton to planktivorous fish increases with increasing size and pigmentation, the predator evasion hypothesis predicts that DVM should be most pronounced in larger and more heavily pigmented individuals and species and, indeed, there is very extensive empirical evidence to show that this is the case (Zaret & Kerfoot, 1975; Wright et al., 1980; Wiebe et al., 1992; Hays et al., 1994). Smaller zooplankton arrive at the surface earlier and leave later than larger forms, consistent with the interaction of light levels and prey size on prey perception by planktivorous fish (Hays, 1995; De Robertis et al., 2000). Furthermore, in order to constrain their near-surface foraging to the hours of darkness, we can use the predator evasion hypothesis to predict that, particularly at high latitudes, there will be marked seasonal changes in the length of near-surface residence by migrants in line with seasonal changes in day and night length. Again, this prediction is supported by extensive empirical evidence, with migrants spending longer at the surface each night during the winter when nights are longer (Hays, 1995; Hays et al., 1995) (Fig. 1).

The predator evasion hypothesis can be used also to make predictions about how DVM should vary between individuals of the same species depending on their nutritional state. For example, for individuals that have built up lipid reserves, there may be less need to come to the surface to feed and so lipid-rich individuals can maintain their reproductive output without increasing their risk of predation by near-surface foraging (Fiksen & Carlotti, 1998; Sekino & Yamamura, 1999). Moreover, this prediction is supported by empirical observations that show how lipid-rich individuals may spend less time at the surface than their lipid-poor co-specifics (Sekino & Yoshioka, 1995; Hays et al., 2001).

Interestingly, instances where mesozooplankton reverse their normal pattern of DVM can also be explained by predator evasion. In reverse DVM, a population occurs nearer the surface during the day and deeper at night, and this behaviour has been associated with high levels of invertebrate predators that use tactile stimuli, rather than vision, to locate their prey, i.e. the normal pattern of DVM appears to be reversed when evasion of invertebrate predators, rather than visually feeding planktivorous fish, is most important (Ohman et al., 1983; Neill, 1990).

While predator evasion is probably the most important ultimate reason for DVM, there are associated consequences of this behaviour. For example, if migrants move between different currents at different depths, then DVM may translate into horizontal movement into new patches of surface water (Hardy, 1936). However, except in the case of some vertical movements that are coupled to tidal flows in coastal regions (Hill, 1998), horizontal movement is probably most often just a consequence of DVM rather than the ultimate reason for DVM. In short, while there are certainly a number of reasons why mesozooplankton undertake DVM, the most likely ultimate reason, is predator evasion.

Implications for higher trophic levels

Invertebrates and fish

With many species of mesozooplankton undertaking daily vertical movements, it is hardly surprising that some predators at higher trophic levels modify their behaviour to exploit this vertically moving food source, while at the same time possibly minimising their own risk of predation from their own visually orientating predators. For example, many carnivorous zooplankton (e.g. copepods of the genus *Euchaeta*, scyphozoan jellyfish and siphonophores) use tactile stimuli, rather than vision, to locate prey and, hence, probably catch mesozooplankton equally well in darkness *versus* daylight. Such predators can simply track the vertical movements of their prey. Indeed, larger pelagic animals, many of them carnivores, are frequently observed on echo-sounders as 'deep scattering layers' (DSLs) or 'sound scattering layers' (SSLs) which can be seen to ascend around dusk and descend around dawn presumably reflecting this predator-prey tracking. Often discrete layers are evident at different depths, each layer composed of different species or developmental stages (e.g. Tarling et al., 2001).

Fish that require high ambient light levels to locate and capture prey may be unable to feed on animals within the SSL, and so confine their foraging to well illuminated shallow depths. However, fish that are able to hunt at low light levels may join the SSLs to feed. For example, many small mesopelagic fish form part of the SSL (e.g. Rasmussen & Giske, 1994), while acoustic tracking of large (25–50 kg)

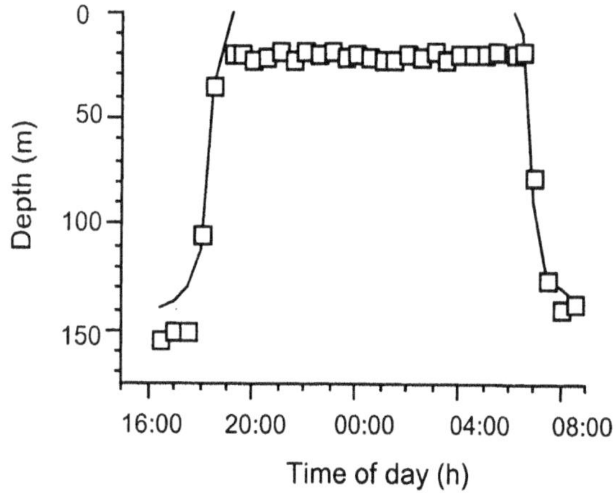

Figure 2. The depth of an acoustically tracked megamouth shark (*Megachasma pelagios*) over 1 day and the estimated depth of the isolume that most closely corresponded with the dawn descent and dusk ascent of the shark. Open squares show depth of the shark, solid line shows depth of the isolume. Adapted from Nelson et al. (1997). This fish followed this isolume, which presumably reflected the depth of its zooplankton prey. At night the isolume 'disappeared', so that ambient light levels experienced by the shark were lower near the surface at night than at depth during the day.

big-eye tuna (*Thunnus obesus*) has shown that there is sometimes a close correspondence between the descent of individuals at dawn and the depth of the SSL (Dagorn et al., 2000). Similar diel vertical movements have been recorded for swordfish (*Xiphias gladius*) (Carey & Robison, 1981). Also, some filter-feeding fish may track the SSL. For example, an acoustically tracked megamouth shark (*Megachasma pelagios*) showed rapid changes in depth around dusk and dawn, presumably associated with this shark attempting to track its zooplankton prey (Nelson et al., 1997) (Fig. 2). However, for all these examples of fish exhibiting diel vertical movements, it would be over simplistic to describe this as an invariant behaviour. On occasion, all species will probably break away from tracking the SSL. For example, tracked bigeye tuna and swordfish have been shown sometimes to move up into shallow water during the day, possibly to re-warm (Carey & Robison, 1981; Dagorn et al., 2000). Similarly, we might expect that vertical movements of such fish might be suspended when prey occur in sufficiently high concentrations near the surface.

Air-breathing vertebrates

Of course, air-breathing vertebrates that feed on animals within SSLs must periodically come to the surface to breathe. This requirement to breathe, means that the commuting costs (i.e. from the surface to the prey) will vary over a diel cycle, being short during the night (when the prey are shallow) and deep during the day (when the prey are deep). At first glance, this appears to suggest that shallow night-time feeding would be preferable for an air-breather feeding within the SSL. However, in addition to the commuting cost, the optimum foraging strategy will also be dictated by the efficiency of prey capture once at the prey depth. For example, even if the prey are very shallow at night, a low commuting cost may translate into a poor rate of prey ingestion if it is too dark to locate and capture prey.

While studies of zooplankton DVM have been reported for a century or more, it is only relatively recently that technological advances have provided the tools to examine diel patterns in the diving performance of air-breathing vertebrates. For example, time-depth recorders (TDRs), which are data loggers that record ambient pressure (and hence depth) at intervals of a few seconds, allow dives to be recorded over periods of weeks or even months. An early description of the first widely used analogue TDR was provided by Gentry & Kooyman (1986), but nowadays miniature digital TDRs, weighing as little as 5 g, are commercially available from several manufacturers. As well as measuring the depth and duration of dives, additional sensor packages allow the swim speed, flipper movements, individual prey ingestion events and prey fields to be recorded (Wilson et al., 1993, 2002a,b; Hooker et al., 2002). TDRs may be interfaced with satellite transmitters which relay dive information from any location in the World via polar orbiting satellites that form part of the Argos tracking system (http://www.argosinc.com/), so that the study animal need not be recaptured for data acquisition to occur. Although the large amounts of data collected by TDRs (up to several Mb) cannot be relayed via the Argos system due to the limited bandwidth that is available (Fedak et al., 2002), onboard data analysis and compression allow dive profiles to be obtained remotely (Fig. 3). These types of technology have (and are currently) being used to explore the diving patterns of certain marine birds, marine mammals and marine reptiles that may feed on prey within the SSL.

Probably the best described interaction between air-breathing divers and vertically migrating prey concerns various species of penguin feeding on fish or euphausiids within the SSL (e.g. Wilson et al., 1993, 2002a; Cherel et al., 1999; Bost el al., 2002). For example, Wilson et al. (1993) recorded the diel

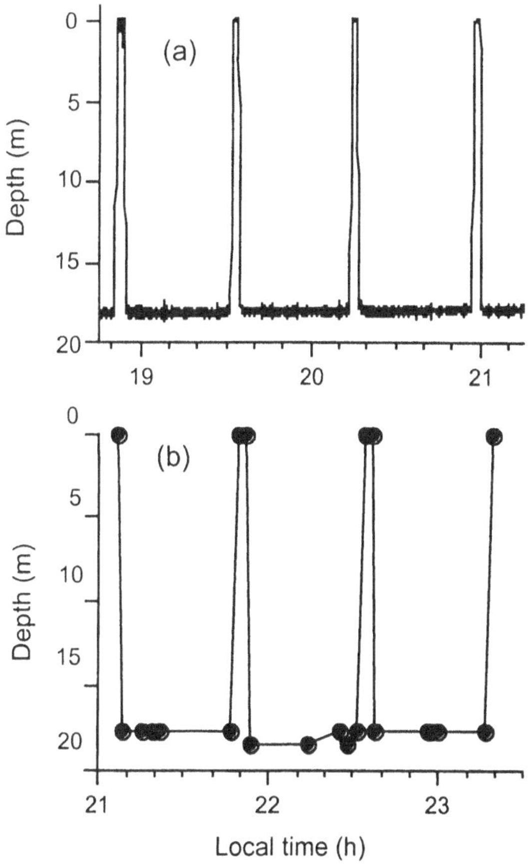

Figure 3. One of the most advanced satellite linked TDRs is the Satellite Relay Data Logger (SRDL) designed by the Sea Mammal Research Unit (SMRU) and originally described by McConnell et al. (1992). In order to allow dive profiles to be relayed via the limited bandwidth of the Argos system, depth is sampled every few seconds and then once a dive is completed there is on-board analysis of the dive profile and only the depth and time of the 5 most important points of inflection during the dive are transmitted, plus the time of the start and end of the dive. In this way the essential elements of the dive profile are transmitted in a most economical way. To illustrate this point, the figure shows the typical U-shaped dive profiles conducted by green turtles (*Chelonia mydas*) at Ascension Island obtained (a) with a time-depth recorder (TDR) sampling depth every 5 seconds during April 2002 and (b) with a SRDL in July 2001. In the latter case, the transmitted times and depths are shown by the symbols. See Hays et al. (2000) for details of this study site and field protocols. Since the dive profiles of planktivorous leatherback turtles and other air breathing planktivores are often very simple in shape (e.g. Southwood et al., 1999), obtaining accurate dive profiles remotely from free-living animals outside the breeding season is now possible.

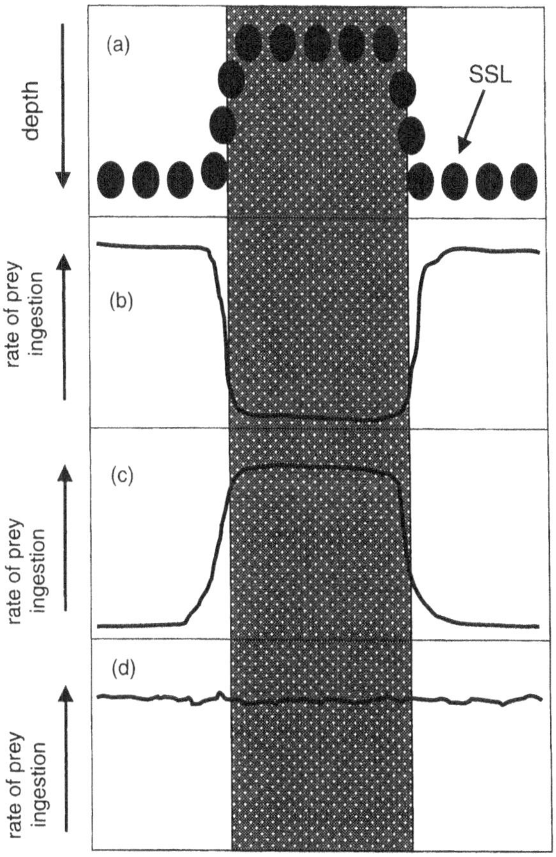

Figure 4. Schematic representation of how the rate of prey ingestion may vary over the diel cycle for various predators feeding within sound scattering layers (SSLs). Shaded area represents night. In (a) the black ovals represent the depth of a SSL. Empirical evidence suggests that penguins often feed best during the day when their prey are deep (b), while fur seals feed best at night when their prey are shallow and then stop diving during the day when the SSL has descended (c). The filter feeding megamouth shark has been shown to follow an isolume, presumably staying within a SSL so that they can feed both day and night (d).

dive patterns in king penguins (*Aptenodytes patagonicus*), African penguins (*Spheniscus demersus*), Adelie penguins (*Pygoscelis adeliae*), gentoo penguins (*P. papua*) and chinstrap penguins (*P. antarctica*). While penguins tended to conduct shallow dives at night and deep dives during the day, there was a distinct diel pattern to prey capture with most prey being captured during daytime deep diving (Wilson et al., 1993). Presumably, this diel pattern is caused by the ambient light levels at prey depths being higher during the day than at night, even though the prey are located deeper during the day. Further support for this inference can be gauged by examining how isolumes approach the surface at dusk and descend at dawn (Fig. 2). Often an isolume will rise towards the surface at dusk, but then light levels everywhere near the surface will be less than this isolume during the middle of the night. This pattern is evident as the isolume 'disappearing' only to re-appear as the sun rises around dawn. For animals in the SSL whose daytime depth is determined by such an isolume, this pattern will mean that despite the animals' vertical migration to the surface, ambient light levels will be lower at the animals' shallow night-time depths than their deep daytime depths.

Diel patterns in diving behaviour have been described for some marine mammals feeding within SSLs. In contrast to the diel pattern of feeding for penguins, deployments of TDRs on fur seals (*Arctocephalus* spp.) suggest that most prey are captured during shallow night-time diving, with individuals not diving during the day when their prey have descended (Croxall et al., 1985; Horning & Trillmich, 1999). During short-term deployments of TDRs on pilot whales (*Globicephala melas*) and pantropical spotted dolphins (*Stenella attenuata*), it was found that during the day individuals were always near the surface but then after dusk they started to dive, possibly in order to feed on animals within an ascending SSL (Baird et al., 2001, 2002). Presumably, differences in diel foraging patterns between different divers may reflect: (i) differences in ambient light levels at prey depths during the diel cycle, (ii) differences in the visual acuity of these different predators at low light levels, (iii) differences in the escape abilities of the different prey, so that ambient light levels are more important for the predator in one case than the other, and (iv) differences in the amplitude of DVM by different prey species so that the commuting costs in reaching the prey vary for these different divers.

In addition to marine mammals and birds, one species of marine reptile, the leatherback turtle (*Dermochelys coriacea*), feeds mainly on plankton and, more specifically, on gelatinous zooplankton such as scyphozoan jellyfish, siphonophores and pyrosomes. Often, these prey form part of the SSL. To date, relatively few records of leatherback diving performance have been obtained, with almost all records focussing on the internesting period when turtles are at their tropical and sub-tropical nesting areas. In some cases, leatherbacks dive to relatively shallow depths (rarely beyond 100 m) and, in these cases, their dives may be constrained by bathymetry and may not reflect their diving behaviour when they are in deep oceanic areas outside the nesting season (Southwood et al., 1999). However, for leatherbacks nesting at St Croix in the Caribbean, much deeper dives have been recorded in the internesting period, with a maximum measured depth of 475 m (Eckert et al., 1986); even deeper dives have been inferred when some dives were deeper than the maximum range of the TDR (Eckert et al., 1989). Furthermore, a diel signal in diving behaviour was evident. At night, more time was spent diving and dives tended to be shallow, while, during the day, less time was spent diving but dives tended to be deeper. These diel patterns suggest an interaction between leatherback diving and vertical movements of SSLs but the details of this interaction remain unclear. It might be, for example, that leatherbacks are mainly shallow night-time feeders and their less frequent deep daytime dives reflect 'prospecting' to see whether the SSL is returning towards the surface. Alternatively, leatherbacks may feed day and night, and simply dive less during the day because deeper dives necessitate periods of recovery at the surface. Furthermore, the pattern of diving by leatherbacks outside the nesting season remains unknown, but the answer to this question is certainly attainable through the deployment of satellite-linked time-depth recorders.

Future studies

After more than a century of studies, there is probably little more to be gained by more simple descriptions of the day and night vertical distributions of zooplankton. However, some investigations of vertical migration are still in their infancy. Firstly, there has been relatively little consideration, particularly in marine environments, of the causes and consequences of individual differences in migration behaviour; for example, related to nutritional state. Secondly, the advent of equipment that allows various aspects of the free-living diving behaviour of vertebrates to be explored, often in remote locations and over time-scales of months, promises a new era of studies of the consequences of zooplankton vertical migration for higher trophic levels.

Acknowledgements

I thank Prof. Ian Boyd and Dr Sascha Hooker for comments on the manuscript. My work on zooplankton vertical migration and the diving behaviour of air-breathing vertebrates has been supported by several grants from the Natural Environment Research Council (NERC).

References

Aksnes, D. L. & J. Giske, 1990. Habitat profitability in pelagic environments. Mar. Ecol. Prog. Ser. 64: 209–215.

Baird, R. W., J. F. Borsani, M. B. Hanson & P. L. Tyack, 2002. Diving and night-time behavior of long-finned pilot whales in the Ligurian Sea. Mar. Ecol. Prog. Ser. 237: 301–305.

Baird, R. W., A. D. Ligon, S. K. Hooker & A. M. Gorgone, 2001. Subsurface and nighttime behaviour of pantropical spotted dolphins in Hawaii. Can. J. Zool. 79: 988–996.

Bollens, S. M. & B. W. Frost, 1989. Zooplanktivorous fish and variable diel vertical migration in the marine planktonic copepod *Calanus pacificus*. Limnol. Oceanogr. 34: 1072–1083.

Bost, C. A., T. Zorn, Y. Le Maho & G. Duhamel, 2002. Feeding of diving predators and diel vertical migration of prey: King penguins' diet *versus* trawl sampling at Kerguelen Islands. Mar. Ecol. Prog. Ser. 227: 51–61.

Carey, F. G. & G. H. Robison, 1981. Daily patterns in the activities of swordfish, *Xiphias gladius*, observed by acoustic telemetry. U.S. Fish. Bull. 79: 277–292.

Cherel, Y., Y. Tremblay, E. Guinard & J. Y. Georges, 1999. Diving behaviour of female northern rockhopper penguins, *Eudyptes chrysocome moseleyi*, during the breeding period at Amsterdam Island (Southern Indian Ocean). Mar. Biol. 134: 375–385.

Cousyn, C., L. De Meester, J. K. Colbourne, L. Brendonck, D. Verschuren & F. Volckaert, 2001. Rapid, local adaptation of zooplankton behavior to changes in predation pressure in the absence of neutral genetic changes. Proc. natl. Acad. Sci. 98: 6256–6260.

Croxall, J. P., I. Everson, G. L. Kooyman, C. Ricketts & R. W. Davis, 1985. Fur seal diving in relation to vertical distribution of krill. J. anim. Ecol. 54: 1–8.

Cushing, D. H., 1951. The vertical migration of planktonic Crustacea. Biol. Rev. 26: 158–192.

Dagorn, L., P. Bach & E. Josse, 2000. Movement patterns of large bigeye tuna (*Thunnus obesus*) in the open ocean, determined using ultrasonic telemetry. Mar. Biol. 136: 361–371.

Dawidowicz, P. & C. J. Loose, 1992. Metabolic costs during predator-induced diel vertical migration of *Daphnia*. Limnol. Oceanogr. 37: 1589–1595.

Dawidowicz, P. & C. J. Loose, 1994. Trade-offs in diel vertical migration by zooplankton – the costs of predator avoidance. Ecology 75: 2255–2263.

De Robertis, A., J. S. Jaffe & M. D. Ohman, 2000. Size-dependent predation risk and the timing of vertical migration in zooplankton. Limnol. Oceanogr. 45: 1838–1844.

Dodson, S. I., R. Tollrian & W. Lampert, 1997. *Daphnia* swimming behavior during vertical migration. J. Plankton Res. 19: 969–978.

Eckert, S. A., D. W. Nellis, K. L. Eckert & G. L. Kooyman, 1986. Diving patterns of two leatherback sea turtles (*Dermochelys coriacea*) during internesting intervals at Sandy Point, St Croix, U.S. Virgin Island. Herpetologica 42: 381–388.

Eckert, S. A., K. L. Eckert, P. Ponganis & G. L. Kooyman, 1989. Diving behavior of leatherback sea turtles (*Dermochelys coriacea*). Can. J. Zool. 67: 2834–2840.

Enright, J. T., 1977. Diurnal vertical migration: adaptive significance and timing. Part 1. Selective advantage: a metabolic model. Limnol. Oceanogr. 222: 856–872.

Fedak, M., P. Lovell, B. McConnell & C. Hunter, 2002. Overcoming the constraints of long range radio telemetry from animals: getting more useful data from smaller packages. Integ. Comp. Biol. 42: 3–10.

Fiksen, O. & F. Carlotti, 1998. A model of optimal life history and diel vertical migration in *Calanus finmarchicus*. Sarsia 83: 129–147.

Forel, E. A., 1878. Faunistische Studien in den Sü*β*wasserseen der Schweiz. Zeitschrift Wissenschaftliche Zooloogie 30: 383–391.

Frost, B. W. & S. M. Bollens, 1992. Variability of diel vertical migration in the marine planktonic copepod *Pseudocalanus newmani* in relation to its predators. Can. J. Fish. aquat. Sci. 49: 1137–1141.

Gentry, R. L. & G. L. Kooyman, 1986. Methods of dive analysis. In Gentry, R. L. & G. L. Kooyman (eds), Fur Seals. Maternal Strategies on Land and Sea. Princeton University Press, New Jersey: 28–40.

Hardy, A. C., 1936. Plankton ecology and the hypothesis of animal exclusion. Proc. linn. Soc. 148: 64–70.

Hays, G. C., 1995. Ontogenetic and seasonal variation in the diel vertical migration of the copepods *Metridia lucens* and *Metridia longa*. Limnol. Oceanogr. 40: 1461–1465.

Hays, G. C., C. A. Proctor, A. W. G. John & A. J. Warner, 1994. Interspecific differences in diel vertical migration of marine copepods: the implications of size, colour and morphology. Limnol. Oceanogr. 39: 1621–1629.

Hays, G. C., A. J. Warner & C. A. Proctor, 1995. Spatio-temporal patterns in the diel vertical migration of the copepod *Metridia lucens* derived from the Continuous Plankton Recorder survey. Limnol. Oceanogr. 40: 469–475.

Hays, G. C., R. P. Harris & R. N. Head, 1997. The vertical nitrogen flux caused by zooplankton diel vertical migration. Mar. Ecol. Prog. Ser. 160: 57–62.

Hays, G. C., C. R. Adams, A. C. Broderick, B. J. Godley, D. J. Lucas, J. D. Metcalfe & A. A. Prior, 2000. The diving behaviour of green turtles at Ascension island. Anim. Behav. 59: 577–586.

Hays, G. C., H. Kennedy & B. W. Frost, 2001. Individual variability in diel vertical migration of a marine copepod: why some individuals remain at depth when others migrate. Limnol. Oceanogr. 46: 2050–2054.

Hill, A. E., 1998. Diel vertical migration in stratified flows: implications for plankton dispersal. J. mar. Res. 56: 1069–1096.

Hooker, S. K., I. L. Boyd, M. Jessopp, O. Cox, J. Blackwell, P. L. Boveng & J. L. Bengtson, 2002. Monitoring the prey-field of marine predators: Combining digital imaging with datalogging tags. Mar. Mamm. Sci. 18: 680–697.

Horning, M. & F. Trillmich, 1999. Lunar cycles in diel prey migrations exert a stronger effect on the diving of juveniles than adult Galapagos fur seals. Proc. R. Soc. Lond. B. 226: 1127–1132.

Kremer, P. & & J. N. Kremer, 1988. Energetic and behavioral implications of pulsed food availability for zooplankton. Bull. mar. Sci. 43: 797–809.

Lampert, W., 1989. The adaptive significance of diel vertical migration of zooplankton. Funct. Ecol. 3: 21–27.

Lampert, W., R-M. Schmitt & P. Muck, 1988. Vertical migration of freshwater zooplankton: test of some hypotheses predicting a metabolic advantage. Bull. mar. Sci. 43: 620–640.

Leech, D. M. & C. E. Williamson, 2001. *In situ* exposure to ultraviolet radiation alters the depth distribution of *Daphnia*. Limnol. Oceanogr. 46: 416–420.

Longhurst, A. & W. G. Harrison, 1989. The biological pump: profiles of plankton production and consumption in the upper ocean. Prog. Oceanogr. 22: 47–123.

McConnell, B. J., C. Chambers & M. A. Fedak, 1992. Foraging ecology of southern elephant seals in relation to bathymetry and productivity of the Southern Ocean. Ant. Sci. 4: 393–398.

McLaren, I. A., 1963. Effect of temperature on growth of zooplankton and the adaptive value of vertical migration. J. Fish. Res. Bd Can. 26: 199–220.

McLaren, I. A., 1974. Demographic strategy of vertical migration by a marine copepod. Am. Nat. 108: 91–102.

Neill, W. E., 1990. Induced vertical migration in copepods as a defence against invertebrate predation. Nature 345: 524–526.

Nelson, D. R., J. N. McKibben, W. R. Strong Jr, C. G. Lowe, J. A. Sisneros, D. M. Schroeder & R. J. Lavenberg, 1997. An acoustic tracking of a megamouth shark, *Megachasma pelagios*: a crepuscular vertical migrator. Env. Biol. Fish. 49: 389–399.

Ohman, M. D., B. W. Frost & E. H. Cohen, 1983. Reverse diel vertical migration – an escape from invertebrate predators. Science 220: 1404–1407.

Pearre, S., 2003. Eat and run? The hunger/satiation hypothesis in vertical migration: history, evidence and consequences. Biol. Rev. 78: 1–80.

Rasmussen, O. I. & J. Giske, 1994. Life-history parameters and vertical distribution of *Maurolicus muelleri* in Masfjorden in summer. Mar. Biol. 120: 649–664.

Ringelberg, J., 1995. Changes in light intensity and diel vertical migration – a comparison of marine and fresh-water environments. J. mar. biol. Ass. U.K. 75: 15–25.

Ringelberg, J., 1999. The photobehavior of *Daphnia* spp. as a model to explain diel vertical migration in zooplankton. Biological Reviews of the Cambridge Philosophical Society 74: 397–423.

Schnetzer, A. & D. K. Steinberg, 2002. Active transport of particulate organic carbon and nitrogen by vertically migrating zooplankton in the Sargasso Sea. Mar. Ecol. Prog. Ser. 234: 71–84.

Sekino, T. & T. Yoshioka, 1995. The relationship between nutritional condition and diel vertical migration of *Daphnia galeata*. Jap. J. Limnol. 56: 145–150.

Sekino, T. & N. Yamamura, 1999. Diel vertical migration of zooplankton: optimum migrating schedule based on energy accumulation. Evol. Ecol. 13: 267–282.

Southwood, A. L., R. D. Andrews, M. E. Lutcavage, F. V. Paladino, N. H. West, R. H. George & D. R. Jones, 1999. Heart rates and diving behavior of leatherback sea turtles in the eastern Pacific Ocean. J. exp. Biol. 202: 1115–1125.

Tarling, G. A., J. B. L. Matthews, P. David, O. Guerin & F. Buchholz, 2001. The swarm dynamics of northern krill (*Meganyctiphanes norvegica*) and pteropods (*Cavolina inflexa*) during vertical migration in the Ligurian Sea observed by an acoustic Doppler current profiler. Deep-Sea Res. I. 48: 1671–1686.

Van Gool, E. & J. Ringelberg, 1998. Light-induced migration behaviour of Daphnia modified by food and predator kairomones. Anim. Beh. 56: 741–747.

Van Gool, E. & J. Ringelberg, 2002. Relationship between fish kairomone concentration in a lake and phototactic swimming by *Daphnia*. J. Plankton Res. 24: 713–721.

Wiebe, P. H., N. J. Copley & S. H. Boyd, 1992. Coarse-scale horizontal patchiness and vertical migration of zooplankton in Gulf Stream warm-core ring 82-H. Deep-Sea Res. 39: 247–278.

Wilson, R. P., K. Puetz, C. A. Bost, B. M. Culik, R. Bannasch, T. Reins & D. Adelung, 1993. Diel dive depth in penguins in relation to diel vertical migration of prey: whose dinner by candlelight. Mar. Ecol. Prog. Ser. 94: 101–104.

Wilson, R. P., Y. Ropert-Coudert & A. Kato, 2002a. Rush and grab strategies in foraging marine endotherms: the case for haste in penguins. Anim. Beh. 63: 85–95.

Wilson, R. P., A. Steinfurth, Y. Ropert-Coudert, A. Kato & M. Kurita, 2002b. Lip-reading in remote subjects: an attempt to quantify and separate ingestion, breathing and vocalisation in free-living animals using penguins as a model. Mar. Biol. 140: 17–27.

Wright, D., W. J. O'Brien & G. L. Vinyard, 1980. Adaptive value of vertical migration: a simulation model arguments for the predation hypothesis. In Kerfoot, W. C. (ed.), Evolution and Ecology of Zooplankton Communities. University Press of New England, New Hampshire: 138–147.

Zaret, T. M. & W. C. Kerfoot, 1975. Fish predation on *Bosmina longirostris*, body size selection *versus* visibility selection. Ecology 56: 232–237.

Zaret, T. M. & J. S. Suffern, 1976. Vertical migration in zooplankton as a predator avoidance mechanism. Limnol. Oceanogr. 21: 804–813.

Hydrobiologia **503:** 171–182, 2003.
M.B. Jones, A. Ingólfsson, E. Ólafsson, G.V. Helgason, K. Gunnarsson & J. Svavarsson (eds), Migrations and Dispersal of Marine Organisms.

Temporal and spatial variability of mobile fauna on a submarine cliff and boulder scree complex: a community in flux

James J. Bell[1] & John R. Turner[2]
[1]*School of Animal and Microbial Sciences, University of Reading, Whiteknights, P.O. Box 228, Reading, Berkshire, RG6 6AJ, U.K.*
[2]*School of Ocean Sciences, University of Wales, Bangor, Menai Bridge, Angelsey, North Wales*
E-mail: j.bell@reading.ac.uk

Key words: circumlittoral zone, infralittoral zone, Lough Hyne, macrofauna, migration

Abstract

The mobile component of a community inhabiting a submarine boulder scree/cliff was investigated at Lough Hyne, Ireland at dawn, midday, dusk and night over a 1-week period. Line transects (50 m) were placed in the infralittoral (6 m) and circumlittoral (18 m) zones and also the interface between these two zones (12 m). The dominant mobile fauna of this cliff consisted of echinoderms (6 species), crustaceans (10 species) and fish (23 species). A different component community was identified at each time/depth interval using Multi-Dimensional Scaling (MDS) even though both species diversity (Shannon–Wiener indices) and richness (number of species) remained constant. These changes in community composition provided indirect evidence for migration by these mobile organisms. However, little evidence was found for migration between different zones with the exception of the several wrasse species. These species were observed to spend the daytime foraging in the deeper zone, but returned to the upper zone at night presumably for protection from predators. For the majority of species, migration was considered to occur to cryptic habitats such as holes and crevices. The number of organisms declined during the night, although crustacean numbers peaked, while fish and echinoderms were most abundant during day, possibly due to predator-prey interactions. This submarine community is in a state of flux, whereby, community characteristics, including trophic and energetic relationships, varied over small temporal (daily) and spatial (m) scales.

Introduction

Migrations of marine organisms may take place in response to biological (e.g. competition, predation, reproduction and changes in food concentration) and physical (e.g. light and increased turbulence or flow rate) factors (Turner & Warman, 1991; Sumich & Erickson, 1992; Crook et al., 2000; Barnes & Crook, 2001; Forward et al., 2003). For example, migration of zooplankton has been considered as a mechanism to avoid visually-orientated predators (Rhode et al., 2001). Most often, migrations are considered to occur over large spatial and temporal scales (e.g. fish shoals and whales), but smaller-scale migrations may occur on time scales ranging from minutes to hours. The length of the migration period often depends on the reason for the initial migration. Scales may also be combined such that organisms might undergo migrations over different times scales throughout their life cycles or even at the same time (i.e. a small migration within a larger migratory pattern). Such migrations are likely to influence community (all species) and assemblage (taxonomically-related species) structure at various scales.

Shallow temperate reefs provide a resource for many mobile animals, including shelter from predators (Anderson, 1994; Barshaw & Spanier, 1994; Ohman & Rajasuriya, 1998), a food source (Palma & Ojeda, 2002) and, in some cases, a surface for the attachment of egg cases (Ellis & Shackley, 1997) or nest building (Uglem & Rosenqvist, 2002). Different organisms may utilise these temperate reefs at different times of the day leading to a community in flux, where ecological characteristics, such as diversity, richness and community composition, may vary throughout the day. Such changes are likely to impact trophic relationships and energy flow through the community. For shallow subtidal communities, certain species (and in some cases assemblages) are well known to show temporal variability in numbers (Turner & Warman, 1991; Nickell & Sayer, 1998) which can be considered important with respect to both seasonal (Worthington et al., 1995; Guidetti & Bussotti, 2000; Kelly, 2001) and daily effects (Turner & Warman, 1991). However, descriptions of entire community differences over both temporal scales are rarer. Such large- and small-scale temporal variability may be an important consideration in assessing an entire community or the contribution of a particular assemblage to an overall community.

In temperate hard substratum environments, the infralittoral zone is usually dominated by macro-algae (Price, 1984; Coleman, 2002) which provides a direct food source for a number of grazing species as well as protection from potential predators. For highly mobile organisms such as benthic fish or crustaceans, migrations may occur between the infralittoral and the deeper circumlittoral zone to forage. However, as well as movements between sublittoral zones, there is the opportunity for within zone movement, most notably to microhabitats such as crevices or holes that act as refuges (Turner & Warman, 1991). This study considers the variation in overall community and assemblage characteristics over daily cycles influenced by small-scale migrations on a boulder scree/cliff complex at Lough Hyne Marine Nature Reserve. The following hypotheses were tested: (1) the overall number of organisms and of component assemblages on the cliff fluctuates between different depth zones over daily cycles, (2) community diversity, based on univariate indices, varies at different times of the day (midday, dusk, night and dawn) within different depth zones (infralittoral, circumlittoral and the interface zone), (3) community composition (based on multivariate analysis) differences can be identified at different times of the day (midday, dusk, night and dawn) within the same, and between different depth zones, and (4) communities migrate together between different depth zones at different times of the day.

Description of study site

Lough Hyne Marine Nature Reserve (est. 1981) is a sheltered, fully marine sea lough on the south-west of Ireland (51°29′ N, 9°18′ W). The lough experiences an unusual tidal regime resulting in essentially unidirectional flow within the lough (Bassindale et al., 1957). This study focused on the boulder scree areas at the Whirlpool Cliff complex (Fig. 1). This site experiences fast (250 cm s^{-1}) water currents which decrease with increasing depth (from 250 to 150 cm s^{-1} between 6 and 12 m at peak inflow). The substratum is composed of areas of loose rock separated by solid substratum. A gravel complex (coarse sediment) extends westward from the base of the cliff/boulder scree area. The infralittoral zone at Whirlpool Cliff extends to approximately 11–12 m and is dominated by kelp, while deeper layers are characterised by an algal turf (predominately red algae). The amount of primary substratum occupied by algae is higher in the infralittoral zone (60–80%) compared to the circumlittoral zone (40–60%). The circumlittoral zone is dominated by sessile forms including sponges, bryozoans, annelids and anthozoans. Fish, crustaceans and echinoderms constitute the majority of the macrofauna.

Materials and methods

Parallel line transects (50 m length) were placed at three depths at the Whirlpool Cliff complex. As the substratum is not homogeneous, transect lines crossed areas of both loose rock and cliff, although the proportions were approximately the same for all transects (50:50) as was habitat complexity (see Turner & Warner, 1991). The first transect was placed within the infralittoral zone at 6 m. The second transect line was placed at 18 m within the circumlittoral zone. The final transect was placed at the interface between the infralittoral and circumlittoral zones (12 m).

Pairs of divers swam along each transect at dawn, midday, dusk and night. Dawn and dusk were defined as the points where the surface light reached approximately 1–1.5% of the midday light levels (determined using a surface and underwater light meter). Night samples were taken a minimum 2 h after dusk, while midday samples were taken at 12:00 h. Divers, swimming along each transect, recorded the abundance of all active, foraging and feeding fish, echinoderms and crustacean species within 2 m either side and 2 m above (important for some fish species) the transect line. Three replicates of observations were made along each transect (each depth) at each time interval (on consecutive days). Work was completed between 30th July and 6th August 2000. Observations at night were

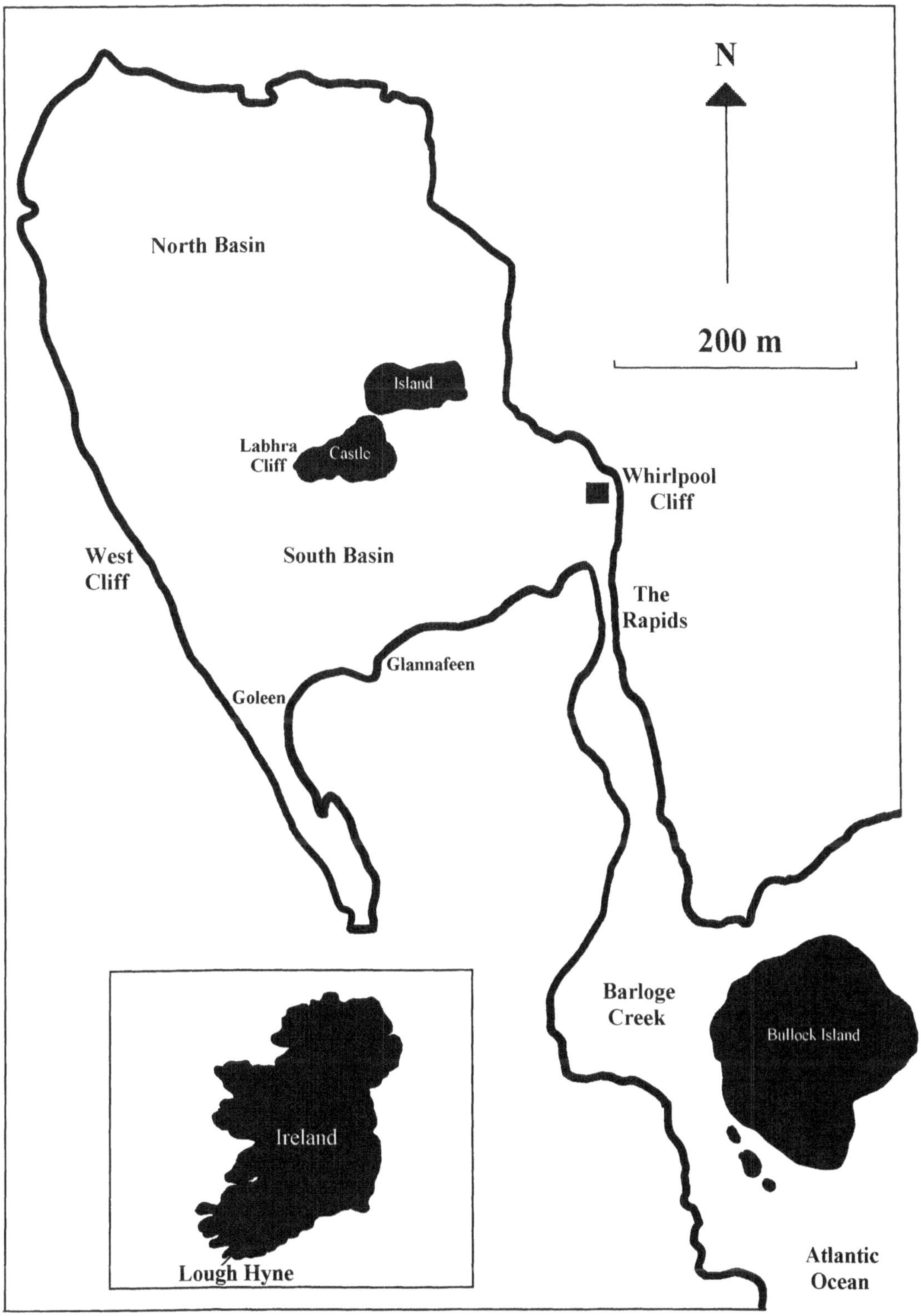

Figure 1. The study site at Lough Hyne Marine Nature Reserve, Co. Cork, Ireland. Whirlpool Cliff is indicated by the filled square.

aided by the use of underwater torches and, therefore, may slightly underestimate abundances of some highly mobile species, particularly fish, although observations of crustaceans and echinoderms are less likely to be affected. Also, the reverse may occur where organisms are drawn to divers torches, however, the latter activity was not observed for any of the assemblages considered.

Patterns of species diversity for overall community composition were described with the Shannon–Wiener (univariate comparisons) information function

$$H' = -\Sigma p_i \ln p_i,$$

(where p_i is the proportion of each species) and by species richness (number of species). Ordination (multivariate comparisons) by non-metric Multi-Dimensional Scaling (MDS in PRIMER V.6, Plymouth Marine Laboratory) was undertaken on a dissimilarity matrix created from a Bray–Curtis similarity analysis to compare communities between different time intervals and the three depth zones. Data were log $(x + 1)$ transformed to reduce the importance of extreme values (i.e. rare species). The use of the MDS technique allows each sample to be displayed in relation to each other sample in 2-D space (being calculated and reduced from multidimensional space). SIMPER analysis (in PRIMER V.6) was used to determine the contribution of each species to the average dissimilarity between different time intervals and depth zones. This method of analysis allows the determination of the species that are most responsible for any differences that occur.

Results

The overall number of organisms in the community, and for the three different assemblages (fish, echinoderms and crustaceans), varied between the different time intervals and transect depths (Fig. 2). For the entire community, the greatest number of organisms (±1 SE) was observed at midday at the lower and middle transect levels (between 189±15 and 173±35 ind. transect^{-1}), and at dusk (96±2 ind. transect^{-1}) along the upper transect. At the lower and middle transects, the overall number of individuals fell at dusk (between 10 and 100 ind. transect^{-1}), with the pattern being more pronounced along the lower transect. Numbers of individuals decreased rapidly between dusk and night to approximately 50 ind. transect^{-1} along the

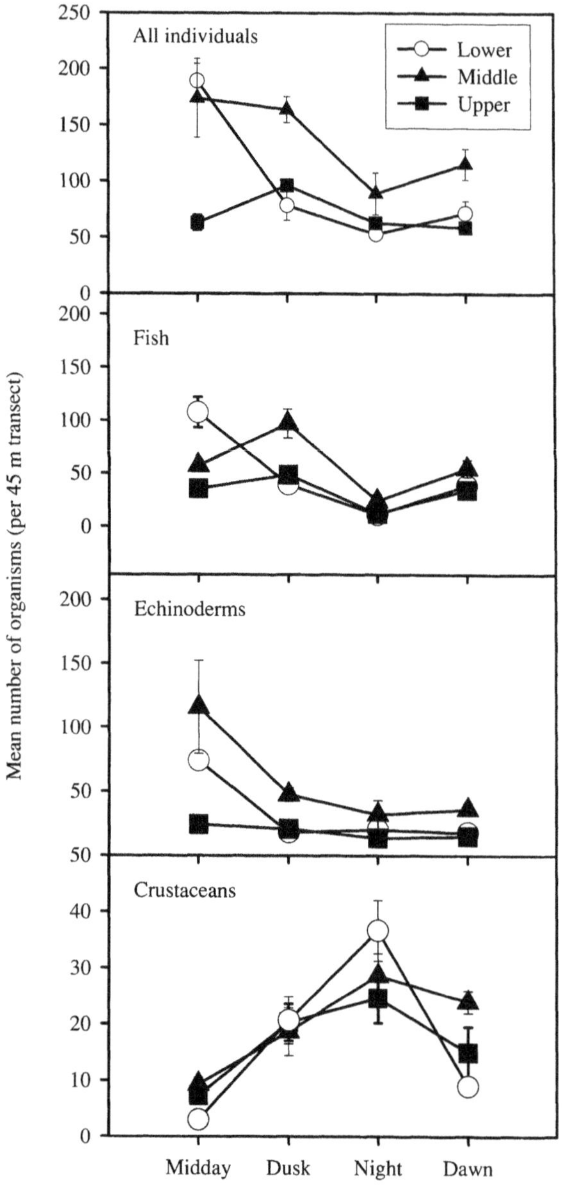

Figure 2. Overall number of individuals and numbers of fish, echinoderms and crustaceans found along three transects (6, 12 & 18 m) at different times of the day (dawn, midday, dusk and night) at Whirlpool Cliff. Standard error bars are shown.

lower and middle transects, but increased at dawn to similar levels seen at dusk.

Individual assemblages showed different patterns of abundance from the overall number of organisms (Fig. 2). Along the lower transect, highest fish numbers were observed at midday (107±14 ind. transect^{-1}), whilst along the upper and middle transects, highest fish numbers were observed at dawn and dusk (between 30 and 100 ind. transect^{-1}). The lowest number of fish was seen at night along all three transects (11–30 ind. transect^{-1}). In contrast, the lowest number of crustaceans was observed at midday with little variation between transect depths (3–10 ind. transects^{-1}) (Fig. 2). Echinoderm abundance was greatest at midday at the lower and middle transect levels (74±5 and 116±37 ind. transect^{-1}), but remained relatively constant between time intervals along the upper transect (between 12 and 27 ind. transect^{-1}). Also, at the lower and middle transect levels, although the greatest number of individuals were found at midday, their numbers decreased rapidly at dusk and remained low until the following midday (between 15–25 and 25–40 ind. transect^{-1} at the lower and middle transect level, respectively).

A total of 38 species (Table 1) were recorded from Whirlpool Cliff during this study. Univariate measures of species richness (Fig. 3) and Shannon–Wiener species diversity (graph not shown) showed no significant variation between time intervals or transect depths (Kruskal–Wallis test, $H < 2.22$, df=3, $P > 0.206$). Therefore, a univariate approach to analysing community differences showed no difference between communities at the different time intervals or depth intervals.

MDS showed differences between sites on the basis of combinations of both time interval and transect depth (Fig. 4). The overall wide separation of samples (i.e. time/depth combinations) indicated that the community composition at each individual time interval/transect combination was different. On the MDS plot, time interval and transect depth could be split along the different axes (Fig. 4B and C). The community composition at both night and midday differed considerably from each other as well as from dusk and dawn communities for any transect depth. For any given transect depth, it was difficult to separate communities at dusk and dawn (Fig. 4). However, each other time interval could be separated further by the different transect depths. Community composition differences occurred between all depth transects, with upper (6 m) and lower (18 m) transects having the most dissimilar communities. Therefore, results of the analysis from the entire community indicated that there was no movement of the entire community between different transects levels at different times of the day since the community at each time/depth interval appeared distinctive.

SIMPER analysis showed the difference between midday and dusk/dawn communities, averaged over time intervals, could be accounted for by essentially echinoderm and fish species (Tables 2 and 3) whereas differences between night and dusk/dawn communities were due essentially to crustacean and fish species. Representatives of all three groups of macro-fauna accounted for the differences between midday and night (Fig. 4). Of the fish species responsible for the various differences between time intervals, wrasse (*Centrolabrus exoletus*, *Ctenolabrus rupestris*, *Labrus mixtus* and *L. bergylta*) and one species of goby (*Thorogobius ephippatus*) were involved in the top eight discriminating species between most time intervals.

When SIMPER analysis (Table 2) was conducted to determine which species contributed to the differences between transect depths (averaged over time interval), echinoderms and fish species accounted for most of the differences. *Echinus esculentus* was the species most responsible for the differences between middle and lower, and upper and lower transects. It was difficult to identify specific assemblages (unlike for time intervals) responsible for most of the community differences since members of all three assemblages were well represented in the SIMPER analysis.

Very few species were observed at all four time intervals, either in the same zone or different zones (Table 1). Many echinoderm and crustacean species were observed only within specific depth zones (e.g. *Echinus esculentus* and *Pagurus prideauxi*) even though, in some cases, variation in numbers was observed between time intervals. Of the echinoderms, only *Asterias rubens* and *Marthasterias glacialis*, showed variation in numbers between time intervals, with significantly higher (Kruskall–Wallis, $H > 10.23$, df=3, $P < 0.001$) numbers being observed at midday. However, no evidence of migration between depth zones was found since reduced numbers within a specific depth zone did not correlate with increased numbers in a another depth zone (Table 1). Many crustacean species showed significantly higher numbers (Kruskall–Wallis, $K > 7.66$, df=3, $P < 0.001$) at specific time intervals. For example, *Galathea strigosa*, *G. squamifera* and *Palaemon serratus* were more abundant at night (Table 1) although, as for most echinoderms, migrations between zones was not evident. However, some fish species showed evidence of migrations between depth zones, with far fewer species exhibiting depth zone-specific distributions. *Labrus mixtus* and *L. bergylta* were abundant across all zones at midday but were not found on the lower transect at dawn and dusk, and not seen at all at night (Table

Table 1. Numbers of individuals (per 50 m transect ±SE) reported from Lough Hyne Marine Nature Reserve along three transects (U=upper, M=middle, L=lower) at four time intervals (Mi=Midday, Du=Dusk, Ni=Night, Da=Dawn) during three consecutive 24 h periods

Species	U/Mi	U/Du	U/Ni	U/Da	M/Mi	M/Du	M/Ni	M/Da	L/Mi	L/Du	L/Ni	L/Da
Fish:												
Scyliorhinus canicula L.	0	0	0	0.3 (+0.3)	0	0	0	0	0	0.3 (±0.3)	0	1 (±1.0)
Scyliorhinus stellaris L.	0	0.3 (±0.3)	0	0	0	0	0	0	0	0	0	0
Conger Conger L.	0	0	0	0	0	0	0	0	0	0	1.7 (±0.7)	0
Syngnathus acus L.	0	0	0	0	0	0	0	0	0	0	0.3 (±0.3)	0
Spinachia spinachia L.	0	0	1.3 (±1)	0	0	0	0	0	0	0	0	0
Pollachius pollachius L.	4.2 (±2.0)	1.0 (±2.0)	3.3 (±2.2)	0.3 (±0.3)	26.0	1.0 (±0)	2.0 (±2.0)	0.3 (±0.3)	0.3 (±0.3)	0	2.3 (±1.7)	2.7 (±0.3)
Trisopterus minutus L.	1.0 (±1.0)	1.3 (±1.1)	2.1 (±1.1)	6.6 (±1.7)	7.0 (±6.0)	12.0 (±8.0)	3.3 (0.7)	7.0 (±3.3)	16.0 (±6.3)	4.0 (±3.0)	1.3 (±1.3)	4.7 (±1.7)
Gaidropsaurus vulgaris (Cloquet)	3.0 (±3.0)	0	0	0	0	0	0	0	0	0	0	0
Zeus faber L.	0	0	0	0	0	0	0	0	0	0	0	0
Labrus bergylta Ascanius	2.3 (±1.7)	5.0 (±4.0)	0	5.0 (±0)	7.6 (±0.9)	1.7 (±1.7)	0.3 (±0.3)	3.7 (±1.7)	4.0 (±2.3)	1.7 (±1.7)	0	1 (±0.3)
Labrus mixtus L.	2.4 (±2.0)	1.0 (±1.0)	0	1.7 (±1.7)	2.7 (±0.7)	2.0 (±2.0)	0	2.0 (±2.0)	4.7 (±1.3)	1 (±1.0)	0	0
Centrolabrus exoletus (L.)	3.0 (±2.3)	3.2 (±3.0)	0	12.0 (±5.8)	7.0 (+5.7)	1.7 (±1.7)	0	2.0 (±2.0)	22.7 (±7.3)	0	0	0
Ctenolabrus rupestris (L.)	15.0 (±6.6)	26.0 (±12.1)	0	7.7 (±8.2)	18.0 (±6.1)	65.0 (±27.0)	0	33.0 (±12.0)	54.0 (±18.0)	18.0 (±8.0)	0.3 (±0.3)	27.0 (±7.0)
Crenilabrus melops (L.)	1.0 (±1.3)	2.0 (±1.5)	0	0.3 (±0.3)	4.0 (0)	0	0	0	2.0 (±2.0)	0	0	0
Parablennius gattorugine (L.)	0	0	0	0	1.7 (±1.7)	0.3 (+0.3)	0	0	0.3 (±0.3)	0	0	0
Gobius niger L.	0	0	0.7 (±0.7)	0	0	0	0.7 (±0.7)	0	0	0.3 (±0.3)	0	0
Gobius paganellus L.	0	1.0 (±1.0)	0	0	0	0.3 (±0.3)	1.0 (±0)	0	2.0 (±2.0)	2.3 (±1.7)	0.3 (±0.3)	0.3 (±0.3)
Gobius cruentatus Gmelin	0	0	0.7 (±0.7)	0	0	0	0	0	0	0.3 (±0.3)	0	0
Thorogobius ephippatus (Lowe)	0	6.6 (±3.4)	4.7 (±3.7)	0.3 (±0.3)	2.7 (±0.5)	12.3 (±6.0)	19.0 (±9.0)	7.7 (±3.3)	0	2.0 (±1.0)	2.0 (±1.0)	1.3 (±0.3)
Gobiusculus flavescens (Fabricius)	3.8 (±2.1)	0.7 (±0.7)	0	0.3 (±0.3)	0	0	0	0	0	0	0	0
Callionymus lyra L.	0	0	0	0	0	0	0	0	0	0	1.0 (±1.0)	0.3 (±0.3)
Taurulas bubalis (Euphrasen)	0	0	0	0	0	0	0	0.3 (±0.3)	0	1.7 (±1.7)	1.7 (±0.7)	1.0 (±1.0)
Eutrigla gurnades (L.)	0	0	0	0	0.3 (±0.3)	0.3 (±0.3)	0	0	0	0	0	0
Crustaceans:												
Cancer pagurus L.	0	0.3 (±0.3)	0	0.3 (±0.3)	0	0.7 (±0.7)	1.3 (±1.3)	2.7 (±1.3)	1.7 (±1.7)	2.7 (±1.7)	0	0.3 (±0.3)
Homarus gammarus (L.)	0	2.0 (±2.0)	1.3 (±1.7)	1.0 (±1.0)	3.3 (±1.7)	1.0 (±1.0)	2.0 (±2.0)	1.7 (±0.3)	1.0 (±0)	1.7 (±1.7)	0.7 (±0.7)	0.7 (±0.7)
Maja squinado (Herbst).	0	0	0	0	1.0 (±1.0)	0.7 (±0.7)	0.3 (±0.3)	0	1.7 (±1.7)	0	0.3 (±0.3)	0.7 (±0.7)
Necora puber (L.)	2.7 (±1.3)	7.7 (±2.3)	3.0 (±1.0)	5.0 (±2.3)	6.7 (±6.7)	13.0 (±12.0)	6.7 (±5.3)	11.3 (±5.7)	2.3 (±1.3)	6.7 (±4.3)	3.0 (±0)	8.3 (±6.7)
Galathea strigosa (L.)	0	0.3 (±0.3	4.0 (±2.3)	0	0	0	4.3 (±2.7)	0.3 (±0.3)	0	0	0	0
Galathea squamifera Leach	0	1.3 (±0.7)	7.7 (±4.3)	0	0	0	4.3 (±1.3)	0	0	0	0	0
Pandalus montagui Leach.	0	13.3 (±7.7)	4.0 (±2.0)	1.0 (±1.0)	0	3.3 (±3.3)	2.0 (±0)	4.7 (±2.3)	0	2.3 (±1.3)	3.3 (±4.7)	0
Inachus sp.	0	0	0.7 (±0.7)	0	0	0	0	0	0	0	0	0
Pagurus prideauxi Leach	0	0	1.0 (±0)	0.3 (±0.3)	0	0	0	0.3 (±0.3)	0.7 (±0.7)	7.0 (±6.0)	5.7 (4.3)	3.0 (±0)
Liocarcinus depurator (L.)	0	0	0	0	0	0	0	0	0	0.3 (±0.3)	4.3 (±3.3)	0.7 (±0.7)
Palaemon serratus (Pennant)	0.3 (±0.3)	0.3 (±0.3)	15.0 (±10.0)	1.3 (±1.3)	0.3 (±0.3)	0	8.0 (±8.0)	3.0 (±0)	0	0	7.3 (±5.7)	1.7 (±1.7)
Echinoderms:												
Echinus esculentus L.	17.0 (±8.0)	20.0 (±10.0)	13.0 (±6.7)	14.7 (±7.7)	42.0 (±24.0)	30.0 (±10.0)	14.0 (±5.0)	21.3 (±9.7)	1.7 (±1.7)	0	0.3 (±0.3)	0.3 (±0.3)
Asterias rubens L.	3.0 (±3.0)	1.3 (±0.7)	0	0	18.0 (±6.0)	0	2.0 (±2.0)	2.7 (±2.7)	44.3 (±22.6)	3.7 (±3.7)	0.7 (±0.7)	2.3 (±1.3)
Marthasterias glacialis L.	4.4 (±1.7)	0.3 (±0.3)	0.7 (±0.7)	0.3 (±0.3)	49.0 (±23.0)	14.0 (±7.0)	10.3 (±5.7)	9.0 (±3.0)	21.7 (±10.3)	7.0 (±0)	7.3 (±4.7)	7.7 (±4.3)
Henricia oculata (Pennant)	0	0.3 (±0.3)	0.3 (±0.3)	0.7 (±0.7)	4.0 (±2.3)	5.0 (±5.0)	2.7 (±1.7)	3.3 (±1.7)	6.3 (±3.3)	6.3 (±3.3)	8.7 (±5.3)	7.7 (±5.3)
Asterina gibbosa (Pennant)	0	0	0	0	0	0	0	0	0	0.3 (±0.3)	0	0

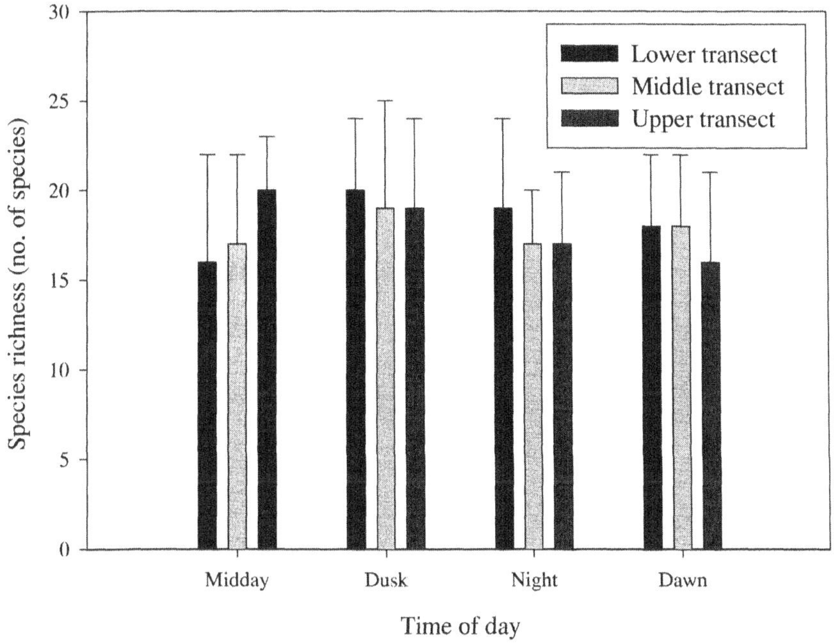

Figure 3. Overall species richness (number of species) found along three transects (6, 12 and 18 m) at different times of the day (dawn, midday, dusk and night) at Whirlpool Cliff. Standard error bars are shown.

1). *Centrolabrus exoletus* and *Ctenolabrus rupestris* occurred mainly along the lower transect at midday, along the upper transect at dusk and dawn, but were not seen at night. *Trisopterus minutus* was found along the lower and middle transects at dusk and dawn, moving down onto the lower transect at midday, was only found in small numbers at night. *Thorogobius ephippatus* was abundant along the middle transect at night, but also occurred at the upper and middle transects at dusk and dawn, with very much reduced numbers at midday (Table 1).

Discussion

The mobile community and number of organisms on the boulder scree/cliff complex at Lough Hyne was not static between time intervals or transects with a different, distinct community being found at each time interval within each depth zone. Therefore, the overall community characteristics of Whirlpool Cliff are likely to fluctuate over daily cycles. The different community composition found at each transect depth/time interval is likely to be a function of a suite of physical (e.g. habitat complexity, flow rate) and biological (e.g. predation, competition and food availability) interactions between the organisms and their environment. Previously, univariate measures have been used to describe community differences for most terrestrial, freshwater and marine systems (Hubálek, 2000). However, this study has shown that univariate measures may not provide a true representation of community structure, since no differences between time intervals or depth intervals were identified with a univariate approach. With a multivariate approach, differences between the different time intervals and transect depth combinations were identified.

The majority of species did not show migration patterns between different zones since their absence from one depth zone could not be correlated with their occurrence in another. However, variability in numbers between different time intervals was evident for most species. This suggests either movements away from the cliff/boulder areas to other locations or migration to cryptic habitats such as under boulders or into crevices. For some of these species, patterns of small-scale migrations are well know and movement away from the open cliff and boulder areas correlates with migration to holes and crevices (e.g. Turner & Warman, 1991). However, some of the fish species (including *Centrolabrus exoletus*, *Ctenolabrus rupestris*, *Labrus mixtus* and *L. bergylta*) were observed to migrate between the circumlittoral and infralittoral zones. Such migrations, presumably, allow

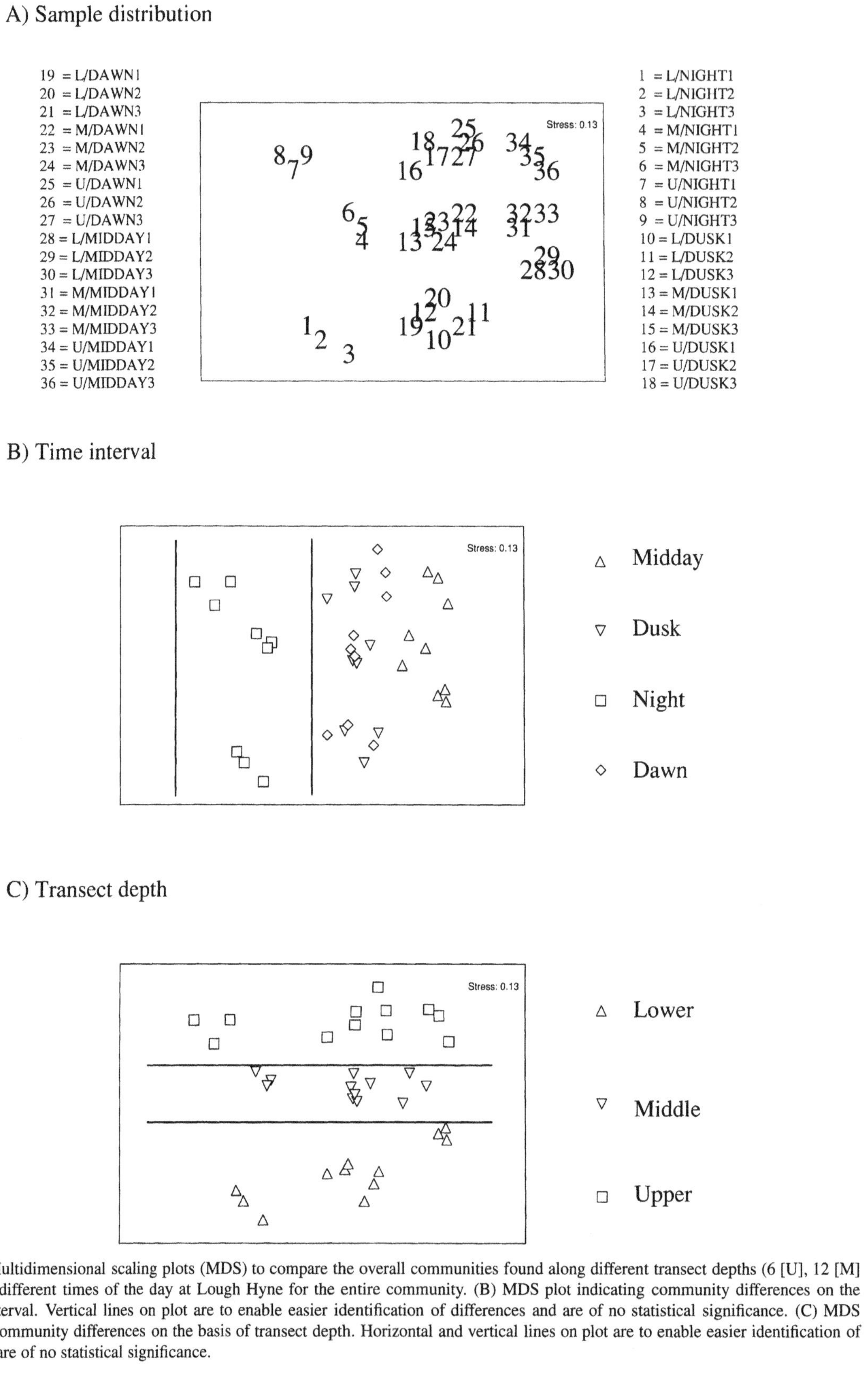

Figure 4. (A) Multidimensional scaling plots (MDS) to compare the overall communities found along different transect depths (6 [U], 12 [M] & 18 m [L]) at different times of the day at Lough Hyne for the entire community. (B) MDS plot indicating community differences on the basis of time interval. Vertical lines on plot are to enable easier identification of differences and are of no statistical significance. (C) MDS plot indicating community differences on the basis of transect depth. Horizontal and vertical lines on plot are to enable easier identification of differences and are of no statistical significance.

Table 2. Results of SIMPER analysis using PRIMER to identify the species that account for the difference between communities identified from Multidimensional Scaling (MDS) at different time intervals (averaged over transect depth). AD=average dissimilarity between time intervals. The contribution (%) of each species to the average dissimilarity between groups is shown in brackets

	Dusk	Night	Dawn
Midday	AD=46.5%	AD=66.4%	AD=45.3%
	Asterias rubens (8.9)	*Ctenolabrus rupestris* (9.9)	*Asterias rubens* (9.6)
	Centrolabrus exoletus (7.2)	*Palaemon serratus* (7.1)	*Marthasterias glacialis* (8.1)
	Pandalus montagui (7.1)	*Asterias rubens* (6.9)	*Centrolabrus exoletus* (6.8)
	Marthasterias glacialis (6.6)	*Centrolabrus exoletus* (6.4)	*Henricia oculata* (6.5)
	Thorogobius ephippatus (6.4)	*Labrus bergylta* (4.9)	*Pollachius pollachius* (5.4)
	Pollachius pollachius (5.2)	*Thorogobius ephippatus* (4.9)	*Labrus bergylta* (5.3)
	Henricia oculata (5)	*Marthasterias glacialis* (4.6)	*Labrus mixtus* (4.5)
	Trisopterus minutus (4.2)	*Pandalus montagui* (4.9)	*Crenilabrus melops* (4.5)
Dusk		AD=56.5%	AD=40.6%
		Ctenolabrus rupestris (13.9)	*Marthasterias glacialis* (8.6)
		Palaemon serratus (9)	*Centrolabrus exoletus* (7)
		Galathea squamifera (6.1)	*Thorogobius ephippatus* (6.7)
		Marthasterias glacialis (4.7)	*Pandalus montagui* (6.6)
		Labrus bergylta (4.5)	*Pagurus prideauxi* (6.2)
		Thorogobius ephippatus (4.2)	*Palaemon serratus* (5.4)
		Pagurus prideauxi (4)	*Ctenolabrus rupestris* (5.1)
		Henricia oculata (3.9)	*Labrus bergylta* (4.9)
Night			AD=55.6%
			Ctenolabrus rupestris (12.1)
			Palaemon serratus (6)
			Centrolabrus exoletus (5.7)
			Thorogobius ephippatus (5.4)
			Galathea squamifera (5.3)
			Marthasterias glacialis (5.2)
			Labrus bergylta (5)
			Trisopterus minutus (4.6)

fish to feed on the abundant invertebrates in the circumlittoral zone during the day (Fjosne & Gjosaeter, 1996; Sayer at al., 1996; Shepherd & Clarkson, 2001) and return to the infralittoral zone for protection within the kelp at dusk (Angel & Ojeda, 2001), returning to the circumlittoral zone to feed the following morning. These wrasse species have also been reported to occupy holes during dark periods (Turner & Warman, 1991), moving out of these holes (spaces between boulders, crevices, caves and cracks) onto open surfaces at dawn. The difference in numbers between wrasse observed in the infralittoral zone at night and circumlittoral zone at midday may be accounted for by migration to holes. Also, these wrasse move out onto the gravel complex and rest by wedging themselves between rocks at night (pers. obs.). Although holes are likely to afford more protection than algae, holes are likely to be limited, forcing individuals to find other areas for protection (i.e. algae and on the nearby coarse sediment areas).

The predominance of crustaceans at night is well documented and is considered as a predator avoidance mechanism, particularly for smaller crustaceans such as *Palaemon serratus*, *Galathea strigosa* and *G. squamifera* (De Grave & Turner, 1997). However, no evidence was available for migration between zones. The main crustacean predators (including wrasse and other fish species) showed the reverse pattern to most crustaceans, being abundant at midday. *Necora puber* was the only species to show a marked increase in

Table 3. Results of SIMPER analysis using PRIMER to identify the species that account for the difference between communities identified from Multi-dimensional Scaling (MDS) at different time intervals (averaged over time interval). AD=average dissimilarity between time intervals. The contribution (%) of each species to the average dissimilarity between groups is shown in brackets

	Upper	Lower
Middle	AD=46.8%	AD=48.3%
	Marthasterias glacialis (9.7)	*Echinus esculentus* (12)
	Ctenolabrus rupestris (7.9)	*Ctenolabrus rupestris* (7)
	Thorogobius ephippatus (6.9)	*Thorogobius ephippatus* (6.3)
	Asterias rubens (6.4)	*Pagurus prideauxi* (5.9)
	Centrolabrus exoletus (5.3)	*Asterias rubens* (5.9)
	Henricia oculata (5.3)	*Centrolabrus exoletus* (4.2)
	Palaemon serratus (5.2)	*Palaemon serratus* (4)
	Trisopterus minutus (4.4)	*Pandalus montagui* (3.9)
Lower	AD=66.7%	
	Echinus esculentus (9.6)	
	Henricia oculata (7.2)	
	Marthasterias glacialis (6.4)	
	Ctenolabrus rupestris (6.2)	
	Centrolabrus exoletus (5.5)	
	Asterias rubens (5.2)	
	Pagurus prideauxi (4.9)	
	Palaemon serratus (4.4)	

abundance during dusk and dawn periods. This species has bright red eyes, which may relate to increased concentration of red pigments used in low-light level adaptation. Small-scale migration patterns over daily cycles may not have been expected for echinoderms given their slow movement compared to crustaceans and fish. However, two asteroid species showed some evidence of migratory patterns. *Asterias rubens* and *Marthasterias glacialis* were only abundant at midday and occurred across all depths but migrated away from the open cliff surfaces at dusk, night and dawn when crustaceans predators were more abundant (Ramsey et al., 2000). Since migration between zones was not evident, movement of these asteroids from the lower transect level may occur onto the adjacent gravel complex to feed on the infaunal bivalves (Anger at al., 1977; Allen, 1983) or beneath boulders (also for other depth zones). However, movement onto the gravel complex makes them vulnerable to attack from the predatory starfish *Luidia ciliaris* (Philippi), which is abundant on the coarse sediment area at the base of the Whirlpool Cliff/boulder scree complex but is not found on the cliff (Greenwood et al., 1999).

Many species recorded have specific habitat requirements and, therefore, are unlikely to show migration patterns. For example, *Echinus esculentus* moves across rock surfaces grazing on algae (Vost, 1983) and is, therefore likely to be restricted to shallower depths where algae are more abundant. Also, the slow moving nature of this species is likely to limit the extent of any migration. Small-scale migration patterns have been described for the purple sea urchin, *Paracentrotus lividis* (Lamarck), within Lough Hyne, but this species inhabits the boulder areas to the north of the lough and migrates beneath boulders (Crook et al., 2000). In most cases, *E. esculentus* inhabits open areas of cliff face, therefore migration under surfaces is not possible. It seems likely that many of the differences between transects depths are explained by the individual habitat requirements of the different species, since some differences existed in the habitat quality of the different depth zones, particularly light and current flow, which decreased with depth. However, differences between time intervals (within depth zones) must represent migrations either between zones or to refuges.

Different communities on the submarine boulder scree/cliff at Lough Hyne were associated with different depth zones over a daily cycle. The majority of species showed migration patterns, however, most did not show vertical migration patterns and most likely moved to cryptic habitats to avoid predators at certain times of the day. Evidence was found for migration of some fish species (particularly wrasse) between zones feeding at the deeper zones during the day and returning to the upper layers for protection in the kelp at night. This submarine community is in a continual state of flux with different taxonomic groups and species dominating different ecological niches at different times of the day and at different depth levels. Such variability creates numerous subcommunities with different ecological characteristics, which should be considered when assessing marine communities.

Acknowledgements

We thank all members of the 2000 University of Wales, Bangor Lough Hyne expedition for helping in data collection and Declan O'Donnell of the Irish Wildlife Service for granting permits for work at Lough Hyne. We also thank University College Cork for use of on-site facilities.

References

Allen, P. L., 1983. Feeding behaviour of *Asterias rubens* on soft bottom bivalves: a study in selective predation. J. exp. mar. Biol. Ecol. 70: 79–90.

Anderson, T. W., 1994. Role of macroalgal structure in the distribution and abundance of a temperate reef fish. Mar. Ecol. Prog. Ser. 113: 279–290.

Angel, A. & F. P. Ojeda, 2001. Structure and trophic organisation of subtidal fish assemblages on the northern Chilean coast: the effect of habitat complexity. Mar. Ecol. Prog. Ser. 21: 81–91.

Anger, K., U. Rogal, G. Schriever & C. Valentin, 1977. In situ investigations on the echinoderm *Asterias rubens* as a predator of soft-bottom communities in the western Baltic sea. Helgol. wiss. Meeresunters. 29: 439–459.

Barnes, D. K. A. & A. C. Crook, 2001. Quantifying behavioural determinants of the coastal European sea-urchin *Paracentrotus lividis*. Mar. Biol. 138: 1205–1212.

Barshaw, D. E. & E. Spanier, 1994. Antipredator behaviours of the Mediterranean clipper lobster, *Scyllarides latus*. Bull. mar. Sci. 55: 375–382.

Bassindale, R., E. Davenport, F. J. Ebling, J. A. Kitching, M. A. Sleigh & J.F. Sloane, 1957. The ecology of Lough Hyne rapids with special reference to water currents. VI. Effects of the rapids on the hydrography of the south basin. Ecology 45: 879–900.

Coleman, M. A., 2002. Small-scale variability in intertidal and subtidal turfing algal assemblages and the temporal generality of these patterns. J. exp. mar. Biol. Ecol. 267: 53–74.

Crook, A. C., M. Long & D. K. A. Barnes, 2000. Quantifying daily migrations in the sea urchin *Paracentrotus lividis*. J. mar. biol. Ass. U.K. 80: 177–178.

De Grave, S. & J. R. Turner, 1997. Activity rhythms of the squat lobsters, *Galathea squamifera* and *G. strigosa* (Crustacea: Decapoda: Anomura) in south west Ireland. J. mar. biol. Ass. U.K. 77: 273–276.

Ellis, J. R. & S. E. Shackley, 1997. The reproductive biology of *Scyliorhinus canicula* in the Bristol Channel, U.K. J. Fish Biol. 51: 361–372.

Fjosne, K. & J. Gjosaeter, 1996. Dietary composition and the potential of food competition between 0-group cod (*Gadus morhua* L.) and some other fish species in the littoral zone. ICES J. mar. Sci. 53: 757–770.

Forward, R. B., R. A. Tankersley & P. N. Pochelon, 2003. Circatidal activity rhythms in ovigerous blue crabs, *Callinectes sapidus*: implications for ebb tide transport during the spawning migrations. Mar. Biol. 142: 67–76.

Greenwood, A., D. K. A. Barnes, R. M. O'Riordan & G. M. Burnell, 1999. Asteroid diversity and abundance in Lough Hyne Ireland. In Candia Carnevalia, M. D. & F. Bonasoro (eds), Proceedings of the Fifth European Conference on Echinoderms, Milan, Italy. A.A. Balkema, Rotterdam: 481–486.

Guidetti, P. & S. Bussotti, 2000. Fish fauna of a mixed meadow composed by the seagrasses *Cymodocea nodosa* and *Zostera noltii* in the Western Mediterranean. Oceanolgia Acta 23: 759–770.

Hubálek, Z., 2000. Measures of species diversity in ecology: an evaluation. Folia Zool. 49: 241–260.

Kelly, S., 2001. Temporal variation in the movement of the spiny lobster *Jasus edwardsii*. Mar. Freshwat. Res. 52: 323–331.

Lowry, M. B. & I. M. Suthers, 1998. Home range, activity and distribution patterns of a temperate rocky-reef fish, *Cheilodactylus fuscus*. Mar. Biol. 132: 569–578.

Nickell, L. A. & M. D. J. Sayer, 1998. Occurrence and activity of mobile macrofauna on a sublittoral reef: diel and seasonal migrations. J. mar. biol. Ass. U.K. 78: 1061–1082.

Ohman, M. C. & A. Rajasuriya, 1998. Relationships between habitat structure and fish communities on coral and sandstone reefs. Environ. Biol. Fish. 53: 19–31.

Palma, A. T. & F. P. Ojeda, 2002. Abundance, distribution and feeding patterns of a temperate reef fish in subtidal environments of the Chilean coast: the importance of the understory algal turf. Rev. Chilena De Hist. Nat. 75: 189–200.

Ramsey, K., J. R. Turner, S. J. Vize & C. A. Richardson, 2000. A link between predator density and arm loss in the starfish *Marthasterias glacialis* and *Asterias rubens*. J. mar. biol. Ass. U.K. 80: 565–566.

Price, J. H., 1984. The distribution of benthic marine algae – a bibliography of the British Isles. Brit. Phycol. J. 19: 385–404.

Rhode, S. C., M. Pawlowski & R. Tollrian, 2001. The impact of ultraviolet radiation on the vertical distribution of zooplankton of the genus *Daphnia*. Nature 412: 69–72.

Sayer, M. D. J., J. W. Treasurer & M. J. Costello, 1996. Wrasse: Biology and use in Aquaculture. Fishing News Books. 283 pp.

Shepherd, S. A. & P. S. Clarkson, 2001. Diet, feeding behaviour, activity and predation of the temperate blue-throated wrasse, *Notolabrus tetricus*. Mar. Freshwat. Res. 52: 311–322.

Sumich, J. L. & C. K. Erickson, 1992. Gray whale migrations. Currents 11: 23–25.

Turner, J. R. & C. G. Warman, 1991. The mobile fauna of the sublittoral cliffs. In Myers, A. A., C. Little, M. J. Costello & J. C. Partridge (eds), The Ecology of Lough Hyne: Proceeding of a Conference 4–5th September 1990. Royal Irish Academy, Dublin: 127–138.

Uglem, I & G. Rosenqvist, 2002. Nest building and mating in relation to male size in Corkwing wrasse, *Symphodus melops*. Environ. Biol. Fish 63: 17–25.

Vost, L. M., 1983. The influence of *Echinus Esculentus* grazing on subtidal algal communities. Brit. Phycol. J. 18: 211–221.

Worthington, D. G., S. E. McNeill, D. J. Ferrell & J. D. Bell, 1995. Large scale variation in the abundance of five common species of decapod sampled from seagrass in New South Wales. Aust. J. Ecol. 20: 515–525.

Hydrobiologia **503:** 183–193, 2003.
M.B. Jones, A. Ingólfsson, E. Ólafsson, G.V. Helgason, K. Gunnarsson & J. Svavarsson (eds), Migrations and Dispersal of Marine Organisms.

Diatom migration and sediment armouring – an example from the Tagus Estuary, Portugal

T. J. Tolhurst[1], B. Jesus[2,3], V. Brotas[2] & D. M. Paterson[3]
[1]*Special Research Centre for Ecological Impacts of Coastal Cities, Marine Ecology Laboratories, A11, University of Sydney, NSW 2006, Australia*
[2]*Instituto de Oceanografia, Faculdade de Ciências da Universidade de Lisboa, Campo Grande, 1749-016, Lisboa, Portugal*
[3]*Sediment Ecology Research Group, Gatty Marine Laboratory, University of St Andrews, Fife, KY16 8LB, Scotland*
E-mail: vbrotas@fc.ul.pt

Key words: microphytobenthos, diatom migration, sediment stability, chlorophyll *a*

Abstract

This study measured sediment stability, colloidal and total carbohydrate and chlorophyll *a* in the upper 2 mm of the sediment over a whole emersion period (0800–1140) in the Tagus estuary, Portugal on the 18th April 2000. Low-temperature scanning electron microscopy (LTSEM) time series images revealed the migration of microalgae throughout the emersion period, including their appearance at the sediment surface at the beginning of emersion and their subsequent return below the surface at the end of emersion. Different species arrived at the sediment surface at different times and there was a slight lag between the appearance of the first cells and the subsequent increase in sediment stability. Increased chlorophyll *a* concentration in the surface sediments preceded the increase in sediment stability, whilst colloidal and total carbohydrate concentrations increased afterwards. Sediment water content decreased during the emersion period. Erosion threshold increased shortly after the microalgal cells appeared at the sediment surface, suggesting that the cells themselves act to 'armour' the sediment surface, retarding erosion. Lack of correlation between sediment stability and factors traditionally considered to control sediment stability (e.g., water content and carbohydrate content) indicates that an important variable or interaction has yet to be identified. One possibility is that the carbohydrate fraction extracted does not measure accurately the binding effectiveness of the carbohydrates in the sediment. We propose feedback and 'critical point' models to explain how the various sediment properties determine sediment stability. The implication is that sediment stability varies in an apparently idiosyncratic and site-specific fashion due to the complex interaction of physical and biological variables. Given the importance of ecological processes in intertidal sediments, the measurement, understanding and modelling of sediment erosion would benefit greatly from the application of ecological methods of experimental design and sampling.

Introduction

The erosion, transport, deposition and consolidation (ETDC) cycle of sediments is controlled by a number of physical and biological processes. In muddy cohesive sediments, such as those found in estuaries, there is often considerable biotic activity. This results in a complex interaction of biological and physical processes, and properties that control the sediment dynamics (Black et al., 2002). Sediment processes (fluxes, erosion, deposition, nutrient cycling) are controlled by the physical sediment properties and are directly or indirectly mediated by the biota. Physical properties include particle size, shape and density, water content, and mineralogy. Biological mediation includes secretion of extracellular polymeric substances (EPS) by organisms such as diatoms, bioturbation (including feeding and tube building), biofiltration, pelletisation and faeces excretion. Due to the number of factors involved, trying to predict cohesive sediment stability from just a few easily measurable parameters has proven difficult (Paterson et al., 2000; Defew et al., 2002). Nevertheless, the two most important factors influencing sediment stability are normally considered to be water and EPS content. In this paper, sediment stability is used to describe the erodibility of the sediment, which is defined by both erosion threshold and rate.

Water content is the only physical variable likely to change over an emersion period, and thus the only one that could influence sediment stability on this time scale. The importance of microbial activity in affecting sediment dynamics has been recognised for some time (Manzenreider, 1983; Grant et al., 1986; Paterson, 1989), organisms such as bacteria, cyanobacteria and diatoms increase sediment stability (Paterson, 1994). Epipelic diatoms can form extensive biofilms in intertidal cohesive sediments, constituting the major taxonomic group in microphytobenthic communities of estuarine ecosystems in Europe (Brotas & Plante-Cuny, 1998; Underwood & Kromkamp, 1999). They are motile, secrete copious quantities of extracellular polymeric substances (EPS) (Hoagland et al., 1993) and are considered often to be the most significant of the stabilising organisms. EPS increase the strength of the inter-particle bonds in cohesive sediments and create cohesion in non-cohesive sediments (Dade et al., 1990; Decho, 1990; Tolhurst et al., 2002). However, recent work has shown that the binding created by EPS extracted from its secreting organism is less than that found with the organism present, leading to the suggestion that the structuring of the EPS by the organism as it moves through the sediment is important in the stabilising process (de Brouwer et al., 2002).

In a number of *in situ* field studies, sediment stability correlated poorly with all measured sediment properties except chlorophyll *a* (Riethmüller et al., 1998, 2000; Paterson et al., 2000; Defew et al., 2002). Lack of correlation with physical factors has been explained by the influence of biological processes that mask the physical properties (Defew et al., 2002). Chlorophyll *a* itself does not affect sediment stability, however, it is a measure of diatom biomass. Poor correlation between sediment stability and EPS, and the good correlation between sediment stability and chlorophyll *a* in a variety of different estuaries, however, is at odds with laboratory findings. Clearly, some factor or process is being overlooked. A possible explanation may be that the diatom cells *themselves* are important in stabilising the sediment. This would also explain why sediment stability is higher in studies with live diatom cells than with extracted EPS.

With this in mind, changes in sediment properties potentially related to sediment stability and the migration of a diatom dominated biofilm were measured over a single emersion period in the Tagus estuary, Portugal, on 18 April 2000. The Tagus estuary (38° 44′ N, 9 08′ W) is an extensive mesotidal estuary, with semi-diurnal tides of amplitudes varying from 1 to 4 m; the intertidal zone covers up to 40% of its total area and consists mainly of mudflats.

Materials and methods

Sediment stability was measured approximately every 20 min over the emersion period with a Cohesive Strength Meter (CSM) vertical jet erosion device (see Tolhurst et al., 1999 and website www.sediments.co.uk for more detailed descriptions). The device consists of a water-filled chamber (30 mm in diameter) pushed into the sediment. The erosive jet of water comes from a downward-directed nozzle in the chamber; the velocity of the jet is increased systematically during each experiment. Bed erosion is calculated from the drop in the transmission of infrared light across the chamber caused by the suspension of sediment. A drop in transmission to 90% was taken as the erosion threshold, which is approximately equivalent to the erosion of 0.01 kg m^{-2} of sediment.

Sediment properties (water content, colloidal and total carbohydrate and chlorophyll *a*) were measured from contact core sediment samples (Honeywill et al., 2002) matched with the CSM measurements. The contact core technique freezes the upper surface 2 mm of sediment using liquid nitrogen and an aluminium core with a plastic collar. It is important to measure on this scale as it represents the extent of both diatom migration and the photic zone within the sediments (Hay et al., 1993). The sediment core was weighed, lyophilised, homogenized and sub-sampled for the different analyses. Sediment water content (W%) was calculated by the following equation:

$$W\% = (Wt_{wet} - Wt_{dry}) / Wt_{wet} * 100,$$

where Wt_{wet} = wet weight of sediment and Wt_{dry} = dry weight of the sediment.

Colloidal carbohydrate (the water soluble fraction of carbohydrates) and total carbohydrate were measured using the Dubois assay (Dubois et al., 1956) and expressed as glucose equivalents using a standard curve. Chlorophyll *a* was measured spectrophotometrically using a dimethyl formamide (DMF) extraction and the following equation (Porra et al., 1989) and expressed as mg m^{-2}:

$$\text{Chlorophyll } a \text{ mg l}^{-1} = 12 (A_{664}–A_{750}) - 3.11(A_{647}–A_{750}),$$

where A_{664}, A_{647}, and A_{750} stand for absorbance at 664, 647 and 750 nm, corresponding respectively to absorbance peaks of chlorophylls *a* and *b*, and to the absorbance due to turbidity. This equation was chosen instead of the more common Lorenzen (1967) equations (which correct for chlorophyll degradation products), as the abundance of chlorophyll *b*-containing organisms varied along the emersion period, whereas the percentage of degradation products was not expected to change over such a short period within the upper layer of sediment.

Aluminium stubs were used to collect sequential samples of the sediment surface during the emersion period; samples were collected more frequently at the beginning and end. Samples were frozen immediately in liquid nitrogen to preserve the structure and viewed using low-temperature scanning electron microscopy (LTSEM). Pictures of the sediment surface were taken to ascertain the distribution of diatoms and other microalgae. Frozen samples were fractured perpendicularly to the surface and LTSEM photographs were taken also at the fracture face, allowing cell migration through the sediment to be viewed (see Paterson [1995] for details of this technique).

Results

On 18 April 2000, the sample site was exposed at 0800 h and tidal water returned at 1140 h. Sediment stability remained low until 0910 h, but increased rapidly between then and 0928 h (Fig. 1). Subsequently, sediment stability remained high until 1112 h when a brief (~5 min) low-intensity rain shower reduced sharply sediment stability to a level similar to that at the beginning of the emersion period. Recovery was rapid as sediment stability returned to a level similar to that before the rain shower at 1135 h. Sediment water content showed its most rapid decrease in the period from the beginning of emersion to just before the marked increase in sediment stability.

The LTSEM images show how the diatoms gradually migrated to the sediment surface at the beginning of emersion and migrated downwards immediately before immersion (Figs 2 and 3). Whilst the appearance of different species at the sediment surface at different times has been documented in the laboratory (Hay et al., 1993), these images are the first to show migration over a complete emersion period and the movement of cells below the sediment surface *in situ*. Large diatom cells from the genus *Plagiotropis*, and a few small *Navicula* spp, appeared at the sediment surface at 0830 h (Fig. 2B); migration continued resulting in a dense covering of cells over the sediment surface (Fig. 3A). LTSEM images taken in 90-degree fracture faces showed the abundant presence of diatom cells, at least as deep as 600 μm, presumably migrating upwards (Fig. 2C and F). *Euglena* spp cells first appeared at the sediment surface at 0917 h (Fig. 3B), and continued to migrate upwards until 1100 h, when microalgal cells covered the entire sediment surface. Apart from the clearly different migration behaviour between diatoms and euglenoids, within the diatom community, there was an apparent change in species composition over exposure time. *Plagiotropis* spp appeared at the beginning, *Staurophora* spp at 0917 h (Fig. 3B) and *Cylindrotheca closterium* Reinmann & Lewin, later in the morning, whereas *Navicula* spp and other small species increased in abundance from 0830 to 0917 h. From 0900 to 1100 h, fracture face images showed practically no cells were present in non-surface layers. Images taken at the surface at 1130 h, 10 min before immersion time, showed some diatom cells in a vertical position, ready to undergo downward migration (Fig. 3D), whereas the depth profile showed many cells migrating downwards (Fig. 3E).

The sudden increase in sediment stability between 0910 and 0928 h occurred at approximately the same time as an increase in chlorophyll *a* in the surface 2 mm (Fig. 4). It is also around this time that the majority of migrating diatom cells reached the surface and just after the first appearance of euglenoids at the sediment surface (Fig. 3B). The colloidal and total carbohydrate concentrations in the top 2 mm did not increase until around 1000 h (Fig. 5), 30 min after the increase in sediment stability. This lag indicates that changes in carbohydrate concentrations are not the cause of the increased stability. Whilst the rain shower had a large effect upon the sediment stability, there was no obvious effect on any of the other sediment properties (including water content).

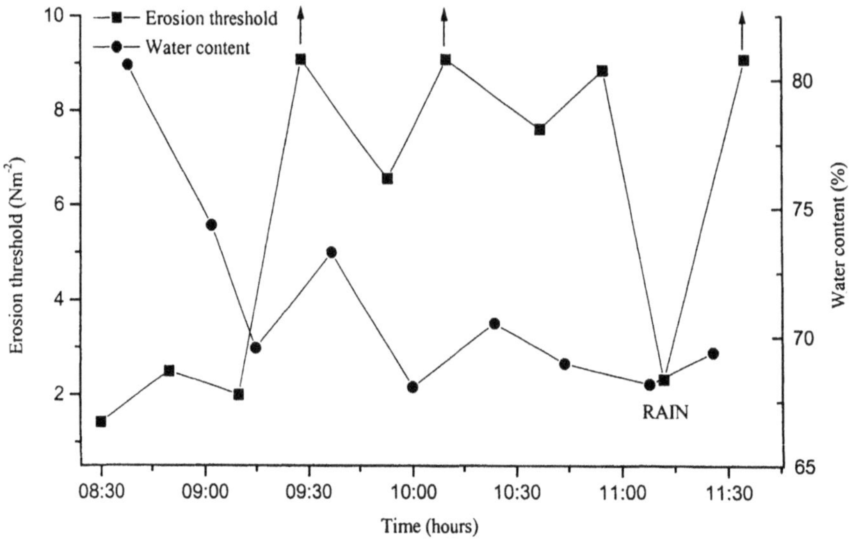

Figure 1. The change in erosion threshold (Nm^{-2}) and associated sediment water content (%) over the low tide emersion period. A brief rain shower resulted in a sharp but transitory decrease in erosion threshold. Arrows indicate where no erosion occurred at the maximum erosion stress of 9 Nm^{-2} and therefore these should be considered minimum values.

Discussion

Relationship between sediment stability and its controlling factors

Much remains unknown about how sediment properties and processes interact to determine sediment stability. Physical parameters affect sediment stability in predictable ways; for example, decreasing the water content of muddy sediments increases the sediment stability. Biotic components are complex, having multiple effects that can vary temporally and spatially. Sediment fauna can simultaneously increase sediment stability (e.g. secretion of EPS, tube formation) and decrease it (e.g. burrow cleaning, feeding). Selective biocide treatments have revealed the net effect of fauna is to destabilise sediments while microbes stabilise sediment (Underwood & Paterson, 1993; de Deckere et al., 2001). When individual effects of the biota are considered under controlled conditions, however, they too behave in a predictable fashion. For example, increasing the EPS content of muddy sediments under laboratory conditions increases sediment stability in a linear fashion (Tolhurst et al., 2002).

Unfortunately, as the present study and others show, when the natural system is investigated the relative importance of both physical and biological components varies spatially and temporally, and many of the relationships apparently disappear (Paterson et al., 2000; de Brouwer et al., 2000; Defew et al., 2002). Clearly, the system is greater than the sum of its parts, and there is a complex interaction between the various properties and processes. We suggest that processes that behave in a linear fashion in isolation in the laboratory could interact *in situ* to generate non-linear responses.

One possibility would be a 'feedback' effect. For example, desiccation, reducing the sediment water content, would increase sediment stability by physically altering the sediment and existing EPS, but could also result in increased secretion of EPS from biota; this would result in increased sediment stability greater than that expected from just the physical effect of changing water content. Another possibility is that the system behaves in a 'critical point' fashion, a controlling factor reaches a threshold that causes a larger than expected change in the variable of interest. In this study, the appearance of diatom cells at the sediment surface is the critical point resulting in a sudden increase in sediment stability. Such interactions explain why natural sediment stability correlates poorly with factors shown experimentally to control it. Different properties change in different ways over the emersion and, at a certain critical point, the net effect of the different controlling factors results in a significant non-linear change in sediment stability. This results in a system that appears to behave idiosyncratically, and creates considerable difficulties for understanding

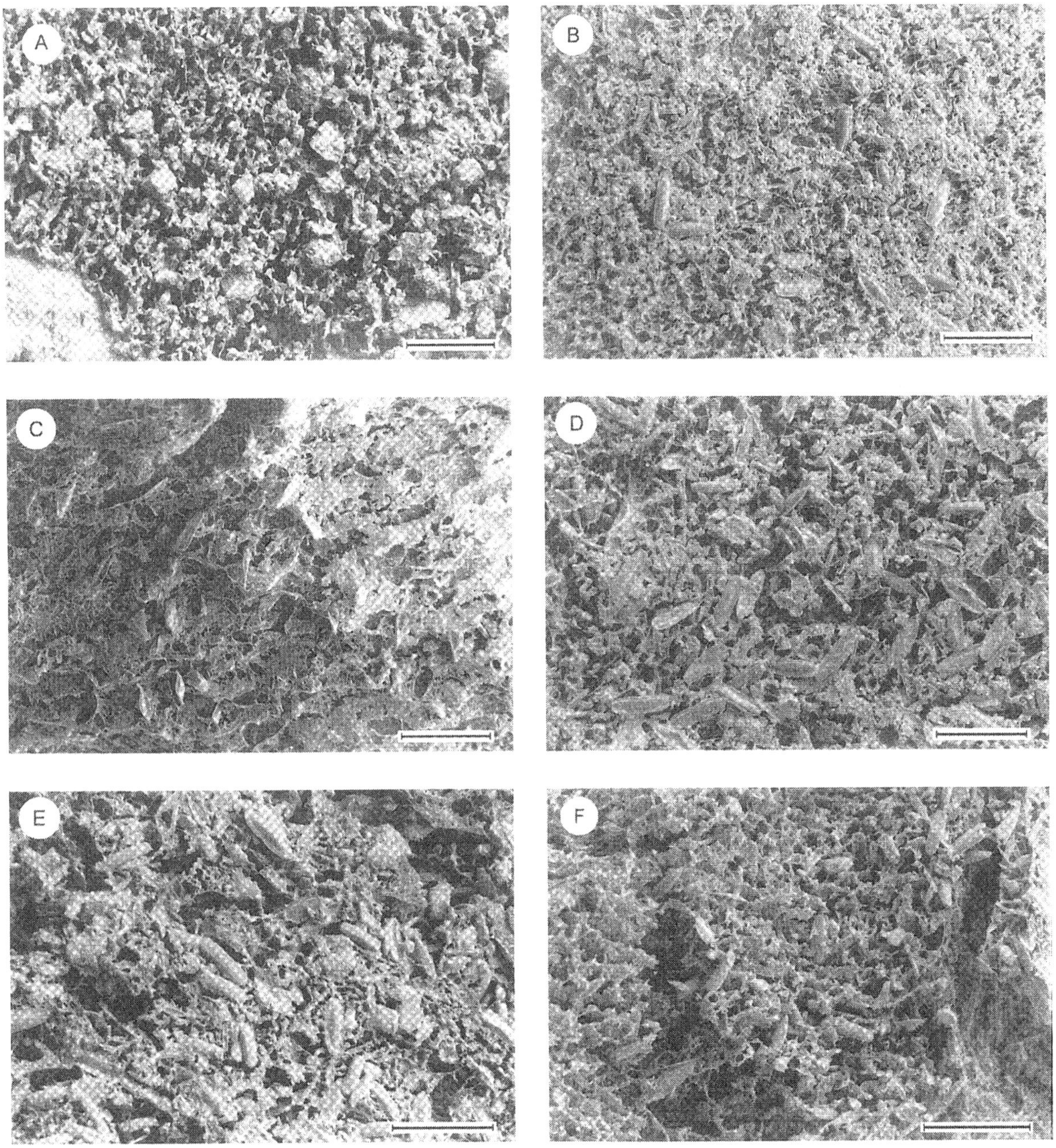

Figure 2. LTSEM images of sediment samples. (A) 0805, sediment surface with no visible microalgae cells; (B) 0830, large diatom cells of *Plagiotropis* arrive at the surface; (C) 0830, fracture face photograph, (taken in depth profile) showing diatoms down to a depth of 600 μm; (D) 8:35 am, arrival of more diatom cells at the sediment surface, (E) 8:40 am, abundant *Navicula* cells; (F) 8:40 am, depth profile, showing large quantities of diatom cells migrating upwards.

and modelling. This is especially true if only a few of the many variables that could potentially influence sediment stability are measured.

The presence of diatoms at the sediment surface results in very high erosion thresholds (Tolhurst et al., 1999). Diatoms have hard silica cell walls and we suggest that the diatom assemblage of cells and EPS strands acts to 'armour' the sediment surface in a manner similar to the scale mail armour of mediaeval knights. The diatoms act as scales and the EPS

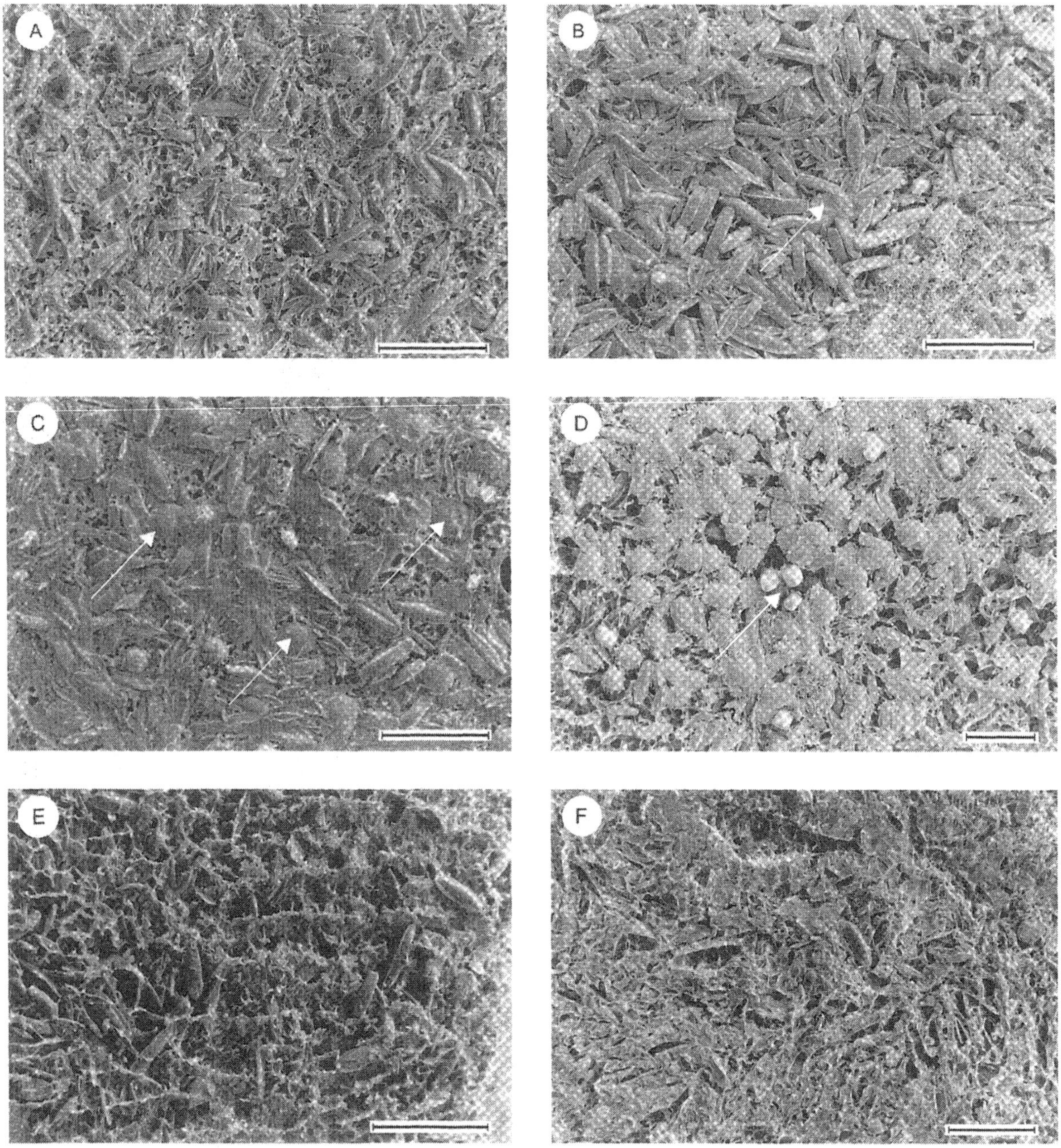

Figure 3. LTSEM images of sediment samples. (A) 0845, the sediment surface is completely covered with diatom cells; (B) 0917, euglenoids start to appear at the surface, as well as the diatom *Staurophora*; (C) 1100, euglenoids are now also very abundant, and new diatom species are visible for the first time, such as *Cylindrotheca closterium*. (D) 1130, some diatom cells in vertical position, preparing to undergo downward migration; (E) 1130, depth profile, the cells are actively migrating downwards; (F) 1136, very few cells are present on sediment surface, prior to immersion.

connects these into strong armour like 'skin' over the sediment surface, retarding erosion and resulting in extremely high erosion thresholds. This armouring is enhanced by the elastic nature of EPS, thick diatom biofilms respond elastically to stress (Paterson, 1989), this helps protect the bed from the cyclic stresses caused by wave action or sweep events (inrushes of high velocity fluid moving downwards towards the bed associated with moving eddy structures), which are correlated with erosion (Clifford et al., 1993).

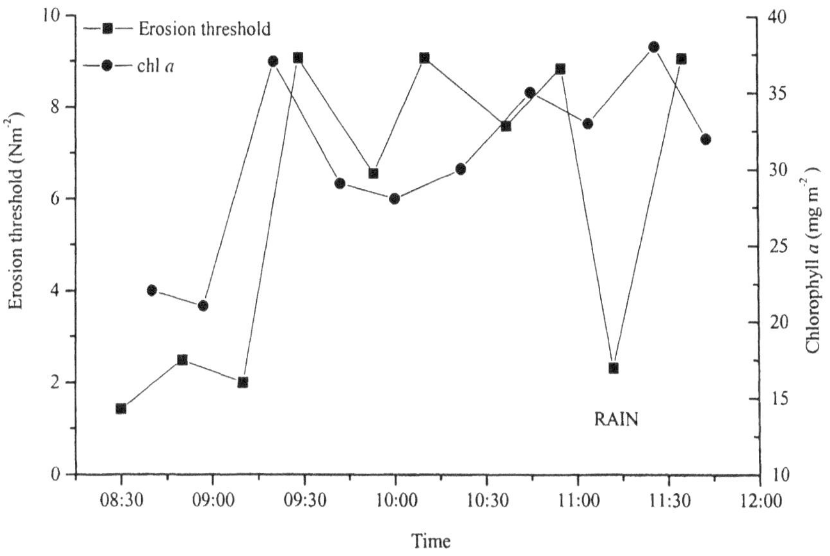

Figure 4. The changes in erosion threshold (Nm^{-2}) and sediment chlorophyll *a* (concentration mg m^{-2}) over the emersion period.

Sediment armouring by diatom cells has important implications for measuring sediment dynamics. The armouring will occur only when the cells are at the sediment surface. Usually, diatoms are considered to migrate away from the sediment surface before the tide returns, thus no sediment armouring would occur in this case. However, some diatom biofilms do not appear to be migratory, remaining at the sediment surface during immersion (T. J. Tolhurst, pers. obs.; Hay et al., 1993). When sediment armouring occurs, different diatoms may stabilise the sediment differently, depending upon their size/migration behaviour as well as EPS secretion. The LTSEM photographs show very clearly that there is a shift in species composition, starting with the diatoms *Plagiotropis* spp and *Navicula* spp appearing at the sediment surface in early morning, followed by euglenoids and other diatom genera (such as *Staurophora* and *Cylindrotheca*) later at mid-morning. This pattern is caused by the different physical and chemical factors that influence migratory behaviour in diatoms (Edgar & Pickett-Heaps, 1984; Paterson, 1986). Sauer et al. (2002) reported the alteration of species composition in response to changes in irradiance and Brotas et al. (2003) observed a shift in cell size during an emersion period. Euglenoids appear at the surface about the time of the increase in sediment stability, however, their appearance at this time is probably merely correlative rather than causative and could be due to an independent factor, such as irradiance.

There is a possibility that we are not measuring the correct aspect of the different components and thus missing important interactions. In a laboratory study, Paterson (1989) found a steady increase in both sediment stability and diatom numbers over emersion in recently-deposited sediment beds. He suggested that the lag between diatoms appearing at the sediment surface and the increase in sediment stability represented a lag period required for the diatoms to produce sufficient mucilage to form an extracellular matrix. Our data do not support that theory, as the total and colloidal carbohydrates increase *after* the increase in sediment stability (Fig. 5). This finding is surprising as there is considerable evidence that the secretion of EPS is the primary mechanism for biostabilisation (Decho, 1990; Dade et al., 1990; Chenu & Gerif, 1991; Tolhurst et al., 2002). However, it should be noted that colloidal carbohydrate is an operational separation of the total carbohydrate pool no matter what its source or type. Given the wide range of molecules detected by the test, it is likely that only a small fraction of the carbohydrate pool is actually contributing to sediment stability. EPS is considered usually to be the fraction most likely to contribute to sediment stability, yet EPS constitutes only approximately 30% of colloidal carbohydrate (Underwood et al., 1995) and, of this 30%, some molecules will be better sediment stabilisers than others. Thus, maybe we are not looking at the right operational fraction in terms of sediment stability and a more specific assay is required

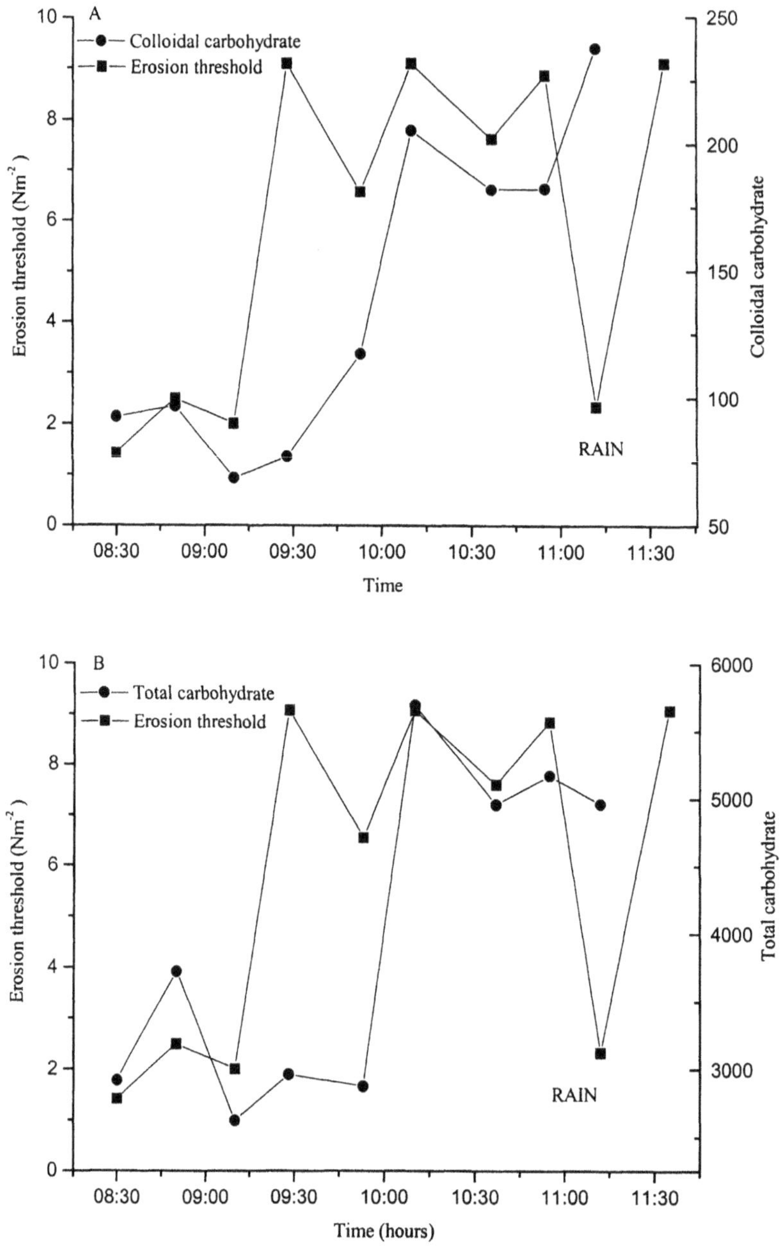

Figure 5. The change in erosion threshold and the associated change in concentration of colloidal carbohydrate (top graph) and total carbohydrate (bottom graph) on the top 2 mm over the emersion period. Carbohydrate units in μg cm^{-2}.

to reveal patterns we may be currently missing. This is important for a variety of reasons, not least of which is that different diatom species secrete different amounts and types of EPS that can have different effects on sediment stability (Holland et al., 1974). As diatoms migrate at different rates (Hay et al., 1993) it may be the timing of the appearance of certain 'stabilising' species at the sediment surface along with physical armouring that is causing the sudden peak in sediment stability. Unfortunately, only very preliminary work has been carried out to investigate how EPS from different diatom species affects sediment stability, but there is evidence that EPS alone is not responsible for all of the stabilising effect seen (Holland et al., 1974; de Brouwer et al., 2002). Clearly, we need to be sure that what we measure is meaningful in terms of the processes we are interested in.

The effect of rain

The rain shower reduced sediment stability to a level similar to that at the beginning of the emersion period yet there were no significant changes in either water content, chlorophyll *a*, or carbohydrate content, which means that changes in these were not directly responsible for the decrease in sediment stability. This supports the findings of Paterson et al. (2000) that rain can significantly reduce intertidal sediment stability during emersion. It is of importance here because the data indicate that sediment stability is controlled largely by an interaction of biological factors, yet a brief rain shower was sufficient to negate the biological effect, highlighting the interaction between biological and physical properties and processes. The exact mechanism at work is unclear, but such rapid changes in sediment stability do not occur at the beginning of seawater immersion, suggesting that the physical impact of the rain drops and the fresh water causes physico-chemical dispersion of clay minerals, causing the rapid and transient drop in stability.

Implications for erosion measuring devices and modelling of sediment stability

Change in sediment properties and stability during tidal emersion has significant implications for modelling and understanding cohesive sediment behaviour. Of particular importance is that large increases in sediment stability measured during emersion do not persist through the subsequent immersion (Paterson, 1989). This means that measuring the sediment stability changes during emersion cannot be considered to accurately represent the changes occurring during immersion. Currently, no erosion devices are capable of measuring sediment stability over immersion on the time scale of minutes that is necessary to detect the changes occurring. Flume-type devices that produce one measure of sediment stability over a few hours collect unrepresentative data, by averaging short-term changes in stability such as those found in this study (Tolhurst et al., 2000). They also change the properties of the sediment during measurement (overlying water will cause diatoms to migrate away from the sediment surface, dissolve EPS, increase sediment water content and change faunal behaviour) and, due to a lack of replication, collect data that are potentially temporally and spatially confounded. This brings into question the relevance of such measurements for modelling natural sediment dynamics.

The range of naturally occurring shear stresses is small, thus a small change in erosion threshold can have a large effect on the erosion of sediment. Ultimately, erosion studies should be able to determine if the sediments in question will be eroded by naturally occurring flows and if erosion does occur, how much. In this respect, it is essential that accurate measurements be made not only during emersion but also immersion without significantly altering the sediment; this presents quite a challenge. LTSEM images show diatoms were still at the sediment surface when the last CSM measurement was taken, but just 6 min later most had migrated away from the surface. If diatoms had been armouring the sediment, we would expect a decrease in sediment stability at this time but, unfortunately, given current technology, taking such measurements as the tide returns in a large estuary such as the Tagus, where the tide returns very rapidly, is potentially dangerous and was, therefore, not attempted. Such an approach would be possible near to the shore in smaller less dynamic systems.

Conclusions

This study begins to address the fine-scale temporal measurement of sediment stability over emersion. The challenge now is to collect replicated measurements over both emersion and immersion to better understand and model the temporal and spatial variation in processes controlling erosion. Sediment stability is a function of the complex interactions of a number of biological and physical properties and processes. These interactions can result in non-linear responses of the system due to feedbacks and 'critical point' interactions. In this study, sediment stability appeared to be controlled primarily by diatom migration. Other factors, such as the carbohydrate and water content of the sediments, did not appear to be significant. We propose, therefore, that the diatom cells act to physically armour the sediment surface protecting it from erosion, resulting in very high erosion thresholds. A rain shower disrupted the stabilising effect of the diatoms resulting in a transitory decrease in sediment stability.

Acknowledgements

This work was supported through the European MAS3-CT98-0166-CLIMEROD project, Anglo-Portuguese Joint Research Program "Treaty of Windsor/CRUP", by FCT (project Sat-Tagis no PDCTM/MAR/15256/99 and POCTI/FEDER), and received additional staff support for T. J. Tolhurst during writing from the Ecological Impacts of Coastal Cities commonwealth special research centre. Bruno Jesus is supported by a PhD grant from FCT (Praxis XXIBD21634/99). The assistance of Irvine Davidson in doing all the LTSEM work is acknowledged. The authors wish to thank Michael Sullivan and an anonymous referee for valuable criticisms and suggestions.

References

Black, K. S., T. J. Tolhurst, S. E. Hagerthey & D. M. Paterson, 2002. Working with natural cohesive sediments. J. Hydraul. Eng. 128: 1–7.

Brotas V. & M. R. Plante-Cuny, 1998. Spatial and temporal patterns of microphytobenthic taxa of estuarine tidal flats in the Tagus estuary using pigment analysis by HPLC. Mar. Ecol. Prog. Ser. 171: 43–57.

Brotas, V., N. Risgaard-Petersen, J. Serôdio, L. Ottosen, T. Dalsgaard & L. Ribeiro. *In situ* measurement of photosynthetic activity and respiration of intertidal benthic microalgal communities undergoing vertical migration. Ophelia 57: 13–26.

Chenu, C. & J. Guérif, 1991. Mechanical strength of clay minerals as influenced by an adsorbed polysaccharide. Soil Sci. Soc. am. J. 55: 1076–1080.

Clifford, N. J., J. R. French & J. Hardisty, 1993. Turbulence: Perspectives on Flow and Sediment Transport. Wiley and Sons Ltd. Chichester. 360 pp.

Dade, W. B., J. D. Davis, P. D. Nichols, A. R. M. Nowell, D. Thistle, M. B. Trexler & D.C. White, 1990. Effects of bacterial exopolymer adhesion on the entrainment of sand. Geomicrobiology. J. 8: 1–16.

de Brouwer, J. F. C., S. Bjelic, E. M. G. T. de Deckere & L. J. Stal, 2000. Interplay between biology and sedimentology in a mudflat (Biezelingse Ham, Westerschelde, The Netherlands). Cont. Shelf Res. 20: 1159–1177.

de Brouwer, J. F. C., G. K. Ruddy, T. E. R. Jones & L. J. Stal, 2002. Adsorption of EPS to sediment particles and the effect on the rheology of sediment slurries. Biogeochemistry 61: 57–71.

Decho, A. W., 1990. Microbial exopolymer secretions in ocean environments: their role(s) in food webs and marine processes. Oceanogr. mar. niol. Ann. Rev. 28: 73–153.

de Deckere, E. M. G. T., T. J. Tolhurst & J. F. C. de Brouwer, 2001. Destabilisation of cohesive intertidal sediments by infauna. Estuar. coast. shelf Sci. 53: 665–669.

Defew, E. C., T. J. Tolhurst & D. M. Paterson, 2002. Site-specific features influence sediment stability of intertidal flats. Hydrol. Earth Syst. Sci. 6: 971–981.

Dubois, M., K. A. Gilles, J. K. Hamilton, P. A. Rebers & F. Smith, 1956. Colorimetric method for determination of sugars and related substances. Ann. Chem. 28: 350–356.

Edgar, L, A. & J. D. Pickett-Heaps, 1984. Diatom locomotion. In Round, F. E. & D. J. Chapman (eds), Progress in Phycological Research. 3: 47–88.

Grant, J., U. U. Bothmann & E. L. Mills, 1986. The interaction between benthic diatom films and sediment transport. Estuar. coast. shelf Sci. 23: 225–38.

Hay, S. I., T. C. Maitland & D. M. Paterson, 1993. The speed of diatom migration through natural and artificial substrata. Diatom Res. 8: 371–384.

Hoagland, K. D., J. R. Rosowski, M. R. Gretz & S. C. Roemer, 1993. Diatom extracellular polymeric substances: function, fine structure, chemistry and physiology. J. Phycol. 29: 537–566.

Holland, A. F., R. G. Zingmark & J. M. Dean, 1974. Quantitive evidence concerning the stabilization of sediments by marine benthic diatoms. Mar. Biol. 27: 191–196.

Honeywill, C., D. M. Paterson & S. E. Hagerthey, 2002. Determination of microphytobenthic biomass using pulse modulated minimum fluorescence. Eur. J. Phycol. 37: 1–8.

Lorenzen, C. J., 1967. Determination of chlorophyll and pheopigments: spectrophotometric equations. Limnol. Oceanogr. 12: 343–346.

Manzenreider, H., 1983. Retardation of initial erosion under biological effects in sandy tidal flats. Leichtweiss Inst, Tech University Braunschweig: 469–79.

Paterson, D. M., 1986. The migratory behaviour of diatom assemblages in a laboratory tidal micro-ecosystem examined by low temperature scanning electron microscopy. Diatom Res. 1: 227–239.

Paterson, D. M., 1989. Short-term changes in the erodibility of intertidal cohesive sediments related to the migratory behaviour of epipelic diatoms. Limnol. Oceanogr. 34: 223–234.

Paterson, D. M., 1994. Microbial mediation of sediment structure and behaviour. In Stal, L. J. & P. Gaumethe (eds), NATO ASI Series, Vol. 635 Microbial Mats. Springer-Verlag Berlin Heidelberg.

Paterson, D. M., 1995. Biogenic structure of early sediment fabric visualised by LTSEM. J. Geol. Soc. 152: 131–140.

Paterson, D. M., T. J. Tolhurst, J. A. Kelly, C. Honeywill, E. M. G. T. de Deckere, V. Huet, S. A. Shayler, K. S. Black, J. de Brouwer & I. Davidson, 2000. Variations in sediment stability and sediment properties across the Skeffling mudflat, Humber estuary, U.K. Conti. Shelf Res. 20: 1373–1396.

Porra, R. J., W. A. Thompson & P. E. Kriedman, 1989. Determination of accurate extinction coefficients and simultaneous equations for assaying chlorophylls *a* and *b* extracted with four different solvents: verification of the concentration of chlorophyll standards by atomic absorption spectroscopy. Biochim. Biophys. Acta 975: 384–394.

Riethmüller, R., J. H. M. Hakvoort, M. Heineke, K. Heymann, H. Kühl & G. Witte, 1998. Relating erosion threshold to tidal flat surface colour. In Black, K. S., D. M. Paterson & A. Cramp (eds), Sedimentary Processes in the Intertidal Zone. Geological Society, London, Special Publications 139: 283–293.

Riethmüller, R., M. Heineke, H. Kühl & R. Keuker-Rüdiger, 2000. Chlorophyll *a* concentration as an index of sediment surface stabilisation by microphytobenthos? Conti. Shelf Res. 20: 1351–1372.

Sauer, J., K. Wenderoth, U. G. Maier & E. Rhiel, 2002. Effects of salinity, light and time on the vertical migration of diatom assemblages. Diatom Res. 17: 189–203.

Tolhurst, T. J., K. S. Black, S. A. Shayler, S. Mather, I. Black, K. Baker & D. M. Paterson, 1999. Measuring the *in situ* erosion shear stress of intertidal sediments with the Cohesive Strength Meter (CSM). Estuar. coast. shelf Sci. 49: 281–294.

Tolhurst T. J., K. S. Black, D. M. Paterson, H. Mitchener, R. Termaat & S. A. Shayler, 2000. A comparison and measurement standardisation of four *in situ* devices for determining the erosion shear stress of intertidal sediments. Conti. Shelf Res. 20: 10–11.

Tolhurst, T. J., G. Gust & D.M. Paterson, 2002. The influence of an extracellular polymeric substance (EPS) on cohesive sediment stability. In Winterwerp, J. C. & C. Kranenburg (eds), Fine Sediment Dynamics in the Marine Environment. Proc. Mar. Sci. 5: 409–425.

Underwood, G. J. C. & D. M. Paterson, 1993. Recovery of intertidal benthic diatoms after biocide treatment and associated sediment dynamics. J. mar. biol. Ass. U. K. 73: 25–45.

Underwood, G. J. C. & J. Kromkamp, 1999.Primary production by phytoplankton and microphytobenthos in Estuaries. Adv. Ecol. Res. 29: 93–153.

Underwood, G. J. C., D. M. Paterson & R. J. Parkes, 1995. The measurement of microbial carbohydrate exoploymers from intertidal sediments. Limnol. Oceanogr. 40: 1243–1253.

Hydrobiologia **503:** 195–209, 2003.
M.B. Jones, A. Ingólfsson, E. Ólafsson, G.V. Helgason, K. Gunnarsson & J. Svavarsson (eds), *Migrations and Dispersal of Marine Organisms.*

Life-cycle strategies and seasonal migrations of oceanic copepods in the Irminger Sea

Astthor Gislason
Marine Research Institute, Skúlagata 4, 101 Reykjavik, Iceland
E-mail: astthor@hafro.is

Key words: copepods, egg production, Irminger Sea, life-cycle, ontogenetic migration, vertical distribution

Abstract

Abundance and seasonal vertical distribution of dominant zooplankters in the Irminger Sea was studied from data collected during four cruises between November 1996 and June 1997. In addition, egg production of *Calanus finmarchicus* was measured during winter, spring and summer 1996–2001. Five taxa constituted >95% of the copepod biomass, *C. finmarchicus*, *Pareuchaeta norvegica*, *C. hyperboreus*, *Oithona* spp. and *Oncaea* spp. A seasonal migration pattern was evident for *C. finmarchicus*, *P. norvegica* and *Oithona* spp.: from December to February, they inhabited the deeper layers, whereas, from April to June, they were most abundant in the upper layers. *Oncaea* spp. also stayed deep during winter and only a very limited part of the population rose to the surface during summer. *C. hyperboreus* remained deep from December to April, but had virtually disappeared in June. Reproduction of *C. finmarchicus* took place in May in the surface layers and was linked to the phytoplankton spring bloom. In contrast, reproduction of *P. norvegica* occurred at depth in February and was uncoupled with the spring bloom. *C. hyperboreus* did not reproduce in the Irminger Sea. Data on *Oithona* spp. and *Oncaea* spp. indicated that the former reproduced between April and June in the upper layers, whereas the latter reproduced year-round at depth. Thus, data on vertical distribution and seasonal stage composition suggested that the dominant copepods are separated, at least partly, at spatial and temporal scales with regard to overwintering and development.

Introduction

The zooplankton community of the Irminger Sea is characterized by low diversity and few species (Beaugrand et al., 2000). Yet, there are several indications that the biological productivity of the area is high. Thus, the Irminger Sea serves as nursery and feeding areas for the large oceanic redfish stock (*Sebastes mentella* Travin) which, with an average stock size of around 2 million tonnes, is utilized by several nations and supports a total fishery of about 110 000 tonnes (ICES, 1999). Redfish is a pelagic species and zooplankton is an important part of its diet, especially for larvae that feed almost exclusively on eggs and young stages of *Calanus finmarchicus* (Gunnerus) (Einarsson, 1960; Bainbridge & McKay, 1968; Magnússon & Magnússon, 1995). A further indication of the biological productivity of the Irminger Sea is the so-called deep scattering layers, a characteristic feature of the Irminger Sea occurring in various intensities throughout a vast area (Magnússon, 1996; Sigurdsson et al., 2002). A great variety of organisms is found in these layers but, excluding the oceanic redfish, the major components are myctophids (several species), viperfish (*Chauliodus sloani* Schneider), jellyfish (several species), cephalopods (several species) and euphausiids (mainly *Meganyctiphanes norvegica* [M. Sars]) (Magnússon, 1996; Sigurdsson et al., 2002). Biomass of organisms found in these layers is unknown but is probably very high (Magnússon, 1996). The Irminger Sea is also important as a feeding area for several baleen whale species (for instance the fin whale and the sei whale) (Sigurjónsson, 1995). Every spring and summer, whales undertake feeding migrations to the area from the wintering/breeding grounds at lower latitudes. During the summer feeding period, zooplankton and small fish form an important component of the food (Sigurjónsson & Víkingsson, 1998). Taken together, data suggest that the secondary productivity of the Irminger Sea is high. With this in mind, it is clearly of interest to study the seasonal dynamics and life-cycle strategies of zooplankton in the area.

The first systematic study of the zooplankton community in the Irminger Sea was made by Hermann Einarsson (1960), who characterized the plankton community of the central Irminger Sea as one dominated by the copepod *C. finmarchicus* and the euphausiid *Thysanoessa longicaudata* (Kröyer). Subsequent studies, the so-called NORWESTLANT surveys, were confined mainly to records of occurrence and distribution (Bainbridge & Corlett, 1968), while Matthews (1968) used data collected during the NORWESTLANT surveys to study the life-cycle of *C. finmarchicus*. However, all these studies were limited as they did not cover the whole water column and were confined mainly to the spring period.

In the present study, I report the seasonal vertical distribution and life-cycle strategies of the dominant copepods in the Irminger Sea, based mainly on data obtained during two cruises in winter, one in spring and one in summer.

Materials and methods

Samples were collected at one station located in the northern part of the Irminger Sea (64° 20′ N, 32° 00′ W, bottom depth ~2500 m, Fig. 1) on the following dates: 4 December 1996, 4 February 1997, 8 April 1997 and 24 June 1997. In February, the station had to be moved 1° eastwards (to 64° 20′ N, 31° 00′ W) because of ice. In addition, egg production rates of *Calanus finmarchicus* were determined at several stations in the northern and western parts of the Irminger Sea during spring and summer 1996–2001 (Fig. 5).

Temperature and salinity were recorded with a CTD (Sea Bird Electronics SBE-9). Seawater samples (1–2 l) for the measurement of chlorophyll *a* were collected from depths of 0, 10, 20, 30 and 50 m, and filtered through GF/C glass fibre filters. Filters were then homogenized in 90% aqueous acetone and the extract measured in a spectrophotometer according to the method described by Strickland & Parsons (1968).

Zooplankton were collected using a Multi Plankton Sampler from HydroBios (0.25 m^2 mouth area, 200 μm mesh size) towed from near the bottom to the surface with a speed of ~1 m s^{-1}. The Multi Plankton Sampler was equipped with five nets that can be opened and closed on command from the ship. By deploying the sampler three times, a total of 13 depth layers was sampled from 2300 m to the surface. The volume of water filtered by the net was measured with HydroBios flowmeters fitted in the mouth of each net.

Zooplankton samples were preserved in 4% neutralized formalin until analysis in the laboratory. As a rule, the entire sample was counted for larger plankton (euphausiids, amphipods, mysids, decapods, chaetognaths and copepods >1 mm prosome length). Except when the sample was very small, the remainder was subsampled with a Motoda splitter (Motoda, 1959) and an aliquot, containing at least 500 individuals, analysed for species composition. Further, the copepods *C. finmarchicus*, *C. hyperboreus* Kröyer and *Pareuchaeta norvegica* (Boeck) were classed to developmental stages. For *P. norvegica*, counts of males and females with spermatophores and/or egg sacs attached to the genital segment were made. A few isolated egg sacs of *P. norvegica* were also observed in the samples. Probably, they became detached during sampling (Bakke, 1977; Hopkins, 1982). The number of isolated egg sacs in the samples was, therefore, taken into account when estimating the number of ovigerous females.

Data on abundance (numbers per m^{-3}) were calculated for each net using the data from the flowmeters. Number per m^{-2} was obtained for each net by multiplying the abundance (number per m^{-3}) by the depth range for that net. Total water column abundances (number per m^{-2}) were calculated from the sum of the abundances (number per m^{-2}) for all nets in a vertically-stratified tow.

Numbers were converted to biomass using stage-specific dry weights from various sources. For *C. finmarchicus, P. norvegica, Metridia lucens* Boeck, and *Pleuromamma robusta* (Dahl), data from Longhurst & Williams (1992) obtained at Ocean Weather Station I were adopted. The biomass of *C. hyperboreus* and *C. glacialis* Jaschnov was calculated using data from Hirche (1997) and Slagstad & Tande (1990), respectively. For *M. longa* (Lubbock), the dry weight of the copepodite stage C5 and adults was calculated using data from Grönvik & Hopkins (1984), while the dry weight for younger copepodite stages was estimated by assuming the same ratio between the biomass of C5 and a particular copepodite stage as for *M. lucens* (Longhurst & Williams, 1992). Biomass of *Oithona* spp. was calculated using information from Tande (1991). *Oithona* spp. was taken to be 50% adults and 50% copepodite stages C5, as the younger copepodite

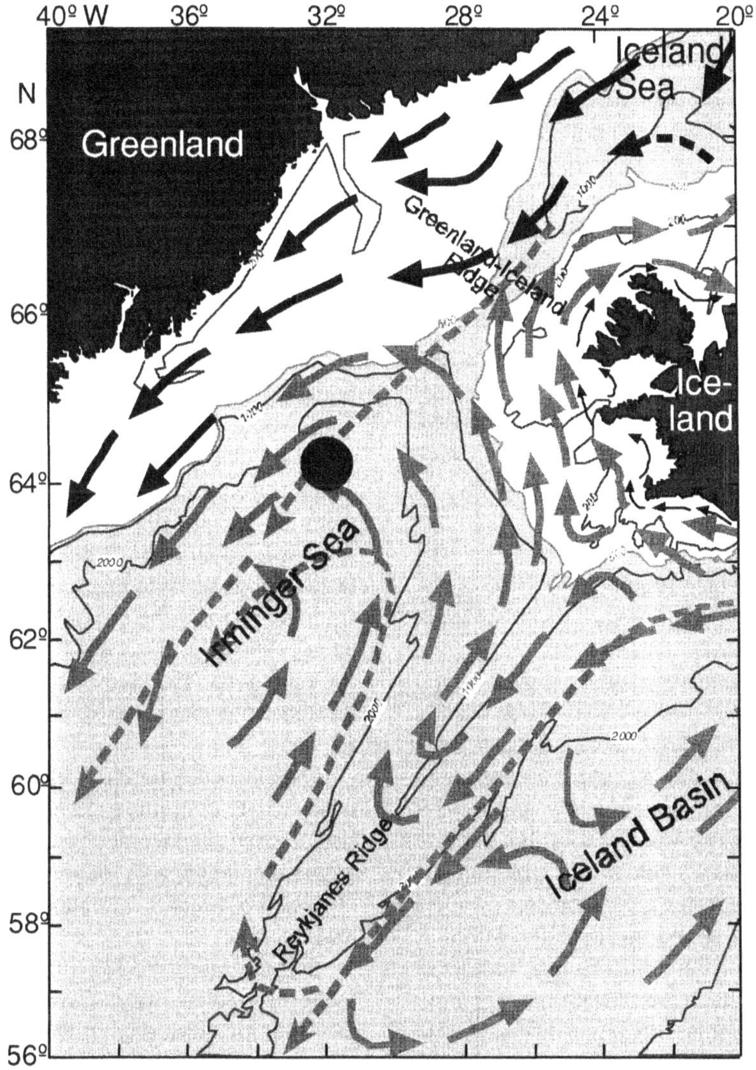

Figure 1. Map of the sea area around Iceland showing the sampling station in the Irminger Sea. Inserted on the figure are the main ocean currents in the upper layers, adapted from Valdimarsson & Malmberg (1999), and the main overflow paths over the Greenland–Scotland Ridge, adapted from Hansen & Østerhus (2000). Grey arrows=Atlantic Water; black arrows=Polar Water; black broken arrow=mixed water; grey broken arrows=main overflow paths.

stages were probably greatly undersampled by the 200 μm mesh nets (Gallienne & Robins, 2001). The dry weight of *Oncaea* spp. was assumed to be the same as for *Oithona* spp. *Pseudocalanus* spp. was taken as 50% adults and 50% copepodite stages C1–5, and the dry weight calculated using information from Davis (1984), assuming 40% of the dry weight as carbon content. The dry weights of *Scolecithricella minor* (Brady) and *Spinocalanus* spp. were assumed to be the same as that of the similarly-sized *Pseudocalanus* spp. Biomass of *Microcalanus* spp. was calculated using information from Tande (1991), assuming the stock consisted of 50% adults and 50% juveniles. Other copepod species were of minor importance in terms of biomass and were excluded from calculations. These calculations provide only crude estimates of biomass but can, at least, be considered conservative values of copepod biomass. No attempt was made to estimate the biomass of other zooplankton groups.

When interpreting the present data, it should be borne in mind that some smaller zooplankton components will inevitably escape through the 200 μm mesh nets. It has been pointed out that the breadth of the zooplankters determines their retainment by plankton nets (Colton et al., 1980; Nichols & Thompson, 1991) and, therefore, the zooplankters, whose breadth is less than the mesh size (200 μm), will be underrepresented in the samples. Accordingly, the younger stages of some copepods caught in the present investigation (*Pseudocalanus* spp., *Scolecithricella minor* and *Spinocalanus* spp.) have been undersampled. For the smallest copepods (*Oithona* spp., *Oncaea* spp. and *Microcalanus* spp.), it is likely that adults were underestimated also. However, in spite of these limitations, it is likely that present material demonstrates the main seasonal changes in the copepod community in the Irminger Sea.

To facilitate comparison of the depth distributions of the dominant copepods, the weighted mean depth (WMD) was calculated according to the equation:

$$WMD = \Sigma(n_i d_i)/\Sigma n_i,$$

where n_i is the abundance (number m^{-3}) of copepods at depth d_i (the midpoint of each depth stratum) (Bollens & Frost, 1989).

To measure egg production of *C. finmarchicus*, separate vertical hauls (0–100 m) were made with a WP-2 net (0.25 m^2 mouth area, 200 μm mesh size) fitted with a nonfiltering codend. On retrieval, contents of the codend were emptied gently into a ~15 l bucket containing surface water. From this bucket, samples were immediately taken with a large pipette and transferred to Petri dishes where females in good condition were quickly sorted out (usually within 1 h after capture) using a stereomicroscope with cold fibre optic light. Females were placed into egg production cylinders of plexiglas (65 mm in diameter, 180 mm in height) with a funnel at the bottom into which the eggs settled. A 330 μm net separated the cylinder from the funnel, which prevented *C. finmarchicus* from eating its own eggs. One female was added to each cylinder and, usually, 20 replicates were done per station. However, due to the paucity of females, experiments were run with fewer animals during the winter cruises (November–February). Egg production chambers were incubated for 24 h in the dark at ambient surface water temperatures. After incubation, eggs were filtered onto a 20 μm screen on which they were counted immediately under a stereomicroscope.

Runge & Plourde (1996) suggested that cannibalism may occur in egg production chambers even if these are fitted with nets to separate eggs and females; however, they also noted that this source of error was most apparent when groups of females were incubated together. As individual females were always incubated in this study and, as experiments were run in a relatively large volume of water, I do not consider that cannibalism will influence the results presented here.

Results

Hydrography

Figure 1 illustrates the basic current pattern around Iceland and in the Irminger Sea. Iceland lies at the boundary of warm and cold currents. The warm Atlantic Water comes from the south and southwest, and the cold Polar and Arctic Water from the north. Deep-water currents transport water into the Irminger Sea from the Iceland Sea and the Iceland Basin (Fig. 1). The circulation in the Irminger Sea is part of the so-called Subpolar Gyre that unites the circulation in the Irminger Sea and the Labrador Sea (Aksnes & Blindheim, 1996). In the central Irminger Sea, there is a cyclonic gyre. The study site of the present study is located somewhat north of the centre of this gyre (Fig. 1).

Over the study period, surface temperatures ranged from ~5 °C (December 1996) to ~7 °C (June 1997) (Fig. 2). Seasonal temperature changes decreased with increasing depth and, below 1000 m, temperatures remained between ~4 °C and ~5 °C throughout the study period.

Surface salinity ranged from ~34.86 (December 1996) to ~35.09 (June 1997) (Fig. 2). The low surface salinities observed in December 1996 (<34.88), reflected ice melting in nearby regions. Salinity below 1000 m depth changed only slightly during the study period (~34.86–34.90).

Based on salinity and temperature measurements, different water masses may be identified (Blindheim et al., 1996; Mortensen et al., 1999). In May and June, Atlantic Water (AW, T~5–7 °C, S>35.00) was evident, mainly above 500 m depth. Throughout the study period, the so-called Labrador Sea Water (LSW, T~3 °C, S<34.88), which originates by deep convection in the Labrador Sea, was evident at ~1500–2000 m depth. In February and March, traces of Denmark Strait Overflow Water (DSOW, T~2.9–3 °C, S>34.88) were found near the bottom. The DSOW overflows the Greenland–Iceland Ridge from the Iceland Sea, and then flows as a bottom current southwards in the Irminger Sea (Fig. 1). Cooling of the AW, and its mixing with the deeper water masses, leads to

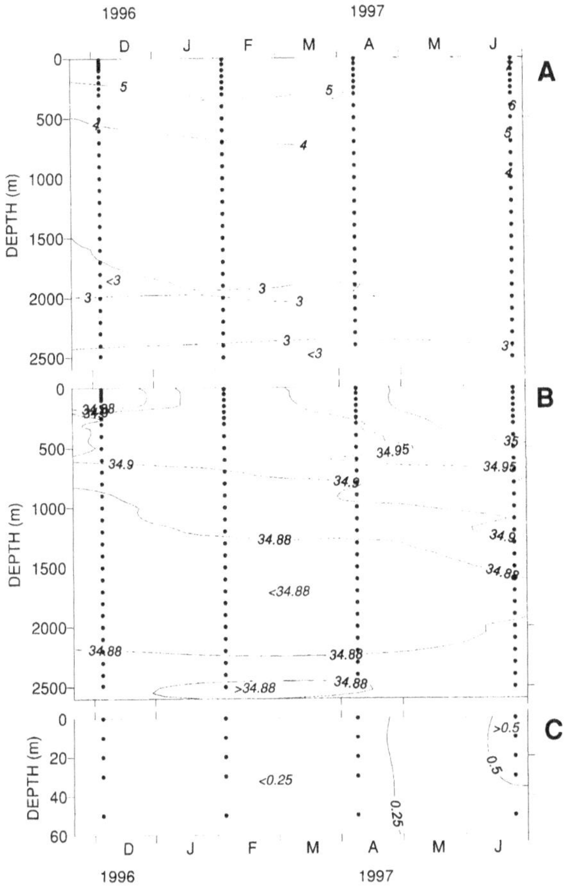

Figure 2. Seasonal variation in temperature (A), salinity (B), and chlorophyll *a* (C) at sampling station in the Irminger Sea from December 1996 to June 1997.

the formation of the so-called Intermediate Water (IW, T~3–5 °C, S~34.88–35.00).

Phytoplankton

During winter (December 1996–February 1997), the concentration of chlorophyll *a* was negligible (<0.08 mg Chl *a* m^{-3}) (Fig. 2C). In April–May, as the surface waters began to warm (Fig. 2A), some very limited phytoplankton growth began (>0.25 mg Chl *a* m^{-3}) (Fig. 2C). In June, the stratification was stronger (Fig. 2A) and the biomass of phytoplankton was somewhat higher, while still very low (>0.5 mg Chl *a* m^{-3}) (Fig. 2C).

Abundance and composition of zooplankton

Copepods were by far the most dominant group in the samples in terms of numbers. During all sampling occasions, they represented >90% of all the animals collected; chaetognaths were the second most abundant group (~1.5%).

Among the copepods, *Calanus finmarchicus* was the dominant copepod (~79%) with *Pareuchaeta norvegica* and *C. hyperboreus* also important, making up ~10% and ~8% of the copepod biomass, respectively (Fig. 3A,C). In terms of numbers, *Oithona* spp. dominated (~39%), whereas *C. finmarchicus* was the second most numerous species (~30%) and *Oncaea* spp. ranked third (~16%) (Fig. 3B,D).

During winter (December–February), the bulk of the copepod biomass was located at 200–1600 m depth in a homogenous environment with respect to both temperature (~3–5 °C) and salinity (~34.85–34.94) (Fig. 4). In spring (April), most of the biomass occurred above 6–800 m depth (T~4–5 °C, S~34.94–34.98), whereas in June, most of the biomass was found in the top 100 m (T~7.5 °C, S~35.10).

Calanus finmarchicus

As Gislason & Astthorsson (2000) reported recently on the biology of *C. finmarchicus* in the oceanic waters southwest of Iceland, using partly the same material dealt with here, I only submit the most important findings in the present paper.

Gislason & Astthorsson (2000) found that, during winter (December–February), *C. finmarchicus* in the Irminger Sea was distributed mainly from ~200 to 1600 m. Ascent to the surface layers had started in April, when the bulk of the stock was found above 800 m depth. The spring ascent continued in May and, by June, the animals were mostly concentrated in the uppermost 100 m. During winter (December–February), the majority of the stock was present as C5 (~90%), with a smaller proportion as C4 (~6%) and females (~3%). In April, adults were prominent in the samples, both females (~23%) and males (~21%), whereas in June, the youngest copepodite stages (C1–3) from the spring spawning made up some 22% of the population, the remainder being manly C4 (~33%) and C5 (~39%).

Data on daily egg production rates of *C. finmarchicus* are shown in Figure 5. During winter, the few females that were found in the uppermost 100 m were immature and did not reproduce (Fig. 5A). In April, when a significant part of the population had emerged (Gislason & Astthorsson, 2000), daily rates of egg production were low (<5 eggs $female^{-1}$ d^{-1}, Fig. 5B). In May, daily rates of egg production rates were highest (>40 eggs $female^{-1}$ d^{-1}, Fig. 5C), whereas in June,

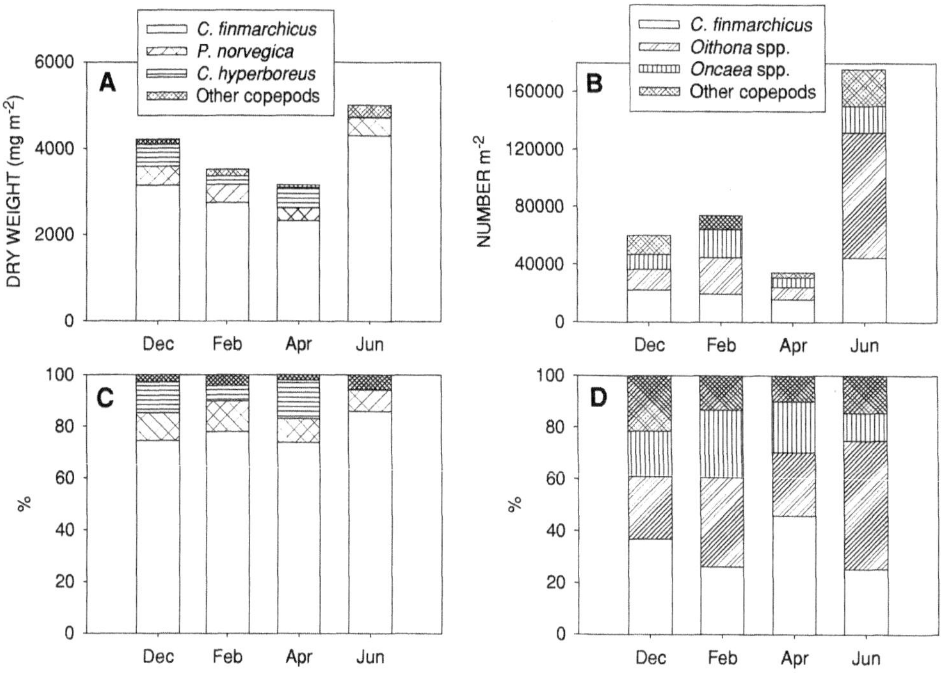

Figure 3. Total depth integrated (0–2300 m) biomass (A), number (B), relative biomass (C), and relative number (D) of copepods at sampling station in the Irminger Sea from December 1996 to June 1997.

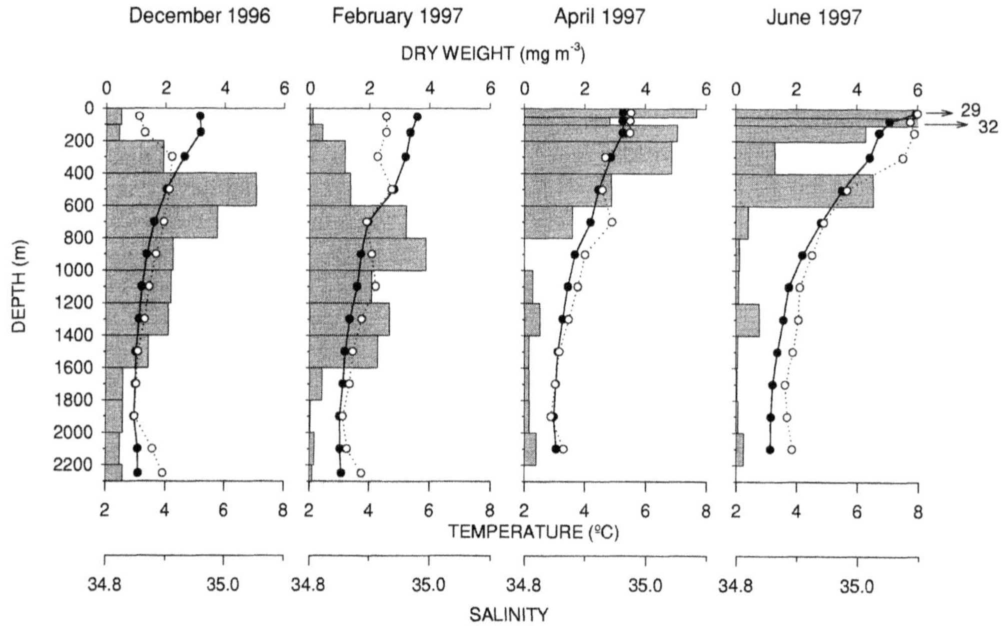

Figure 4. Vertical distribution of copepod biomass at sampling station in the Irminger Sea from December 1996 to June 1997. Profiles for temperature (filled circles) and salinity (open circles) obtained by simultaneous CTD casts are also shown.

daily egg production rates in the oceanic Irminger Sea had decreased again (<20 eggs female^{-1} d^{-1}, Fig. 5D).

On the basis of the seasonal vertical distribution and stage composition (Gislason & Astthorsson 2000), and daily egg reproduction rates (Fig. 5), it is concluded that *C. finmarchicus* in the Irminger Sea repro-

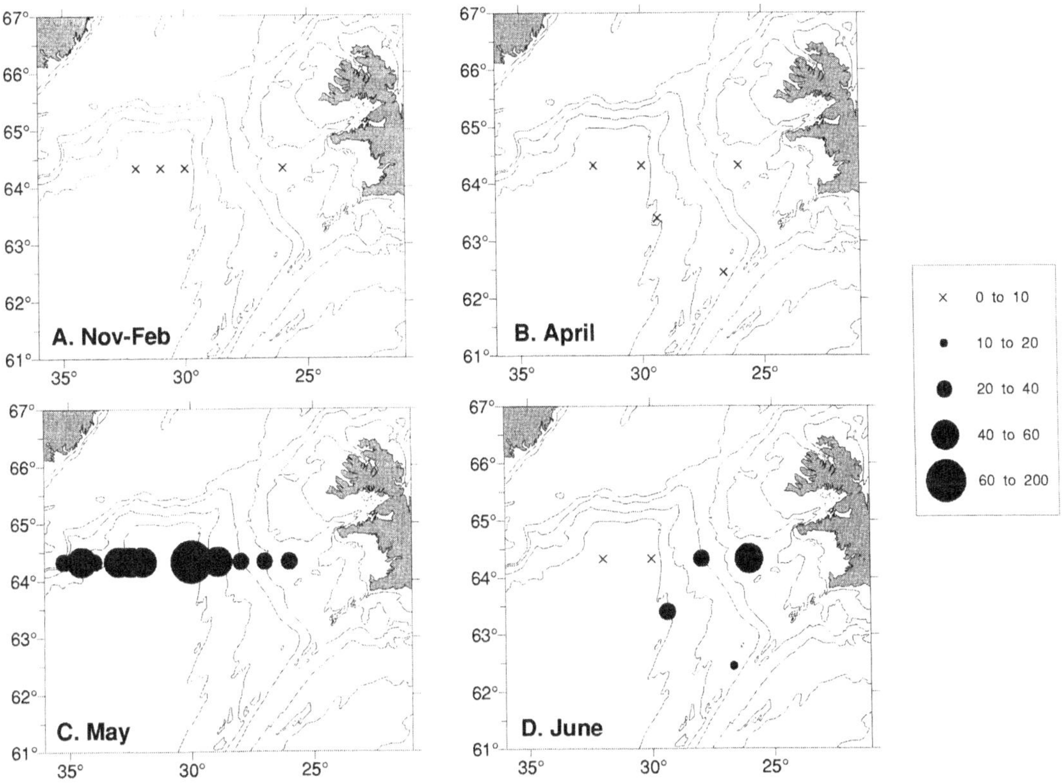

Figure 5. Calanus finmarchicus. Egg production rates (eggs female^{-1} day^{-1}) in the Irminger Sea; November 1996 and February 1997 (A), April 1997 (B), May 2000, 2001 (C), and June 1997 (D).

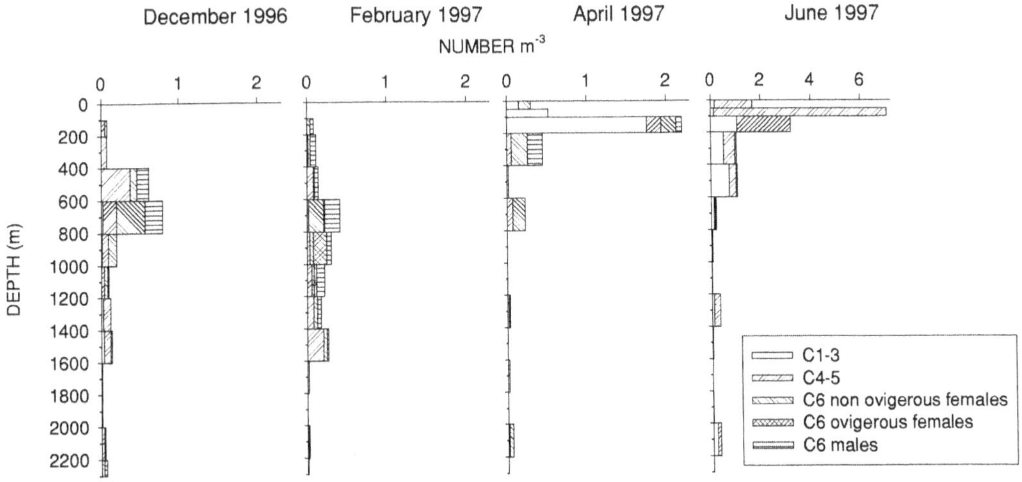

Figure 6. Pareuchaeta norvegica. Vertical distribution at sampling station in the Irminger Sea from December 1996 to June 1997. Note the change in abundance scales between April and June.

duces mainly in the upper layers of the ocean (0–100 m), with peak reproduction in May.

Pareuchaeta norvegica

Pareuchaeta norvegica ranked second in biomass and fourth in copepod numbers (Fig. 3). During winter, the main part of the population was found between 400 and 1600 m depth (Fig. 6). Ascent to the upper layers occurred sometime between February and April, when the highest concentrations were observed in the 100–200 m depth interval. In June, concentrations were highest above 200 m depth.

During winter and spring, the depth-integrated biomass of *P. norvegica* decreased steadily from ~450 mg dry weight m^{-2} in December to ~300 mg dry weight m^{-2} in April (Fig. 3A), while numbers fluctuated between ~350 and ~400 individuals m^{-2} (Fig. 7A). After April, biomass and numbers increased to a maximum in June (~400 mg dry weight m^{-2} and ~1300 individuals m^{-2}, respectively).

From December to February, the proportion of stages C4–5 decreased while that of adults increased, suggesting that juveniles were developing into adults during this time (Fig. 7B). A high proportion of males in the population indicated breeding activity (Mauchline, 1994) and this ratio was highest in February (~30%, Fig. 7B). The frequency of ovigerous females was also highest in February (~15%, Fig. 7B), and few males and females with spermatophores were observed in samples during this time. Taken together, this suggests that the main reproductive activity occurred around February. Young developmental stages from this reproduction appeared in April (Fig. 7A,B), when the population became dominated by juvenile stages C1–3 (~50%). In June, the percentage of stages C4–5 was highest (~65%), suggesting that offspring from the winter spawning were recruited to the older stages (Fig. 7A,B).

On the basis of the description above, it is concluded that *P. norvegica* reproduced mainly at depth (~400–1600 m) in February, and nauplii and/or youngest copepodite stages swim or drift up to the surface layer after the eggs hatch. It is suggested further that growth to the more advanced copepodite stages (C4–5) takes place in the surface layers, mainly between April and June.

Calanus hyperboreus

Calanus hyperboreus was the third most important copepod species in the Irminger Sea in terms of biomass but fifth in terms of numbers (Fig. 3). From December to April, almost the whole population stayed below 400 m depth (Fig. 8). In June, the whole stock had virtually disappeared.

During winter and spring, biomass fluctuated between ~200 and ~500 mg dry weight m^{-2} (Fig. 3), and numbers between ~130 and ~300 individuals m^{-2} (Fig. 9A). The stage composition showed little temporal trend (Fig. 9B). The youngest copepodite stages (C1 and C2) were never found.

From these data it is concluded that *C. hyperboreus* does not reproduce in the Irminger Sea.

Oithona *spp.*

Two species belonging to genus *Oithona* (*O. similis* Claus and *O. spinirostris* Claus) were found in the samples; *O. similis* was more numerous, while *O. spinirostris* was also regularly found. As they were usually counted as belonging to one group, they are treated together in the present analysis. *Oithona* spp. was the most abundant taxon encountered during this study, contributing ~39% of copepod numbers (Fig. 3D), however, due to small size, it only contributed ~2% of the biomass.

From December to April, the population was found mainly between the surface and 800 m depth (Fig. 10A). In June, the bulk of the population aggregated in the top 100 m. During winter and spring, the depth-integrated biomass fluctuated between ~17 and ~50 mg dry weight m^{-2}, and numbers between ~8000 and ~25 000 individuals m^{-2} (Fig. 3A,B). After April, numbers and biomass increased markedly to a maximum in June (~170 mg dry weight m^{-2} and ~87 000 individuals m^{-2}, respectively).

As the developmental stages of *Oithona* spp. were not separated, it is difficult to evaluate to what extent the increase after April reflects reproduction. However, since the seasonal pattern is very clear, with low values during the winter and one main increase from April to June (Figs 3B and 10A), most likely the main breeding activity occurred sometime between April and June. As judged by the vertical distribution (Fig. 10A), spawning probably took place in the upper layers.

Oncaea *spp.*

Oncaea spp. comprised two species (*O. borealis* G. O. Sars and *O. conifera* Giesbrecht) of which, the former was more numerous. For the same reasons as with *Oithona* spp., they were treated together in the present analysis.

Animals belonging to the genus *Oncaea* were found mainly between 400 and 1600 m depth during winter (Fig. 10B). A small part of the population was found in relatively shallow waters during spring and summer (0–200 m), while the main part remained at depth. The depth-integrated abundance values fluctuated irregularly between ~13 and ~38 mg dry weight

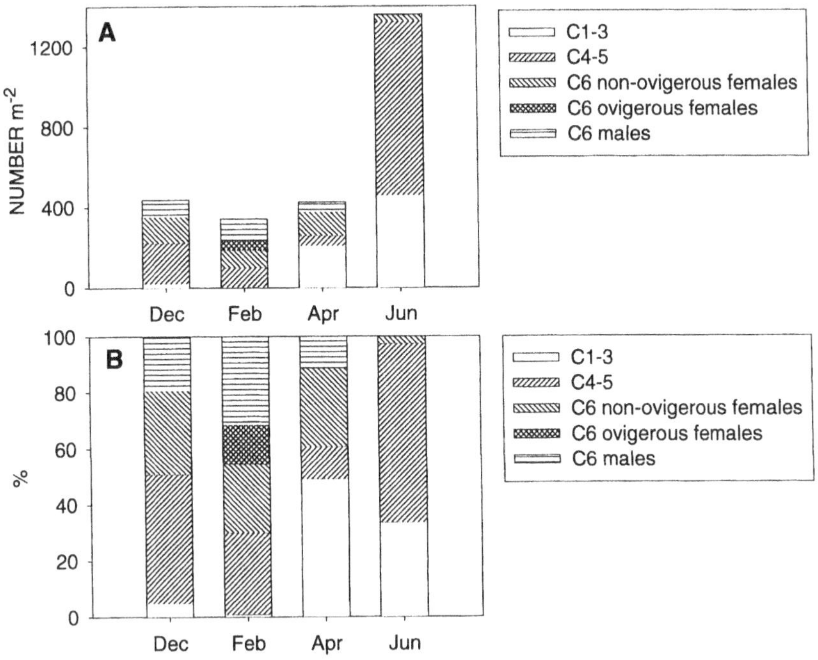

Figure 7. Pareuchaeta norvegica. Total depth integrated (0–2300 m) number (A) and relative number (B) of copepodite stages at sampling station in the Irminger Sea from December 1996 to June 1997.

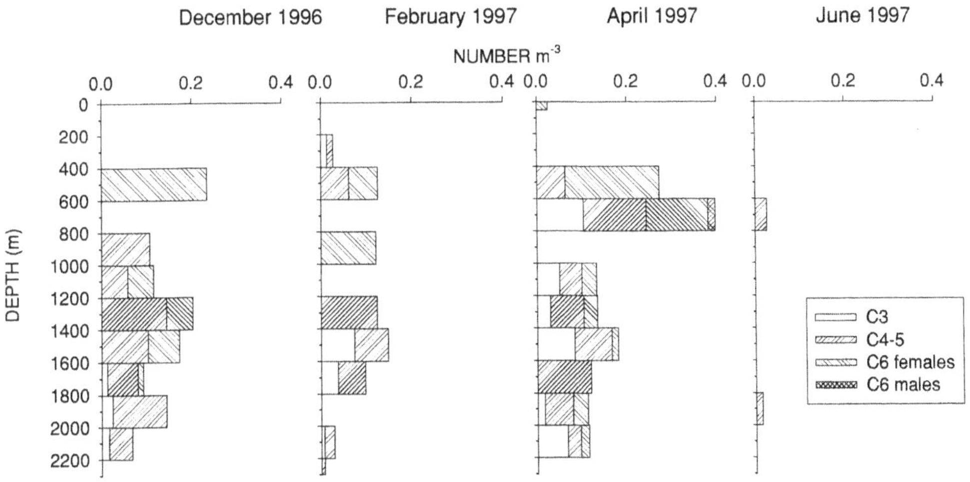

Figure 8. Calanus hyperboreus. Vertical distribution at sampling station in the Irminger Sea from December 1996 to June 1997.

m^{-2} (biomass), and ~7000 and ~19 000 individuals m^{-2} (numbers) (Fig. 3A,B).

As with *Oithona* spp., the stages of *Oncaea* spp. were not separated and, therefore, it is difficult to evaluate when reproduction took place. However, the fact that abundance was rather stable during the study period (Figs 3D and 10B), suggests that animals were reproducing continuously throughout the year.

Discussion

Very low chlorophyll values observed from April to June suggest that the main phytoplankton growth period had not started in June (Fig. 2). However, the broad spacing between sampling dates should be kept in mind when interpreting these data with respect to the dynamics of the phytoplankton. Thus, it is likely that, due to the long time interval between sampling,

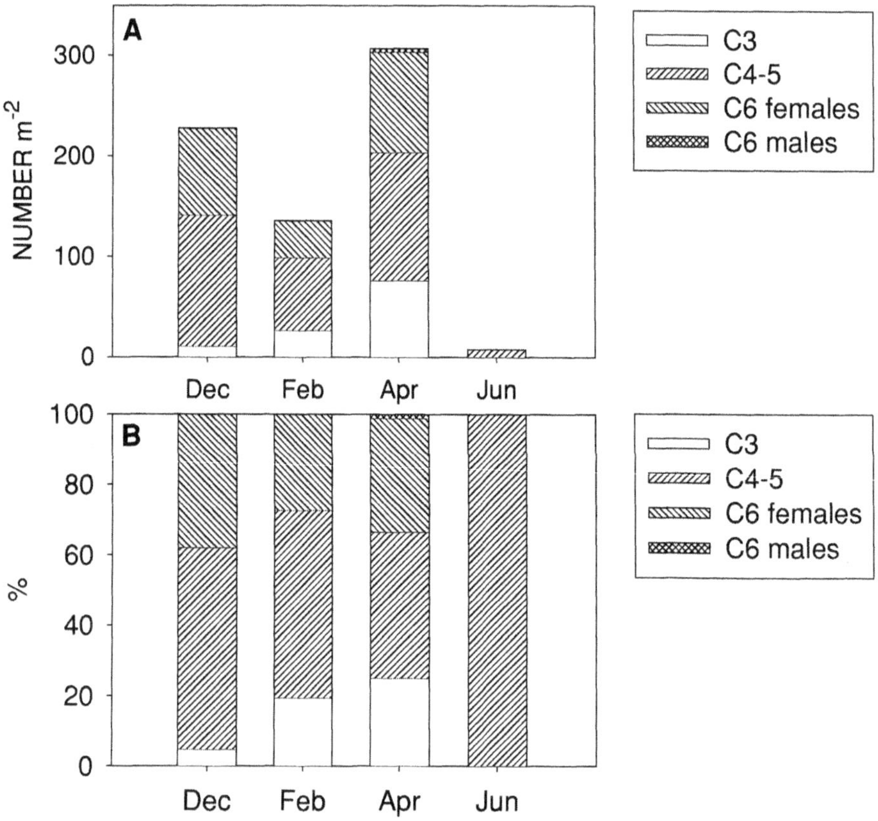

Figure 9. Calanus hyperboreus. Total depth integrated (0–2300 m) number (A) and relative number (B) of copepodite stages at sampling station in the Irminger Sea from December 1996 to June 1997.

the spring bloom was not detected in this study. Results from the Icelandic spring survey in 1997 showed a vigorous phytoplankton growth in the oceanic areas west of Iceland in mid May 1997 (Anonymous, 1999). Based on this, and the low chlorophyll values observed in April and June (Fig. 2), it is concluded that the spring bloom at the study site started in April, culminated in May and declined in June. This is supported by long-term studies showing that, in the offshore waters west of Iceland, the spring bloom generally starts in late April and culminates in mid May (Gudmundsson, 1998).

In the present material, copepods were by far the most dominant group, accounting for 90% or more of the plankton, which is in accordance with earlier zooplankton investigations in the Irminger Sea (Einarsson, 1960; Bainbridge & Corlett, 1968). The most numerous copepods of the present study were also reported among the most abundant ones in earlier investigations of the Irminger Sea (Einarsson, 1960; Bainbridge & Corlett, 1968). Similarly, from the northern North Atlantic Ocean (Ocean Weather Station I), Williams (1988) reported zooplankton to be dominated by copepods, with *Calanus finmarchicus* as the single-most abundant species. In the Labrador Sea (Sameoto, 1984), the Iceland Sea (Gislason & Astthorsson, 1998) and the Norwegian Sea (Østvedt, 1955), copepods were also the dominating group.

The dominant copepods exhibited clear differences in seasonal abundance, vertical distribution and age structure, reflecting differences between the life histories of these species. All species stayed deep during winter and some rose to the surface during spring (*C. finmarchicus, Pareuchaeta norvegica, Oithona* spp.), while others did not (*Oncaea* spp., *C. hyperboreus*) (Fig. 11).

In the Irminger Sea, *C. finmarchicus* overwinters in deep water and rises to the surface mainly during March and April (Gislason & Astthorsson, 2000). This is in accord with the general behaviour of this species in other regions, with arrested development and diapause at depth during the winter, and termin-

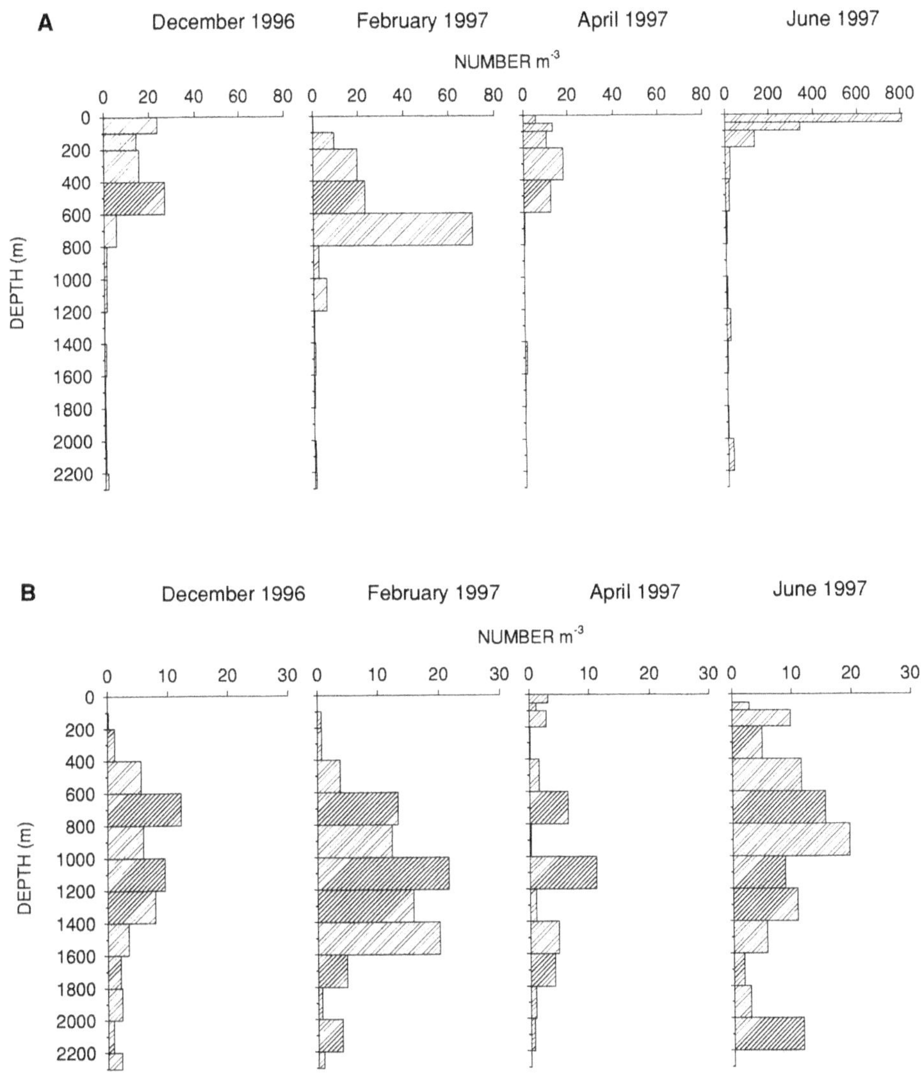

Figure 10. Vertical distribution of *Oithona* spp. (A), and *Oncaea* spp. (B) at sampling station in the Irminger Sea from December 1996 to June 1997. Note the change in abundance scales between April and June in (A).

ation of diapause and spring ascent in late winter and spring (see Hirche, 1996a for review). Reproduction took place mainly in May in the surface layers at similar time as the phytoplankton bloom (see discussion above and Fig. 5). The close coupling between the spawning of *C. finmarchicus* and phytoplankton growth has been documented previously by several workers (see Hirche, 1996b for review).

P. norvegica also stayed at depth during winter, rising to the upper layers during spring (Figs 6 and 11). However, in contrast to *C. finmarchicus*, spawning of *P. norvegica* was uncoupled to the spring bloom. Liberation of nauplii took place in the deeper layers, probably mainly in February and March, and early copepodites drifted or migrated to the upper layers in March and April to utilize the biological production there. *P. norvegica* is a tactile predator and *Calanus* is an important part of its diet (Auel, 1999). It is, therefore, likely that these copepods, even though they spawn at different times and in different habitats, are involved in a strong predator–prey relationship. This would apply, especially during spring and summer, when their vertical distributions generally overlap and both species are active (Fig. 11). During winter, in spite of their similar vertical distributions, the immobility of the overwintering *C. finmarchicus* may

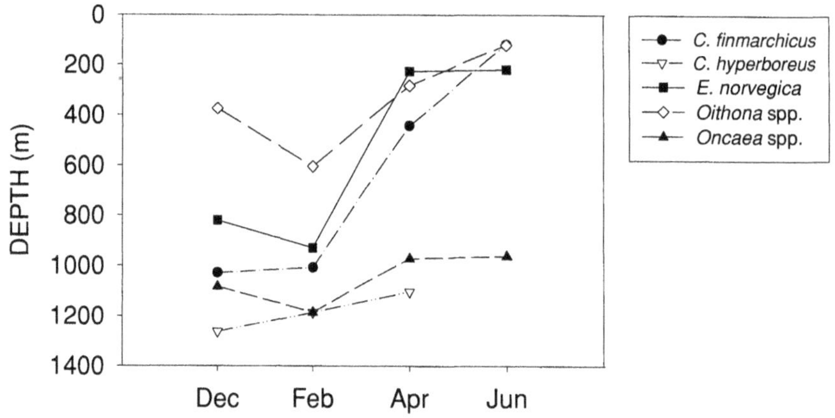

Figure 11. Weighted mean depth of *Euchaeta norvegica, Calanus hyperboreus, Oithona* spp. and *Oncaea* spp. at sampling station in the Irminger Sea from December 1996 to June 1997. The weighted mean depth of *C. finmarchicus* from a wider area of the Irminger Sea (Gislason & Astthorsson, 2000) is shown for comparison.

reduce the probability of encounter with *P. norvegica* (Hairston, 1987). Similar to our findings, Williams (1988) and Gislason & Astthorsson (1992) reported *P. norvegica* in the Iceland Basin south of Iceland to reproduce at depth in late winter prior to the spring bloom. This was reported for this species by Mauchline (1994) from the Rockall Through, farther south in the North Atlantic Ocean, and by Østvedt (1955) from the Norwegian Sea.

Oithona spp. had a similar ontogenetic migration pattern to *C. finmarchicus* and *P. norvegica*, staying deep during winter and rising to the surface in spring (Fig. 11). However, *Oithona* spp. had a shallower winter distribution than *C. finmarchicus* and *P. norvegica* (Fig. 11). As with *C. finmarchicus*, the main growth of *Oithona* spp. took place probably in the surface layer in May and June. However, as *Oithona* spp. is mainly omnivorous (Sabatini & Kiörboe, 1994; Svensen & Kiörboe, 2000) while *C. finmarchicus* is essentially herbivorous (Irigoien et al., 1998; Mayer-Harms et al., 1999), although non-algal components such as microzooplankton may complement phytoplankton in the diet under certain conditions (Ohman & Runge, 1994; Harris, 1996; Harris et al., 2000), they were probably not competing for food. *O. similis* (the most abundant of the two *Oithona* species of the present study) may be the most abundant and ubiquitous copepod in the world's oceans (Sabatini & Kiörboe, 1994; Svensen & Kiörboe, 2000; Gallienne & Robins, 2001), yet, very little information exists on its seasonal depth distribution and life-cycle strategy. From the Greenland Sea, Richter (1994) reported *Oithona* spp. as a mesopelagic species (0–500 m) with the main breeding period in early summer in the surface layer (0–100 m). This is similar to present observations in the Irminger Sea.

Most *Oncaea* spp. stayed below 600 m depth during the whole study period, and only a limited part of the population rose to the surface during summer (Figs 10B and 11). The biology of *Oncaea* spp. is poorly understood but, like *Oithona* spp., it is probably mainly omnivorous (Paffenhöfer, 1993). As with *Oithona* spp., very little is known about its life-cycle. In the Greenland Sea, *Oncaea* spp. maintained high stocks in deep waters year-round and the breeding season was more extended than that of *Oithona* spp. (Richter, 1994). Present data from the Irminger Sea, with limited seasonal variations in total integrated abundance (Figs 3B and 10B), suggest that for *Oncaea* spp., in contrast to the other indigenous copepods considered here, the breeding activity is not confined to a particular season of the year.

There is limited information on the feeding habits of *Oithona* spp. and even less on those of *Oncaea* spp. From the available literature, it appears as *Oithona similis* (the most abundant of the *Oithona* species of the present investigation) prefer motile food (Lampitt, 1978; Turner & Granéli, 1992) whereas species belonging to genus *Oncaea* feed on aggregates and detritus (Turner, 1984; Metz, 1998; Fortier et al., 2001) although small copepods may also be taken (Hopkins et al., 1993). Diet may be a factor in the deeper depth distribution of *Oncaea* spp. compared to *Oithona* spp (Fig. 10).

C. hyperboreus appears the least adapted copepod in the Irminger Sea. Thus, it did not reproduce successfully in the region (Fig. 8). *C. hyperboreus* is known to be associated with water of Arctic origin, with a main distribution in the Labrador Sea, the northern Norwegian Sea, the Greenland Sea and the Arctic Ocean (Jaschnov, 1972; Conover, 1988; Hirche, 1991). In the Irminger Sea, it is probably an expatriate, being transported to the area either from the north by the East Greenland Current or the west by the Labrador Sea Water. Paucity of *C. hyperboreus* in samples taken in June is puzzling (Fig. 8), but may be related to its large size making it more conspicuous to visual planktivorous predators, such as myctophids and oceanic redfish.

Avoidance of visual predators may be the ultimate reason why zooplankton avoids the upper layers during winter (Hirche, 1996a; Kaartvedt, 1996, 2000; Dale et al., 1999). The abundant visual predators in the Irminger Sea are oceanic redfish and various mesopelagic fish species of the deep-scattering layers. In the Irminger Sea, the oceanic redfish is generally registered in highest abundance at a depth of 500–800 m (Sigurdsson et al., 1999; Magnússon, 2000), whereas the deep-scattering layers are located mainly from 400–500 m to 700–800 m depth (Magnússon, 1996). All the copepods considered here, except *Oithona* spp., were centred below this depth during winter (Fig. 11), suggesting that predatory threat from the visually-hunting fish may have an effect on the depth distribution of these copepods. On the other hand, the shallower depth distribution of *Oithona* spp. during winter may be related to its much smaller size and transparency, making it less susceptible to visual predation.

As stated in the Introduction, the purpose of this study was to present an overview of the seasonal vertical distribution and life-cycle strategies of the dominant copepods in the Irminger Sea. Results show that the most abundant copepods differed in migratory behaviour, life-cycles and trophic positions, which may provide a mechanism by which they share the pelagic habitat. As discussed above, predators may also shape the life history of the copepods. Our knowledge of the role of predators in the life history of copepods is, however, still limited (e.g. Dale et al., 1999), and, therefore, it is important for future studies to get a better understanding of how predators affect the structure and function of planktonic communities.

Acknowledgements

I am grateful to the crew on the RV 'Bjarni Sæmundsson' and colleagues at the Marine Research Institute for their help in sampling. I thank A. R. Bödvarsdóttir and H. Pétursdóttir for analysing the zooplankton samples, Kristín Valsdóttir for carrying out the chlorophyll measurements, and H. Valdimarsson and S.-A. Malmberg for helping with the interpretation of the physical data. I also thank two anonymous reviewers for valuable comments on an earlier version of the manuscript. The study was funded by the European commission through the TASC project (Contract No. MAS3-CT95-0039) and the Icelandic Republic Fund.

References

Aksnes, D. L. & J. Blindheim, 1996. Circulation patterns in the North Atlantic and possible impact on the population of *Calanus finmarchicus*. Ophelia 44: 7–28.

Anonymous, 1999. Þættir úr vistfræði sjávar. (Environmental conditions in Icelandic waters 1996 and 1997. In Icelandic, English summary). Hafrannsóknastofnun Fjölrit 73. 48 pp.

Auel, H., 1999. The ecology of arctic deep-sea copepods (Euchaetidae and Aetideidae). Aspects of their distribution, trophodynamics and effect on the carbon flux. Ber. Polarforsch. 319: 1–97.

Bainbridge, V. & J. McKay, 1968. The feeding of cod and redfish larvae. ICNAF Spec. Publ. 7: 187–217.

Beaugrand, G., P. C. Reid, F. Ibañez & B. Planque, 2000. Biodiversity of North Atlantic and North Sea copepods. Mar. Ecol. Prog. Ser. 204: 299–303.

Bainbridge, V. & J. Corlett, 1968. The zooplankton of the NORWESTLANT surveys. ICNAF Spec. Publ. 7: 101–122.

Bakke, J. L. W., 1977. Ecological studies on the deep-water pelagic community of Korsfjorden, western Norway. Population dynamics of *Euchaeta norvegica* (Crustacea, Copepoda) from 1971 to 1974. Sarsia 63: 49–66.

Blindheim, J., E. Buch, E. Fogelqvist, T. Tanhua & S. Østerhus, 1996. The R/V Johan Hjort 1994 NORDIC WOCE cruise: on hydrography and tracers. ICES CM 1996/O: 20. 26 pp.

Bollens, S. M. & B. W. Frost, 1989. Predator-induced diel vertical migration in a planktonic copepod. J. Plankton Res. 11: 1047–1065.

Colton, J. B., J. R. Green, R. R. Byron & J. L. Frisella, 1980. Bongo net retention rates as effected by towing speed and mesh size. Can. J. Fish. aquat. Sci. 37: 606–623.

Conover, R. J., 1988. Comparative life histories in the genera *Calanus* and *Neocalanus* in high latitudes of the northern hemisphere. Hydrobiologia 167/168: 127–142.

Dale, T., E. Bagøien, W. Melle & S. Kaartvedt, 1999. Can predator avoidance explain varying overwintering depth of *Calanus* in different oceanic water masses? Mar. Ecol. Prog. Ser. 179: 113–121.

Davis, C. S., 1984. Predatory control of copepod seasonal cycles on Georges Bank. Mar. Biol. 82: 31–40.

Einarsson, H., 1960. The fry of *Sebastes* in Icelandic waters and adjacent seas. Rit Fiskid. 2: 1–67.

Fortier, M., L. Fortier, H. Hattori, H. Saito & L. Legendre, 2001. Visual predators and the diel vertical migration of copepods under arctic sea ice during the midnight sun. J. Plankton Res. 23: 1263–1278.

Gallienne, C. P. & B. Robins, 2001. Is *Oithona* the most important copepod in the world's oceans? J. Plankton Res. 23: 1421–1432.

Gislason, A. & O. S. Astthorsson, 1992. Zooplankton collected by sediment trap moored in deep water south of Iceland. Sarsia 77: 219–224.

Gislason, A. & O. S. Astthorsson, 1998. Seasonal variations in biomass, abundance and composition of zooplankton in the subarctic waters north of Iceland. Polar Biol. 20: 85–94.

Gislason, A. & O. S. Astthorsson, 2000. Winter distribution, ontogenetic migration and rates of egg production of *Calanus finmarchicus* southwest of Iceland. ICES J. mar. Sci. 57: 1727–1739.

Grönvik, S. & C. C. E. Hopkins, 1984. Ecological investigations on the plankton community of Balsfjorden, Northern Norway. Generation cycle, seasonal vertical distribution, and seasonal variations in body weight and carbon and nitrogen content of the copepod *Metridia longa* (Lubbock). J. exp. mar. Biol. Ecol. 80: 93–107.

Gudmundsson, K., 1998. Long-term variation in phytoplankton productivity during spring in Icelandic waters. ICES J. mar. Sci. 55: 635–643.

Hairston, N. G. Jr., 1987. Diapause as a predator avoidance adaptation. In Kerfoot, W. C. & A. Sih (eds), Predation: Direct and Indirect Impacts. University Press of New England, Hanover: 281–290.

Hansen, B. & S. Østerhus, 2000. North Atlantic–Norwegian Sea exchanges. Prog. Oceanogr. 45: 109–208.

Harris, R. P., 1996. Feeding ecology of *Calanus*. Ophelia 44: 85–109.

Harris, R. P., X. Irigoien, R. N. Head, C. Rey, B. H. Hygum, B. W. Hansen, B. Niehoff, B. Meyer-Harms & F. Carlotti, 2000. Feeding, growth and reproduction in the genus *Calanus*. ICES J. mar. Sci. 57: 1708–1726.

Hirche, H.-J., 1991. Distribution of dominant calanoid species in the Greenland Sea during late fall. Polar Biol. 11: 351–362.

Hirche, H.-J., 1996a. Diapause in the marine copepod, *Calanus finmarchicus* – a review. Ophelia 44: 129–143.

Hirche, H.-J., 1996b. The reproductive biology of the marine copepod, *Calanus finmarchicus* – a review. Ophelia 44: 111–128.

Hirche, H.-J., 1997. Life cycle of the copepod *Calanus hyperboreus* in the Greenland Sea. Mar. Biol. 128: 607–618.

Hopkins, C. C. E., 1982. The breeding biology of *Euchaeta norvegica* (Boeck) (Copepoda: Calanoida) in Loch Etive, Scotland: assessment of breeding intensity in terms of seasonal cycles in the sex ratio, spermatophore attachment, and egg-sac production. J. exp. mar. Biol. Ecol. 60: 91–102.

Hopkins, T. L., T. M. Lancraft, J. J. Torres & J. Donnelly, 1993. Community structure and trophic ecology of zooplankton in the Scotia Sea marginal ice zone in winter (1988). Deep-Sea Res. 1: 81–105.

ICES, 1999. Report of the North-Western Working Group. Advisory Committee on Fisheries Management. ICES CM 1999/ACFM: 17. 329 pp.

Irigoien, X., R. Head, U. Klenke, B. Meyer-Harms, D. Harbour, B. Niehoff, H.-J. Hirche & R. Harris, 1998. A high frequency time series at Weathership M, Norwegian Sea, during the 1997 spring bloom: feeding of adult female *Calanus finmarchicus*. Mar. Ecol. Prog. Ser. 172: 127–137.

Jaschnov, W. A., 1972. On the systematic status of *Calanus glacialis, Calanus finmarchicus* and *Calanus helgolandicus*. Crustaceana 22: 279–284.

Kaartvedt, S., 1996. Habitat preference during overwintering and timing of seasonal vertical migration of *Calanus finmarchicus*. Ophelia 44: 145–156.

Kaartvedt, S., 2000. The life history of *Calanus finmarchicus* in the Norwegian Sea in relation to planktivorous fish. ICES J. mar. Sci. 57: 1819–1824.

Lampitt, R. S., 1978. Carnivorous feeding by a small marine copepod. Limnol. Oceanogr. 23: 1228–1231.

Longhurst, A. & R. Williams, 1992. Carbon flux by seasonal vertical migrant copepods is a small number. J. Plankton Res. 11: 1495–1509.

Magnússon, J., 1996. The deep scattering layers in the Irminger Sea. J. Fish Biol. 49: 182–191.

Magnússon, J., 2000. Djúpkarfi og úthafskarfi. (Oceanic redfish. In Icelandic). Námsgagnastofnun og Hafrannsóknastofnunin, Reykjavík. 7 pp.

Magnússon, J. & J. V. Magnússon, 1995. Oceanic redfish (*Sebastes mentella*) in the Irminger Sea and adjacent waters. Sci. Mar. 59: 241–254.

Matthews, J. B. L., 1968. On the acclimatization of *Calanus finmarchicus* (Crustacea, Copepoda) to different temperature conditions in the North Atlantic. Sarsia 34: 371–382.

Mauchline, J., 1994. Seasonal variation in some population parameters of Euchaeta species (Copepoda: Calanoida). Mar. Biol. 120: 561–570.

Metz, C., 1998. Feeding of *Oncaea curvata* (Poecilostomatoida, Copepoda). Mar. Ecol. Prog. Ser. 169: 229–235.

Meyer-Harms, B., X. Irigoien, R. Head & R. Harris, 1999. Selective feeding on natural phytoplankton by *Calanus finmarchicus* before, during, and after the 1997 spring bloom in the Norwegian Sea. Limnol. Oceanogr. 44: 154–165.

Mortensen, J., H. Valdimarsson & S.-A. Malmberg, 1999. Flædi sjógerda í Nordurhöfum (Advection of water masses in the Nordic Seas. In Icelandic). Hafrannsóknastofnun Fjölrit 73: 28–30.

Motoda, S., 1959. Devices of simple plankton apparatus. Mem. Fac. Fish. Hokkaido Univ. 7: 73–94.

Nichols, J. H. & A. B. Thompson, 1991. Mesh selection of copepodite and nauplius stages of four calanoid copepod species. J. Plankton Res. 13: 661–671.

Ohman, M. D. & J. A. Runge, 1994. Sustained fecundity when phytoplankton resources are in short supply: Omnivory by *Calanus finmarchicus* in the Gulf of St. Lawrence. Limnol. Oceanogr. 39: 21–36.

Østvedt, O. J., 1955. Zooplankton investigations from Weathership M in the Norwegian Sea, 1948–1949. Hvalrd. Skr. 40: 1–93.

Paffenhöfer, G. A., 1993. On the ecology of marine cyclopoid copepods (Crustacea, Copepoda). J. Plankton Res. 15: 37–55.

Richter, C., 1994. Regional and seasonal variability in the vertical distribution of mesozooplankton in the Greenland Sea. Ber. Polarforsch. 154: 1–87.

Runge, J. A. & S. Plourde, 1996. Fecundity characteristics of *Calanus finmarchicus* in coastal waters of eastern Canada. Ophelia 44: 171–187.

Sabatini, M. & T. Kiørboe, 1994. Egg production, growth and development of the cyclopoid copepod *Oithona similis*. J. Plankton Res. 16: 1329–1351.

Sameoto, D. D., 1984. Vertical distribution of zooplankton biomass and species in northeastern Baffin Bay related to temperature and salinity. Polar Biol. 2: 213–224.

Sigurdsson, Th., H.-J. Rätz, A. Pedchenko, V. Mamylov, J. Mortensen, C. Stransky, S. Melnikov, K. Drevetnyak & Y.

Bakay, 1999. Report on the joint Icelandic/German/Russian trawl-acoustic survey on pelagic redfish in the Irminger Sea and adjacent waters in June/July 1999. ICES CM 1999, ACFM: 17, 38 pp.

Sigurdsson, Th., G. Jónsson & S. Sveinbjörnsson, 2002. Djúpsjávarlódningar í sudvesturdjúpi. (Deep scattering layers southwest of Iceland. In Icelandic). Ægir 95: 17–18.

Sigurjónsson, J., 1995. On the life history and autecology of North Atlantic rorquals. In Blix, A. S., L. Walløe & Ø. Ulltang (eds), Whales, Seals, Fish and Man. Elsevier Science: 425–441.

Sigurjónsson, J. & G. Víkingsson, 1998. Seasonal abundance of and estimated food composition by cetaceans in Icelandic and adjacent waters. J. Northw. Atl. Fish. Sci. 22: 271–287.

Slagstad, D. & K. S. Tande, 1990. Growth and production dynamics of *Calanus glacialis* in arctic pelagic food web. Mar. Ecol. Prog. Ser. 63: 189–199.

Strickland, J. D. H. & T. R. Parsons, 1968. A practical handbook of seawater analysis. Bull. Fish. Res. Bd Can. 167. 311 pp.

Svensen, C. & T. Kiörboe, 2000. Remote prey detection in *Oithona similis*: hydromechanical versus chemical cues. J. Plankton Res. 22: 1155–1166.

Tande, K. S., 1991. *Calanus* in North Norwegian fjords and in the Barents Sea. In Sakshaug, E., C. C. E. Hopkins & N. A. Øritsland (eds), Proceedings of the Pro Mare Symposium on Polar Marine Ecology, Trondheim, 12–16 May 1990. Polar Res. 10: 389–407.

Turner, J. T. & E. Granéli, 1992. Zooplankton feeding ecology: grazing during enclosure studies of phytoplankton blooms from the west coast of Sweden. J. exp. mar. Biol. Ecol. 157: 19–31.

Turner, J. T., 1984. The feeding ecology of some zooplankters that are important prey of larval fish. NOAA Tech. Rep. NMFS 7: 1–28.

Valdimarsson, H. & S. A. Malmberg, 1999. Near-surface circulation in Icelandic waters derived from satellite tracked drifters. Rit Fiskid. 16: 23–39.

Williams, R., 1988. Spatial heterogenety and niche differentiation in oceanic zooplankton. Hydrobiologia 167/168: 151–159.

Hydrobiologia **503:** 211–221, 2003.
M.B. Jones, A. Ingólfsson, E. Ólafsson, G.V. Helgason, K. Gunnarsson &
J. Svavarsson (eds), *Migrations and Dispersal of Marine Organisms.*

Seasonality of harpacticoids (Crustacea, Copepoda) in a tidal pool in subarctic south-western Iceland

María Björk Steinarsdóttir[1], Agnar Ingólfsson[1] & Emil Ólafsson[2]
[1]*Institute of Biology, University of Iceland, Sturlugata 7, IS-101 Reykjavík, Iceland*
E-mail: mariab@hi.is
[2]*Department of Zoology, University of Stockholm, SE-106 91 Stockholm, Sweden*

Key words: Harpacticoida, seasonality, reproduction, migration, high latitudes, Iceland

Abstract

Seasonality of phytal harpacticoids was studied in a tidal pool in the lower rocky intertidal in south-western Iceland. Samples of the red alga *Chondrus crispus* were taken monthly from January 1999 to January 2000. Multidimensional scaling ordination showed a clear seasonal pattern in phytal species composition and abundance of individual species. Of more than 70 species recorded, *Zaus spinatus* accounted for one third of the total abundance. Many species were found only sporadically and analysis of the data was confined primarily to the eleven most common species (*Zaus spinatus, Ectinosoma* spp., *Tisbe* spp., *Ameira longipes, Mesochra* sp. 2 and *Mesochra* sp. 1, *Dactylopusia vulgaris, Amphiascoides debilis, Amphiascus* sp. 1, *Parastenhelia spinosa* and *Amphiascus minutus*). Four showed no significant variation in total abundance through the year, while the remainder had more or less pronounced peaks in summer. Most species (on which data was sufficient) had a prolonged ($\geq$ 6 months) to a continuous breeding season, as indicated by the presence of ovigerous females. This contrasts with most shallow-water temperate and subarctic invertebrates which appear to have more restricted breeding seasons. Our results extend to higher latitudes the general conclusion that many temperate phytal harpacticoids have a long, often continuous, breeding season. The reasons for this are unclear but, presumably, involve access to continuous food supply. There is strong evidence that variations in abundance of several species in the tidal pool were not due only to recruitment, but also to migration to the pool in summer and emigration in autumn.

Introduction

Harpacticoid copepods are small crustaceans, mostly living in or around various types of substrata including sediments and plants. A suite of species is associated with marine algae where they occupy several habitats, i.e. interstices of finely divided epiphytes or other algal structures, the surfaces of the fronds themselves and the sediment accumulated at the bases of fronds (Moore, 1973; Pallares & Hall, 1974; Hicks, 1977).

At high latitudes, reproduction of marine invertebrates is generally confined to six months or less of the year (e.g. Moore, 1972). This pattern is shown clearly in crustaceans from shallow waters around subarctic Iceland (Astthorsson, 1987, 1990a,b; Astthorsson & Gislason, 1991, 1992; Gislason & Astthorsson, 1995; Leifsson, 1997). At lower latitudes, prolonged or continuous breeding becomes increasingly more frequent (e.g. Moore, 1972), and most phytal crustaceans, for example of the genera *Gammarus* and *Jaera*, have shorter breeding periods in the subarctic than in the temperate zone (Jones & Naylor, 1971; Steele & Steele, 1972, 1975; Skadsheim, 1984). We would, therefore, expect most phytal harpacticoids in Icelandic waters to exhibit relatively short breeding seasons. However, Hicks (1979) reported that the majority of phytal harpacticoids (of over 70 species) at temperate latitudes showed continuous breeding through the year. Although Hicks (1979) concluded that the majority of temperate harpacticoids living in or above sediments had shorter (i.e. non-continuous) reproductive periods than the phytal ones, several species living in sediment in subarctic Alaska have been found to show continuous reproduction (Schizas & Shirley, 1996). Thus, in general, harpacticoids, perhaps especially phytal species, have unusually long reproductive periods at temperate and even subarctic latitudes. Hicks (1979) suggested that the pattern in phytal harpacticoids may be linked to seasonally unlimited food.

Generally, harpacticoids reach maximum abundance in the warmer months of the year (review by Hicks & Coull, 1983) and we would expect this pattern in Iceland. For several phytal species, seasonal migrations have been suggested as explanations, at least partial, for these patterns (e.g. Hicks, 1979; Johnson & Scheibling, 1987), many species are good swimmers (Ólafsson et al., 2001). The primarily purpose of this study was to test whether the pattern of reproduction seen for phytal harpacticoids at temperate latitudes holds true for subarctic south-western Iceland. Average sea surface temperatures in this area range from about 1 °C in February to about 11 °C in August (Gunnarsson & Ingólfsson, 1995). We also investigated seasonal patterns in abundance of individual species and assessed the importance of migrations, to any variations observed.

Materials and methods

Samples of the red alga *Chondrus crispus* Stackhouse were collected from a 4 × 4 m section of a fully saline large (approx. 25 × 55 m) tidal pool in the lower rocky intertidal zone in Seltjarnarnes, near Reykjavík, south-western Iceland (64° 10′ N, 22° 01′ W). At low tide, the surface of the pool was at an elevation of 1.14 m above chart datum, just below the level of mean low-water neap tide. Two, 4 m long rulers formed a grid. Coordinates of each sample were obtained by drawing random numbers from 0 to 40 for each ruler. Ten samples were taken monthly from January 1999 to January 2000. Whole plants were taken, sufficient numbers for each sample to fit loosely into a 155 ml plastic container, to which a little seawater was added. Samples were preserved in the laboratory in 5% formalin solution; one sample was lost in this process. Preserved samples were rinsed through a 63 μm sieve and *Chondrus crispus* was weighed wet. Samples ($N = 129$) weighed 13.5 g on average (SD = 3.63). Small pieces of other algal species, often found in the samples, were identified and weighed. These additional algal species made up 3% of total algal weight. Harpacticoids, sieved from the algae, were counted and identified to species when ever possible, and classified into non-ovigerous females, ovigerous

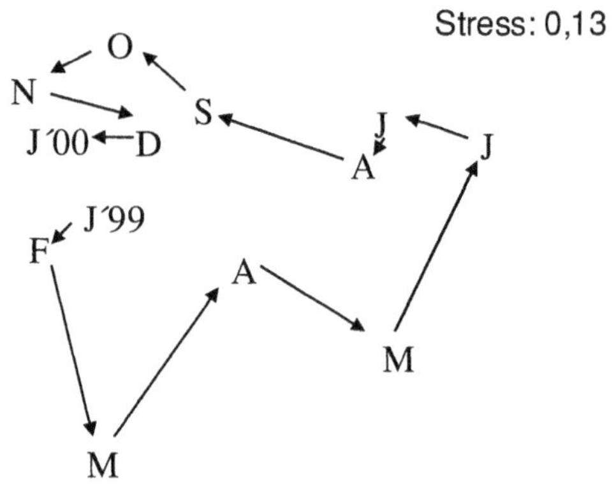

Figure 1. Non-metric multidimensional scaling ordination of harpacticoid species abundance data based on average values for each month. Arrows indicate the route of changes in harpacticoid assemblage structure in successive months from January 1999 (J′99) to January 2000 (J′00).

females, males, and copepodite stages I–V (stage I is the youngest).

Abundances of all harpacticoid species were subjected to non-metric multidimensional scaling ordination (MDS) using the Bray-Curtis similarity measure on fourth root data. The ordination was done using the PRIMER 5.2.4. statistical package developed at the Plymouth Marine Laboratory (England). Associations between various variables were assessed with Spearman correlations (r_s) and regression analyses. Species diversity was assessed by Hurlbert's expected number of species for samples of 19 individuals ($E_{(S19)}$) using the BioDiversity Professional Beta 1 program. Twenty samples contained fewer than 19 individuals and were omitted from analyses. Variations in abundance from January 1999 to January 2000 were investigated by one-way analysis of variance (ANOVA) on densities (number/100 g algae), $\log_{10}$-transformed to reduce heterogeneity. Bonferroni probabilities were used throughout. Trend lines were fitted by Distance Weighted Least Squares (DWLS) in SYSTAT®.

Results

More than seventy species of harpacticoids, belonging to 16 families, were found in this study (Table 1). Family Laophontidae had most species (15), followed by Thalestridae (10) and Diosaccidae (10). Species of family Laophontidae, however, were never very frequent or abundant (Table 1). Many species were found only sporadically (in less than 10 of the 129 samples) and in low abundance (Table 1). Further analysis was

Table 1. List of harpacticoid species (or taxa) found in the tidal pool, Seltjarnarnes, their frequency (both in numbers, F, and %, F%) and mean number (M) of individuals per 100 g of algae, with standard error (SE), from January 1999 to January 2000. $N = 129$. The occurrence of species in monthly samples is indicated with x

Species	F	F (%)	M	SE	J	F	M	A	M	J	J	A	S	O	N	D	J
Ectinosomatidae																	
Ectinosoma spp.	118	91.5	91.1	9.1	x	x	x	x	x	x	x	x	x	x	x	x	x
Halectinosoma sp. A	4	3.1	0.3	0.2				x				x	x				x
Microsetella norvegica (Boeck)	8	6.2	0.5	0.2	x	x				x				x	x		
Harpacticidae																	
Harpacticus chelifer (O.F. Müller)	13	10.1	2.6	0.8					x	x	x						
Harpacticus obscurus T. Scott	2	1.6	0.2	0.1							x	x					
Harpacticus uniremis Kröyer	18	14.0	3.0	1.1				x	x	x	x						
Zaus abbreviatus Sars	6	4.7	0.4	0.2						x	x		x				
Zaus spinatus Goodsir	115	89.1	299.1	52.3	x	x	x	x	x	x	x	x	x	x	x	x	x
Zaus sp. 1	1	0.8	0.1	0.1					x								
Zaus sp. 2	1	0.8	0.1	0.1					x								
Tisbidae																	
Tisbe spp.	61	47.3	100.3	20.3			x	x	x	x	x	x	x	x	x	x	x
Scutellidium hippolytes (Kröyer)	28	21.7	3.5	0.9	x		x		x	x	x	x		x			
Scutellidium sp. 1	3	2.3	0.3	0.2							x	x					
Scutellidium sp. 2	5	3.9	1.2	1.0							x	x	x				
Sacodiscus littoralis (Sars)	5	3.9	0.3	0.1		x							x	x	x		
Tegastidae																	
Tegastes sp.	22	17.1	4.4	1.1						x	x	x					
Thalestridae																	
Thalestris longimana Claus	1	0.8	0.1	0.1	x												
Parathalestris clausi (Norman)	41	31.8	3.9	0.7	x	x		x			x	x	x	x	x	x	x
Parathalestris intermedia Gurney	22	17.1	1.5	0.3	x	x	x	x	x		x	x	x		x	x	x
Diarthrodes major (T. & A. Scott)	30	23.3	4.6	1.2	x	x	x	x	x		x	x	x		x	x	
Diarthrodes nobilis (Baird)	24	18.6	1.8	0.4	x	x		x			x	x	x	x	x	x	
Diarthrodes sp. 1	2	1.6	0.1	0.1											x		x
Dactylopusia glacialis Sars	1	0.8	0.1	0.1					x								
Dactylopusia vulgaris Sars	78	60.5	42.0	11.1	x	x	x	x	x	x	x	x	x	x	x	x	x
aff. *Dactylopusia signata* (Willey)	1	0.8	0.1	0.1					x								
Paramenophia sp. Lang	1	0.8	0.1	0.1						x							
Parastenheliidae																	
Parastenhelia spinosa (Fischer)	65	50.4	18.2	4.4	x	x	x	x	x	x	x	x	x	x	x	x	x
Diosaccidae																	
cf. *Robertsonia* sp.	1	0.8	0.1	0.1			x										
Amphiascus minutus (Claus)	74	57.4	15.8	2.3	x	x	x	x	x	x	x	x	x	x	x	x	x
Amphiascus tenuiremis (Brady & Robertson)	1	0.8	0.1	0.1		x											
Amphiascus sp. 1	58	45.0	19.6	3.4	x	x		x	x	x	x	x	x	x	x	x	x
Dactylopodamphiascopsis latifolius (Sars)	1	0.8	0.1	0.1						x							
Paramphiascopsis longirostris (Claus)	13	10.1	2.1	0.8								x	x	x		x	x
Amphiascoides debilis (Giesbrecht)	85	65.9	28.7	3.8	x	x	x	x	x	x	x	x	x	x	x	x	x
Paramphiascella sp. 1	1	0.8	0.1	0.1				x									
Diosaccidae sp. 2	1	0.8	0.2	0.2			x										
Diosaccidae sp. 3	2	1.6	0.1	0.1									x			x	

Continued on p. 214

Table 1. contd.

Species	F	F (%)	M	SE	J	F	M	A	M	J	J	A	S	O	N	D	J
Ameiridae																	
Ameira longipes Boeck	110	85.3	92.6	17.0	x	x	x	x	x	x	x	x	x	x	x	x	x
Ameira sp. 2	1	0.8	0.1	0.1							x						
Ameira sp. 3	1	0.8	0.05	0.05										x			
Proameira simplex (Norman & T. Scott)	2	1.6	0.3	0.3		x	x										
Nitocra typica Boeck	2	1.6	0.2	0.1		x											
Ameiridae sp. 1	3	2.3	0.3	0.2				x	x			x					
Ameiridae sp. 2	7	5.4	0.6	0.3				x	x			x	x			x	
Canthocamptidae																	
Mesochra sp. 1	109	84.5	50.0	4.8	x	x	x	x	x	x	x	x	x	x	x	x	x
Mesochra sp. 2	107	82.9	85.1	14.8	x	x	x	x	x	x	x	x	x	x	x	x	x
Itunella muelleri (Gagern)	4	3.1	0.3	0.2										x	x	x	x
Canthocamptidae sp. 2	5	3.9	0.4	0.2	x	x			x						x		
Canthocamptidae sp. 3	1	0.8	0.1	0.1				x									
Orthopsyllidae																	
Orthopsyllus major Klie	1	0.8	0.04	0.04		x											
Paranannopidae																	
Danielssenia typica Boeck	1	0.8	0.1	0.1					x								
Huntemanniidae																	
Pontopolites typicus T. Scott	5	3.9	0.3	0.1			x	x		x	x	x					
Rhizothricidae																	
Rhizothricidae sp. 1	1	0.8	0.1	0.1						x							
Argestidae																	
aff. *Eurycletodes (O). similis* (T. Scott)	1	0.8	0.1	0.1								x					
Laophontidae																	
Laophonte thoracica Boeck	2	1.6	0.1	0.1										x		x	
Laophonte setosa Boeck	27	20.9	5.9	1.6		x	x	x	x		x	x	x	x		x	x
aff . *Laophonte serrata* (Claus)	1	0.8	0.05	0.05										x			
cf. *Laophonte baltica* Klie	4	3.1	0.4	0.2				x	x			x	x				
Laophonte inopinata T. Scott	23	17.8	3.9	1.0	x	x	x	x		x	x	x	x	x	x	x	x
Laophonte depressa T. Scott	1	0.8	0.1	0.1									x				
Laophonte sp. 3	3	2.3	0.2	0.1									x		x	x	
Laophonte sp. 4	3	2.3	0.2	0.1		x	x			x							
Heterolaophonte stroemi (Baird)	7	5.4	0.5	0.2	x	x					x	x	x				
Heterolaophonte littoralis (T. & A. Scott)	4	3.1	0.3	0.1				x		x		x					
Heterolaophonte longisetigera Klie	23	17.8	2.2	0.7	x						x	x	x	x	x	x	x
Heterolaophonte sp. 3	4	3.1	0.2	0.1					x		x						
Paralaophonte perplexa (T. Scott)	17	13.2	1.2	0.3	x	x		x			x	x	x			x	x
Paralaophonte sp. 1	32	24.8	4.4	1.0					x	x	x	x	x	x	x	x	x
Onychocamptus horridus (Norman)	5	3.9	0.3	0.1							x	x				x	
Normanellidae																	
Normanella aff. *minuta* (Boeck)	5	3.9	0.4	0.2	x		x					x	x			x	
Unidentified species and copepodites			57.5	9.5													

confined mostly to the eleven most abundant taxa (see Table 1 for authorities of species names). *Ectinosoma* spp. had the highest frequency (in 91% of samples), followed by *Zaus spinatus* (89%), *Ameira longipes* (85%), *Mesochra* sp. 1 (84%), *Mesochra* sp. 2 (83%), *Amphiascoides debilis* (66%), *Dactylopusia vulgaris* (60%), *Amphiascus minutus* (57%), *Parastenhelia spinosa* (50%), *Tisbe* spp. (47%) and *Amphiascus* sp.

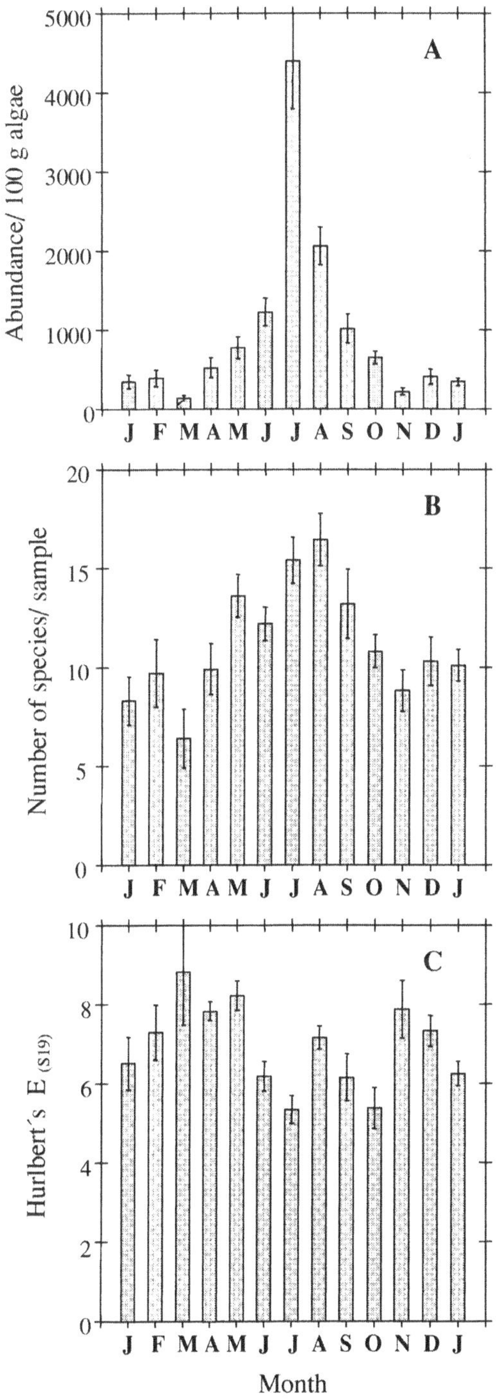

Figure 2. (A) Average monthly abundance of harpacticoids per 100 g of algae, with standard errors. (B) Average monthly numbers of species per sample, with standard errors. (C) Average monthly $E_{(S19)}$ values, with standard errors. The time scale runs from January 1999 to January 2000.

1 (45%). These 11 taxa were also the most abundant, containing almost 90% of the total number of harpacticoid individuals; *Z. spinatus* accounted for one third of the number of individuals.

The MDS ordination showed a clear cyclic seasonal pattern in harpacticoid assemblage structure (Fig. 1). Harpacticoids were most abundant during summer, especially in July and August, and the number of species was also highest then, while the seasonal pattern of species diversity is less clear (Fig. 2). Number of species correlated significantly with the number of individuals ($r_S = 0.82$, $p < 0.0005$, $N = 129$). Hurlbert's index showed a positive correlation with the number of species ($r_S = 0.43$, $p < 0.001$, $N = 109$) but no significant correlation with the total number of harpacticoid individuals ($r_S = -0.17$, $p > 0.05$, $N = 109$). There was no correlation between weight of algae and number of harpacticoid individuals ($r_S = 0.12$, $p > 0.05$, $N = 129$) or species ($r_S = 0.10$, $p > 0.05$, $N = 129$).

Virtually no ovigerous female *D. vulgaris* and *Mesochra* sp. 2 were observed in the tidal pool, although non-ovigerous female *Mesochra* sp. 2 were often common (Table 2, Fig. 3). Ovigerous female *Ectinosoma* spp. were also rare, but were found in most months of the year (Table 2). The pattern for *Tisbe* spp. is unclear, as their occurrence was virtually confined to 4 months during summer (May–August) when ovigerous females were common. Of the remaining common species, *A. longipes*, *A. debilis*, *Mesochra* sp. 1, and *Z. spinatus* appeared to reproduce continuously (Fig. 3); occasional gaps (Fig. 3) being due probably to low sample sizes (Table 2). The pattern for *Amphiascus* sp. 1 is ambiguous, but ovigerous females were observed in most months (Table 2) and breeding may be continuous. In all species except *A. longipes*, however, the frequency of ovigerous females appeared to peak at certain times of the year. For *A. debilis* highest numbers occurred in May and June, for *Mesochra* sp. 1 from March to June, for *Z. spinatus* from April to June and in September, and for *Amphiascus* sp. 1 in April and May and August and September (Table 2). Ovigerous female *P. spinosa* were found from February to August (the gap in March being due probably to insufficient data) and in January the following year, indicating a prolonged breeding season (>6 months). *Amphiascus minutus* had the most distinct breeding season, ovigerous females being confined to a 6-month period (April–September) (Fig. 3, Table 2).

Occurrence of identifiable copepodite stages CI and CII in the pool was rather sporadic for most

Table 2. Total number of females (TF) and proportion of ovigerous females (%O) of total number of females of common species of harpacticoids, from January 1999 to January 2000

		J	F	M	A	M	J	J	A	S	O	N	D	J
Ameira longipes	TF	23	58	4	49	72	43	266	99	28	23	17	55	54
	%O	8.7	10.3	0	28.6	27.8	9.3	9.4	5.1	14.3	8.7	17.6	14.5	25.9
Amphiascoides debilis	TF	8	12	8	53	73	19	33	51	45	5	8	22	14
	%O	37.5	0	25.0	43.4	58.9	63.2	42.4	9.8	13.3	0	12.5	9.1	14.3
Amphiascus minutus	TF	11	21	7	22	12	3	1	7	25	18	2	16	22
	%O	0	0	0	36.4	66.7	33.3	0	42.9	12.0	0	0	0	0
Amphiascus sp. 1	TF	2	15	0	13	42	7	17	50	19	2	1	9	1
	%O	0	26.7	0	53.8	54.8	14.3	11.8	34.0	57.9	100	0	33.3	0
Dactylopusia vulgaris	TF	1	9	1	22	10	14	56	12	1	1	1	1	6
	%O	0	0	0	0	10.0	7.1	0	8.3	0	0	0	0	0
Ectinosoma spp.	TF	84	129	18	115	163	32	25	51	58	96	56	164	243
	%O	0	1.6	5.6	7.8	9.8	0	4.0	3.9	12.1	6.3	1.8	1.2	5.8
Mesochra sp. 1	TF	42	43	30	86	52	16	37	90	79	65	31	66	96
	%O	11.9	18.6	56.7	68.6	67.3	56.3	35.1	27.8	22.8	13.8	0	21.2	19.8
Mesochra sp. 2	TF	12	8	12	56	11	32	201	437	249	95	26	34	29
	%O	0	0	0	3.6	9.1	0	0	0	0	0	0	0	0
Parastenhelia spinosa	TF	8	16	6	28	7	4	3	35	12	3	5	22	7
	%O	0	6.3	0	42.9	42.9	25.0	33.3	17.1	0	0	0	0	14.3
Tisbe spp.	TF	0	0	1	1	15	47	210	113	1	6	0	3	0
	%O	0	0	100	0	46.7	36.2	32.9	35.4	0	33.3	0	0	0
Zaus spinatus	TF	62	48	19	14	49	84	302	159	243	168	17	4	6
	%O	16.1	20.8	31.6	57.1	28.6	47.6	25.2	18.2	49.0	35.1	11.8	25.0	0

species, but they usually occurred when ovigerous females were present (Fig. 3). In the case of *D. vulgaris* and *Mesochra* sp. 2, their occurrence indicated that the breeding season was longer than the scant records of ovigerous females showed (Fig. 3).

All the most abundant taxa, except *Tisbe* spp. and *Amphiascus* sp. 1, were recorded each month (Table 1, Figs 4 and 5). *Ectinosoma* spp., *P. spinosa, A. minutus* and *Mesochra* sp. 1 showed no significant variation in abundance through the year (Table 3, Fig. 4), while the remainder had more or less pronounced peaks in summer (Table 3, Fig. 5). These peaks did not coincide, *D. vulgaris, Tisbe* spp. and *A. longipes* being most abundant during summer, *Z. spinatus* during summer to mid autumn, *Mesochra* sp. 2 in summer to early autumn, and *A. debilis* and *Amphiascus* sp. 1 from mid spring to early autumn (Fig. 5). Some of the less abundant species, e.g. *Harpacticus chelifer, H. uniremis* and *Tegastes* sp. were recorded in spring and summer only (Table 1).

There was no significant variation in the abundance of copepodites of *A. minutus, Amphiascus* sp. 1 and *Mesochra* sp. 1 through the year (Table 3). In other common species, copepodites were most abundant in summer, while copepodites of *Ectinosoma* spp. and *Z. spinatus* were abundant to mid autumn.

Discussion

Many species in the tidal pool were found only sporadically and some are probably not of phytal habits. Thus, species of *Robertsonia, Eurycletodes* and *Halectinosoma* are known to live in soft sediments (reviewed by Hicks & Coull, 1983) and 14 of the recorded species have been found in tidal sediments in Iceland, where one, *Laophonte setosa*, was common (E. Ólafsson, unpublished Icelandic Science Council report, 1994). *Heterolaophonte littoralis* was one of the most common species found on the epiphytic alga *Pilayella littoralis* (L.) in North Wales (Jarvis & Seed, 1996) but was very rare in our tidal pool. Species of *Harpacticus* were not common in the tidal pool. This was surprising, as they may be common in the littoral zone elsewhere (Colman, 1939; Ohm, 1964; Hicks, 1980) and are amongst the most common species on

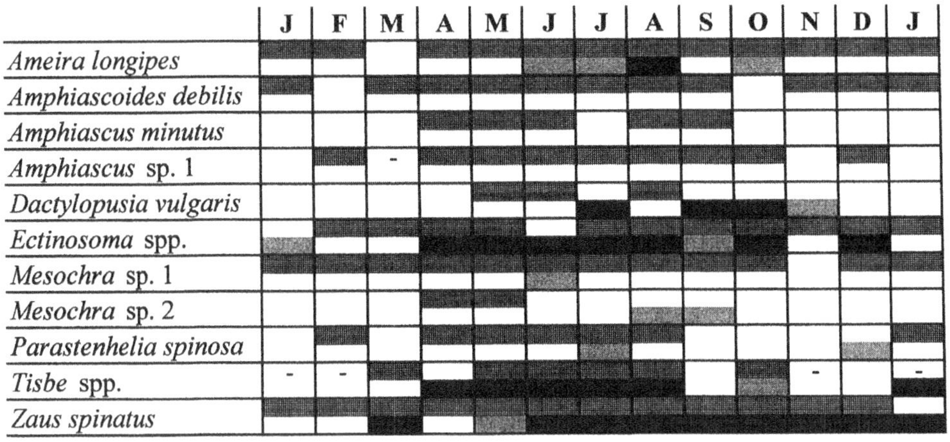

Figure 3. Occurrence of ovigerous females (hatched) and copepodite stages CI (dark shade) and CII (light shade), of common species of harpacticoids, from January 1999 to January 2000. – = no females found in the samples. Note that many CI and CII copepodites found could not be identified to species.

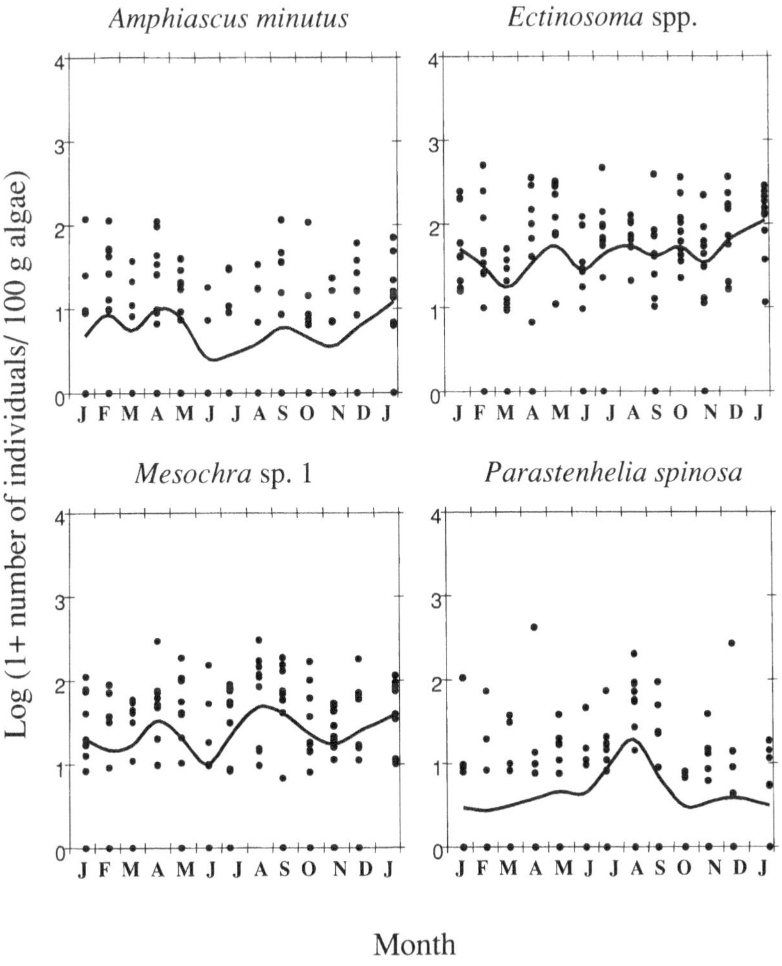

Figure 4. Density of common species (or taxa) of harpacticoids that show no significant seasonal variation during the sampling period (January 1999 to January 2000). Each dot represents a sample (note that dots are in some cases superimposed on each other). Lines are fitted by Distance Weighted Least Squares (DWLS).

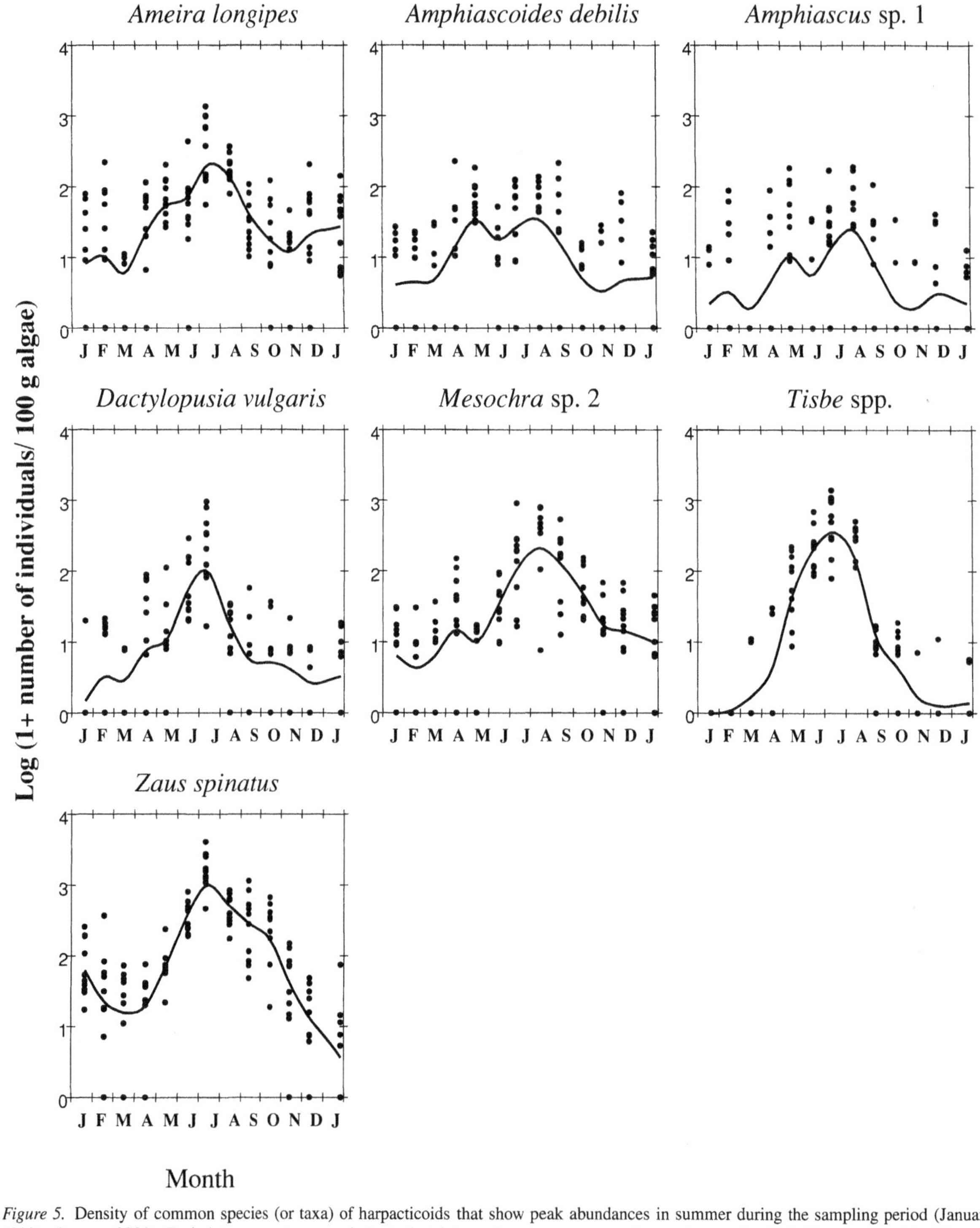

Figure 5. Density of common species (or taxa) of harpacticoids that show peak abundances in summer during the sampling period (January 1999 to January 2000). Each dot represents a sample (note that dots are in some cases superimposed on each other). Lines are fitted by Distance Weighted Least Squares (DWLS).

Table 3. Results from one-way ANOVAs on log-transformed densities (numbers/100 g algae) of common species of harpacticoids during 13 sampling dates from January 1999 to January 2000. F-ratios are shown and Bonferroni probabilities indicated

	Total abundance	Ovigerous females	Non-ovigerous females	Males	Copepodites
Ameira longipes	9.36 ***	3.10 **	5.39 ***	13.79 ***	16.60 ***
Amphiascoides debilis	4.60 ***	8.31 ***	2.83 *	5.39 ***	7.89 ***
Amphiascus minutus	2.13 NS	6.59 ***	1.59 NS	1.98 NS	1.18 NS
Amphiascus sp. 1	5.53 ***	4.37 **	3.77 ***	4.86 ***	1.95 NS
Dactylopusia vulgaris	11.97 ***	0.87 NS	5.58 ***	7.14 ***	26.98 ***
Ectinosoma spp.	2.43 NS	2.23 NS	3.00 *	2.52 NS	5.95 ***
Mesochra sp. 1	1.96 NS	4.91 ***	2.12 NS	3.72 ***	1.10 NS
Mesochra sp. 2	11.68 ***	1.66 NS	10.84 ***	19.75 ***	9.87 ***
Parastenhelia spinosa	2.07 NS	2.58 *	1.54 NS	2.31 NS	8.64 ***
Tisbe spp.	71.09 ***	21.58 ***	26.23 ***	30.07 ***	64.42 ***
Zaus spinatus	20.85 ***	10.47 ***	15.17 ***	16.77 ***	22.95 ***

* $P<0.05$.
** $P<0.01$.
*** $P<0.001$.

floating seaweed in Iceland (Ólafsson et al., 2001). Rare species may prefer algae other than *Chondrus crispus* but we have no information to support this suggestion. Some of the common species found on *Chondrus* in the pool (i.e. *P. spinosa, Z. spinatus, D. vulgaris,* and *A. minutus,* and species of *Tisbe* and *Mesochra*) have been found abundantly on other algae elsewhere (Hicks, 1980; Johnson & Scheibling, 1987). Many factors can potentially influence the abundance and species richness of harpacticoids, e.g. the diversity and heterogeneity of the habitat (Coull et al., 1978; Gunnill, 1982; Jenkins et al., 2002).

Several common species in the tidal pool appeared to reproduce throughout the year. This is generally consistent with other studies on phytal harpacticoids at lower latitudes (reviewed by Hicks, 1979; Hall & Bell, 1993). *Amphiascus minutus* had the most restricted breeding season (approx. 6 months). Our results, however, do not support entirely the results of Johnson & Scheibling (1987) who concluded that most harpacticoids from phytal habits in Nova Scotia did not show continuous breeding; they did imply continuous reproduction for *Tisbe* spp. and *Amphiascopsis* sp. Gurney. Almost no comparisons involving the same species at different latitudes can be made due to scarcity of data. However, in England, *Z. spinatus* was found to breed continuously (Hicks, 1980) as in the present study.

Prolonged and often continuous breeding of a large proportion of phytal harpacticoids at temperate and high latitudes is remarkable, and contrasts with what is known for most other invertebrates at these latitudes (e.g. Moore, 1972), although similar results have been obtained for several harpacticoids in sediment at high latitudes (Schizas & Shirley, 1996). Continuous breeding implies feeding all year round and, thus, a continuous food source. Sieburth & Tootle (1981) showed seasonal differences in microbial composition on algae (*Chondrus crispus* being one of the algae species), but there was no significant difference in total coverage of microorganisms through the year which could provide food for harpacticoids (cf. Hicks & Coull, 1983). Most species in the tidal pool achieved maximum abundance, as elsewhere, during summer. (e.g. Hicks & Coull, 1983; Webb & Parsons, 1988) Ólafsson & Elmgren (1997) found, that the harpacticoid *Microarthridion littorale* (Poppe), which reproduced throughout the year, showed a sharp increase in the number of ovigerous females shortly after the settlement of spring bloom of phytoplankton (April), long before temperatures started to rise, indicating food as the controlling factor.

Many benthic harpacticoids are good swimmers (see Ólafsson et al., 2001) and are common colonizers of floating seaweed (Ingólfsson, 1998), indicative of their power of mobility. All common species in the tidal pool except one (*Mesochra* sp. 2) have been recorded from floating seaweed (Ólafsson, et al., 2001; Steinarsdóttir, unpublished data). Some migrations, especially seasonal, to and away from the pool are, therefore, likely, and are strongly supported by our data. Some species appeared in the pool in spring and disappeared in autumn (e.g. *Harpacticus chelifer* and

H. uniremis). Similar seasonal migrations of *Harpacticus* spp. to and from intertidal algae may be inferred from data given by Colman (1939) from England, Johnson & Scheibling (1987) from Nova Scotia, and by Kito (1977) from Japan. Both *H. chelifer* and *H. uniremis* colonize floating seaweed in spring and mostly disappear in autumn (Ólafsson et al., 2001; Steinarsdóttir, unpublished data). *Tisbe* spp. are suddenly present in the pool in high numbers in May, but have virtually disappeared by September. Other species reach high numbers in spring and summer, and these increases are difficult to explain only by *in situ* recruitment. The best examples are *Mesochra* sp. 2 and *D. vulgaris* which reproduced hardly at all in the pool, but copepodites, non-ovigerous females and males were abundant in summer. Seasonal migrations have been noted in other harpacticoid species. Thus, *Thalestris longimana* colonizes algae in spring and reproduces, while the copepodites overwinter mostly on sand (Hicks, 1979). *Parathalestris croni* (Kröyer) females settle on floating algae in spring and summer to lay their eggs while being planktonic at other times, as are the copepodites (Ingólfsson & Ólafsson, 1997).

Kuebler & Dudgeon (1996) examined the morphological plasticity of *Chondrus crispus* in response to temperatures representative of winter and summer surface seawater in the Gulf of Maine and showed growth of morphologically more complex thalli at summer temperatures. Thus, increase in abundance and species richness of harpacticoids in the tidal pool during summer could be due partly to increased complexity of the habitat. Antibiotic production of *C. crispus* is at a maximum during winter (Hornsey & Hide, 1976), which may deter some species from this biotope at that season. Predation may also control or modify abundance of a local population (Hicks & Coull, 1983). Harpacticoids are one of the main food items of lumpsucker fry (*Cyclopterus lumpus* L.) under 12 mm in length. However, lumpsucker fry reach maximum numbers in the pool during summer (Ingólfsson & Kristjánsson, unpublished data), when harpacticoids are also most abundant.

We conclude that our results extend to higher latitudes the general conclusion of Hicks (1979), that many temperate phytal harpacticoids have a long, often continuous, breeding season, although the reasons for this are obscure. The abundance of harpacticoids in the tidal pool was highest in the summer months. While migration from other unknown areas clearly contributed to this in many cases, *in situ* reproduction and other factors undoubtedly also played a role.

Acknowledgements

We thank Magnea Karlsdóttir for assistance in the field and Samuel Gomez for assistance with the identification of the two *Mesochra* species. We also thank two anonymous reviewers for their helpful comments. Acknowledgements are due to N. McAleece, P.J.D. Lambshead, G.L.J Paterson and J.D. Gage for the use of the program BioDiversity Professional Beta 1. This study was supported in part by a grant from Lýðveldissjóður.

References

Astthorsson, O. S., 1987. Records and life history of *Praunus flexuosus* (Crustacea: Mysidacea) in Icelandic waters. J. Plankton Res. 9: 955–964.

Astthorsson, O. S., 1990a. Ecology of the euphausiids *Thysanoessa raschi*, *T. inermis* and *Meganyctiphanes norvegica* in Ísafjord-deep, northwest-Iceland. Mar. Biol. 170: 147–157.

Astthorsson, O. S., 1990b. Aspects of the biology of *Mysis oculata* on the west coast of Iceland. Ann. Zool. Fenn. 27: 247–251.

Astthorsson, O. S. & A. Gislason, 1991. Seasonal abundance and distribution of Caridea larvae in Ísafjord-deep, north-west Iceland. J. Plankton Res. 13: 91–102.

Astthorsson O. S. & A. Gislason, 1992. Investigation on the ecology of the zooplankton community in Ísafjord-deep, northwest Iceland. Sarsia 77: 225–236.

Colman, J., 1939. On the faunas inhabiting intertidal seaweeds. J. mar. biol. Ass. U. K. 24: 129–183.

Coull, B. C., S. S. Bell, A. M. Savory & B. W. Dudley, 1978. Zonation of meiobenthic copepods in a southeastern United States salt march. Estuar. coast. mar. Sci. 9: 181–188.

Gislason, A. & O. S. Astthorsson, 1995. Seasonal cycle of zooplankton southwest of Iceland. J. Plankton Res. 17: 1959–1976.

Gunnarsson, K. & A. Ingólfsson, 1995. Seasonal changes in the abundance of intertidal algae in southwestern Iceland. Bot. Mar. 38: 69–77.

Gunnill, F. C., 1982. Effects of plant size and distribution on the numbers of invertebrate species and individuals inhabiting the brown alga *Pelvetia fastigiata*. Mar. Biol. 69: 263–280.

Hall, M. O. & S. S. Bell, 1993. Meiofauna of the seagrass *Thalassia testudinum*: population characteristics of harpacticoid copepods and associations with algal epiphytes. Mar. Biol. 116: 137–146.

Hicks, G. R. F., 1977. Observations on substrate preference of marine phytal harpacticoids (Copepoda). Hydrobiologia 56: 7–9.

Hicks, G. R. F., 1979. Pattern and strategy in the reproductive cycles of benthic harpacticoid copepods. In Naylor, E. & R. G. Hartnoll (eds), Cyclic Phenomena in Marine Plants and Animals. Pergamon press, Toronto: 139–147.

Hicks, G. R. F., 1980. Structure of phytal harpacticoid copepod assemblages and the influence of habitat complexity and turbidity. J. exp. mar. Biol. Ecol. 44: 157–192.

Hicks, G. R. F. & B. C. Coull, 1983. The ecology of marine meiobenthic harpacticoid copepods. Oceanogr. mar. biol. Ann. Rev. 21: 67–175.

Hornsey, I. S. & D. Hide, 1976. The production of antimicrobial compounds by british marine algae II. Seasonal variation in production of antibiotics. Br. Phycol. J. 11: 63–67.

Ingólfsson, A. 1998. Dynamics of macrofaunal communities of floating seaweed clumps off western Iceland: a study of patches on the surface of the sea. J. exp. mar. Biol. Ecol. 231: 119–137.

Ingólfsson, A. & E. Ólafsson, 1997. Vital role of drift algae in the life history of the pelagic harpacticoid *Parathalestris croni* in the northern North Atlantic. J. Plankton Res. 19: 15–27.

Jarvis, S. C. & R. Seed, 1996. The meiofauna of *Ascophyllum nodosum* (L.) Le Jolis: characterization of the assemblages associated with two common epiphytes. J. exp. mar. Biol. Ecol. 199: 249–267.

Jenkins, G. P., G. K. Walker-Smith & P. A. Hamer, 2002. Elements of habitat complexity that influence harpacticoid copepods associated with seagrass beds in a temperate bay. Oecologia 131: 598–605.

Johnson, S. C. & R. E. Scheibling, 1987. Reproductive patterns of harpacticoid copepods on intertidal macroalgae (*Ascophyllum nodosum* and *Fucus vesiculosus*) in Nova Scotia, Canada. Can. J. Zool. 65: 129–141.

Jones, M. B. & E. Naylor, 1971. Breeding and bionomics of the British members of the *Jaera albifrons* group of species (Isopoda: Asellota). J. Zool., Lond. 165: 183–199.

Kito, K., 1977. Phytal animals in the *Sargassum confusum* region in Oshoro bay, Hokkaido: phenology of harpacticoid copepods. J. Fac. Sci. Hokkaido Univ. Series VI, 20: 691–696.

Kuebler, J. E. & S. R. Dudgeon, 1996. Temperature dependent change in the complexity of form of *Chondrus crispus* fronds. J. exp. mar. Biol. Ecol. 207: 15–24.

Leifsson, B. J., 1997. Life cycles, breeding periods and vertical distribution of *Idotea granulosa* Rathke and *I. pelagica* Leach (Crustacea, Isopoda) on Icelandic shores. Sarsia 83: 1–13.

Moore, H. B., 1972. Aspects of stess in tropical marine environment. Adv. mar. Biol. 10: 217–269.

Moore, P. G., 1973. The kelp fauna of northeast Britain. II. Multivariate classification: turbidity as an ecological factor. J. exp. mar. Biol. Ecol. 13: 127–163.

Ohm, G., 1964. Die Besiedlung der *Fucus*-Zone der Kieler Bucht und der westlichen Ostee unter besonderer Berücksichtigung der Microfauna. Kieler Meeresforsch. 20: 30–64.

Ólafsson, E. & R. Elmgren, 1997. Seasonal dynamics of sublittoral meiobenthos in relation to phytoplankton sedimentation in the Baltic sea. Estuar. coast. shelf. Sci. 45: 149–164.

Ólafsson E., A. Ingólfsson & M. B. Steinarsdóttir, 2001. Harpacticoid copepod communities of floating seaweed: controlling factors and implications for dispersal. Hydrobiologia 453/454: 189–200.

Pallares, R. E. & M. A. Hall, 1974. Análisis bioestadístico-ecológico de la fauna de copepodos asociados a los bosques de *Macrocystis pyrifera*. (Conclusions). Physis (Buenos Aires) 33: 409–432.

Schizas, N. V. & T. C. Shirley, 1996. Seasonal changes in structure of an Alaskan intertidal meiofaunal assemblage. Mar. Ecol. Prog. Ser. 133: 115–124.

Sieburth, J. McN. & J. L. Tootle, 1981. Seasonality of microbial fouling on *Ascophyllum nodosum* (L.) LeJol., *Fucus vesiculosus* L., *Polysiphonia lanosa* (L.) Tandy and *Chondrus crispus* Stackh. J. Phycol. 17: 57–64.

Skadsheim, A., 1984. Life cycles of *Gammarus oceanicus* and *G. salinus* (Amphipoda) in the Oslofjord, Norway. Holarct. Ecol. 7: 262–270.

Steele, D. H. & V. J. Steele, 1972. The biology of *Jaera* spp. (Crustacea, Isopoda) in the northwestern Atlantic. 1. *Jaera ischiosetosa*. Can. J. Zool. 50: 205-211.

Steele, D. H. & V. J. Steele, 1975. The biology of *Gammarus* (Crustacea, Amphipoda) in the northwestern Atlantic. XI. Comparison and discussion. Can. J. Zool. 53: 1116–1126.

Webb, D. G. & T. R. Parsons, 1988. Empirical analysis of the effect of temperature on marine harpacticoid copepod development time. Can. J. Zool. 66: 1376–1381.

Hydrobiologia **503:** 223–236, 2003.
M.B. Jones, A. Ingólfsson, E. Ólafsson, G.V. Helgason, K. Gunnarsson &
J. Svavarsson (eds), Migrations and Dispersal of Marine Organisms.

Spatio-temporal distribution of recruits (0 group) of *Merluccius merluccius* and *Phycis blennoides* (Pisces, Gadiformes) in the Strait of Sicily (Central Mediterranean)

F. Fiorentino, G. Garofalo, A. De Santi, G. Bono, G. B. Giusto & G. Norrito
CNR-IRMA (Istituto di ricerche sulle Risorse Marine e l'Ambiente), Sezione di Mazara del Vallo dell' IAMC (Istituto per l'Ambiente Marino e Costiero), Via L. Vaccara, 61 91026 – Mazara del Vallo (TP), Italy
E-mail: fabiof@irma.pa.cnr.it

Key words: recruitment, nurseries, *Merluccius merluccius*, *Phycis blennoides*, Strait of Sicily, Mediterranean

Abstract

Aspects of the recruitment of hake (*Merluccius merluccius* L., 1758) and greater fork beard (*Phycis blennoides* Brunnich, 1768) in the Strait of Sicily (Central Mediterranean) are presented. Data were collected from 1994 to 1999 during the international bottom trawl survey program MEDITS. In view of the available literature on juvenile growth of these two species in the Mediterranean area, a length-based criterion was adopted to separate fish belonging to the 0 group (recruits). Recruit density indices (R/km^2) by haul were calculated based upon the likely variability of recruit growth among years and used to study abundance variability and spatial pattern of recruitment and to identify the main nursery areas. Although there was inter-annual variability, two stable areas for *M. merluccius* were identified on the eastern side of the Adventure Bank and the Malta Bank at depths ranging between 100 and 200 m. The main nursery areas for *P. blennoides* were deeper (from 200 to 400 m) and two stable nursery areas were identified on the western and eastern side of the Adventure Bank; other nurseries were found in the easternmost part of the Strait in 1998 and 1999. Recruitments of the two species were significantly correlated, with the strongest recruitment occurring in 1998 and 1999 for each species.

Introduction

Description of space and time patterns of different age groups of fish populations is one of the basic elements for understanding stock dynamics. Many authors have identified spawning, nursery and feeding areas, and migratory patterns, as prerequisites for effective stock assessment (Harden-Jones, 1968; Cushing, 1975; Jennings et al., 2001). Research trawl surveys, used to estimate abundance and demography of many demersal stocks, provide useful information of the distribution of fished species. Trawl surveys carried out in many part of the Mediterranean Sea, within national (for example the GRUND program for the Italian Seas in Relini, 2000) and international frameworks (the MEDITS program for the Mediterranean in Bertrand et al., 2002) are allowing increased understanding of the recruitment of hake (*Merluccius merluccius*), the main demersal resource in the Mediterranean (Fiorentini et al., 1997). These studies have focused on timing and depth range of recruitment (Orsi Relini et al., 1989a), juvenile growth (Orsi Relini et al., 1989b; Orsi Relini et al., 1992; Recasens, 1992; Papaconstantinou et al., 1992; Morales-Nin & Aldebert, 1997), diel migration of juveniles from the bottoms into the water column (Orsi Relini et al., 1997), and mapping the nursery grounds (Orsi Relini et al., 1988; Ardizzone & Corsi, 1997; Ardizzone et al., 1999; Lembo et al., 2000; Maynou et al., 2003). Although these studies provide the basic information on the dynamics of the 0-group hake, they have not considered the inter-annual variability of recruitment processes either in terms of year class strength or spatial distribution of recruits.

Despite the importance of the recruitment process in the dynamics of fished species, knowledge on the early life stages and reproductive phases of most of the demersal species living in the Mediterranean is fragmentary compared with available information on Atlantic stocks (Fogarty, 2000).

According to the literature, *M. merluccius* has a prolonged spawning period in the Mediterranean (Oliver & Massutí, 1995; Papaconstantinou & Stergiou, 1995; Colloca, 1999; Orsi Relini et al., 2002), however, few data are available on the localization of spawning areas. Aggregation of mature adults was reported in the outer shelf-upper slope of the Adriatic Sea (Zupanovic, 1968), and between 100 and 200 m in the Gulf of Tunis (Boulhal, 1973). Eggs were found between 50 and 200 m depth in the Adriatic (Karlovac, 1965) and pelagic larvae in the outer shelf-upper slope along the Catalan coast (Sabatés, 1990). Change from the pelagic to the benthic habitat occurs when young individuals are about 3 cm TL (total length) (D'Ancona, 1956). Orsi Relini et al. (1989a) reported that the distance between spawning areas and nurseries, covered through larval dispersion due to currents, may be considerable, as for other species of genus *Merluccius*, such as *M. productus* (Bailey, 1981; Babcock Hollowed & Bailey, 1989). Settlement is a long process with more than one peak per year (Orsi Relini et al., 2002).

Little is known of the spawning period, eggs and larvae of *P. blennoides* in the Mediterranean region. Reproduction occurs from late summer to early winter (Massutí et al., 1996; Belcari & Biagi, 1999). According to D'Ancona (1933), the change from pelagic to benthic habitat occurs when young fish are between 3 and 4.6 cm TL. In contrast to *M. merluccius*, the Mediterranean populations of *P. blennoides* have a spring discrete recruitment (Belcari & Biagi, 1999; Ragonese et al., 2002); however small fish, between 3 and 8 cm TL, may be caught in winter (Massutí et al., 1996).

The present study aims to provide information on aspects of the recruitment process of two very important commercial gadoids, the hake (*Merluccius merluccius*) and greater fork beard (*Phycis blennoides*), in the Strait of Sicily (Central Mediterranean). This area is one of the most important demersal fishing grounds in the Mediterranean. This work reports recruitment variability and the spatial distribution of the 0 group through a time series, comparing patterns observed in the two species. Patterns are discussed taking into account key biotic and abiotic factors affecting the dynamics of young stages in fish populations.

Study site

The Strait of Sicily has a complex bottom morphology (Fig. 1). Along the southern coasts of Sicily, the shelf is widest in the westernmost (Adventure Bank) and Easternmost sectors (Malta Bank). These large banks are separated by the narrow shelf in the central part. The shape of slope is extremely irregular, incised by many canyons, trenches and steep declives. From an oceanography point of view, the Strait of Sicily is characterised by a two-layer flux model. The upper layer, called "Modified Atlantic Water" (MAW), is identified by water with a relatively low salinity flowing from the western to the eastern basin. The lower layer, called "Levantine Intermediate Water" (LIW), is identified by water with a relatively high salinity flowing in the opposite direction, like an undercurrent (Onken & Sellschopp, 1998). According to Robinson et al. (1991), the MAW motion, called "Atlantic-Ionian Stream" (AIS), shows a rather steady mean path. It enters the Strait from the western boundary along the Adventure Bank, comes close to the southern coast of Sicily, and then moves further from the coast, once it encounters the Malta Bank. The complex bathymetry influences the current features in the region. Measurements have shown that the mesoscale signal prevails in the upper layer, with the presence of eddies, meanders, and filaments.

Materials and methods

Analyses were based on six surveys carried out once a year from 1994 to 1999 within the framework of the MEDITS programme (Bertrand et al., 2002). Sampling was carried out in June, with the exception of 1998 and 1999 when the trawl surveys lasted from late May to early June. Surveys included trawlable areas from a depth of 10 m to 800 m, covering about 44,500 km^2. Each area was sampled with a stratified sampling design, where the number of hauls per stratum was proportional to the stratum surface. A total of 56 hauls were conducted on the same site during each survey (Fig. 1). During each survey, hauls were performed using the MEDITS standardized sampling protocols (MEDITS, 1998). The sampling gear was a bottom trawl with stretched mesh size of 20 mm in the cod-end (GOC 73) and a vertical opening between 2.0 and 2.5 m. The duration of hauls was 30 min over the shelves (up to 200 m), and 60 min along the slopes. The area swept in each tow was calculated

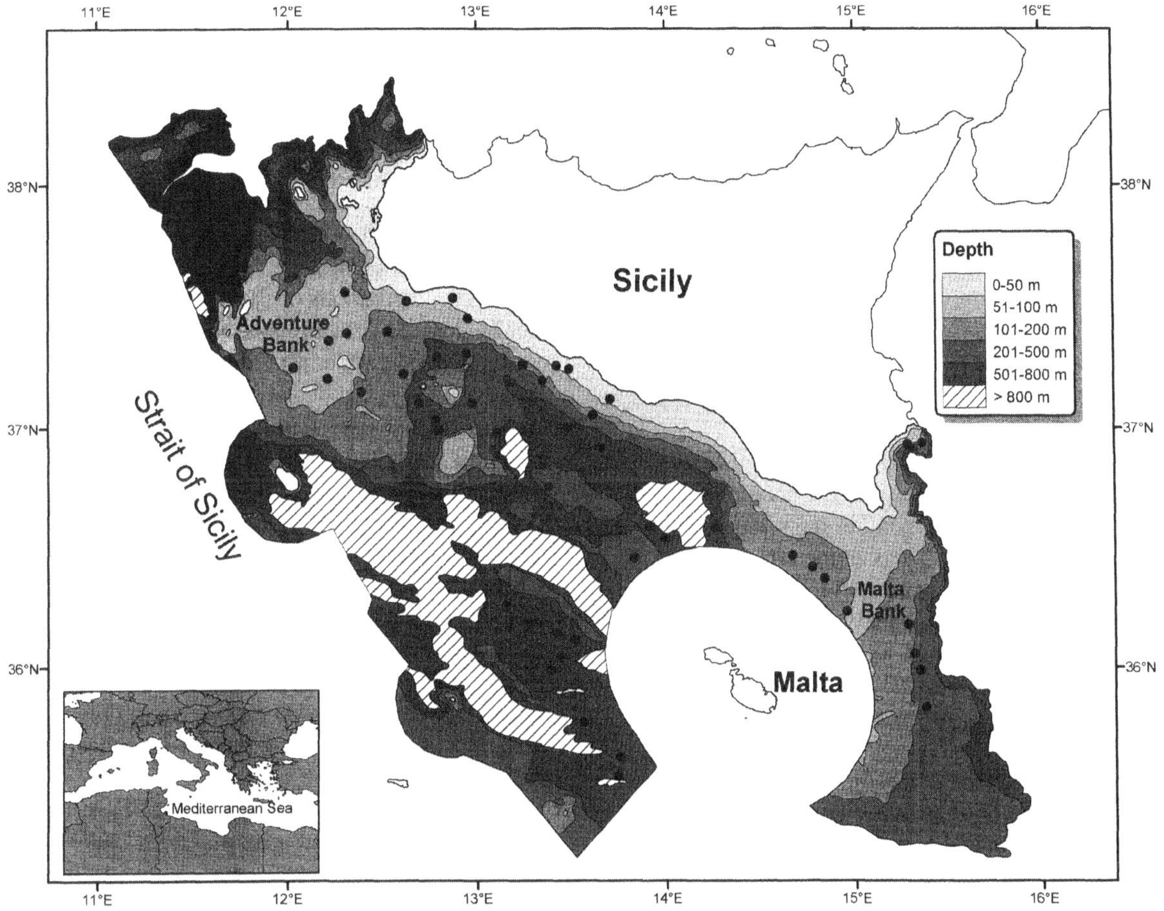

Figure 1. Map showing the extension of the study area, the bathymetric strata and the 56 fixed hauls.

according to the formula proposed by Fiorentini et al. (1999) based on the measures of horizontal opening detected by SCANMAR equipment.

The entire catch of each haul was identified in terms of number and weight per species. For the target species (which included the hake and greater forkbeard), total length of each fish was measured to the nearest 0.5 cm.

Identification of recruits

Recruits (R), corresponding to fish younger than 1 year (0 group), were identified in length/frequency distributions (LFD) for each haul. This was done to take into account the likely variability of the recruit length among years (Orsi Relini et al., 1992; Campana, 1996; Ottersen & Loeng, 2000). To separate the recruits from the older groups, the use of a fixed cut-off for all the years was avoided. Abundance indices of recruits (R/km^2) by haul were calculated according to the following 'length-based' three-step procedure (Levi et al., 2003, modified): 1. preparing 'pooled strata' mean LFD, standardised by unit surface according to the procedure reported in SAMED (2002), for each survey; 2. analysing 'pooled strata' LFD for each survey to estimate mean length ($\bar{l}$), the corresponding standard deviation (sd) and number of individuals of the 1st component from the poly-modal LFD, using the Bhattacharya method as implemented in the MPA routine in the FiSAT package (Gayanilo et al., 1996). On the basis of recent papers on hake (Fiorentino et al., 2000) and greater fork beard (Ragonese et al., 2002), juvenile growth rate in the Mediterranean, a maximum mean length at 12 months (i.e. the end of the 1st year of life), was used as the threshold for assigning a putative age to the 1st component. Therefore, all 'Gaussian' groups with a mean length lower than 14.4 cm were considered belonging to age 0 group for both the gadoids;

and 3. calculating the abundance as number of recruits for each haul for all specimens whose length was below $\bar{l}$ + 1 sd. This arbitrary threshold was chosen to limit the inclusion of specimen of the age 1 group in the estimation of the 0 group abundance. On the basis of swept area, the number of R per haul was transformed into R per km^2, assuming a catchability coefficient equal to 1. These values form the input for the spatial analysis.

Identification of nurseries

The nursery areas were identified through a GIS approach.

For a more effective detection of the discrete nurseries in the area, it was decided to combine the density index of recruits with a index of 'recruit exclusive richness' (RER), calculated as the ratio between the number of recruits and species total catch. This latter index contributes to the identification of areas mostly occupied by the 0 group, excluding areas with older individuals for each species.

Distribution maps of density and RER index were generated for each year by ordinary kriging interpolation. Each pair of GIS layers was overlaid and queried to identify the areas characterised by the co-occurrence of the following criteria: density index higher than the 3rd quartile of the yearly distribution and RER index equal or greater than 70%. The newly generated maps allowed location of areas which fulfilled the above criteria. Such areas were considered the most relevant nurseries in the Strait of Sicily for a given year. Also, in this phase, the procedure adopted allowed the likely inter-annual variability of the recruitment process to be consider. Thus, to discriminate nurseries from other areas, use of a constant threshold of recruit density was avoided. The strength of recruitment was estimated by integrating the interpolated densities over the whole investigated area.

To assess the temporal stability of the observed spatial patterns, an index of persistence was defined as follows. Based on a binary classification, areas were scored with values of 1 (nurseries) or 0 (otherwise). By overlaying the series of reclassified nursery maps, the sum of the scores by pixel divided by the number of years allowed the assessment of the degree of persistence of the nurseries. The index ranged between 0 and 1, where 1 indicates 'stable nursery' and 0 'nursery absence'. Visualisation was restricted to persistence indices of 0.5 or more. All analyses were carried out using ArcView 8 by ESRI.

Results

The length–frequency distributions of each species showed that most belonged to the 0 group (Figs 2 and 3). The analysis indicated also some variability among years in the mean length of the 0 group, ranging from 9.0 (1996) to 12.2 cm (1995) for *M. merluccius*, and between 10.3 (1999) and 12.8 cm (1994) for *P. blennoides* (Table 1).

Yearly distributions of the hake recruitment index highlighted the presence of several areas of recruit concentration which appeared variable in terms of location, extension and density (Fig. 4). However, identification of nurseries based on the criteria described previously, resulted in the delimitation of a few well defined patches in which the young of the year were highly and exclusively concentrated. Two large nurseries were evident each year on the Adventure Bank and Malta Bank, while minor nurseries were found occasionally in the central part of the area. Overall, hake nurseries covered a fairly constant surface over time (mean surface $\approx$ 2200 km^2; sd $\approx$ 375). Mean density of recruits in the nursery areas varied between 460 and 1260 individuals per km^2, with a mean of 990 (CV= 31%).

Yearly maps of recruit density for *P. blennoides* showed a more defined spatial pattern with two extended areas of recruit concentration located on both the western and eastern side of the Adventure Bank (Fig. 5). Only in the final two years (1998 and 1999) relevant nurseries were found along the eastern border of the Malta Bank. Different from hake, large inter-annual variability of the nursery surface was observed. The minimum surface (about 2230 km^2) corresponded to the lowest recruitment for greater fork beard (about 1 950 000 recruits), while the largest area (about 10 950 km^2) corresponded to the highest abundance (about 5 700 000 recruits). In nurseries recruit mean densities ranged between 160 and 430 individuals per km^2, with a mean value across years of 290 (CV= 35%).

Estimation of recruit abundance for the total area showed that hake recruitment was more stable than that of greater fork beard (Fig. 6). Hake recruitment ranged between 3 746 000 and 9 337 000 (mean=7 420 000; sd=1 964 000), while *P. blennoides* varied between 1 952 000 and 5 697 000 (mean=3 846 000; sd=1 545 000) and showed an increasing trend over time. The recruitment strength of the two species was significantly correlated (r= 0.844;

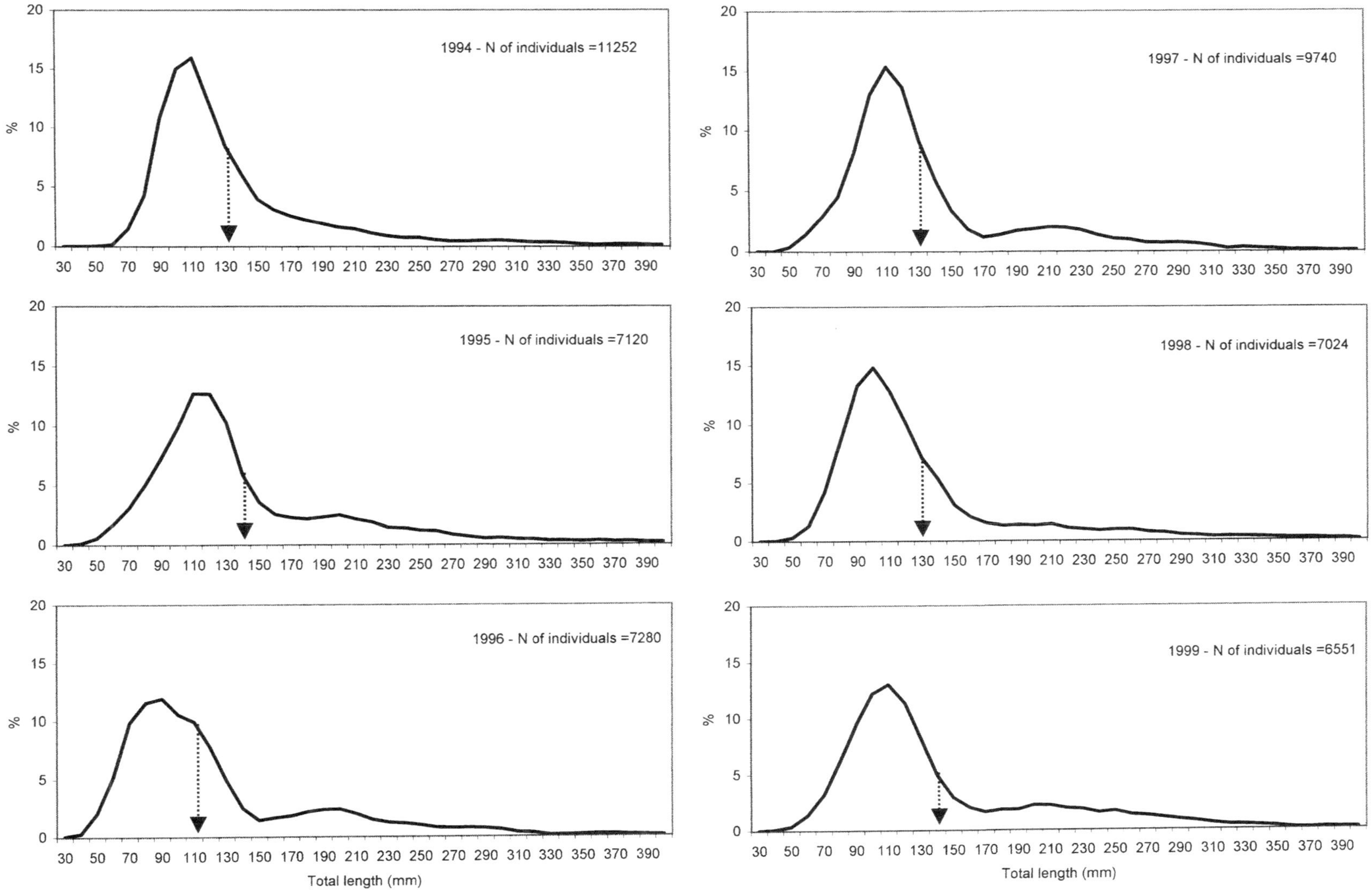

Figure 2. Length–frequency distribution of *M. merluccius* per survey. The cut-off used for the recruit identification are overimposed.

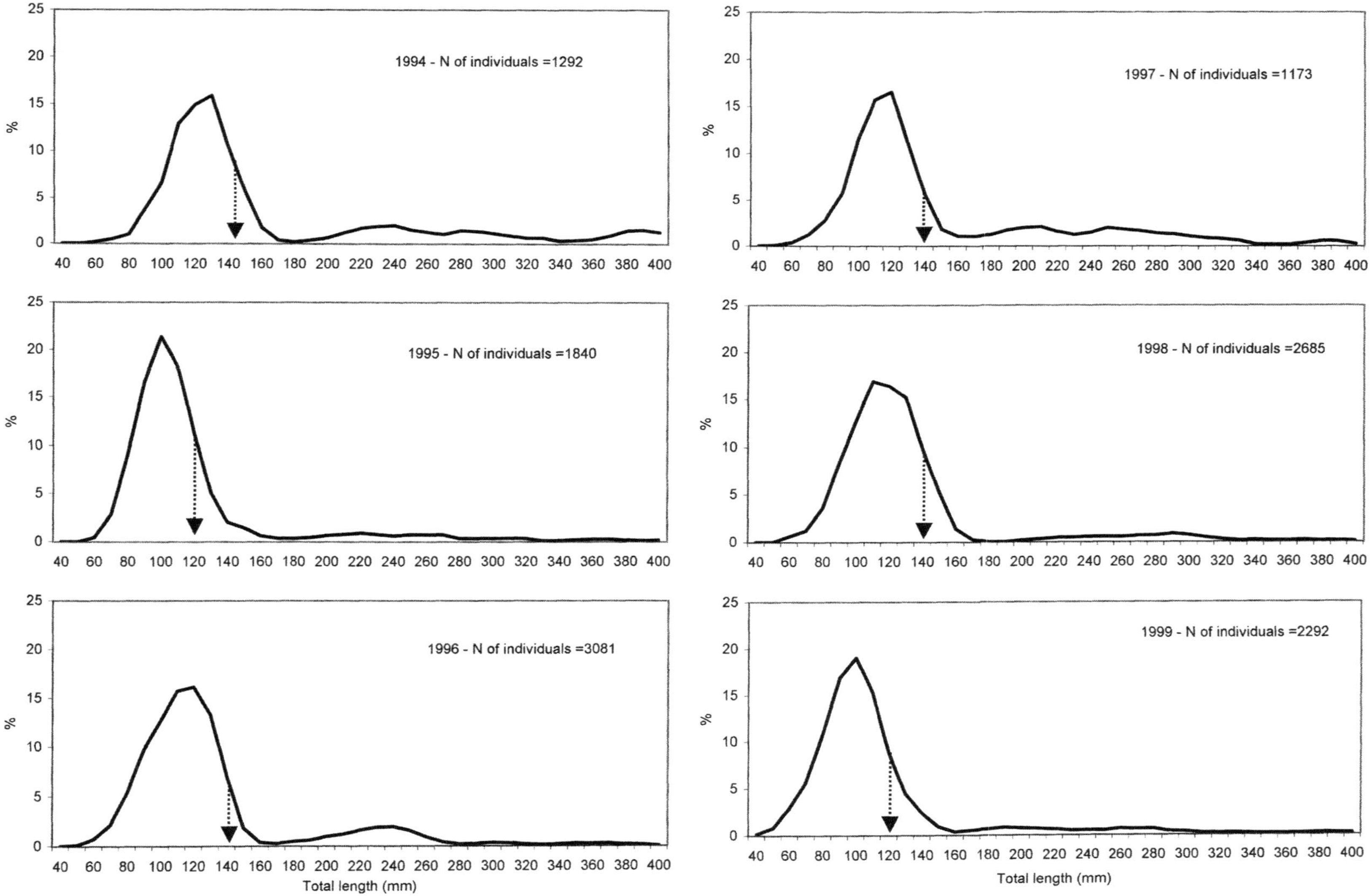

Figure 3. Length–frequency distribution of *P. blennoides* per survey. The cut-off used for the recruit identification are overimposed.

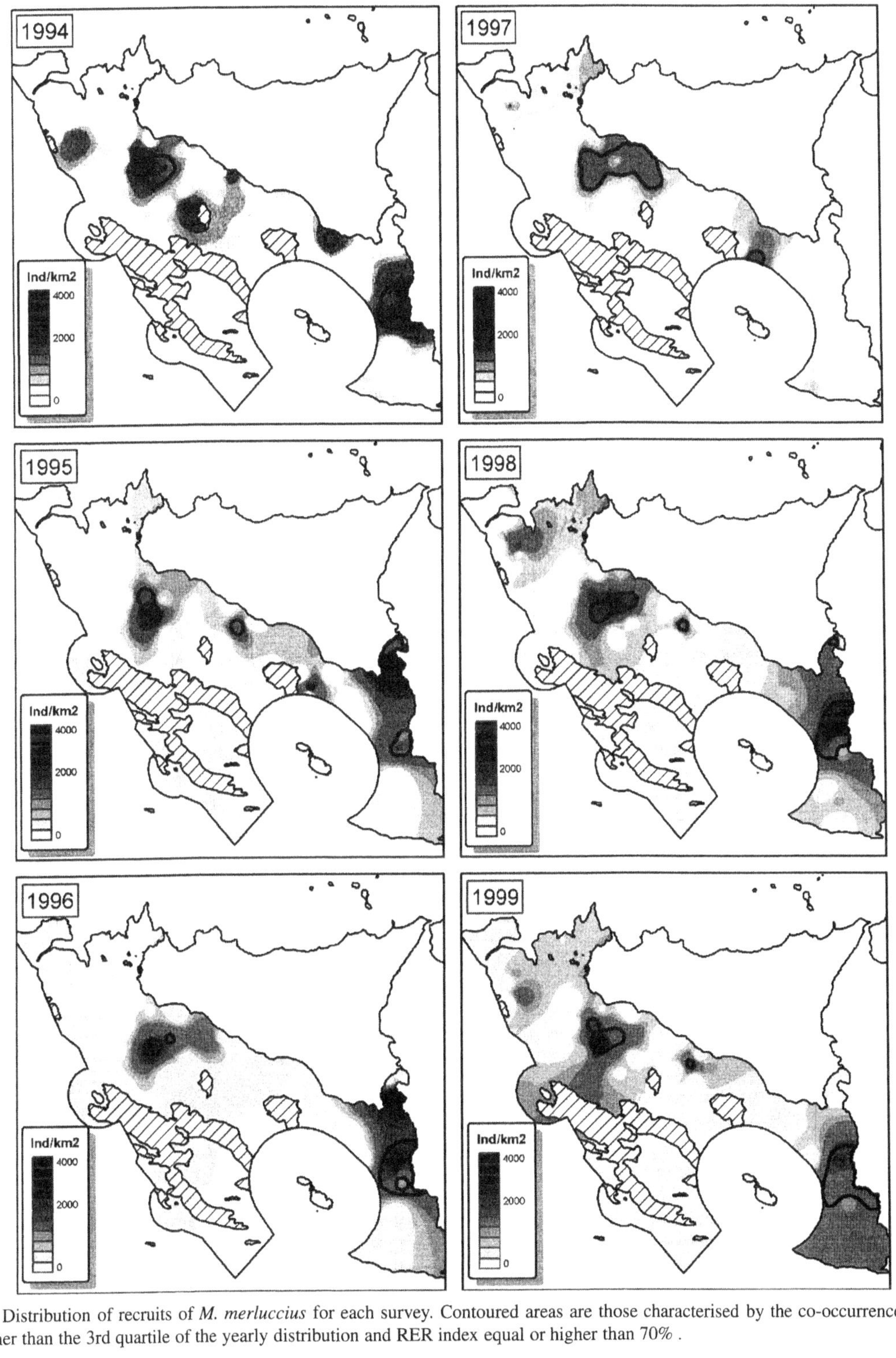

Figure 4. Distribution of recruits of *M. merluccius* for each survey. Contoured areas are those characterised by the co-occurrence of density index higher than the 3rd quartile of the yearly distribution and RER index equal or higher than 70% .

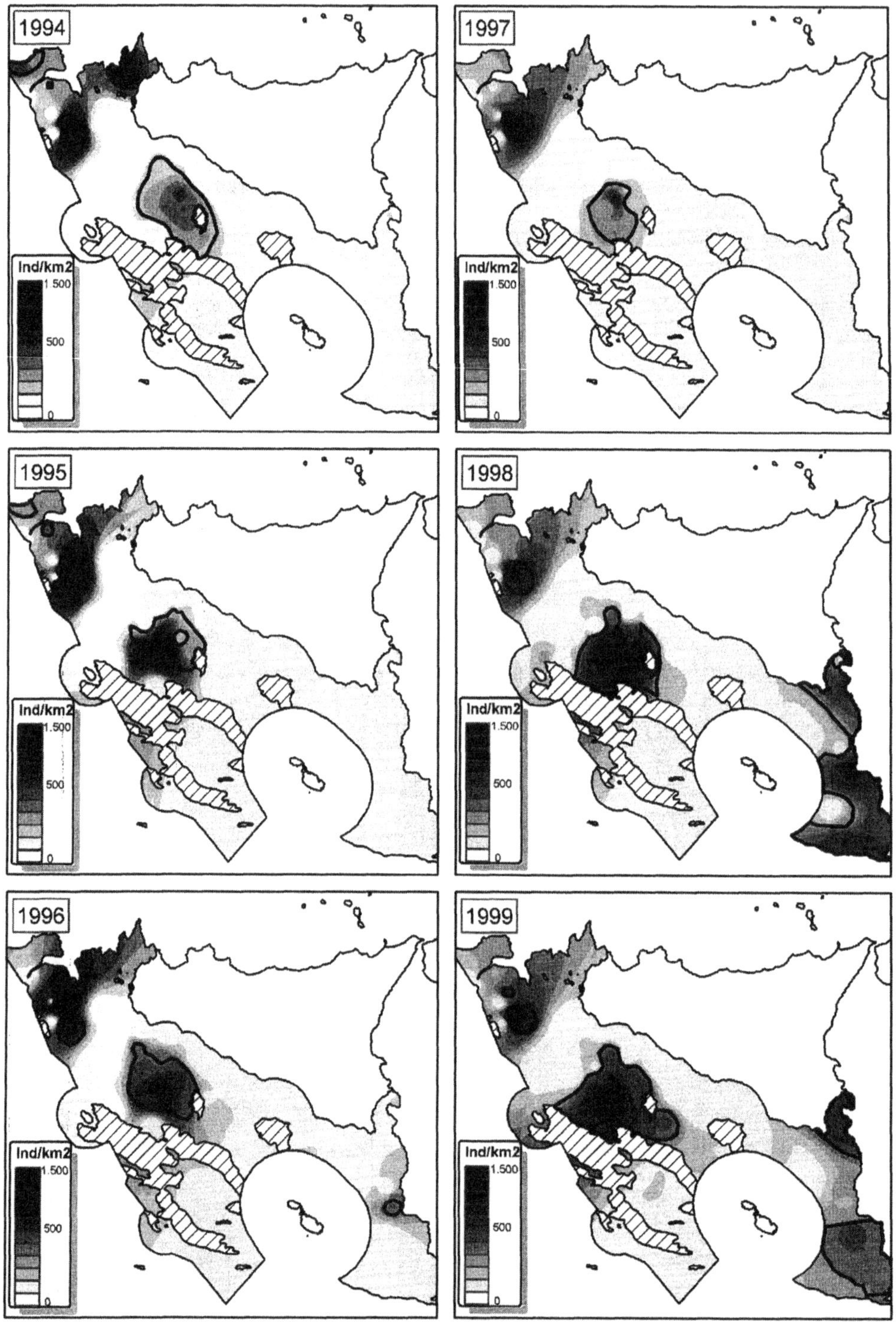

Figure 5. Distribution of recruits of *P. blennoides* per each survey. Contoured areas are those characterised by the co-occurrence of density index higher than the 3rd quartile of the yearly distribution and RER index equal or higher than 70%.

Table 1. Characteristics of the first modal group of *M. merluccius* and *P. blennoides* in standardised LFD. Mean R = density in N per 10 km^2; mean TL = mean length in cm; sd = standard deviation; cut-off size in cm

Year	Merluccius merluccius				Phycis blennoides			
	N	Mean	sd	Cut-off	*N*	Mean	sd	Cut-off
1994	6739	11.06	1.45	12.51	883	12.83	1.7	14.53
1995	3988	12.22	1.76	13.98	1561	10.75	1.61	12.36
1996	3734	9.05	1.6	10.65	2554	11.75	2.06	13.81
1997	6115	11.64	1.8	13.44	776	11.99	1.7	13.69
1998	4616	10.68	1.83	12.51	2399	12.26	2.03	14.29
1999	4582	11.33	2.62	13.95	1943	10.31	1.83	12.14
Mean	**4962**	**11**	**1.84**	**12.89**	**1686**	**11.65**	**1.82**	**13.43**

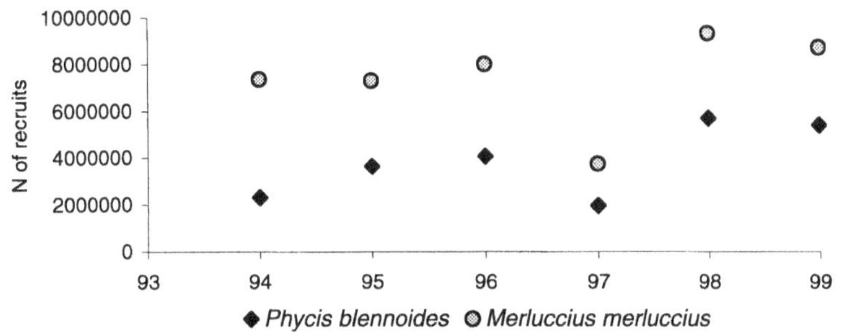

Figure 6. Time series of recruit abundance of *M. merluccius* and*P. blennoides* in the Strait of Sicily, from 1994 to 1999.

$p<0.01$), suggesting that species fluctuation might be synchronised.

Finally, analysis of nursery persistence throughout the study period showed that there were only two areas, which may be considered as stable for *M. merluccius* (Fig. 7). The first, located on the eastern side of the Adventure Bank, presented a core of maximum stability between 100 and 200 m depth, with two surrounding areas of lower persistence oriented along two orthogonal directions. The other, located on the eastern side of the Malta Bank, was characterised by a lower stability and extended to deeper bottoms. Two stable nurseries were identified also for *P. blennoides* (Fig. 8); they were located at deeper bottoms, mainly from 200 to 400 m, on the western and eastern side of the Adventure Bank.

Discussion and conclusions

Approaches followed in this paper avoided to use of fixed thresholds of size and density for recruits and nurseries identification, respectively. Despite the observed variability among years, results for each species showed the occurrence of consistently 'stable' areas, characterized by an almost exclusive presence of high concentrations of recruits, at least during spring. Nursery presence for hake in the southeastern side of the Strait of Sicily had been already identified by Lembo et al. (2000).

The estimated strength of recruitment should be considered mainly as a relative measure for many reasons. It is well known that hake recruits feed mainly on small crustaceans (Euphasiacea), caught in the water column at night (Orsi Relini et al., 1989a; Colloca, 1999). This behaviour could affect severely estimation of recruit densities in the nurseries. According to Orsi Relini et al. (1997), trawling, carried out between 08:00 and 14:00 solar time, sampled about 80% of recruits present at a given site. Our hauls were carried out during the day and, since the maximum recruit vulnerability to trawling occurs at midday with a progressive reduction towards sunrise and sunset, our estimation of spring recruitment strength in hake should be considered an underestimation. Furthermore, hake settlement is a lengthy process that, like reproduction, has more than one peak. The presence of more than one pulse of recruitment in hake, implies that our

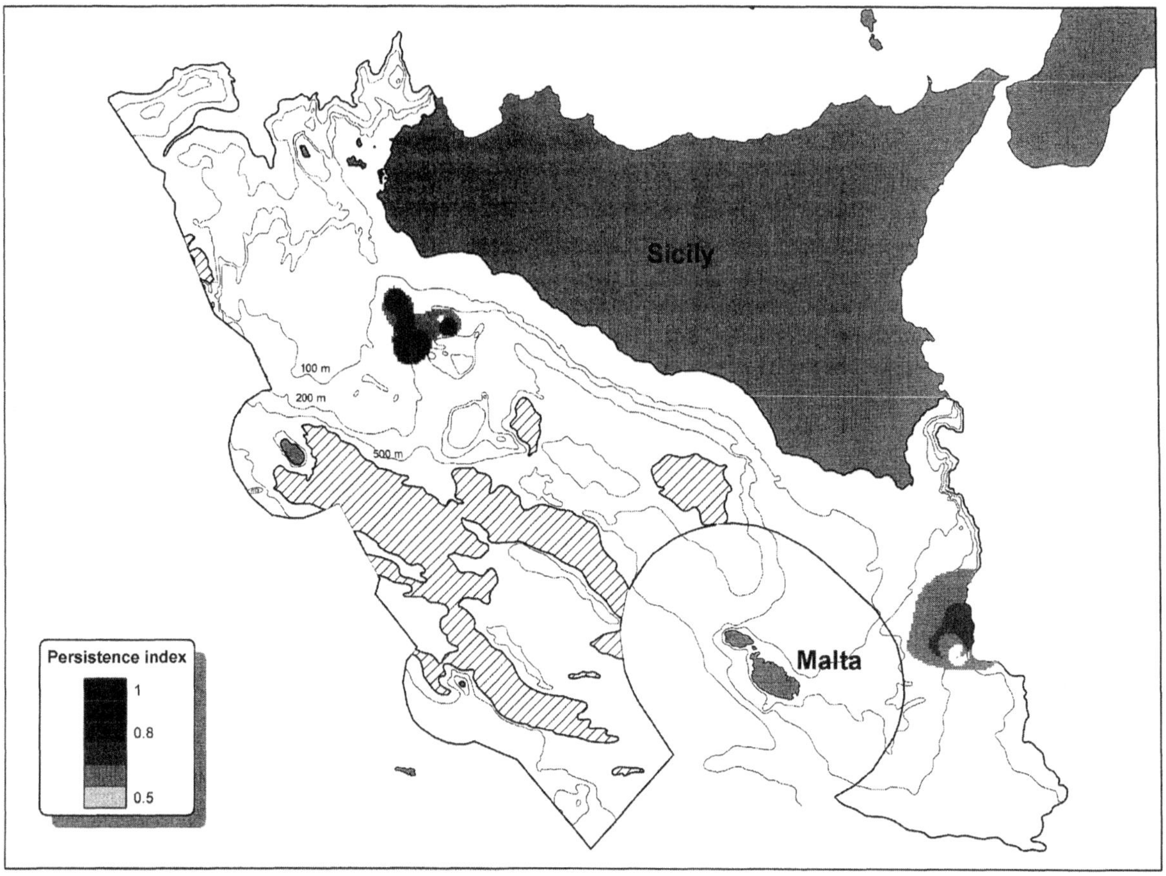

Figure 7. Areas showing stable presence of *M. merluccius* recruits throughout the examined period. The index of persistence ranges between 0 and 1, where 1 indicates stable nursery and 0 complete lack of nursery. Visualization was restricted to persistence index of 0.5 or more.

nursery identification, and corresponding estimates of recruit abundance in spring, give only a partial picture of the recruitment processes in the area, underestimating the annual amount of recruits per year. Since hake recruits can be found throughout the year in the Strait of Sicily, further studies on the seasonal variability in abundance and spatial distribution of recruits should be carried out.

Evidence suggests that hake recruits settling in different seasons may behave in different ways. According to Orsi Relini et al. (1989a), the smallest hake recruits (5 cm tl modal length) settled in the Ligurian Sea at about 100 m mainly in spring-early summer. Another settlement of very small hake, although less abundant than the previous one, occurred in winter at about 50 m, when the seawater was well mixed and inshore transport was greater. From the first settlement zone, the young fish spread over wider areas, forming nurseries (50–250 m). This expansion occurs towards both shallow and deep waters in winter, and only towards the deepest level in summer–autumn. In certain areas and years, 0 group fish may be found at greater depths, i.e. between 300 and 400 m in the Eastern Ligurian Sea (Orsi Relini et al., 1988).

More recently, differences in depth affected by recruitment were confirmed by Maynou et al. (2003) in the Catalan Sea. Recruits were found between 60 and 160 m depth in autumn and winter, while in spring and summer, their depth range extended down to 300 m. Investigating the seasonal variability in hake nurseries off the Catalan Sea, Maynou et al., *loc. cit.* found areas where recruits were present all year round, representing the nursery core. These areas were encircled by larger ones in which the presence of hake recruits might vary seasonally. Recruit density in the core and the surrounding areas was similar, although the contribution of the latter was more important than the former in producing the total recruitment for the Catalan Sea.

Many recent studies outside the Mediterranean have shown a positive correlation between recruitment

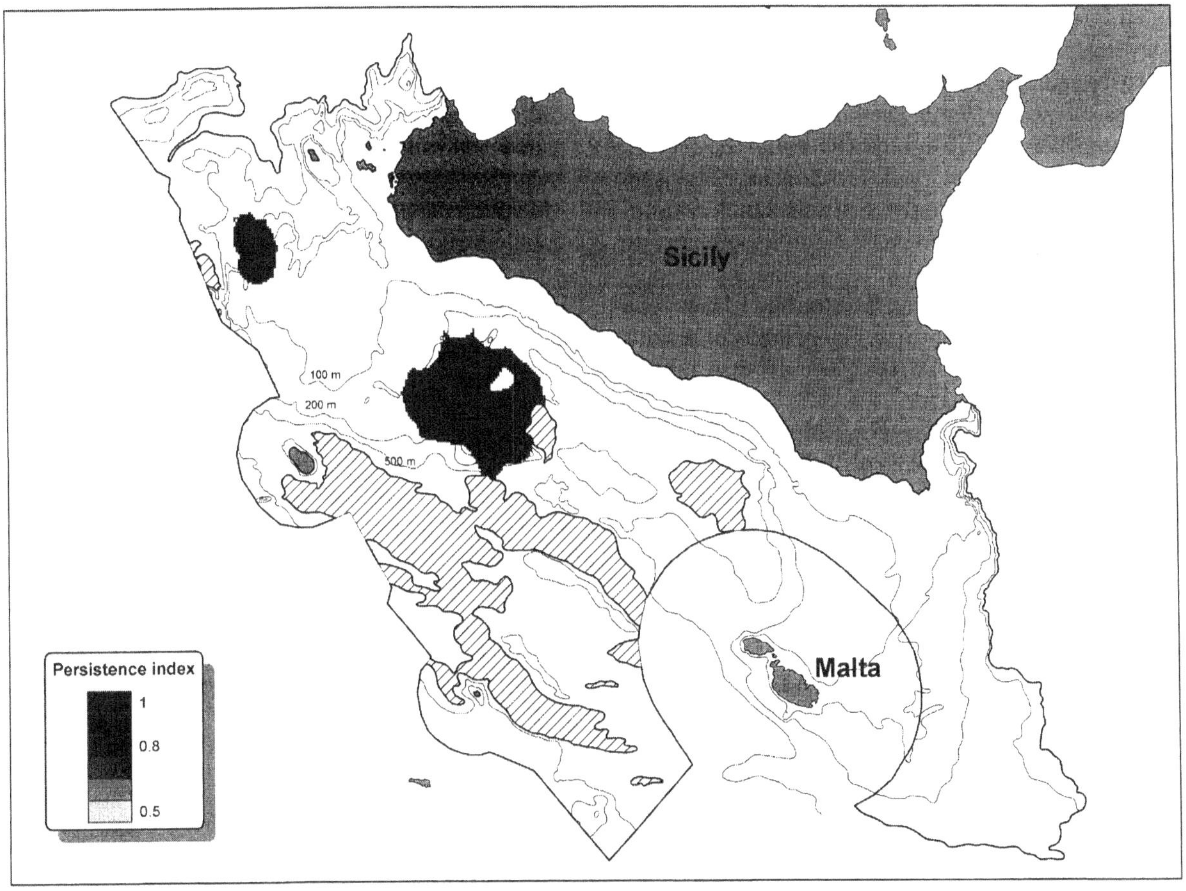

Figure 8. Areas showing stable presence of *P. blennoides* recruits throughout the examined period. The index of persistence ranges between 0 and 1, where 1 indicates stable nursery and 0 complete lack of nursery. Visualization was restricted to persistence index of 0.5 or more.

strength and the surfaces occupied by the nurseries (Petigas & Poulard, 1989; Rijnsdorp et al., 1992; Van der Veer et al., 2000). According to Casey & Pereiro (1995), the distribution of *M. merluccius* recruits in the Atlantic Ocean varied with the year-class strength, contracting in less abundant years and expanding during the most abundant ones. Conversely, Sanchez & Gil (2000) reported that hake nurseries in the Bay of Biscay remained generally stable in location and surface, occupying well-defined patches in localised areas on the continental shelf, although hake recruit densities varied considerably over time.

According to our results, the examined species behaved in different ways. No obvious variation of the nursery surface in function of the year class strength was detected for hake. This finding may be due to little variability in abundance of hake recruits in the time series. On the other hand, the greater fork beard nurseries showed a significant increase in the area occupied as a function of year-class strength. Another remarkable result was the small variation of recruitment strength in the two species due, probably, to the shortness of available time series. The ratio between the strongest and the weakest year class varied between 2.5 in hake and 2.9 in greater fork beard.

As indicated by Ursin (1982), recruitment of gadoids in the North Sea (1963–1975) ranged by a minimum factor of 6 in cod and saithe to a maximum of 100 in haddock. This level of year-to-year variability was confirmed by Rothschild (2000), who reported the existence of decadal and multi-decadal changes in recruitment in a longer time series for North Sea cod, whiting and haddock (1920–1994). However, Alvarez et al. (2001) pointed out that *M. merluccius* showed rather low variability in recruitment strength compared with other gadoids, with the coefficient of variation ranging between 22% in the northern ICES stock and 60% in southern ICES stock.

It is well known that the production of fished species, and the variation in abundance of these species in space and time, is driven by physical and biological processes, but very few studies are available for Mediterranean demersal resources. Recently, Levi et al. (2003) investigated the stock-recruitment relationship for red mullet in the Strait of Sicily, including environmental information in terms of sea surface temperature (SST) anomaly as a proxy for oceanographic processes affecting recruitment. Results showed that, for a given level of spawning stock, higher levels of recruitment corresponded to SST, being warmer than average during the early life stages.

Unfortunately, the heavy exploitation of gadoids in the Mediterranean (Farruggio & Papaconstantinou, 1998), together with the supposed low vulnerability of adults to fine mesh trawling (Caddy, 1991; Abella et al., 1997), makes it difficult to study the adult fraction of these stocks. However, our results suggest that oceanographical features play a major role, with the eddies and the frontal system, produced by the AIS, influencing and maintaining the spatial structure of nurseries in time. Very recently, Garcia Lafuente et al. (2002) pointed out the importance of frontal systems and eddies in influencing the spawning, larval drift and recruitment of anchovy in the Strait of Sicily. Sanchez & Gil (2000), studying the coupling of hake recruitment with hydrographical structures in the Bay of Biscay, reported that the size and location of nurseries over the continental shelf is influenced by mesoscale hydrological anomalies (eddies). These eddies appear to retain larvae and juveniles, and sustain the feeding behavior of the recruits. Thus, these aggregations are located in areas of the continental shelf where the anomalies are repeated to a greater or lesser extent every year. Another result, suggesting the importance of environmental factors in determining the strength of recruitment, is the high correlation between recruit abundance in the two species, with 1997 being the poorest year class and 1998 the best for both hake and greater fork beard.

Information on the time persistence of nurseries allows management measures, based on seasonal closures, to be implemented, which may have an effect comparable to measures aimed at increasing mesh size (Caddy, 1993). Within this framework, the delimitation of 'stable' nursery areas, taking into account not only the highest presence of recruits/juveniles but also their exclusive occurrence in the area, can be crucial for providing advice to fisheries management, especially when the nursery areas are included in the fishing grounds. Indeed, nurseries of the investigated species are situated mainly between depths of 100 and 200 m for the hake; those for the greater fork beard were found at depths greater than 200 m. This is in agreement with the literature concerning both the hake (Orsi Relini et al., 2002) and the greater fork beard (Massutí et al., 1996). These habitats, which are commonly exploited by trawlers, are less vulnerable than those closer to the coast shallow areas, but could be damaged by dumping and excessive trawling.

Acknowledgements

This study was carried out within the framework of two UE-funded projects managed by Ifremer, Ieo, Sibm and Ncmr: MEDITS (Mediterranean International Trawl Surveys) and SAMED (Stock Assessment in the Mediterranean). L. Orsi Relini, S. Ragonese and G. Lembo, together with all the colleagues involved in these programs, are warmly thanked for their support.

References

Abella, A., J. F. Caddy & F. Serena, 1997. Do natural mortality and availability decline with age? An alternative yield paradigm for juvenile fisheries, illustrated by the Hake *Merluccius merluccius* fishery in the Mediterranean. Aquat. Living Res. 10: 257–269.

Alvarez, P., L. Motos, A. Uriarte & J. Egana, 2001. Spatial and temporal distribution of European hake, *Merluccius merluccius* (L.), eggs and larvae in relation to hydrographical conditions of the Bay of Biscay. Fish. Res. 50: 111–128.

Ardizzone, G. D. & F. Corsi (eds), 1997. Atlas of Italian Demersal Fishery Resources. Trawl Survey 1985–1987. Biol. Mar. Medit. 4: 568 pp.

Ardizzone, G. D., F. Corsi & S. Agnesi (eds), 1999. Atlante delle Risorse Ittiche Demersali Italiane Triennio 1994–1996. Ministero per le Politiche Agricole. Roma

Bailey, M. K., 1981. Larval transport and recruitment of Pacific Hake *Merluccius productus*. Mar. Ecol. Prog. Ser. 6: 1–9.

Babcock Hollowed, A. & M. K. Bailey, 1989. New perspectives on the relationship between recruitment of Pacific Hake *Merluccius productus* and the ocean environment. Can. Spec. Publ. Fish. aquat. Sci. 108: 207–220.

Bertrand, J. A., L. Gil de Sola, C. Papaconstantinou, G. Relini & A. Souplet, 2002. The general specifications of the MEDITS surveys. Sci. mar. 66 (Suppl. 2): 9–17.

Belcari, P. & F. Biagi, 1999. *Phycis blennoides*. In Relini G., J. A. Bertrand & A. Zamboni (eds), Synthesis of Knowledge on Bottom Fishery Resources in Central Mediterranean (Italy and Corsica). Biol. Mar. Medit. 6: 189–196.

Boulhal, M., 1973. Le merlu de côtes nord de la Tunisie: Etude économique et biologique (reproduction, sex ratio et répartition bathymétrique). Bull. Inst. Océanogr. Pêche Salammbô 2: 579–603.

Caddy, J. F., 1991. Death rates and time intervals: is there an alternative to the constant natural mortality axiom? Rev. Fish Biol. Fish. 1: 109–138.

Caddy, J. F., 1993. Some future perspectives for assessment and management of Mediterranean Fisheries. Sci. mar. 57: 121–130.

Campana, S. E., 1996. Year class strength and growth rate in young Atlantic cod *Gadus morhua*. Mar. Ecol. Prog. Ser. 135: 21–26.

Casey, J. & J. Pereiro, 1995. European hake (*M. merluccius*) in the North-east Atlantic. In Alheit, J. & T. J. Pitcher (eds), Hake, Fisheries, Ecology and Markets. Chapman & Hall, London. Fish & Fisheries Ser. 15: 125–147.

Colloca, F., 1999. *Merluccius merluccius*. In Relini, G., J. A. Bertrand & A. Zamboni (eds), Synthesis of Knowledge on Bottom Fishery Resources in Central Mediterranean (Italy and Corsica). Biol. mar. Medit. 6: 259–270.

Cushing, D. H., 1975. Marine Ecology and Fisheries. University Press, Cambridge: 278 pp.

D'Ancona, U., 1933. Uova, larve e stadi giovanili di Teleostei. Fauna e Flora del Golfo di Napoli 38: 232–238.

D'Ancona, U., 1956. Gadidae: uova, larve e stadi giovanili dei teleostei. Pubbl. Stn. Zool. Napoli 38: 19–31.

Farruggio, H. & C. Papaconstantinou, 1998. The status of fisheries resources of the Mediterranean. In Gaps in Mediterranean Fishery Science. CIESM Workshop Ser. 5: 13–24.

Fiorentini, L., J. F. Caddy & J. I. De Leiva, 1997. Long and short trends of Mediterranean fishery resources. FAO GFCM Stud. Rev. 69: 72 pp.

Fiorentini, L., G. Cosimi, A. Sala, I. Leonori & V. Palumbo, 1999. Efficiency of the bottom trawl used for Mediterranean international trawl survey (MEDITS). Aquat. Living Resour. 12: 187–205.

Fiorentino, F., S. Gangitano, S. Ragonese, P. Rizzo & D. Levi, 2000. How the Mediterranean hake grow? Does the available von Bertalanffy parameters describe growth of *Merluccius merluccius* adequately? Working Group on Demersals of Sub-Committee on Stock Assessment of the SAC of GFCM. Sète, France, 21–23 March, 2000: ED/P/FF-SG-SR-PR-DL/2/0300/REL: 18 pp (mimeo).

Fogarty, M. J. (ed.), 2000. Recruitment Dynamics of Exploited Marine Populations: Physical-Biological Interactions. Part 1. ICES J. mar. Sci. 57: 189–464.

Gayanilo, F. C. Jr., P. Sparre & D. Pauly, 1996. FAO-ICLARM Stock Assessment Tools (FISAT) User's Manual. FAO - Computerized Information Series – Fisheries: 126 pp.

García Lafuente, J., A. García, S. Mazzola, L. Quintanilla, J. Delgado, A. Cuttitta & B. Patti, 2002. Hydrographic phenomena influencing early life stages of the Sicilian Channel anchovy. Fish. Oceanogr. 11: 31–44.

Harden-Jones, F. R., 1968. Fish Migration. Edward Arnold, London: 325 pp.

Jennings, S., M. J. Kaiser & J. D. Reynolds, 2001. Marine Fisheries Ecology. Backwell Science, Oxford: 417 pp.

Karlovac, J., 1965. Contribution a la connaissance de l'écologie du merlu, *Merluccius merluccius* L., dans le stade planctonique de sa vie in Adriatique. Rapp. P.V. Reun. CIESM 18: 461–464.

Lembo, G., T. Silecchia, P. Carbonara & M. T. Spedicato, 2000. Nursery areas of *Merluccius merluccius* in the Italian Seas and in the east side of the Adriatic Sea. Biol. mar Medit. 7: 98–116.

Levi D., M. G. Andreoli, A. Bonanno, F. Fiorentino, G. Garofalo, S. Mazzola, G. Norrito, B. Patti, G. Pernice, S. Ragonese, G. B. Giusto & P. Rizzo, 2003. Embedding sea surface temperature anomalies in the stock recruitment relationship of red mullet (*Mullus barbatus* L. 1758) in the Strait of Sicily. Sci. mar. 67 (Suppl. 1): 259–268.

Maynou, F., J. Lleonart & J. E. Cartes, 2003. Seasonal and spatial variability of hake (*Merluccius merluccius* L.) in the NW Mediterranean. Fish. Res. 60: 65–78.

Massutí, E., B. Morales-Nin & D. Lloris, 1996. Bathymetric distribution and recruitment patterns of *Phycis blennoides* (Pisces:Gadidae) from the slope of the north-western Mediterranean. Sci. mar. 60: 481–488.

Morales-Nin, B. & Y. Aldebert, 1997. Growth of juvenile *Merluccius merluccius* in the Gulf of the Lions (NW Mediterranean) based on otolith microstructure and length–frequency analysis. Fish. Res. 30: 77–85.

Oliver, P. & E. Massutí, 1995. Biology and fisheries of western Mediterranean hake (*M. merluccius*). In Alheit, J. & T. J. Pitcher (eds), Hake, Fisheries, Ecology and Markets. Chapman & Hall, London. Fish & Fisheries Ser. 15: 181–201.

Onken, R. & J. Sellschopp, 1998. Seasonal Variability of Flow Instabilities in the Strait of Sicily. J. Geophys. Res. (C Oceans) 103: 24799–24820.

Orsi Relini, L., F. Fiorentino & M. Cappanera, 1988. Mapping the nurseries of *Merluccius merluccius* in the Ligurian Sea. Rapp. Comm. int. Mer. Médit. 31: 275.

Orsi Relini, L., F. Fiorentino & A. Zamboni, 1989a. Nurseries del Nasello Mediterraneo: dove, quando, perche'. Nova Thalassia 10 (Suppl. 1): 407–416.

Orsi Relini, L., M. Cappanera & F. Fiorentino, 1989b. Spatial-temporal distribution and growth of *Merluccius merluccius* recruits in the Ligurain Sea. Observations on the 0 group. Cybium 13: 263–270.

Orsi Relini, L., F. Fiorentino & A. Zamboni, 1992. Growth of the Mediterranean hake. Experiences gained in the Ligurian Sea. In Colombo G., I. Ferrari, V. U. Ceccherelli & R. Rossi (eds), Proceedings of the 25th European Marine Biology Symposium, Ferrara, 1990. Olsen and Olsen, Fredensborg: 307–315.

Orsi Relini, L., A. Zamboni, F. Fiorentino & G. Relini, 1997. Vulnerabilità luce-dipendente del nasello (*Merluccius merluccius*) giovanile. Biol. mar. Medit. 4: 256–262.

Orsi Relini, L, C. Papaconstantinou, S. Jukic-Peladic, A. Souplet, L. Gil de Sola, C. Piccinetti, S. Kavadas & M. Rossi, 2002. Distribution of the Mediterranean hake populations (*Merluccius merluccius smiridus* Rafinesque, 1810) based on six years monitoring by trawl surveys: some implications for management. Sci. mar. 66 (Suppl. 2): 21–38.

Ottersen, G. & H. Loeng, 2000. Co-variability in early growth and year-class strength of Barents Sea cod, haddock, and herring: the environmental link. ICES J. mar. Sci. 57: 339–348.

Papaconstantinou, C., G. Petrakis & A. Fourtouni, 1992. A preliminary study on the fishery biology of hake in the Aegean Sea. FAO Fish. Rep. 477: 139–149.

Papaconstantinou, C. & K. I. Stergiou, 1995. Biology and fisheries of eastern Mediterranean hake (*M. merluccius*). In Alheit, J. & T. J. Pitcher (eds), Hake, Fisheries, Ecology and Markets. Chapman & Hall, London. Fish & Fisheries Ser. 15: 140–180.

Petigas, P. & J. Ch. Poulard, 1989. Applying stationary geostatistics to fisheries: a study on hake of the Gulf of Biscay. ICES Doc. 1989/G:62: 21 pp (mimeo).

Ragonese, S., F. Fiorentino, P. Rinelli & S. Greco, 2002. A procedure to evaluate the effect of lag time in studying length structure and growth rate of young fish. The case of *Phycis blennoides*, Brunnich, 1768 (Pisces; Gadiformes) in the Central Mediterranean. Sci. mar. 66 (Suppl. 2): 253–260.

Recasens, L., 1992. Dinamica de poblacion i pesqueria del lluc (*Merluccius merluccius*) al Golfo de Lleon i la mar Catalana. Ph.D. Thesis, Universitat de Barcelona: 398 pp.

Relini, G., 2000. Demersal trawl surveys in Italian Seas: a short review. Actes de Colloques IFREMER 26: 76–93.

Rijnsdorp, A. D., F. A. Van Beek, S. Flatman, R. M. Milner, J. D. Riley, M. Giret & R. De Clerk, 1992. Recruitment in sole stocks, *Solea solea* (L.) in the northeast Atlantic. Neth. J. Sea Res. 29: 173–192.

Robinson, A. R., M. Golnaraghi, W. G. Leslie, A. Artegiani, A. Hecht, E. Lazzoni, A. Michelato, E. Sansone, A. Theocharis & U. Unluata, 1991. The Eastern Mediterranean General Circulation: Features, Structure and Variability. Dyn. Atmos. Oceans 15: 215–240.

Rothschild, B. J., 2000. Fish stocks and recruitment: the past thirty years. ICES J. mar. Sci. 57: 191–201.

Sabatés, A., 1990. Distribution pattern of larval fish populations in the Northwestern Mediterranean. Mar. Ecol. Prog. Ser. 59: 75–82.

SAMED, 2002. Stock Assessment in the Mediterranean SAMED. Final Report. European Commission. DG Fisheries (DG XIV). Project n 99/047.

Sanchez, F. & J. Gil, 2000. Hydrographic mesoscale structures and Poleward Current as a determinant of hake (*Merluccius merluccius*) recruitment in southern bay of Biscay. ICES J. mar. Sci. 57: 152–170.

Ursin, E., 1982. Stability and variability in the marine ecosystem. Dana 2: 51–67.

Van der Veer, H. W., R. Berghahm, J. M. Miller & A. D. Rijnsdorp, 2000. Recruitment in flat fish, with special emphasis on North Atlantic species: Progress made by Flatfish Symposia. ICES J. mar. Sci. 57: 202–215.

Zupanovic, S., 1968. Study of hake (*Merluccius merluccius* L.) biology and population dynamics in the Central Adriatic. FAO GFCM Stud. Rev. 32: 1–24.

Hydrobiologia **503:** 237–244, 2003.
M.B. Jones, A. Ingólfsson, E. Ólafsson, G.V. Helgason, K. Gunnarsson &
J. Svavarsson (eds), Migrations and Dispersal of Marine Organisms.

Growth aspects of *Flustra foliacea* (Bryozoa, Cheilostomata) in laboratory culture

Jens Kahle[1], Gerd Liebezeit[1] & Gisela Gerdes[2]
[1]*Forschungszentrum Terramare, Schleusenstrasse 1, D-26382 Wilhelmshaven*
[2]*Carl von Ossietzky University of Oldenburg, Institute for Chemistry and Biology of the Marine Environment, Marine Station, Schleusenstrasse 1, D-26382 Wilhelmshaven*
E-mail: jens.kahle@terramare.de

Key words: bryozoa, *Flustra foliacea*, growth rate, biomass production

Abstract

Flustra foliacea is a cheilostome marine bryozoan which, after initial horizontal growth during the first year, changes to erect growth thereafter. Appraising this slow-growing species for biotechnological use, required the development of cultivation methods for economical biomass production. Vegetative reproduction via cuttings, as employed in the horticulture of various plants, was tested. Fronds of *F. foliacea* were cut into pieces (10–150 mg fresh weight) and fixed in longitudinally cut silicon tubes. Bryozoa were fed with a mixture of *Isochrysis galbana*, *Dunaliella tertiolecta* and *Cryptomonas* sp. yielding growth rates of 0.3% $\pm$ 0.02 day^{-1}of the initial fresh weight. Growth of the colony occurred firstly at the original growth margin, but later also at the transections, including the side fixed in the silicon tube, opposite to the growth margin. After three months of growth, the two layers of zooids split at the growth margin, forming two monolayers. This growth feature, not described previously for *Flustra foliacea,* was observed under experimental and natural conditions, and is interpreted as the beginning of one type of ramification. According to our findings, vegetative reproduction of *F. foliacea* under laboratory conditions is possible and might be an alternative to natural resources regarding biomass of this species.

Introduction

The erect-growing anascan bryozoan, *Flustra foliacea* (Linneus) has been one of the target species of the Lower Saxonian Marine Biotechnology Programme 1998–2002. This species produces natural products such as a.o. brominated alkaloids (Carle & Christophersen, 1980, 1981; Wulff et al, 1981, 1982a, b; Laycock et al., 1986; Holst et al., 1994a, b), which may act as muscle relaxants (Sjöblom et al., 1983). Since the natural resources of these organisms are not sufficient for large-scale production of the above-mentioned metabolites, laboratory-enhanced biomass production is desirable. General information on *Flustra* is given by Tyler-Walters & Ballerstedt (2002). Under natural conditions, *F. foliacea* colonies grow from a settled, newly-metamorphosed larva in the form of an encrusting mat. If laterally developed lobes at the growth margin meet, growth is deflected upwards. The two layers of zooids grow synchronously, forming a bilaminar, erect frond at 90 ° to the original encrusting mat. Branching of the erect fronds varies between branches and colonies (Stebbing, 1971; Silén, 1981). Stebbing (1971) reported that, in England, growth began in late February/early March and stoped in October/November, with a slight check in growth in August; no growth occurred over winter (Eggleston, 1963; cited in Stebbing, 1971). *F. foliacea* colonies regularly reach 6 years of age, although 12-year old specimens have been reported (Stebbing, 1971; Ryland, 1976). In our study, *Flustra* was grown in common fish tanks and in a specially designed bioreactor. Here, we report our results for vegetative and generative reproduction of *Flustra foliacea* for *in vitro* biomass production.

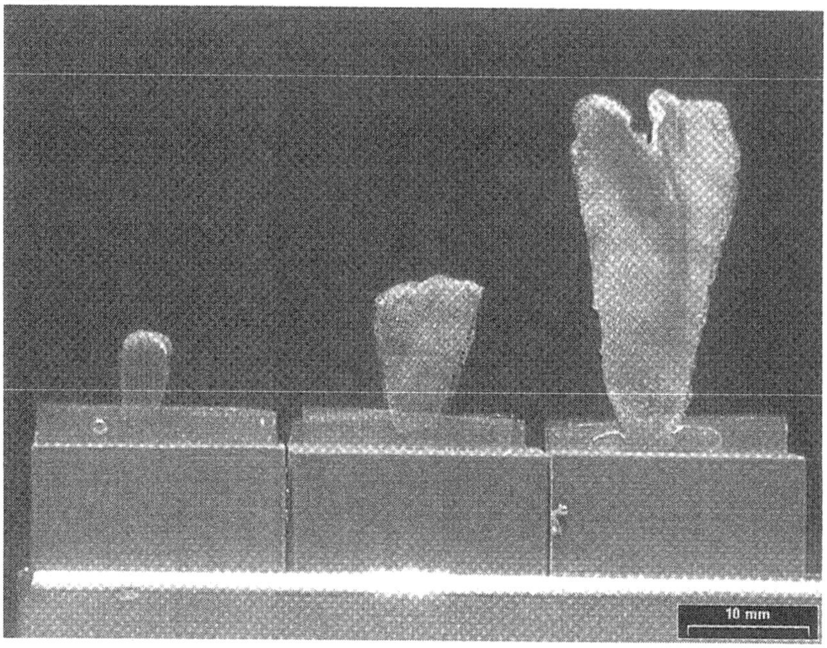

Figure 1. Flustra foliacea cuttings fixed in a silicon tube embedded in PVC blocks.

Table 1. Chemical properties of seawater used for the cultivation of *Flustra foliacea*

	Ammonia [mg L^{-1}]	Nitrite [mg L^{-1}]	Nitrate [mg L^{-1}]	Salinity PSU	pH
Mean	0.12	0.04	20.45	32.23	8.02
±	0.21	0.02	10.06	2.80	0.07

±: standard deviation, number of independent determinations = 4 (over 6 months).

Materials and methods

Original colonies of *Flustra foliacea* were obtained by SCUBA diving from the Steingrund, about 6 n.m. NE of Heligoland, southern North Sea, and by dredging north of the East Frisian island of Spiekeroog (52° 52.50′ N, 007° 44′ E). Initially, colonies were kept intact for the production of larvae. Loose material and colonies still attached to smaller stones were transferred under continuous seawater flow to a constant temperature room (9 °C), and kept in flow-through 20 l aquaria. Food consisted of a mixture of the autotrophic flagellates *Cryptomonas* sp., *Dunaliella tertiolecta* and*Isochrysis galbana*, and the heterotrophic flagellate *Oxyrrhis marina*. To feed *Flustra*, the experimental water was enriched with algae twice a week. Directly after feeding, concentrations of about 10 000 cells m l $^{-1}$ were measured. Considering the water volume of the system (250 l), the introduced biomass [5 kg fresh weight (fw)] of *Flustra,* and a twice weekly feeding regime the food supply, resulted in about 1.4 × 10^3 cells g^{-1} day^{-1}. To achieve higher growth rates in a second set up (one 15 l aquarium with 25 g of *Flustra*), food was supplied daily in concentrations of up to 9 × 10^4 cells l^{-1}, leading to a food supply of 5.4 × 10^7 cells g^{-1} day^{-1}. Good water quality, as indicated by nitrate concentrations <40 mg l^{-1} (Table 1), was maintained by the use of a biofilter and skimmer, additional to regular water changes of 50% of the water. Water flow in the aquaria was enhanced by small submerged rotary pumps (40 l min^{-1}). To encourage vegetative reproduction, cuttings of *Flustra* colonies were fixed in a silicon tube embedded in PVC blocks (Fig. 1). Monthly documentation of growth was achieved by weighing the cuttings and by computer-based measurements of height (mm) and size (mm^2) using the software package Analysis Pro. Growth rates were determined as the slope of the linear regression for the increase in biomass and on the basis of differences between the monthly measurements of the height. Free-swimming larvae were released by adult colonies, indicating good culture conditions. For settlement, larvae were taken up in a pipette and transferred to petri dishes at 9 °C and 4 °C.

Table 2. Actual growth measurements of *Flustra foliacea* cuttings

Day	Front size of the cuttings [mm^2]				Number of zoids *				FW of the cuttings [mg]			
	F_1	F_2	F_3	F_4	F_1	F_2	F_3	F_4	F_1	F_2	F_3	F_4
0	22	71	71	84	411	1330	1331	1573	11	42	39	47
26	22	72	72	84	419	1351	1348	1571	11	44	41	50
42	23	73	73	86	429	1376	1370	1618	12	46	43	52
63	24	75	78	88	452	1455	1411	1650	13	50	47	56
92	26	82	84	97	491	1572	1544	1821	14	52	51	59
125	n.a.	n.a.	n.a.	n.a.	n.a.	n.a.	n.a.	n.a.	(15)	(55)	(54)	(63)
Absolute growth**	4	11	13	13	80	242	214	248	3	10	12	12
Mean ± SD (n = 4)		10 ± 4				196 ± 78				9 ± 4		
% increase	20	16	18	16	20	18	16	16	27	25	30	26
Mean ± SD (n = 4)		17 ± 2				17± 2				27 ± 2		

* Number of zooids was calculated from the frond size and the mean size of a single zooid = 0.11 ± 0.01 mm^2 (n = 20) × 2 with respect to the double layered structure of the fronds of *Flustra foliacea*.
** Measurements of the front size were only applicable for 92 days, due to the divergent growth of the layers (see text for details).
() for comparability, day 125 was not included in the calculations of the absolute growth and % increase.

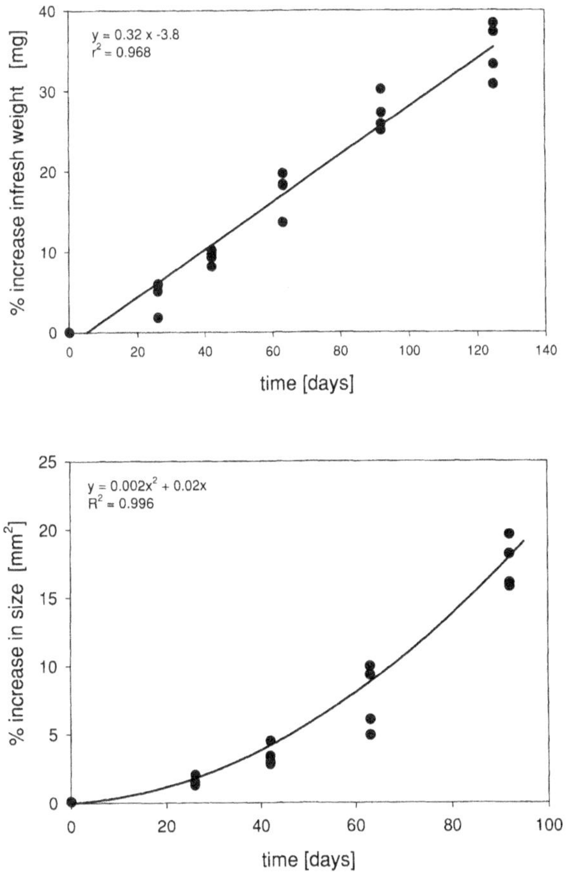

Figure 2. Relative increase in size and fresh weight of *Flustra foliacea* cuttings.

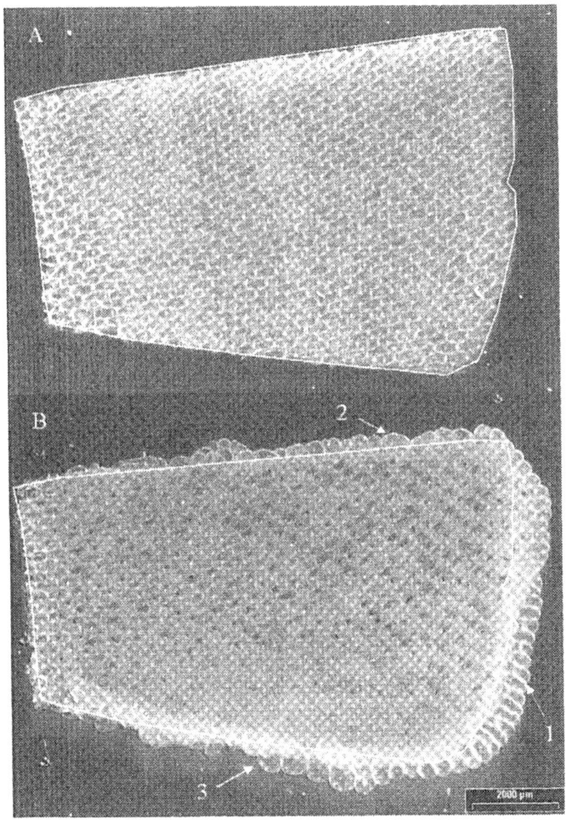

Figure 3. Photo documentation of the growth for *Flustra foliacea* cuttings over a period of 125 days (A: day 0; B: day 125). Growth occurred first at the original growth margin (arrow 1), but later also at the transections (arrows 2 and 3).

Results

Flustra foliacea has been maintained continuously in the laboratory for 3 years. Survival and growth was controlled largely by ambient physical conditions, i.e. temperature, currents, and food supply. Whereas hardly any measurable growth occurred when feeding *Flustra* colonies twice a week, cuttings fed daily increased measurably in weight and size after 1 month (Fig. 2). Growth of the cuttings, related to the initial fresh weight, led to a daily mean growth rate of 0.3% (± 0.02) of the initial weight. Measurements of the growth in height led to an daily increase of 0–0.3 mm. Actual growth measurements of the cuttings size and fresh weight are given in Table 2, next to the calculated numbers of zoids. Growth of the colonies occurred first at the original growth margin, but later also at the transections (Fig. 3). After 3 months of growth, the two layers of zooids split at the growth margin and formed two monolayers (Fig. 4). The divergent growth proceeded while monolayers bend to such an extent that growth margins of each monolayer might eventually meet and form two bilayers (Fig. 5). Splitting up into two monolayers was also observed for *Flustra* colonies under natural conditions, sampled north of the island of Spiekeroog. For these samples, a change in the direction of growth was also observed (Fig. 6). Next to growth at the margin and transections, growth also occurred at the base fixed in the silicon tube (Fig. 7). Colonies collected in August 2001 released, under laboratory conditions, small quantities of larvae, during February and March 2002, and these settled actively on petri dish surfaces. Budding of the settled ancestrula over the first month is documented in Figure 8.

Discussion

Our results demonstrate that *Flustra foliacea* has the potential for *in vitro* biomass production, achieving sexual as well as vegetative reproduction under the documented experimental conditions. Vegetative reproduction via cuttings seems to be a promising procedure, since repair mechanisms work efficiently enough to keep small cuttings of 10 mg fresh weight actively growing. Since bryozoans are filter feeders, fastening the cuttings is essential to allow currents across the surface to bring food to the organisms. The way of securing the cuttings, shown in Figure 1, has several advantages for the experimental work, including easy handling, no chemical interference due to any kind of glue etc., cuttings can be individually marked and investigated and space saving, since distance between the cuttings can be adjusted to their size. Tracking the growth of the cuttings shows that weight increases earlier than size (Fig. 2), which may be explained by the increase of body mass of the existing zooids before budding of new zooids results in a measurable growth in size. The growth experiment was performed with colonies showing initially few polypides and high quantities of brown bodies, which decreased rapidly in number during the first month of the experiment.

Comparison of the growth rate (0.3% of the initial weight day^{-1}) obtained here with other published data is difficult because literature values report height in combination with annual lines found on the frond. Stebbing (1971) reported a mean increment in frond height of 16.8 mm/y, Eggleston (1972) noted that annual lines were usually between 2 and 3 cm apart in Isle of Man specimens, and Menon (1975) found that Heligoland specimens reached an average of 21.2 mm in height at 2 y and an average of 79.3 mm after 8 y. Growth in height of the cuttings in our experiment indicated a daily growth rate of between 0 and 0.03 mm, with a mean daily growth rate of 0.009 mm day^{-1} (n = 16). Extrapolating this growth for a year, a maximum increase in height of about 11 mm seems possible, but the mean of 3.3 mm indicates less growth in the laboratory than under natural conditions. In our experiments, cuttings grew at the growth margin as well as at the transections (Fig. 3), therefore the actual height is not a reliable indicator for growth. Extraplation for 1 year is highly speculative, since the increase in height, size or number of zoids is not linear throughout the experiment, and increased after a lag-phase of about one month (Table 2, Fig. 2). For this reason, growth rates for height were determined at different stages of the experiment leading to the reported minimum and maximum rates. Biomass increased almost linearly throughout the experiment and is, therefore, a better basis for extrapolation. Data presented by Stebbing (1971) indicated exponential growth for *Flustra foliacea* on a time scale of 6 years whereas growth in height was linear. Differences in the patterns of growth regarding height and weight between present and published data are likely to be due to different time scales. As can be inferred from Table 2, the smallest cutting F1 shows lowest absolute growth, however, the relative increase in biomass is comparable to that of the larger cuttings. From this, it

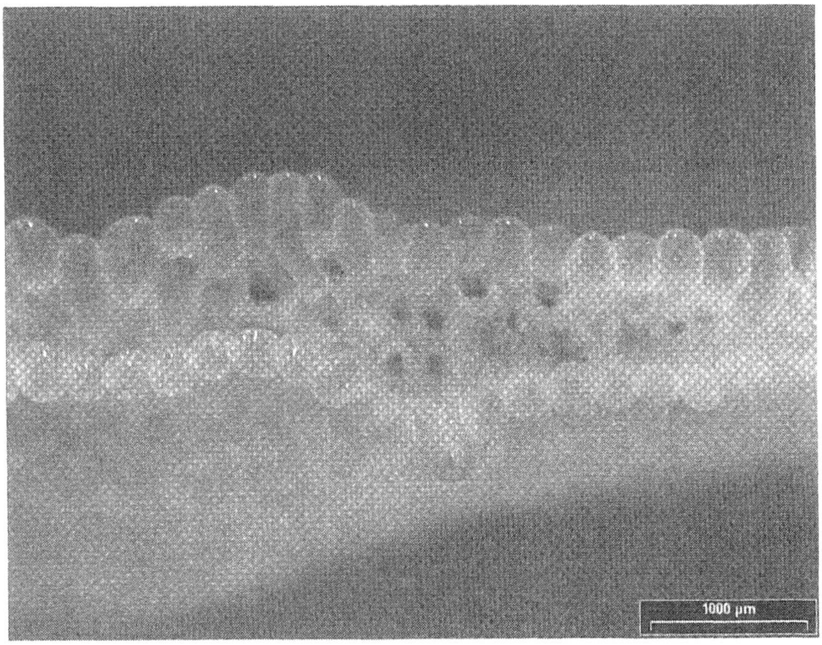

Figure 4. After three months of growth the frond of *Flustra foliacea* split up at the growth margin forming two monolayers.

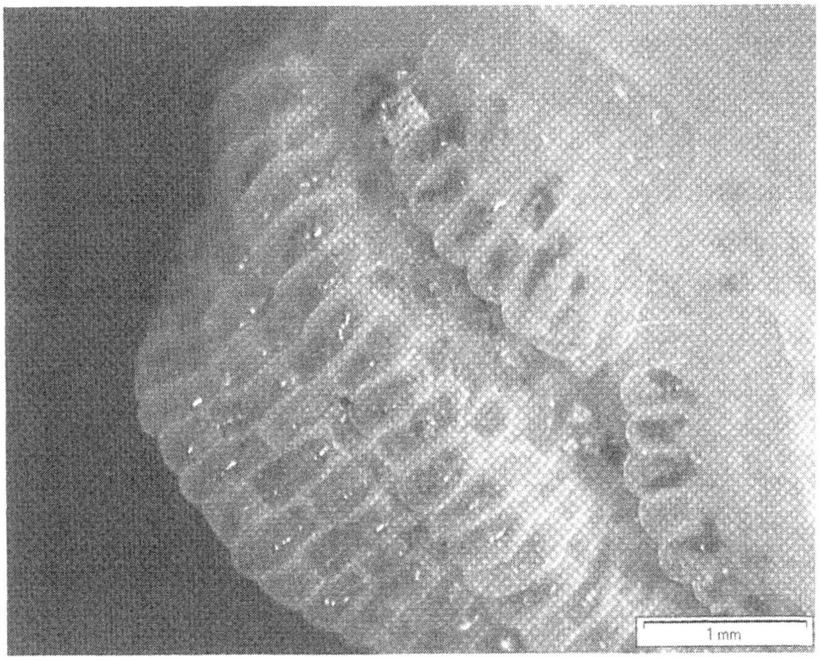

Figure 5. Extended divergent growth of *Flustra foliacea.*

is concluded that the absolute increase of biomass is related closely to the initial weight of the cutting, and this is in agreement with Stebbing (1971). In addition to different time scales, seasonality will also bias comparison. Since cuttings were cultivated under constant experimental conditions (food, temperature, salinity and photoperiod), it is questionable whether colonies will show a growth check in August and stop growing over the winter (Eggleston 1963, cited in Stebbing, 1971).

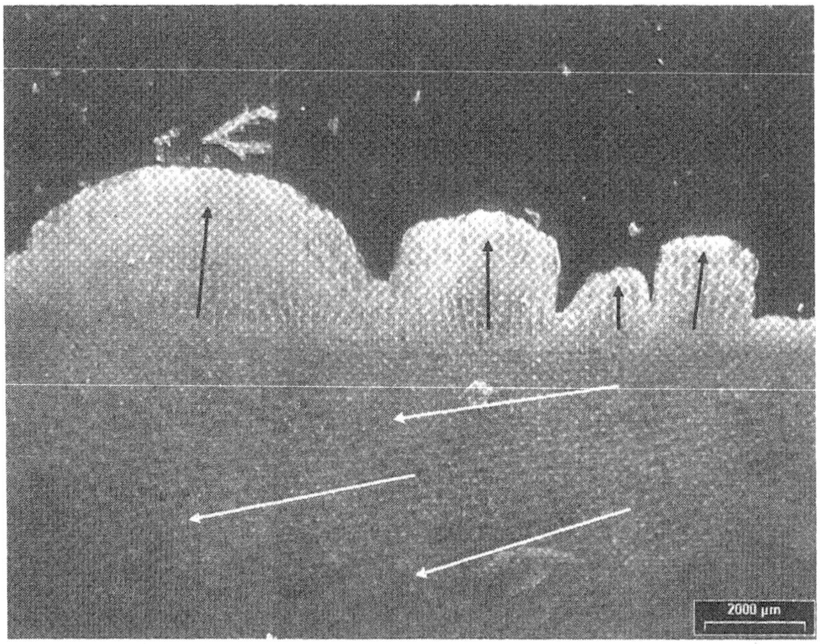

Figure 6. Change in the direction of growth (arrows) and irregular growth patterns of the growth margin in field samples of *Flustra foliacea*.

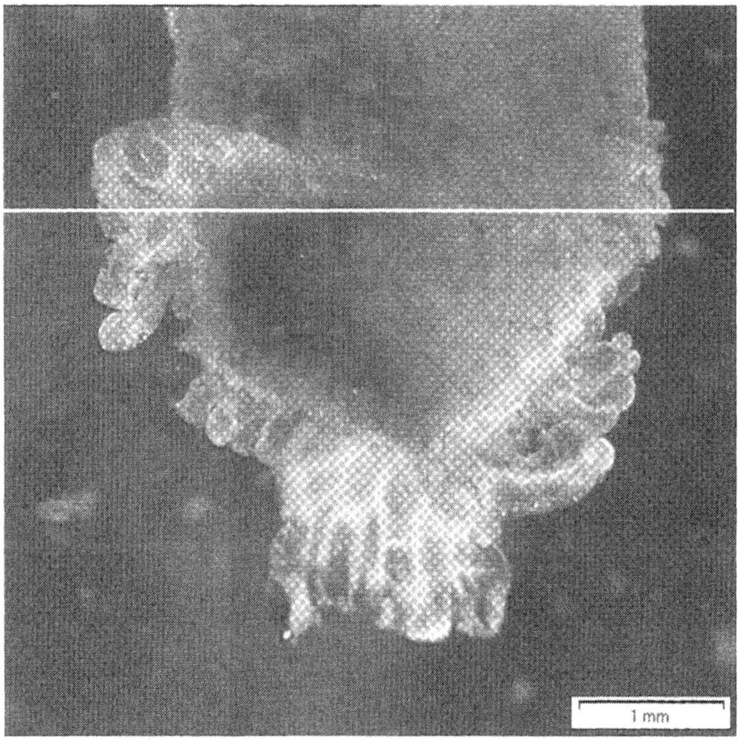

Figure 7. Budding of *Flustra foliacea* at the basis, fixed in the silicon tube (white line marks the original edge of the silicon tube).

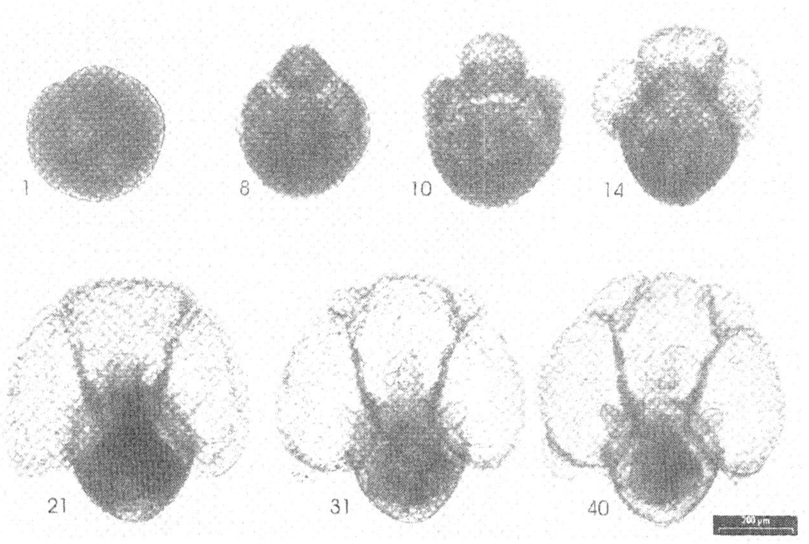

Figure 8. Growth of a *Flustra foliacea* larva during the first month (numbers indicate the days after settlement of the larva).

Growth of the cuttings occurred not only at the original growth margin but also at the transections (Fig. 3). This finding is in line with observations by Stebbing (1971), who found repair of the colony in areas where one layer of the bilamellar frond was missing. In field samples, we occasionally found fronds with zooids orientated at an angle of 90 ° (Fig. 6). It is assumed, that this change in the growing direction results from colonial repair of a ripped frond, comparable to the transections in our experiment. Growth disturbance resulting in two monolayers (Figs 4 and 5), as observed in our experiments and in field samples, has not been reported in the literature. Since field samples were collected in June, it is unlikely that this particular growth feature was the result of food deficiency because food for filter feeders such as *Flustra* is plentiful in June. Additionally, under experimental conditions, divergent growth might result in the formation of two bilayers (Fig. 5). If this interpretation is correct, divergent growth will lead eventually to bilateral ramification of the frond, resulting in branches at right angles to the original frond. Irregular growth patterns at the growth margin will also result in ramification of the frond within the same layer (Fig. 6). Next to growth at the margin and transections, adherent growth also occurred at the base fixed in the silicon tube (Fig. 7). This feature is of special interest for culture purposes, since it would allow a second attachment of the adult colony.

Larvae released by adult colonies were transferred to petri dishes for settlement. Larvae are generally positively phototactic on release and swim only for short periods, however in species in which light stimulus is not critical, larvae may delay metamorphosis for 12 h or more (Hayward & Ryland, 1998). In our experiments, larvae did not settle at 9 °C over a period of 3 days but, after transferring the larvae to 4 °C, they settled within 15 min. In addition to reported sensitivity to surface contour, chemistry and the proximity of conspecific colonies (Hayward & Ryland, 1998), our findings suggest a high sensitivity to temperature. Larvae cultured a 4 °C developed as shown in Figure 8. In the first week after settlement of the larva, lateral walls were built up separating three daughter zoids. Up to day 21, the daugher zooids grew to their final size and polipide rudiments were built up. At the same time, the polipide rudiment of the mother zooid developed to a fully functional polipide. First feeding of the larva was noticed on day 25. On day 28, further lateral walls became visible, forming two buds of the middle daugther zooids. The outer daughter zooids also form each lateral walls, separating buds, around day 40.

Comparison of generative an vegetative cultivation of *F. foliacea* for culture purposes favours the vegetative reproduction because the higher number of zooids at the start of the culture will result in a higher biomass production for a given time span. Additionally, vegetative cultivation omitts the one year of horizontal growth occuring in generative reproduction.

Acknowledgements

This work has been supported financially by the State of Lower Saxony through its Marine Biotechnology Programme. We are grateful to two anonymous reviewers and Malcome Jones for helpful comments on an earlier version of the manuscript.

References

Carle, J. S. & C. Christophersen, 1980. Marine alkaloids. 2. Bromo alkaloids from the marine bryozoan *Flustra foliacea*. Isolation and structure elucidation. J. Org. Chem. 45: 1586–1589.

Carle, J. S. & C. Christophersen, 1981. Marine alkaloids. 3. Bromo-Substituted alkaloids from the marine bryozoan *Flustra foliacea*, flustramine C and flustraminol A and B. J. Org. Chem. 46: 3440–3443.

Eggleston, D., 1972. Patterns of reproduction in marine Ectoprocta off the Isle of Man. J. Nat. Hist. 6: 31–38.

Hayward, P. J. & Ryland, J. S., 1998. Cheilostomatous Bryozoa. Part 1. Aeteoidea -Cribrilinoidea. Synop. Br. Fauna, New Ser. 10: 1–366.

Holst, P. B., U. Anthoni, C. Christophersen & P. H. Nielsen, 1994a. Marine alkaloids, 15. Two alkaloids, flustramine E and debromoflustramine B, from the marine bryozoan *Flustra foliacea*. J. Nat. Prod. 57: 997–1000.

Holst, P. B., U. Anthon, C. Christophersen, P. H. Nielsen & K. Bock, 1994b. A racemic diterpene from the marine bryozoan *Flustra foliacea*, natural product or artefact? Acta Chem. Scand. 48: 765–768.

Laycock, M. V., J. L. C. Wright, J. A. Findlay & A. D. Patil, 1986. New physostigmine related bromoalkaloids from the marine bryozoan *Flustra foliacea*. Can. J. Chem. 64: 1312–1316.

Menon, N. R., 1975. Observations on growth of *Flustra foliacea* (Bryozoa) from Helgoland waters. Helgoländer wiss. Meeresuntersuch. 27: 263–267.

Ryland, J. S., 1976. Physiology and ecology of marine bryozoans. Adv. mar. Biol. 14: 285–443.

Silén, L., 1981. Colony structure in *Flustra foliacea* (Linnaeus) (Bryozoa, Cheilostomata). Acta Zool. (Stockholm.) 62: 219–232.

Sjöblom, T., L. Bohlin & C. Christophersen, 1983. Studies of Swedish marine organisms. II. Muscle-relaxant alkaloids from the marine bryozoan *Flustra foliacea*. Acta Pharm. Suec. 20: 415–419.

Stebbing, A. R. D., 1971. Growth of *Flustra foliacea*. Mar. Biol. 9: 267–273.

Tyler-Walters, H. & S. Ballerstedt, 2002. *Flustra foliacea*. Hornwrack. Marine Life Information Network: Biology and Sensitivity Key Information Sub-programme [on-line]. Plymouth: Marine Biological Association of the United Kingdom. [cited 02/07/02]. Available from: <http://www.marlin.ac.uk>.

Wulff, P., J. S. Carle & C. Christophersen, 1981. Marine alkaloids. Part 4. A formamide, flustrabromine, from the marine bryozoan *Flustra foliacea*. J. Chem. Soc. Perkin Trans.1: 2895–2898.

Wulff, P., J. S. Carle & C. Christophersen, 1982a. Marine alkaloids 5. Flustramide A and 6-bromo-N_b-methyl-N_b-formyltryptamine from the marine bryozoan*Flustra foliacea*. Comp. Biochem. Physiol. 71B: 523–524.

Wulff, P., J. S. Carle & C. Christophersen, 1982b. Marine alkaloids 6. The first naturally occurring bromo-substituted quinoline from *Flustra foliacea*. Comp. Biochem. Physiol. 71B: 525–526.

Hydrobiologia **503:** 245–250, 2003.
M.B. Jones, A. Ingólfsson, E. Ólafsson, G.V. Helgason, K. Gunnarsson & J. Svavarsson (eds), Migrations and Dispersal of Marine Organisms.

Distributional pattern of rays (Pisces, Rajidae) in the Strait of Sicily in relation to fishing pressure

G. Garofalo[1], M. Gristina[1], F. Fiorentino[1], F. Cigala Fulgosi[2], G. Norrito[1] & G. Sinacori[1]
[1] *CNR-IRMA (Istituto di ricerche sulle Risorse Marine e l'Ambiente), Sezione di Mazara del Vallo dell' IAMC (Istituto per l'Ambiente Marino e Costiero), Via L. Vaccara 61, 91026 Mazara del Vallo (TP), Italy*
[2] *Università di Parma, Italy*
E-mail: garofalo@irma.pa.cnr.it

Key words: rays, spatial distribution, abundance trend, Strait of Sicily, Mediterranean

Abstract

Rays are amongst the most vulnerable of exploited marine fish due to their life-history strategy (low fecundity, high age at maturity, slow growth). In this paper, we examine the distributional pattern of the thornback ray (*Raja clavata* Linnaeus, 1758), brown ray (*Raja miraletus* Linnaeus, 1758) and other species of Rajidae in the Strait of Sicily, in the context of changed fishing pressure. In the last decade, most of the Mazara del Vallo trawl fleet (one of the largest and most active in the Mediterranean) has moved southeastwards onto the African continental shelf looking for new fishing grounds. Data are provided from four scientific trawl surveys conducted over a wide area of the Strait of Sicily (covering part of the African shelf) in autumn 1997, 1998, 2000 and 2001. In addition, we analysed research survey data collected along the Sicilian continental shelf between 1985 and 2002. Analysis of distributional patterns and mean catch rates, over the 4-year survey series, revealed low abundances of all ray species on the Sicilian shelf. Data also suggested a remarkable decline in ray abundance on the African shelf. However, data from 1985 to 2002 reported an increasing abundance trend of *Raja clavata* and *Raja miraletus* on the Adventure Bank, located on the southwest side of the Sicilian shelf. Analyses of trawling effort for this latter area indicated that fishing effort has decreased gradually in the last 10 years.

Introduction

Rays are slow-growing species with low fecundity and late maturity, hence they are particularly vulnerable to exploitation (Holden, 1973; Hoenig & Gruber, 1990; Walker & Hislop, 1998; Smith et al., 1999), even when caught as by-catches. However, individual species react to exploitation in different ways, some being more resilient than others (Dulvy et al., 2000; Jukic-Peladic et al., 2001). According to Ragonese et al. (2003), 16 out of the 17 species reported for the Mediterranean Sea occur in the Strait of Sicily; some are very rare (i.e. *Raja africana* Capapé, 1977), while others are caught frequently by trawlers.

The Strait of Sicily is an intensively fished area. The Mazara del Vallo trawl fleet (about 180 vessels; mean GRT = 23.4), one of the largest in the Mediterranean, has exploited the trawlable inshore and offshore grounds of the Strait of Sicily for the last 35 years (Levi et al., 1998). Important changes occurred in this fleet during the late 1980s and early 1990s when many old vessels, primarily used for coastal trawling, were demolished and larger trawlers, for fishing in distant waters, were built (Andreoli et al., 1994). Changes of the fleet, and technological improvements in navigation and fishing equipment, lead to a southward expansion of the fishery. New fishing grounds, included the north African shelf, were forced by the overexploitation of resources in the Italian side of the Strait of Sicily (Levi et al., 1998). At present, the southern limit of the Mazara del Vallo fleet operation area is about 50 nm off the Libyan coast, while the eastern border is delineated by the line that links Malta to the Libyan coast. The spatial and temporal distribution of the fishing effort exerted by this fleet is patchy,

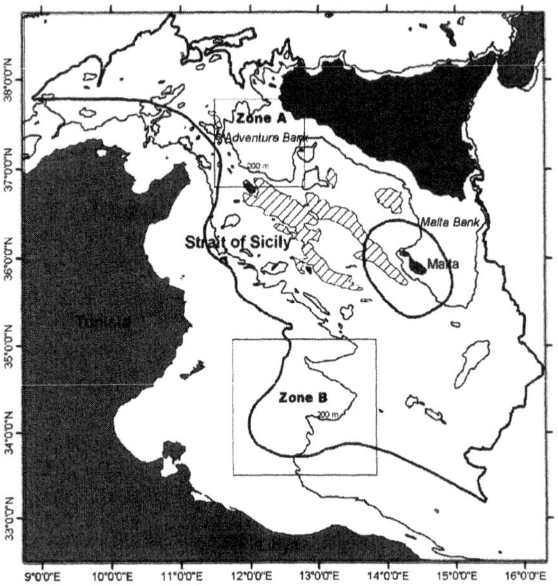

Figure 1. Extension of the study area (bold line) and location of Zone A and Zone B. The 200 m depth contour and water depths over 800 m (hatched areas) are shown.

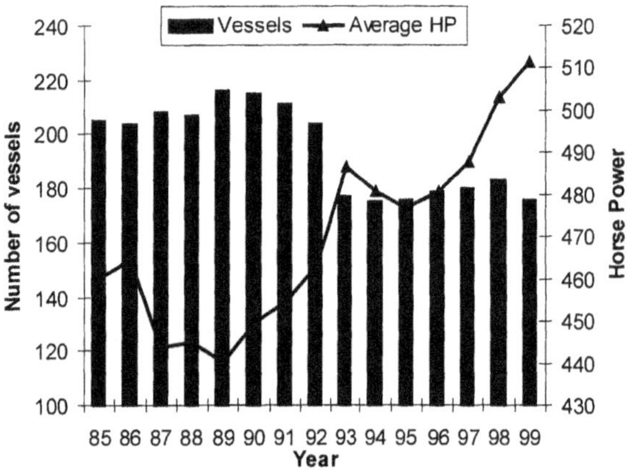

Figure 2. Development of main technical characteristics of the Mazara del Vallo trawling fleet since 1985.

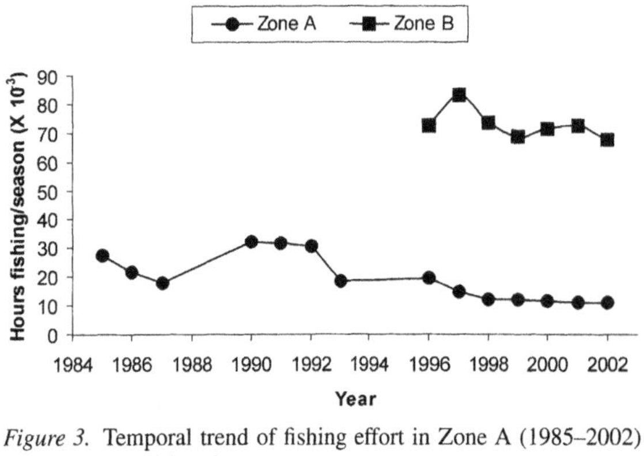

Figure 3. Temporal trend of fishing effort in Zone A (1985–2002) and Zone B (1996–2002).

with some zones fished intensively all year around and others lightly fished (European Commission, 2000).

Here, we evaluate the spatial distribution and abundance of thornback ray (*Raja clavata* Linnaeus, 1758), brown ray (*Raja miraletus* Linnaeus, 1758) and other species of Rajidae in the context of the described changes in exploitation. In particular, abundance trends have been analysed for two areas affected by fleet displacement: the Adventure Bank (southwest side of the Sicilian shelf) and the southeastern fishing grounds, located on the African shelf.

Materials and methods

Abundance data were obtained from scientific trawl surveys carried out on the Italian side of the Strait of Sicily, excluding the Maltese Exclusive Fishing Zone (MEFZ), in the framework of the Italian Programmes for the evaluation of Demersal Resources (Levi et al., 1998; IRMA-CNR, 1999). Surveys were conducted in the periods 1985–86, 1990–91 1994–98 and 2000–2002. Between 1997 and 2001, the surveyed area was extended to include part of the African shelf. A depth-stratified random sampling design was adopted. The same commercial vessel, which utilised an 'Italian Bottom Trawl' with a 28 mm codend mesh size, was chartered each year. We decided to aggregate into a single taxon all species but thornback ray (*Raja clavata*) and brown ray (*Raja miraletus*). This was due to inconsistencies in the identification of Raijdae and, in particular, for juveniles. Records of thornback and brown rays have been collected consistently since 1985. Abundance for each taxon, in terms of density and biomass indices, was standardised to 1 km^2 using the swept-area method.

Fishing-effort assessment was based on data from the Harbour Office and verified with interviews to crews and captains of the Mazara del Vallo trawlers (Andreoli et al., 1995; European Commission, 2000). Two zones were identified in the study area with different exploitation patterns (Fig. 1). Adventure Bank (Zone A), historically exploited by short-range fishery, showed a reduction of fishing pressure over the last 10 years. The other, Zone B, was located on the African shelf about 50 nm off the Libyan coast, where a large displacement of trawlers occurred in the same period. Time series of effort were compiled for these two zones, using the total number of trawling hours as estimates.

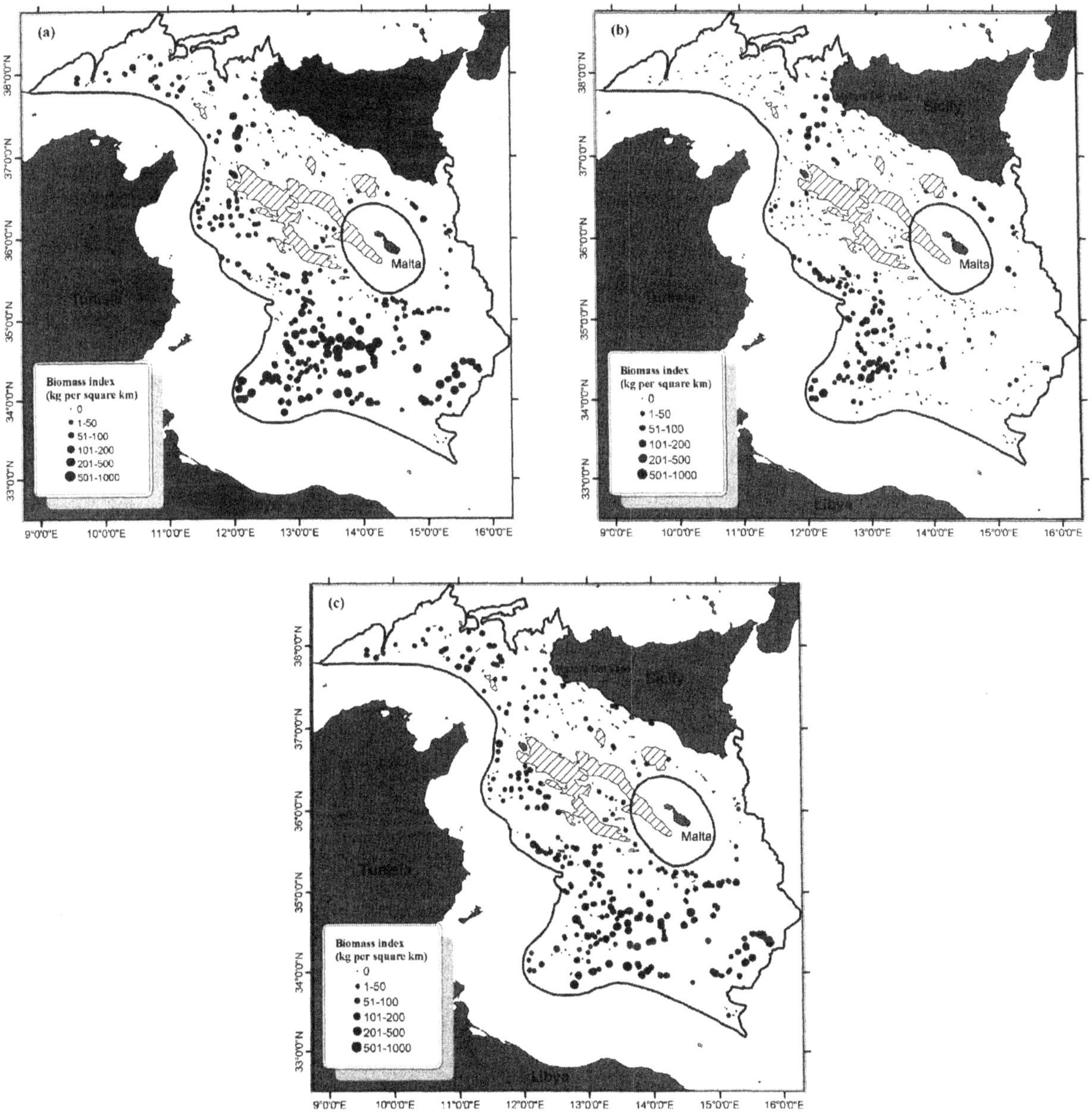

Figure 4. Distribution maps of (a) *Raja clavata*, (b) *Raja miraletus*, and (c) other Rajidae in the Strait of Sicily in the period 1997–2001. Circle radius for each value is proportional to biomass index.

For both species, time series of abundance was calculated restricting analyses to the shelf. Annual means were computed on log-transformed data to reduce the influence of occasionally high catch rates. Comparison of the two zones was based on data collected from 1997 to 2001. For Zone A only, data were available to study longer temporal trends (1985–2002). Mean distribution maps were produced for each taxon pooling the records of surveys from 1997 to 2001 and using a 'proportional symbol' plot. Circles of radius proportional to the biomass index were used. The GIS software ArcView 8 by ESRI was utilized.

Results

The study area and location of Zones A and B are shown in Figure 1. The area is about 100 000 km^2 and Zone B is about 180 nm south the port of Mazara del

Vallo. Following the structural changes in the Mazara del Vallo fleet in the early 1990s (Fig. 2), fishing effort in Zone A has declined gradually (Fig. 3); 30% reduction of fishing effort was observed in this zone. Values fell from over 30 000 fishing h in autumn 1990 to about 11 000 fishing h in autumn 2002. In Zone B, the area most affected by the southward expansion of the fleet, the current effort level is around 65 000 fishing h in autumn (Fig. 3). Although no effort data are available prior to 1994, it may be assumed that intensive exploitation in this area started in the early 1990s, when the distant fishery was developed. However, exploitation of this area is relatively recent compared to Zone A.

The distributional pattern of the three taxa of Rajidae in the Strait of Sicily over the period 1997–2001 is shown in Figure 4. Low abundances of all taxa were evident in the northern part of the Strait. Most of the biomass was located offshore in the central part of the Strait of Sicily, south of Malta. Sparse catches of *R. clavata* were recorded in the central and eastern part of the Strait on the Sicilian side. Moreover, on the Adventure Bank this species was restricted to the western side. *Raja miraletus* exhibited a discrete distribution, occuring mainly in the African shelf and Adventure Bank, being less abundant in the Malta Bank with the exclusion of MEFZ. Similarly, *R. miraletus* appeared absent from the coastal waters off the central Sicilian shelf. The other rays were located mainly in offshore waters of the Strait of Sicily, except for some sparse low catches recorded in the Sicilian shelf. However, in some areas, there was a total absence of this group of species.

Analyses of the annual average biomass of Rays (all species combined), indicated a severe decline in Zone B between autumn 1997 and 2001 (Fig. 5). Biomass and density indices decreased by about 50%. *Raja clavata* showed a negative trend in Zone B, declining from about 45 kg/km^2 to about 20 kg/km^2 (Fig. 6). Although the biomass of *R. miraletus* decreased markedly (about 30%) from 1997 to 2000, it increased in 2001. In Zone A, the biomass levels of each taxon appeared lower than in Zone B although they remained stable over the period 1997–2001. It is worth noting that the dominance ratio between *R. clavata* and *R. miraletus* in the two zones was inverted, *R. miraletus* was more abundant than *R. clavata* in Zone A. An increasing biomass trend for both *R. miraletus* and *R. clavata* emerged from analysis of the longer temporal series available for the Adventure Bank (Fig. 7); while *R. miraletus* showed a clear increasing trend throughout the whole period, *R. clavata* decreased until 1990 and increased thereafter.

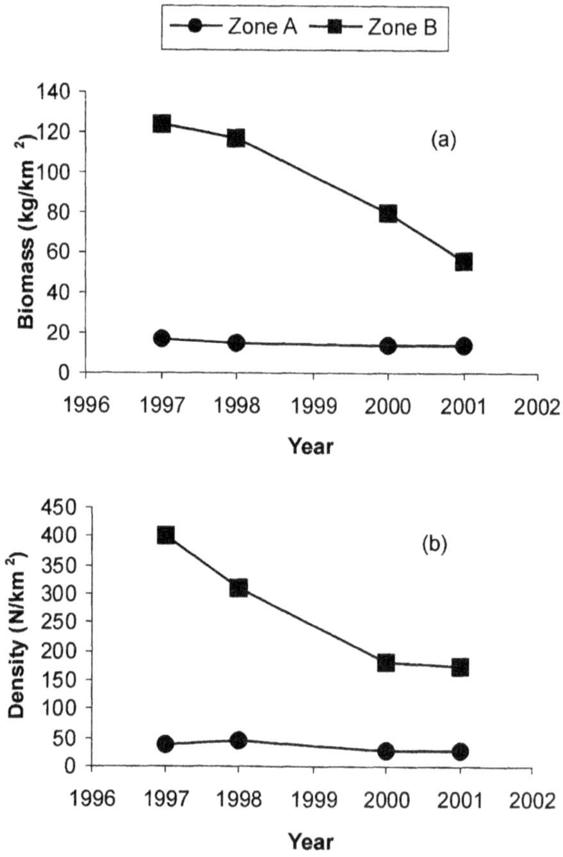

Figure 5. Annual means of all Rays in Zone A and Zone B for the 1997–2001 bottom trawl surveys: (a) weight and (b) number.

Discussion

Distributional patterns and mean catch rates from 1997 to 2001 indicated that highest abundances of *Raja* sp. were found on the African side of the Strait of Sicily. Yet, this area has been subjected to intensive trawling exploitation in the last decade, and a warning decline of ray abundance is occurring. At the same time, rays appeared to be highly depleted in the northern part of the Strait, an area exploited intensively since the 1970s. Nevertheless, there were trends of increasing abundance for *R. miraletus* and *R. clavata* in the Adventure Bank (Zone A) from 1985 to 2002. These findings are in agreement with the decreasing trend of fishing effort in the same area. Some differences have been reported for the temporal pattern of *R. clavata* and *R. miraletus* and these, probably, reflect their spe-

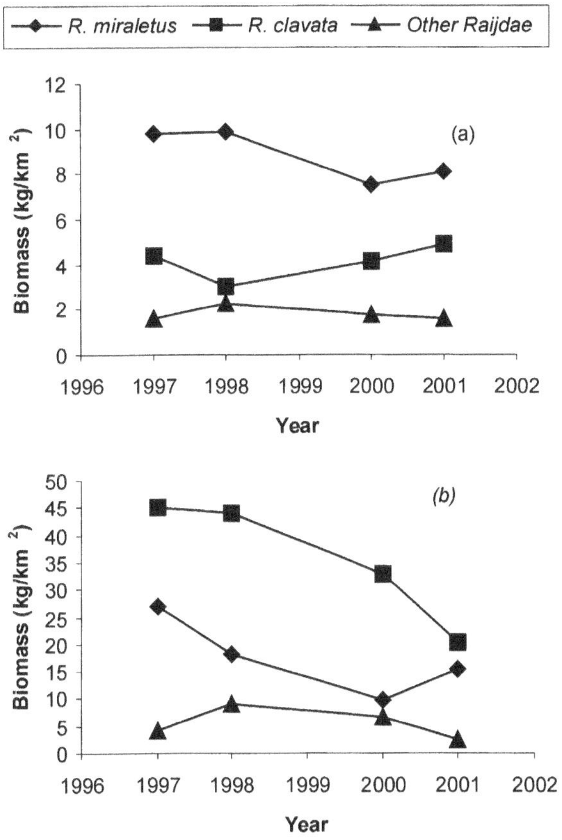

Figure 6. Annual mean biomass indices of *Raja clavata*, *Raja miraletus* and other Rajidae between 1997 and 2001 in (a) Zone A and (b) Zone B.

cific sensitivity to fishing exploitation. *Raja clavata* is larger, slower growing and later maturing than *Raja miraletus* (Relini et al., 1999), making it more susceptible to fishing (Walker & Hislop, 1998; Jennings et al., 1999; Stevens et al., 2000). Similar observations were reported by other authors for the Mediterranean; *Raja clavata* stock in the Gulf of Lions was strongly depleted between 1983 and 1997 by high trawling pressure (Bertrand et al., 2000). Conversely, *R. miraletus* has maintained a stable abundance in the Adriatic Sea in the last 50 years, while other rays have disappeared or become rare (Jukic-Peladic et al., 2001). The dominance ratio between the two species was opposite in the two investigated areas, with *R. miraletus* being more abundant than *R. clavata* in the Adventure Bank, whereas the latter dominated the African shelf. Assuming that the fishing pattern in the Strait of Sicily will remain unchanged in future, observed trends and literature evidence suggest that a switch may occur in the abundance ratio of the two species in the African

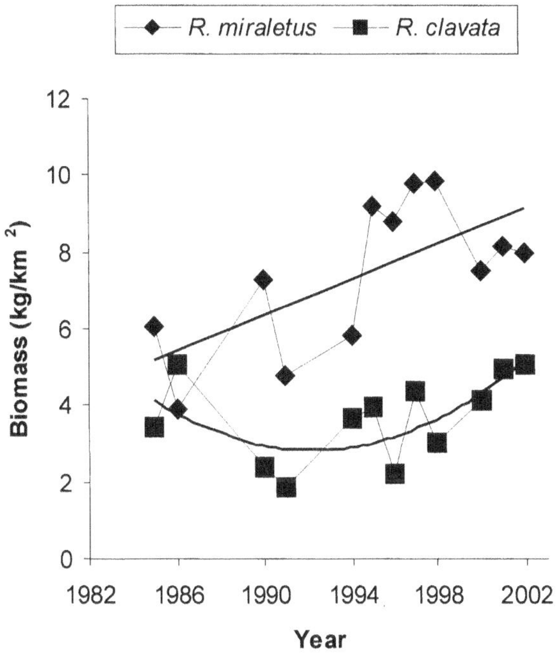

Figure 7. Mean biomass trends of *Raja clavata* and *Raja miraletus* in Zone A from 1985 to 2002.

shelf. *Raja miraletus* may become dominant due to its life-history characteristics. On the other hand, it is difficult to predict the recovery of Rays on Adventure Bank in the long term if trawling effort continues to decrease. At present, little is known about the recovery dynamics of impacted communities, particularly in open-sea banks (Kaiser et al., 2002). These preliminary results highlight the need for further investigation to clarify the role of sensitive species in the resistance and recovery of ecosystems exploited by fisheries.

References

Andreoli, M. G., D. Levi, L. Cannizzaro & G. Garofalo, 1994. Demersal landing and trawling effort data along the southern coast of Sicily (Program Minipestat). GFCM Central Mediterranean Consultation, Tunis, 8–12 Nov. 1994. 22 pp.

Andreoli, M. G., N. Campanella, L. Cannizzaro, G. Garofalo, G. B. Giusto, P. Jereb, D. Levi, G. Norrito, S. Ragonese, P. Rizzo & G. Sinacori, 1995. Sampling statistics of Southern Sicily trawl fisheries (MINIPESTAT): Data Report N.T.R.-I.T.P.P. Special Publication No 4 (Vol. II). 436 pp.

Bertrand, J. A., Y. Aldebert & A. Souplet, 2000. Temporal variability of demersal species in the Gulf of Lions from trawl surveys (1983–1997). In Bertrand J. A. & G. Relini (eds), Demersal Resources in Mediterranean. Proceedings of the Symposium held in Pisa, 18–21 March 1998. Actes Colloq. IFREMER 26: 153–164.

Dulvy, N. K., J. D. Metcalfe, J. Glanville, M. G. Pawson & J. D. Reynolds, 2000. Fishery stability, local extinctions and shifts in community structure in skates. Conserv. Biol. 14: 283–293.

European Commission, 2000. Analysis of trawls' discard operation in the central and eastern Mediterranean sea. Commission of the European Communities, DGXIV, Project No. 97/0044, Final Report: 161 pp.

Hoenig, J. M. & S. H. Gruber, 1990. Life-history patterns in the elasmobranchs: implications for fisheries management. In Pratt, H. L., S. H. Gruber & T. Taniuchi (eds), Elasmobranchs as Living Resources: Advances in the Biology, Ecology, Systematics and the Status of the Fisheries. Proceeding of the Second United States-Japan workshop East-West Center, Honolulu, Hawaii, 9–14 December 1987, NOAA Tech. Rep. NFMS 90. 518 pp.

Holden, M. J., 1973. Are long-term sustainable fisheries for elasmobranches possible? Rapp. P.-V. Réun. Cons. Int. Explor. Mer 164: 360–367.

IRMA-CNR, 1999. Results of experimental groundfish investigations in the Strait of Sicily (Mediterranean Sea) during 1996–'99 within the frame of the italian law 41/8: final report – Vol. 1–2. Gruppo Nazionale Valutazione Risorse Demersali (GRUND), Unità Operativa no 11, IRMA-CNR, Mazara (TP), Italy: 97 pp.

Jennings, S., S. P. R. Greenstreet & J. D. Reynolds, 1999. Structural change in an exploited fish community: a consequence of differential fishing effects on species with contrasting life histories. J. anim. Ecol. 68: 617–627.

Jukic-Peladic, S., N. Vrgoc, S. Krstulovic-Sifner, C. Piccinetti, G. Piccinetti-Manfrin, G. Marano & N. Ungaro, 2001. Long-term changes in demersal resources of the Adriatic Sea: comparison between trawl surveys carried out in 1948 and 1998. Fish. Res. 53: 95–104.

Kaiser, M. J., J. S. Collie, S. J. Hall, S. Jennings & I. R. Poiner, 2002. Modification of marine habitats by trawling activities: prognosis and solutions. Fish Fish. 3: 114–136.

Levi, D., S. Ragonese, M. G. Andreoli, G. Norrito, P. Rizzo, G. B. Giusto, S. Gancitano, G. Sinacori, G. Bono, G. Garofano & L. Cannizzaro, 1998. Results of experimental groundfish investigations in the Strait of Sicily (Central Mediterranean Sea) during 1985–1997 within the frame of the Law 41/82. Biol. mar. Medit. 5: 130–139.

Ragonese, S., F. Cigala Fulgosi, M. L. Bianchini, G. Norrito & G. Sinacori, 2003. Annotated check list of the skates (chondrichthyes, Rajidae) in the Strait of Sicily (Central Mediterranean). Biol. mar. Medit. 10: (in press).

Relini, G., J. Bertrand & A. Zamboni (eds), 1999. Synthesis of the Knowledge on Bottom Fishery Resources in Central Mediterranean (Italy and Corsica). Biol. mar. Medit. 6 (Suppl.1). 868 pp.

Smith, S. E., D.W. Au & C. Show, 1999. Intrinsic rebound potentials of 26 species of Pacific sharks. Mar. Freshwat. Res. 49: 663–678.

Stevens, J. D., R. Bonfil, N. K. Dulvy & P. A. Walker, 2000. The effects of fishing on sharks, rays and chimaeras (chondrichthyans), and implications for marine ecosystems. ICES J. mar. Sci. 57: 476–494.

Walker, P. A. & J. R. G. Hislop, 1998. Sensitive skates or resilient rays? Spatial and temporal shifts in ray species composition in the central and north-western North Sea between 1930 and the present day. ICES J. mar. Sci. 55: 392–402.

Hydrobiologia **503:** 251–262, 2003.
M.B. Jones, A. Ingólfsson, E. Ólafsson, G.V. Helgason, K. Gunnarsson & J. Svavarsson (eds), Migrations and Dispersal of Marine Organisms.

Distribution of tintinnid species from 42° N to 43° S through the Indian Ocean

M. Modigh[1], S. Castaldo[1], M. Saggiomo[2] & I. Santarpia[1]
[1]*Zoological Station "A. Dohrn", Villa Comunale, 80121 Naples, Italy*
[2]*Department of Zoology, University of Naples "Federico II", Via Mezzocannone 8, 80134 Naples, Italy*
E-mail: modigh@szn.it

Key words: tintinnids, biogeography, diversity, oligotrophic

Abstract

Taking advantage of the transfer of the Italian R/V "Italica" from Italy to New Zealand for the Italian Research Expedition in the Antarctic, surface sampling of plankton was carried out from 42°N to 43°S from 14 November to 16 December 2001. Collaboration between several Italian research groups resulted in the *M*editerranean, *I*ndian and *P*acific *O*cean *T*ransect (MIPOT), aimed to merge oceanographic measurements with remotely sensed data and to study latitudinal changes in plankton communities. We present preliminary results on the tintinnids (abundance and species composition) along the MIPOT1 transect. Tintinnid distributions and food vacuole contents are discussed in relation to chlorophyll concentrations and contribution of phytoplankton size classes. Generally, chlorophyll concentrations were well below 0.5 mg m^{-3} and tintinnid numbers were only a few tens per liter. Eighty-six tintinnid species belonging to 27 genera were recorded. Highest diversity and number of tintinnid species occurred in the warm-water regions, from the Mediterranean Sea to the Tropic of Capricorn. Shannon diversity index ranged from 1.0 (Tasman Sea) up to 2.9 (tropical Indian Ocean). Each of the oceanic provinces showed a well defined tintinnid assemblage, a tintinnid 'fingerprint' characterizing each area. Data are compared to previous records for the different areas and the persistence in time of typical tintinnid species is discussed.

Introduction

Tintinnid ciliates are a common component of planktonic communities and occur in all the oceans of the world (Pierce & Turner, 1992, 1993). Tintinnids, and pelagic ciliates in general, are important consumers of pico- and nano-plankton (Rassoulzadegan et al., 1988; Bernard & Rassoulzadegan, 1993) and are, in turn, exploited by higher trophic levels (e.g. Stoecker & Sanders, 1985). The importance of protozoan grazing and recycling in the carbon budget of the oceans has been established (e.g. Sherr et al., 1986; Paffenhöfer, 1998). The 'changing paradigm' of planktonic food webs has focused much attention on microbial compartments, and tintinnids have been identified as ideal organisms for the study of changes in the structure and composition of microzooplankton (Thompson et al., 1999; Dolan, 2000).

Pierce & Turner (1993) reviewed tintinnid distributions, yet few recent studies provide quantitative data on tintinnid diversity in coastal (Cariou et al., 1999; Dolan & Gallegos, 2001; Modigh & Castaldo, 2002) and offshore waters (Dolan & Marrasè, 1995; Thompson et al., 1999; Dolan, 2000; Pitta et al., 2001). In contrast to other microorganisms, such as the benthic ciliates, for which local and global diversity is similar, i.e. "everything is everywhere" (Fenchel et al., 1997), most tintinnid taxa are not cosmopolitan (Pierce & Turner, 1993) and latitudinal trends in tintinnid species diversity have been described (Dolan & Gallegos, 2001). Tintinnid biogeographic distributions in relation to hydrography are well known for several oceanic provinces (Pierce & Turner, 1993). Tintinnids may be used as tracers of upwelling (Balech, 1972), oceanic currents and water masses; when other characteristic features of the water has been lost due to heating or mixing, the occurrence of specific tintinnid taxa may give information on the origin of the current or water mass (Zeitzschel, 1982; Kato & Taniguchi 1993; Pierce & Turner, 1993; Thompson et al., 1999). In addition, for sites where repeated observations of tintinnid assemblages are available, the dominant tintinnid species show recurrent patterns of occurrence (e.g. Rassoulzadegan, 1979; Verity, 1987; Stelfox et al., 1999; Thompson et al., 1999; Modigh & Castaldo, 2002).

Lorica morphology is used for tintinnid taxonomy, as other criteria such as physiological and biogeographical data and, in particular, cytological characters of the cytosome, are available only for a very small number of species (Laval-Peuto & Brownlee, 1986). Tintinnid lorica preserve well and may be collected easily by fine-meshed plankton nets, but taxonomic determinations based exclusively on features of the lorica must take into account the variability in lorica size and shape reported for several species. Nevertheless, some conservative lorica characteristics are particularly useful for tintinnid taxonomic work, e.g. lorica general shape, texture and oral diameter (Laval-Peuto & Brownlee, 1986). However, a global synthesis of tintinnid biogeography is still lacking primarily due to a poor understanding of tintinnid taxonomy and the proliferation of species and synonyms (Pierce & Turner, 1993).

Taking advantage of the transfer of the Italian R/V "Italica" from Italy to New Zealand for the XVII Italian Research Expedition in the Antarctic, surface sampling of plankton was carried out from 42° N to 43° S from 14 November to 16 December 2001. Cooperation between several Italian research groups resulted in the *M*editerranean, *I*ndian and *P*acific *O*cean *T*ransect (MIPOT), aimed to merge oceanographic measurements with remotely sensed data and to study latitudinal changes in plankton communities. The present paper deals exclusively with tintinnid assemblages along the MIPOT1 transect, as related to chlorophyll concentrations and continuous temperature and salinity recordings. We present preliminary results on tintinnid assemblages in surface waters in the Eastern Mediterranean, the Red Sea, the Arabian Sea, the Indian Ocean and the Tasman Sea. We define tintinnid assemblages for each of the oceanic provinces in terms of abundance, species composition, diversity and equitability, and compare our findings to literature records of tintinnid genera and species along the transect. We also discuss the relationships between phytoplankton biomass concentrations and tintinnid numbers and vacuole content.

Materials and methods

Surface chlorophyll and microzooplankton sampling was carried out on board R/V "Italica" during the MIPOT1 cruise between Italy and New Zealand. The ship's track crossed different oceanic provinces from 42° N to 43° S and from 16° E to 170° E, passing through the Eastern Mediterranean, the Red Sea, the Arabian Sea, the Indian Ocean and the Tasman Sea (Fig. 1). The ship's transfer was non-stop and sampling was performed during navigation from −5 m depth. A peristaltic pump provided water for continuous recording of temperature and salinity by means of a CTD SEACAT 19, and a membrane pump was used for chlorophyll and microzooplankton sampling. For the first 2 days of the transect, CTD recordings were not available due to technical problems. Water samples for chlorophyll *a* (chl*a*) analyses ($N = 347$) were collected every 2 h and the spectrofluorometric analyses carried out on board within 24 h from sampling (Neveux & Panouse, 1987). Phytoplankton size fractions (micro- >20 μm, nano- 2 μm to 20 μm, pico- <2 μm phytoplankton) were separated by serial filtration (for details see Modigh & Castaldo, 2002).

The division of oceanic provinces was based on well-known geographical and hydrographical (major oceanic currents) settings (Sverdrup et al., 1942; Wyrtki, 1973 for the Indian Ocean), and supported by surface recordings of environmental variables.

Daily microzooplankton samples (40–120 liters of seawater) were collected onto a net of 20 μm mesh size; 20 samples are presented here. Sample collection continued until initial clogging of the net (20–60 min), thus, the microzooplankton sample was collected along approximately 4–15 nautical miles, depending on the ship's speed and the volume of water collected. The net samples were preserved in borate formol at 4% final concentration, and kept dark and refrigerated till analysis. A Zeiss inverted microscope equipped with phase contrast was employed for examination at 320×. The entire bottom of the sedimentation chamber was scanned, the concentrated samples corresponded to a surface water sample of 2–8 l.

The taxonomic authority was Kofoid & Campbell (1929, 1939). In addition, Balech (1959) and other taxonomic literature (Daday, 1887; Brandt, 1907; Jörgensen, 1924) were consulted. Original drawings and first description of species from the older literature were useful to resolve doubt. Kofoid & Campbell (1929, 1939) performed a thorough revision of tin-

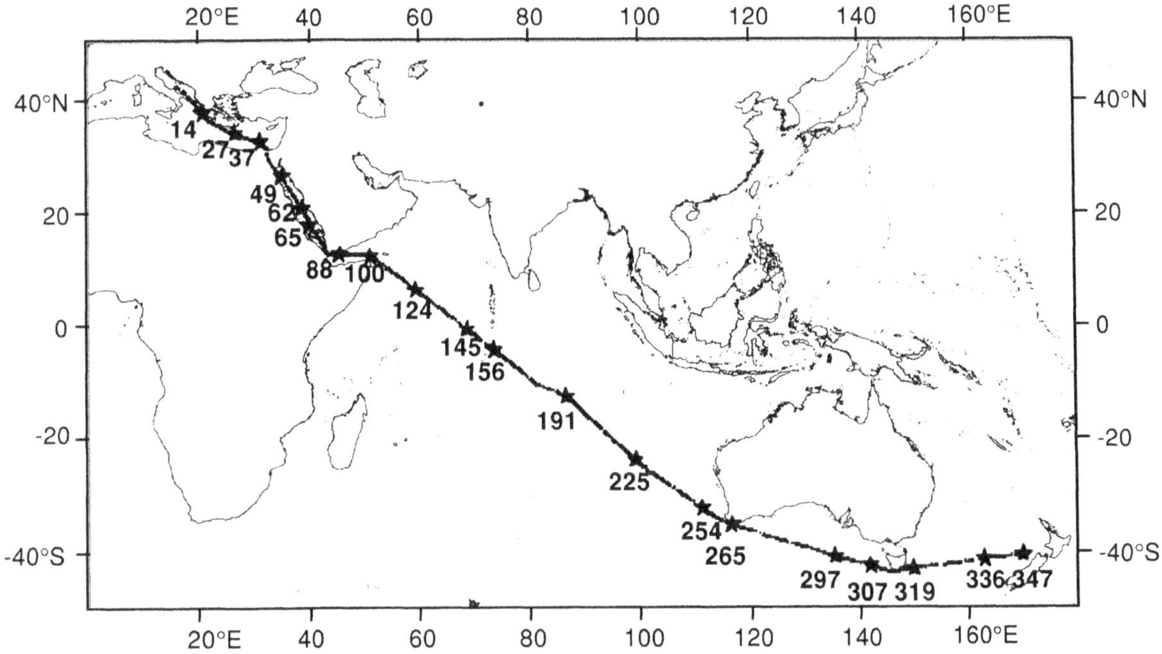

Figure 1. MIPOT1, November-December 2001: ship's route and location of tintinnid samplings (★).

tinnid taxonomy and their works are still the only general scheme for tintinnid taxonomy. However, several tintinnid taxa need to be revised as, in some cases, definition of species was based upon minor differences in lorica morphology. Such revisions have been made for some species, e.g. Balech (1959) considered several species of *Dadayiella* as described by Kofoid & Campbell (1929) to be all *Dadayiella ganymedes*; Zeitzschel (1969) examined the genera *Epiplocyloides, Epiplocylis, Epiorella.* For these genera, we followed the more recent and conservative nomenclature. Finally, a few specimens of *Coxliella*-forms represented another case when we did not follow the taxonomy of Kofoid & Campbell; *Coxliella* may not be a genus but rather a particular lorica shape assumed during the life cycle of several tintinnid taxa (Laval-Peuto & Brownlee, 1986). Therefore, the *Coxliella*-forms were counted as undetermined tintinnids and were not included in the calculation of species diversity.

Shannon diversity index was calculated for each sample; we included also the tintinnids determined to the genus level when no uncertainties could arise with other taxa recorded in the same sample. This was the case either because no other species of the same genus occurred in the same sample or the most conservative features of the lorica, such as general shape and oral diameter, were clearly different from congeneric species in the same sample. When these conditions were not met, specimens were not included in the calculations of diversity.

Analyses of diversity of tintinnid assemblages were performed using the "Biodiversity Professional" program (beta release; McAleece N., Lambshead P.J.D., Paterson G.L.J., Gage J.D., 1997; from the Scottish Association for Marine Science web site http://www.sams.ac.uk/dml/projects/benthic/ bdpro).

Pearson's correlation coefficients were calculated between tintinnid numbers and the environmental variables measured. In addition, for examination of tintinnid vacuole content, daily microscope observations of microzooplankton samples were carried out within a few hours from sampling with an inverted epifluorescence microscope from the Red Sea to New Zealand.

Results

The MIPOT transect covered approximately 85° from north to south and 154° in west-east direction. Temperature varied from 12.6 till 31.7 °C and salinity from 33.09 till 42.80 psu over the entire transect (Fig. 2), but a much smaller range of these variables characterized

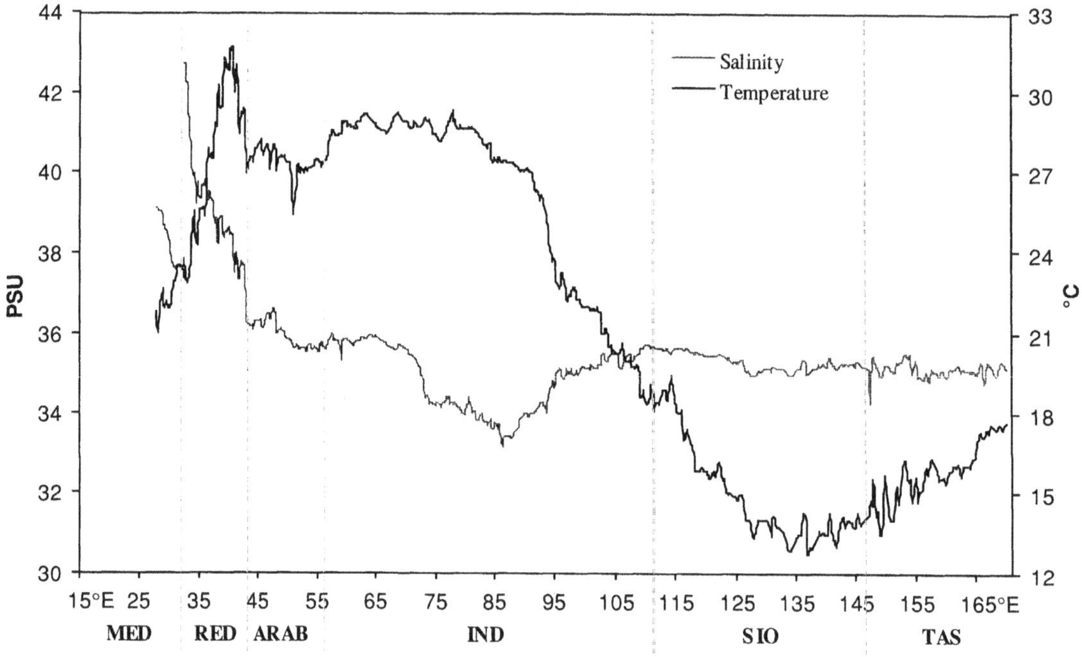

Figure 2. Temperature and salinity records (-5m) along the MIPOT 1 transect.

Table 1. Oceanic regions sampled along the MIPOT1 transect. Surface temperature and salinity ranges; tintinnid numbers: range and (mean ± SD)

	Latitude	Longitude	Temperature (°C)	Salinity (psu)	Tintinnids (Cells/l)
Mediterranean Sea	42° 13′ N–32° 33′ N	16° 10′ E–31° 12′ E	20.9–23.6	37.5–39.2	29.7–58.2 (45.6±14.5)
Red Sea	29° 45′ N–13° 17′ N	32° 31′ E–43° 04′ E	22.7–31.7	37.1–42.8	19.1–53.5 (30.9±19.6)
Arabian Sea	12° 58′ N–08° 52′ N	43° 11′ E–55° 56′ E	25.0–28.3	35.3–36.6	145.0–316.0 (230.5±120.9)
Indian Ocean	08° 37′ N–32° 13′ S	56° 16′ E–110° 48′ E	18.2–29.6	33.1–36.0	13.7–76.0 (41.7±24.4)
Southern Indian Ocean	32° 29′ S–43° 46′ S	111° 12′ E–146° 23′ E	12.6–19.9	34.8–35.7	2.1–20.8 (9.5±8.2)
Tasman Sea	40° 37′ S–43° 40′ S	147° 03′ E–170° 04′ E	13.4–17.8	33.9–35.7	8.6–496.5 (10.5±2.7)*

each oceanic province (Table 1). Temperatures were >20 °C from the Mediterranean to the Tropic of Capricorn, and varied within a narrow range within each of the warm-water regions, except for the Red Sea where temperature increased and salinity decreased from north to south. A sharp decrease in surface temperatures was recorded east of 90° E in the Southern Indian Ocean (SIO) due to the Western Australian Current carrying cold waters from the south. Higher temperatures were recorded in the Tasman Sea due to a warm water current, the Eastern Australian Current, which enters the Tasman Sea from the north.

Chlorophyll concentrations were generally low (<0.5 mg chl*a* m^{-3}) (Fig. 3). Lowest values occurred in the Indian Ocean (0.07±0.03 mg chl*a* m^{-3}; $N = 138$) and in the Eastern Mediterranean Sea (0.07

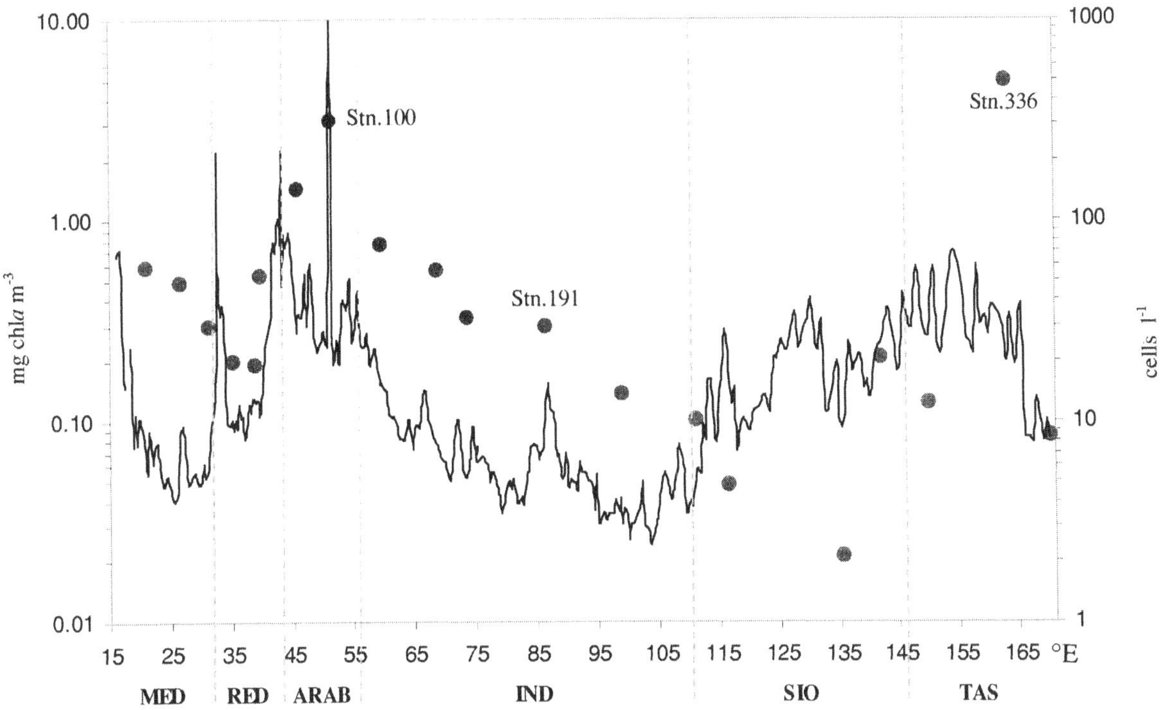

Figure 3. Chlorophyll *a* concentrations (-5m) along the MIPOT 1 transect.

±0.02 mg chl*a* m^{-3}; $N = 26$, not including a few high values up to 0.74 mg chl*a* m^{-3} recorded in the Southern Adriatic Sea); highest concentrations occurred in the Red and Arabian Seas (0.36±0.30 and 0.41±0.16 mg chl*a* m^{-3}, respectively). Maximum chl*a* values, up to 10.4 mg chl*a* m^{-3}, occurred in the Gulf of Aden (Arabian Sea). Most of the phytoplankton biomass, in terms of chlorophyll, was due to the picophytoplankton, from about 50% in the Tasman Sea to about 74% in the tropical Indian Ocean. The contribution of microphytoplankton was lowest (average 7–8%) in the Mediterranean Sea and in the Indian Ocean, and the largest contribution of this size fraction (average 22%) was recorded in the Tasman Sea.

Tintinnid abundances were in the order of a few tens per liter at most stations; high numbers occurred in the Arabian Sea and at one station off the western coast of New Zealand (Fig. 3). Maximum tintinnid numbers occurred at stations where a sharp drop in surface temperature and salinity were recorded. In particular, at Stn.100 (Arabian Sea) surface temperature decreased from 27.39 to 25.05 °C over 360 km and salinity decreased from 35.88 to 35.54 (Fig. 2). After this patch, temperature values were 27.30 °C and salinity around 35.8. The maximum chl*a* concentration (10.4 mg chl*a* m^{-3}) was recorded at Stn.100, and tintinnid abundance was 316 ind. l^{-1}. Maximum tintinnid numbers (497 ind. l^{-1}) were recorded at Stn.336 in the Tasman Sea. This station was located within an area of 250 km where temperature and salinity decreased from 16.1 to 15.4 °C and from 35.2 to 34.9, respectively; chl*a* concentration was 0.20 mg chl*a* m^{-3} (Fig. 3).

Eighty-six tintinnid taxa were identified to the species (95%) or the genus level (2.1%); undetermined tintinnids were 2.9% (Table 2). The number of species recorded for each oceanic province was highest in the oligotrophic regions; 45 species were found in the Indian Ocean, 35 in the Mediterranean, 20 in the Red and 23 in the Arabian Sea, and only 18 and 13 species were recorded in the SIO and Tasman Sea respectively. The diversity index ranged from 1.5 on average in the SIO and the Tasman Sea to 2.4 in the Mediterranean Sea and the Indian Ocean (Fig. 4). Maximum number of species, 25 species belonging to 17 genera, and highest diversity (H'=2.9) occurred at Stn.191, sampled on 1 December 2001 in an area where a tropical cyclone had passed 24 h prior to the ship's transit. Chl*a* concentrations were up to 0.15 mg chl*a* m^{-3}, while 0.03 mg chl*a* m^{-3} was recorded before and after

Table 2. Tintinnid species recorded along the MIPOT1 transect and area of occurrence for each species

	Mediterranean Sea	Red Sea	Arabian Sea	Indian Ocean	S. Indian Ocean	Tasman Sea	No. of observations
Amphorella amphora	•	•	•	•		•	13
Amphorella quadrilineata				•			1
Amphorella laackmanni	•						1
Ascampbeliella torulata	•		•	•			8
Canthariella sp.	•						1
Climacocylis scalaroides			•	•			3
Codonella aspera	•			•			2
Codonellopsis ecaudata			•				2
Codonellopsis orthoceras				•			1
Codonellopsis sp.						•	1
Craterella urceolata			•				1
Cymatocylis sp.		•					1
Dadayiella curta		•					1
Dadayiella ganymedes	•	•	•	•			10
Dictyocysta elegans				•		•	4
Dictyocysta mitra	•				•	•	4
Dictyocysta reticulata			•				2
Epiplocylis acuminata					•		1
Epiplocylis blanda	•			•			2
Epiplocylis bruhni				•			1
Epiplocylis pacifica	•						1
Epiplocylis sp. 1		•					1
Epiplocylis sp.2					•		1
Epiplocylis sp.3					•		1
Epiplocyloides freymandli					•	•	2
Epiplocyloides reticulata		•	•	•			4
Eutintinnus apertus	•	•		•	•		10
Eutintinnus elegans	•						2
Eutintinnus elongatus				•			1
Eutintinnus fraknoii	•	•	•	•	•	•	9
Eutintinnus haslae				•		•	2
Eutintinnus lusus-undae	•	•	•	•		•	10
Eutintinnus macilentus	•	•		•			4
Eutintinnus medius				•			1
Eutintinnus stramentus	•	•	•	•			10
Eutintinnus tubulosus	•	•	•	•	•	•	10
Eutintinnus sp.		•		•			2
Favella markusovszkyi				•			1
Metacylis annulifera			•				1
Parundella longa				•			1
Petalotricha pacifica			•				1
Petalotricha sp.1					•		1
Proplectella claparedei	•			•	•		5
Proplectella ovata	•						1
Proplectella perpusilla				•			1
Proplectella praelonga				•			1

Continued on p. 257

Table 2. contd.

	Mediterranean Sea	Red Sea	Arabian Sea	Indian Ocean	S. Indian Ocean	Tasman Sea	No. of observations
Protorhabdonella curta	•	•	•	•			7
Protorhabdonella simplex	•	•	•	•	•		11
Protorhabdonella sp.1				•			1
Protorhabdonella sp.2		•					1
Rhabdonella amor	•			•	•		3
Rhabdonella brandti	•						1
Rhabdonella cornucopia		•		•			3
Rhabdonella hebe			•	•			4
Rhabdonella inflata	•			•			2
Rhabdonella poculum			•				2
Rhabdonella quantula			•	•			4
Rhabdonella spp.		•		•	•	•	5
Rhabdonellopsis intermedia				•			3
Rhabdonellopsis longicaulis				•			2
Rhabdonellopsis minima				•			1
Rhabdonellopsis sp.1					•		1
Salpingella acuminata			•	•	•		3
Salpingella attenuata	•						1
Salpingella curta	•			•			2
Salpingella decurtata	•		•	•			5
Salpingella laminata	•						1
Salpingella lineata				•			1
Salpingella spp.		•					2
Steenstrupiella gracilis				•			3
Steenstrupiella intumescens	•						1
Steenstrupiella steenstrupii	•	•		•	•	•	14
Tintinnopsis cylindrica				•			1
Tintinnopsis parvula				•			1
Undella clevei	•						1
Undella hemisferica	•						1
Undella minuta			•	•			3
Undella ostenfeldii				•		•	2
Undella turgida		•				•	2
Xystonella lanceolata					•		1
Xystonella lohmanni					•		1
Xystonella longicauda	•						3
Xystonella treforti				•			1
Xystonellopsis brandti	•						2
Xystonellopsis heroica	•						1

this patch (Fig. 3) whereas tintinnid abundance was similar to the values in the other Indian Ocean stations.

For 43 species, only a single specimen was encountered and, in the following description of regional occurrence of tintinnid taxa, only the taxa recorded in at least three stations are included. No species occurred in all samples, but five species showed a widespread occurrence and may be defined as cosmopolitan: *Amphorella amphora, Eutintinnus tubulosus, Steenstrupiella steenstrupii, E. fraknoi* and *E. lusus-undae* (in rank order). *Eutintinnus* was the most widely distributed genus represented by 11 species (Fig. 5A) including *Eutintinnus haslae*, described for the first time in the tropical Pacific and Indian Oceans (Taniguchi & Hada, 1981).

Several genera and species occurred throughout the warm-water regions, from the Mediterranean to the Tropic of Capricorn, but were not encountered south of 23° S. These warm-water taxa included the genera *Ascambelliella, Dadayiella, Protorhabdonella* and *Salpingella* (Fig. 5), and thirteen species e.g. *A. torulata, D. ganymedes, E. apertus, E. macilentus, E. stramentus, P. curta, P. simplex, Rhabdonella amor* and *S. decurtata*. However, the definition of *Salpingella* and *Protorhabonella* as warm-water taxa was based on the species encountered in this study; these very species-rich genera are in fact cosmopolitan (Pierce & Turner, 1993).

Some species occurred in only one oceanic province, e.g. *Xystonella longicauda* (Fig. 6A) and the genus *Xystonellopsis*, represented by two species *X. brandti* and *X. heroica*, occurred exclusively in the Mediterranean Sea; *Rhabdonella cornucopia* occurred in the Red Sea; and *Epiplocyloides reticulata* in the Red and Arabian Seas (Fig. 6B); *Rhabdonella quantula* and *R. hebe* in the Arabian Sea and Indian Ocean (Fig. 6C); and *Steenstrupiella gracilis* solely in the Indian Ocean (Fig. 6D). No species was encountered exclusively in the SIO and Tasman Sea.

Equitability was fairly constant (0.77–0.97) at most stations; the minimum value (0.57) occurred at Stn.336 in the Tasman Sea due to maximum abundances of the two cosmopolitan species *Amphorella amphora* and *Eutintinnus tubulosus*.

Tintinnid numbers were positively correlated ($p < 0.05$) with total chl*a* ($r = 0.54$), and with nano- ($r = 0.53$) and pico- ($r = 0.58$) phytoplankton concentrations, but not with microphytoplankton concentrations ($r = 0.41$). No significant correlation occurred between tintinnid numbers and temperature, salinity or contribution of different phytoplankton size classes. However, temperature showed a positive correlation with tintinnid species number ($r = 0.72$, $p < 0.01$) and with the Shannon index ($r = 0.50$, $p < 0.05$).

Epifluorescence microscope observations of tintinnid vacuole content (number of specimens examined=496) showed the presence of fluorescent algal prey in 19% of tintinnid vacuoles. Amongst these, red fluorescent prey (eukaryotic algae), generally 2–5 μm in diameter, occurred in 55%, yellow fluorescent (cyanobacteria) in 30%, and 15% contained both types of prey.

Discussion

Oligotrophic conditions prevailed during most of the MIPOT1 transect as indicated by low chl*a* concentrations and low-abundance, high-diversity tintinnid assemblages. Most tintinnid taxa showed a clear regional distribution and a well-defined tintinnid assemblage characterized each oceanic province. Several of the typical species had been reported previously for the same area indicating their persistence in time. Such is the case for *Xystonella* and *Xystonellopsis* defined as typical for the Mediterranean Sea (Krþiniæ, 1988). Similarly, *Epiplocyloides reticulata* was first described from the Red Sea and appears to be typical for this area (Brandt, 1907; Kofoid & Campbell, 1929; Halim, 1969). Review of literature records indicates a sort of tintinnid 'fingerprint' may characterize each oceanic or coastal area (Modigh & Castaldo, 2002 and references therein). However, 'typical' does not correspond with 'endemic' species and most of the tintinnid species reported in this study for specific regions show a more widespread occurrence, e.g. the species encountered in the Red and Arabian Seas and Indian Ocean have been reported also for other sub-tropical and tropical areas. In addition, some atypical species were encountered in the tropical Indian Ocean and may be used as tracers of oceanic currents. The neritic species *Favella markusovszkyi, Tintinnopsis cylindrica* and *T. parvula* were recorded at Stn.'s 145, 156 and 191; *Tintinnopsis* and *Favella* are common in coastal areas and have been reported as the most abundant tintinnid genera in Indian coastal waters (Naidu & Krishnamurthy, 1985; Srinivasan et al., 1988).

A sharp decrease in species number and species diversity occurred in the Southern Indian Ocean and the Tasman Sea; the turning point for tintinnid assemblages was the Tropic of Capricorn. The considerably lower diversity and minor number of tintinnid species recorded south of 23° S compared to the findings in the warm water regions is in agreement with the well-established finding of lower diversity at higher latitudes (Rohde, 1992; Dolan & Gallegos, 2001).

Low tintinnid abundance characterized most of our samples; tintinnid numbers were similar to previous reports from the different areas of this study (Chang, 1990; Dennett et al., 1999; Dolan, 2000; Pitta et al., 2001). Our samples refer only to the surface layer and do not permit any extrapolation for tintinnid assemblages in the water column. Data on tintinnid assemblages are very scanty particularly for the Red Sea

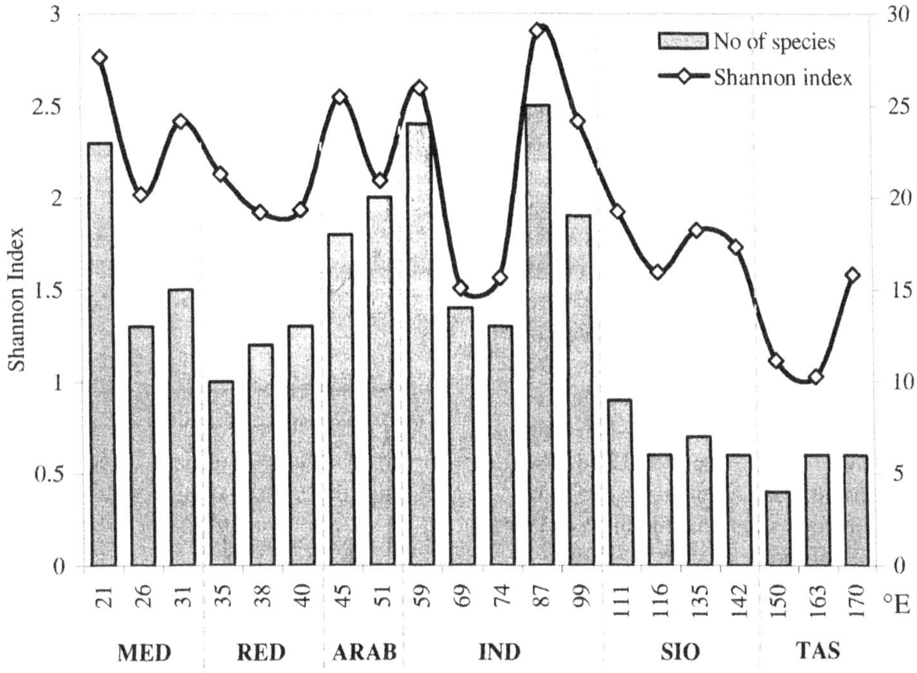

Figure 4. Number of tintinnid species (□) and Shannon diversity index (-◊-) at all stations.

and the Indian Ocean, but available data suggest no significant differences in abundance and species composition in the water column for several of the areas of this study as well as for other offshore areas (off New Zealand: Chang, 1990; Eastern Mediterranean: Dolan, 2000; Pitta et al., 2001; NW Indian Ocean: Stelfox et al., 1999; SW Atlantic: Thompson et al., 1999). On the other hand, higher tintinnid numbers at the surface but similar species composition along the water column were reported for the Arabian Sea (Zeitzschel, 1969; Leakey et al., 1996).

Maximum tintinnid numbers occurred at stations where a sharp drop in surface temperature and salinity was recorded, suggesting that mixing had occurred. In this study, three different situations associated with mixing were observed: high chl*a* and high tintinnid concentrations, unaltered chl*a* and maximum tintinnid numbers and, finally, a five-fold increment in chl*a* concentrations but unaltered tintinnid numbers. Maximum chl*a* concentrations were recorded in the Arabian Sea (Stn.100) associated with a relative maximum of tintinnid numbers. However, at Stn.336 in the Tasman Sea tintinnid numbers were 50 fold, but chl*a* concentrations were similar to the values recorded over this area. At both these stations, increase in tintinnid numbers was due to the sharp increase of a few tintinnid species. Such response of tintinnid assemblages to mixing events has been reported frequently for coastal and offshore waters (e.g. Thompson et al., 1999; Modigh & Castaldo, 2002). The species most frequently observed to become very abundant were, *Epiplocyloides reticulata* and *Rhabdonella poculum* in the Arabian Sea and *Amphorella amphora* and *Steenstrupiella steenstrupii* in the Tasman Sea.

A third mixing event occurred at Stn.191, resulting from the recent passage of a tropical cyclone. Chl*a* concentrations were five-fold but tintinnid numbers were unaltered with respect to the values recorded before and after this patch. At this station, maximum number of tintinnid species and maximum diversity was observed. We presume that the different relationship between chl*a* concentrations and tintinnid abundance recorded at these three stations was due to the different timing of our sampling with respect to the mixing event. The short-time interval between the passage of the cyclone and our sampling (24 h) would not have allowed any perceptible tintinnid growth; the increment in number of species suggests an initial stage of the tintinnid assemblage response. The high tintinnid numbers, and the extremely high phytoplankton biomass (10.4 mg chl*a* m^{-3}) recorded at Stn.100, suggest that the tintinnid assemblage had sufficient time to benefit from the abundant algal resource. Finally, the high tintinnid numbers at Stn.336 and low chl*a* concentration (0.2 mg chl*a* m^{-3}) may have represented the end of a tintinnid bloom, when algal biomass had been consumed and the high tintinnid abundance would probably crash.

On the basis of these observations, the relationship between chl*a* and tintinnid concentrations was expected to be fairly complex. Lynn & Montagnes (1991), pooling data available in literature, found a positive correlation ($r = 0.41$) between pelagic ciliate concentrations and phytoplankton biomass. In this study, a significant positive correlation was found between chl*a* concentrations and tintinnid numbers when calculations were performed on the entire MIPOT1 dataset. However, no significant correlation ($r = 0.33$) occurred if we cancelled one single data point, Stn.100 (maximum chl*a* values and a relative maximum in tintinnid numbers). Correlations only suggest an association between parameters, and a prey–predator relationship may be either positive or negative, depending on factors such as the timing of sampling and the extent of grazing pressure. The lack of correlation between tintinnid abundance and phytoplankton biomass in the oligotrophic regions of this study might be explained by observations of tintinnid vacuole content.

Most tintinnid food vacuoles did not contain fluorescent (autotrophic) prey and, in fact, tintinnids have been shown to feed also on heterotrophic pico- and nano-plankton (Verity, 1991). Experiments conducted in the Arabian Sea showed the major importance of ciliate grazing on heterotrophic nano- and pico-flagellates that in turn prey upon the picophytoplankton (Burkill et al., 1993; Reckerman & Veldhuis, 1997). In this study, when fluorescent prey did occur within tintinnid vacuoles, nano-eucaryotic algae were the most frequently observed, thus confirming tintinnid preferential feeding on these algae (Bernard & Rassoulzadegan, 1993; Pitta et al., 2001). In contrast, the heterotrophic dinoflagellates, by far more abundant than the tintinnids at all stations, had mainly yellow fluorescent cyanobacteria in their food vacuoles (unpublished data). Cyanobacteria are a major component of picoplankton in oligotrophic areas (e.g. Burkill et al., 1993) and the tintinnid preference for the less abundant eukaryotic algae may have limited their numbers.

In addition, no correlation was found between tintinnid numbers and the microphytoplankton chl*a* concentrations. Large tintinnids occurred mainly in the oligotrophic warm-water regions where the con-

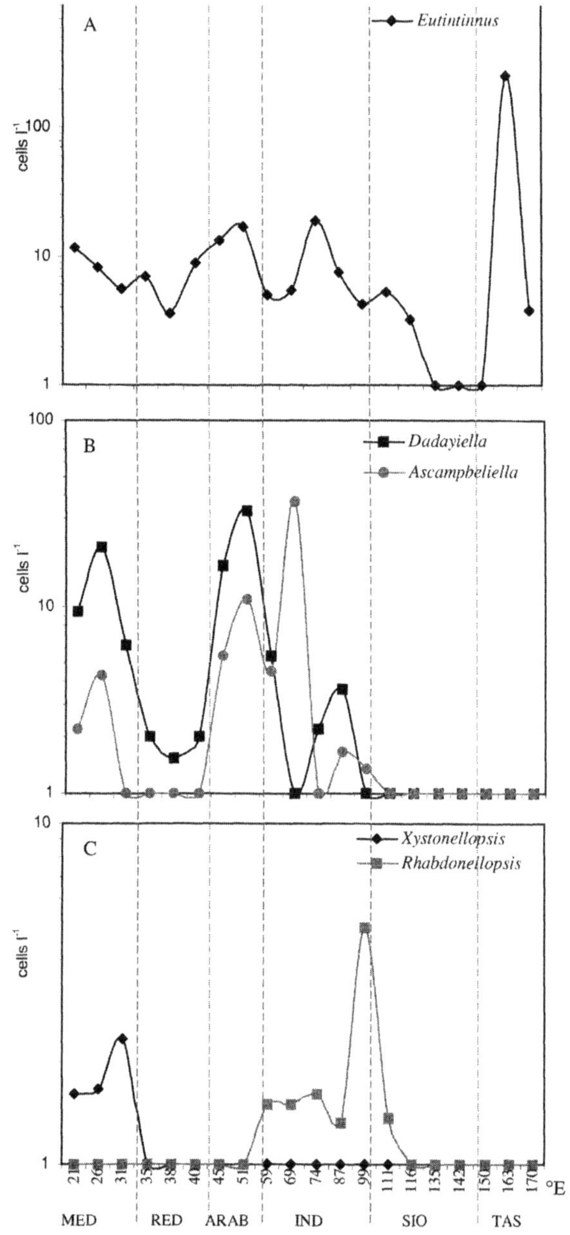

Figure 5. Occurrence of some tintinnid genera: A) cosmopolitan, B) warm-water, C) regional distributions.

tribution of microphytoplankton was very low. On the other hand, microphytoplankton attained relatively higher concentrations in the SIO and Tasman Sea where tintinnids were generally of smaller size and the microphytoplankton would not be appropriate prey as most tintinnids feed on prey not exceeding 45% of their oral diameter (Spittler, 1973). The counterintuitive finding of larger tintinnids where prey items are

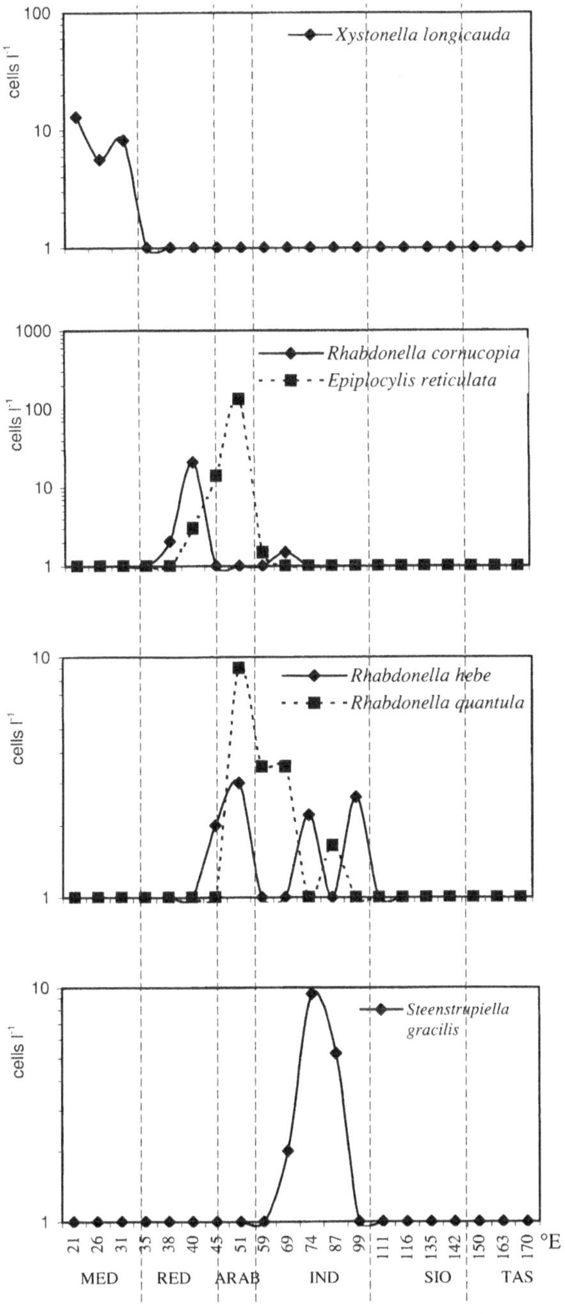

Figure 6. Regional occurrence of some characteristic tintinnid species.

smaller, i.e. in the oligotrophic oceans, has already been reported (Cariou et al., 1999; Dolan, 2000). Predation rather than competition for resources has been hypothesized as a major factor for selecting plankton morphotypes (Verity & Smetacek, 1996) and such hypothesis has been specifically suggested also for the selection of tintinnid species (Capriulo et al., 1982; Cariou et al., 1999). In this study, *Eutintinnus,* characterized by a large transparent lorica, was the most widespread and species-rich genus, and transparency is a good predator-avoiding device (Capriulo et al., 1982). However, the high diversity and low abundance of tintinnid assemblages in the oligotrophic regions of this study is probably determined both by predation pressure and by scarce food resources. From this first analysis of the MIPOT1 data, we conclude that well-defined tintinnid assemblages characterized each oceanic region and that comparison of our data with previous records from the different areas indicate the area-specific tintinnid species persist in time. A second transect, MIPOT2, was carried out from New Zealand to Italy in March 2002. The analysis of this second data set will be helpful to further define tintinnid assemblages associated with different environmental settings.

References

Balech, E., 1959. Tintinnoinea del Mediterráneo. Trab. Inst. Esp. Ocean. 28: 1–88.

Balech, E., 1972. Los tintinnidos indicadores de afloramientos de aguas (Ciliata). Physis, Buenos Aires 31: 519–528.

Bernard, C. & F. Rassoulzadegan, 1993. The role of picoplankton (cyanobacteria and plastidic picoflagellates) in the diet of tintinnids. J. Plankton Res. 15: 361–373.

Brandt, K., 1907. Die Tintinnodeen der Plankton-Expedition. Systematischer Teil. Kiel und Liepzig: 1–488.

Burkill, P. H., R. J. G. Leakey, N. J. P. Owens & R.F.C. Mantoura, 1993. *Synechococcus* and its importance to the microbial food-web of the northwestern Indian Ocean. Deep Sea Res. II 40: 773–782.

Capriulo, G. M., K. Gold & A. Okubo, 1982. Evolution of the lorica in tintinnids: a possible selective advantage. Ann. Inst. Océanogr. Paris 58(S): 319–324.

Cariou, J. B., J. R. Dolan & S. Dallot, 1999. A preliminary study of tintinnid diversity in the NW Mediterranean Sea. J. Plankton Res. 21: 1065–1075.

Chang, H. F., 1990. Quantitative distribution of microzooplankton off Westland, New Zealand. N.Z. J. mar. Freshwat. Res. 24: 187–195.

Daday, E., 1887. Monographie der Familie der Tintinnodeen. Mitt. a d Zoolog. Station Neapel Bd VII: 473–591.

Dennett, M. R., D. A. Caron, S. A. Murzov, I. G. Polikarpov, N. A. Gavrilova, L. V. Georgieva & L. V. Kuzmenko, 1999. Abundance and biomass of nano- and microplankton during the 1995 Northeast Monsoon and Spring Intermonsoon in the Arabian Sea. Deep Sea Res. II 46: 1691–1717.

Dolan, J. R., 2000. Tintinnid ciliate diversity in the Mediterranean Sea: longitudinal patterns related to water column structure in late spring–early summer. Aquat. Microbiol. Ecol. 22: 69–78.

Dolan, J. R. & C. L. Gallegos, 2001. Estuarine diversity of tintinnids (planktonic ciliates). J. Plankton Res. 23: 1009–1027.

Dolan, J. R. & C. Marrasé, 1995. Planktonic ciliate distribution relative to a deep chlorophyll maximum: Catalan Sea, N. W. Mediterranean, June 1993. Deep Sea Res. I 42: 1965–1987.

Fenchel, T., G. F. Esteban & B. J. Finlay, 1997. Local versus global diversity of microorganisms: cryptic diversity of ciliated protozoa. Oikos 80: 220–225.

Halim, Y., 1969. Plankton of the Red Sea. Oceanogr. mar. biol. Ann. Rev. 7: 231–275.

Jörgensen, E., 1924. Mediterranean Tintinnidae. Rep Danish Oceanographical Expeditions 1908–1910 to the Mediterranean and adjacent seas II (J3): 1–110.

Kato, S. & A. Taniguchi, 1993. Tintinnid ciliates as indicator species of different water masses in the western North Pacific Polar Front. Fish. Oceanogr. 2: 166–174.

Kofoid, C. A. & A. S. Campbell, 1929. A conspectus of the marine and fresh-water ciliata belonging to the suborder Tintinnoinea, with descriptions of new species principally from the Agassiz expedition to the Eastern Tropical Pacific. 1904–1905. Univ. Calif. Publ. Zool. 34: 1–403.

Kofoid, C. A. & A. S. Campbell, 1939. The Ciliata: the Tintinnoinea. Bull. Mus. Comp. Zool. Harvard Coll. 84: 1–473.

Kršinić, F., 1988. The family *Xystonellidae* (Ciliophora, Tintinnina) in the Adriatic Sea. J. Plankton Res. 10: 413–429.

Laval-Peuto, M. & D. C. Brownlee, 1986. Identification and systematics of the tintinnina (Ciliophora): evaluation and suggestions for improvement. Ann. Inst. Océanogr. Paris 62: 69–84.

Leakey, R. J.G., P. H. Burkill & M. A. Sleigh, 1996. Planktonic ciliates in the northwestern Indian Ocean: their abundance and biomass in waters of contrasting productivity. J. Plankton Res. 18: 1063–1071.

Lynn, D. H. & D. J. S. Montagnes, 1991. Global production of heterotrophic marine planktonic ciliates. In Reid P. C., C. M. Turley & P. Burkill H. (eds), Protozoa and their Role in Marine Processes. NATO ASI Ser. G25 Springer Verlag, Berlin: 281–307.

Modigh, M. & S. Castaldo, 2002. Variability and persistence in tintinnid assemblages at a Mediterranean coastal site. Aquat. Microbiol. Ecol. 28: 299–311.

Naidu, D. W. & K. Krishnamurthy, 1985. Biogeographical distribution of tintinnids (Protozoa; Ciliata) from Porto Novo waters. Bull. Nat. Inst. Oceanogr. 18: 417–423.

Neveux, J. & M. Panouse, 1987. Spectrofluorometric determination of chlorophylls and phaeophytins. Archiv Hydrobiol. 109: 567–581.

Paffenhöfer, G. A., 1998. Heterotrophic protozoa and small metazoa: feeding rates and prey consumer interactions. J. Plankton Res. 20: 121–133.

Pierce, R. W. & J. T. Turner, 1992. Ecology of planktonic ciliates in marine food webs. Rev. aquat. Sci. 6: 139–181.

Pierce, R. W. & J. T. Turner, 1993. Global biogeography of marine tintinnids. Mar. Ecol. Prog. Ser. 94: 11–26.

Pitta, P., A. Giannakourou & U. Christaki, 2001. Planktonic ciliates in the oligotrophic Mediterranean Sea: longitudinal trends of standing stocks, distributions and analysis of food vacuole content. Aquat. Microbiol. Ecol. 24: 297–311.

Rassoulzadegan, F., 1979. Évolution annuelle des Ciliés pélagiques en Méditerranée nord-occidentale. II. Ciliés Oligotriches. Tintinnides (Tintinnina). Inv. Pesq. 43: 417–448.

Rassoulzadegan, F., M. Laval-Peuto & R. W. Sheldon, 1988. Partitioning of the food ration of marine ciliates between pico- and nanoplankton. Hydrobiologia 159: 75–88.

Reckermann, M. & M. J. W. Veldhuis, 1997. Trophic interactions between picophytoplankton and micro- and nanozooplankton in the western Arabian Sea during the NE monsoon 1993. Aquat. Microbiol. Ecol. 12: 263–273.

Rohde, K., 1992. Latitudinal gradients in species diversity: the search for the primary cause. Oikos 65: 514–527.

Sherr, E. B., B. F. Sherr & G. A. Paffenhöfer, 1986. Phagotrophic protozoa as food for metazoans: a 'missing' trophic link in marine pelagic food webs? Mar. Microb. Food Webs 1: 61–80.

Spittler, P., 1973. Feeding experiments with tintinnids. Oikos 15: 128–132.

Srinivasan, A., R. Santhanam & G. Jegatheesan, 1988. Biomass and seasonal distribution of planktonic tintinnids of Pullavazhi Estuary, southeast coast of India. Ind. J. mar. Sci. 17: 131–133.

Stelfox, C. E., P. H. Burkill, E. S. Edwards, R. P. Harris & M. A. Sleigh, 1999. The structure of zooplankton communities, in the 2–2000 μm size range, in the Arabian Sea during and after the SW monsoon, 1994. Deep Sea Res. II 46: 815–842.

Stoecker, D. K. & N. K. Sanders, 1985. Differential grazing by *Acartia tonsa* on a dinoflagellate and a tintinnid. J. Plankton Res. 7: 85–100.

Sverdrup, H. U., M. W. Johnson & R. H. Fleming, 1942. The Oceans. Their Physics, Chemistry, and General Biology. Prentice-Hall, Inc. New York: 1087 pp.

Taniguchi, A. & Y. Hada, 1981. *Eutintinnus haslae* n. sp., a new ciliated protozoa from the Tropical Pacific and Indian Oceans. La mer 19: 18–22.

Thompson, G. A., V. A. Alder, D. Boltovskoy & F. Brandini, 1999. Abundance and biogeography of tintinnids (Ciliophora) and associated microzooplankton in the Southwestern Atlantic Ocean. J. Plankton Res. 21: 1265–1298.

Verity, P. G., 1987. Abundance, community composition, size distribution, and production rates of tintinnids in Narragansett Bay, Rhode Island. Estuar. coast. shelf Sci. 24: 671–690.

Verity, P. G., 1991. Measurement and simulation of prey uptake by marine planktonic ciliates fed plastidic and aplastidic nanoplankton. Limnol. Oceanogr. 36: 729–750.

Verity, P. G. & V. Smetacek, 1996. Organism life cycles, predation, and the structure of marine pelagic ecosystems. Mar. Ecol. Prog. Ser. 130: 277–293.

Wyrtki, K., 1973. Physical oceanography of the Indian Ocean. In Zeitzschel, B. (ed.), The Biology of the Indian Ocean. Ecological Studies 3, Springer Verlag, Berlin: 18–36.

Zeitzschel, B., 1969. Tintinnen des westlichen Arabischen Meeres, ihre Bedeutung als Indikatoren für Wasserkörper und Glied der Nahrungskette. Meteor Forschungsergeb. Reihe D 4: 47–101.

Zeitzschel, B., 1982. Zoogeography of pelagic marine protozoa. Ann. Inst. Océanogr. Paris 58(S): 91–116.

GPSR Compliance

The European Union's (EU) General Product Safety Regulation (GPSR) is a set of rules that requires consumer products to be safe and our obligations to ensure this.

If you have any concerns about our products, you can contact us on ProductSafety@springernature.com

In case Publisher is established outside the EU, the EU authorized representative is:

Springer Nature Customer Service Center GmbH
Europaplatz 3
69115 Heidelberg, Germany

Batch number: 10762189

Printed by Printforce, the Netherlands